SURFACE AREA

Right prism: $T = L + 2B$, where $L = h \cdot P$

Regular pyramid: $T = L + B$, where $L = \frac{1}{2} \cdot \ell \cdot P$

Right circular cylinder: $T = 2\pi r^2 + 2\pi rh$

Right circular cone: $T = \pi r^2 + \pi r \ell$

Sphere: $S = 4\pi r^2$

VOLUME

Daniel Marsh (728-6348)

Right prism: $V = Bh$

Pyramid: $V = \frac{1}{3}Bh$

Right circular cylinder: $V = \pi r^2 h$

Right circular cone: $V = \frac{1}{3}\pi r^2 h$

Sphere: $V = \frac{4}{3}\pi r^3$

TRIGONOMETRY

$$\sin \theta = \frac{\text{opposite}}{\text{hypotenuse}} \qquad \cos \theta = \frac{\text{adjacent}}{\text{hypotenuse}} \qquad \tan \theta = \frac{\text{opposite}}{\text{adjacent}}$$

SELECTED TRIGONOMETRIC RELATIONSHIPS

$\sin^2 \theta + \cos^2 \theta = 1$

$A_\triangle = \frac{1}{2}bc \sin \alpha$

$\dfrac{\sin \alpha}{a} = \dfrac{\sin \beta}{b} = \dfrac{\sin \gamma}{c}$

$c^2 = a^2 + b^2 - 2ab \cos \gamma$

π radians $= 180°$

ELEMENTARY GEOMETRY
FOR COLLEGE STUDENTS

DANIEL C. ALEXANDER
Parkland Community College

GERALYN M. KOEBERLEIN

 Wm. C. Brown Publishers

Book Team

Editor *Earl McPeek*
Developmental Editor *Theresa Grutz*
Visuals/Design Freelance Specialist *Barbara J. Hodgson*
Developmental Visuals/Design Consultant *Marilyn A. Phelps*

Wm. C. Brown Publishers

President *G. Franklin Lewis*
Vice President, Publisher *George Wm. Bergquist*
Vice President, Operations and Production *Beverly Kolz*
National Sales Manager *Virginia S. Moffat*
Group Sales Manager *Vincent R. Di Blasi*
Vice President, Editor in Chief *Edward G. Jaffe*
Marketing Manager *Liz Robbins*
Advertising Manager *Amy Schmitz*
Managing Editor, Production *Colleen A. Yonda*
Manager of Visuals and Design *Faye M. Schilling*
Production Editorial Manager *Julie A. Kennedy*
Production Editorial Manager *Ann Fuerste*
Publishing Services Manager *Karen J. Slaght*

WCB Group

President and Chief Executive Officer *Mark C. Falb*
Chairman of the Board *Wm. C. Brown*

Production and design by Bookman Productions

Line art by Illustrious, Inc.

Cover photo of Dallas skyline by George Glod/SUPERSTOCK

Library of Congress Catalog Card Number: 91–72230

ISBN 0–697–11067–2

Printed in the United States of America by Wm. C. Brown Publishers,
2460 Kerper Boulevard, Dubuque, IA 52001

10 9 8 7 6 5 4 3 2 1

We thank our families for their sacrifices during the time when this book was written and taken through its developmental stages. We thus dedicate this finished product to our spouses, Mary Alexander and Richard Koeberlein, and to our children, Matthew, Phillip, and Sarah Alexander and Alice and Laura Koeberlein.

CONTENTS

Preface   ix

1

PRELIMINARY CONCEPTS   1

1.1   Reasoning in Mathematics   **1**
1.2   Statements in Mathematics   **8**
1.3   Algebraic Expressions   **12**
1.4   Formulas and Equations   **20**
1.5   Inequalities   **26**
A Look Beyond: The Nature of Proof   **32**
Summary   **33**
Review Exercises   **34**

2

LINE AND ANGLE RELATIONSHIPS   36

2.1   Some Undefined Terms and Initial
       Definitions   **36**
2.2   Angles and Their Relationships   **44**
2.3   Introduction to Geometry Proofs   **50**
2.4   The Formal Proof of a Theorem   **58**
A Look Beyond: Historical Sketch of Euclid   **63**
Summary   **64**
Review Exercises   **64**

3

PARALLEL LINES   66

3.1   The Parallel Postulate, Transversals, and
       Special Angles   **66**
3.2   Indirect Proof   **74**
3.3   Proving Lines Parallel   **78**

3.4   Applications of Parallels to Triangles and
       Other Polygons   **85**
A Look Beyond: Non-Euclidean Geometries   **93**
Summary   **95**
Review Exercises   **95**

4

TRIANGLES AND QUADRILATERALS   98

4.1   Congruent Triangles   **98**
4.2   CPCTC and HL   **106**
4.3   Isosceles Triangles   **113**
4.4   Quadrilaterals and Parallelograms   **120**
4.5   More Quadrilaterals   **127**
4.6   Inequalities in a Triangle   **134**
A Look Beyond: Historical Sketch of
Archimedes   **140**
Summary   **140**
Review Exercises   **141**

5

SIMILAR TRIANGLES   145

5.1   Quadratic Equations and Solutions   **145**
5.2   Ratio and Proportions   **152**
5.3   Similar Polygons and Triangles   **159**
5.4   Sigments Divided Proportionally   **168**
5.5   Similar Right Triangles and the Pythagorean
       Theorem   **177**
5.6   Special Right Triangles   **185**
A Look Beyond: An Unusual Application of Similar
Triangles   **192**
Summary   **193**
Review Exercises   **194**

CIRCLES   197

6.1 Circles and Related Segments and Angles 197
6.2 More Angle Measures in the Circle 207
6.3 Line and Segment Relationships in the Circle 216
6.4 Some Constructions and Inequalities for the Circle 223
6.5 Locus and Concurrency 229
A Look Beyond: The Value of π 239
Summary 240
Review Exercises 240

AREAS OF POLYGONS AND CIRCLES   244

7.1 Area and Initial Postulates 244
7.2 More with Perimeter and Area 255
7.3 Regular Polygons and Area 262
7.4 The Area of a Circle 270
A Look Beyond: Another Look at the Pythagorean Theorem 276
Summary 278
Review Exercises 278

ANALYTIC GEOMETRY   281

8.1 The Rectangular Coordinate System 281
8.2 Graphs of Linear Equations and Slope 289
8.3 Equations of Lines 297
8.4 Circles and Parabolas 303
8.5 Preparing to Do Analytic Proofs 310

8.6 Analytic Proofs 320
A Look Beyond: The Banach-Tarski Paradox 325
Summary 326
Review Exercises 327

SURFACES AND SOLIDS   329

9.1 Prisms and Pyramids 329
9.2 Cylinders and Cones 339
9.3 Polyhedrons and Spheres 346
9.4 A Three-Dimensional Coordinate System 352
A Look Beyond: Historical Sketch of René Descartes 358
Summary 359
Review Exercises 359

INTRODUCTION TO TRIGONOMETRY   361

10.1 Sine Ratio and Applications 361
10.2 Cosine Ratio and Applications 369
10.3 Tangent and Other Ratios 376
10.4 More Trigonometric Relationships 385
A Look Beyond: Radian Measure of Angles 395
Summary 397
Review Exercises 397

APPENDIX: Summary of Constructions, Postulates, and Theorems and Corollaries 400

ANSWERS: Selected Odd-Numbered Exercises and Proofs 407

INDEX 434

 **PREFACE**

Purpose

Elementary Geometry for College Students is written for college students who have not previously completed a course in geometry and for those who need a refresher course. Note that students could be concurrently enrolled in a beginning or intermediate-level algebra class. We, as authors, see the ultimate purpose of our work as helping students to learn and apply the principles of geometry.

Goals

Specific outcomes for users of this textbook are:

- An introduction to the essentials of geometry for students planning to use geometry in a vocation
- The preparation necessary to enable transfer students to further study mathematics and disciplines that require a knowledge of geometry
- Exposure to the step-by-step development of a logical mathematical system
- The maintenance and application of certain algebraic skills

Style

This textbook was prepared and carefully developed by geometry teachers to be read by students. We made every attempt to present material in this book in much the same way we present it in the classroom. The book is more complete than most in that we **explain** what we are doing, **anticipate** where we are going, and **recall** where we have been. A second difference is that we present many paragraph proofs (common at the college level) rather than the "two-column" proofs usually featured in secondary-level geometry texts. Our method of proof should help improve students' writing styles in all classes, not just mathematics classes, making their transition to college mathematics a bit smoother.

Content

This textbook parallels the goals of most secondary-level geometry textbooks, adheres to the standards the National Council of Teachers of Mathematics recommends, and presents sufficient geometry content to prepare the potential teacher of geometry. The content also follows the recommendations of the Illinois Mathematical Association of Community Colleges and the Mathematical Association of America (Illinois section). Rather than include a separate chapter based on transformational geometry, that material has been carefully interwoven throughout the book.

Features

- Interesting chapter-opening art relevant to each chapter introduces the principal notion of each chapter.
- A chapter outline in color on each chapter-opening page gives students an overview of what they will be studying.
- Within chapters, a thorough set of computational, applied, and theoretical exercises follows each section. More challenging section exercises are identified by an asterisk.
- Answers to selected odd-numbered exercises and proofs are at the end of the book.
- Color enhances the instructional value of many of the illustrations within the text, examples, exercises, and end-of-chapter material. Color is also used to emphasize and highlight important pedagogical features, including definitions, warnings, theorems, and postulates.
- A color ▲ symbol clearly indicates the end of each example and proof.
- Definitions of primary importance are boxed; other important terms are boldfaced.

- A convenient triple-number system is used for theorems, corollaries, and lemmas to indicate the order in which these items are found in the book. For example, Theorem 5.4.2 is the second theorem in Section 5.4; Corollary 5.4.3 indicates the corollary to Theorem 5.4.2.

- Tables summarizing and relating concepts are used where appropriate, and tables for the trigonometric sine, cosine, and tangent ratios are provided.

- A Look Beyond section completes the text of each chapter; these sections provide sketches that are interesting, sometimes historical, and always informative.

- A Chapter Summary lists the important concepts covered within each chapter.

- A Look Back section within each Chapter Summary presents an overview of the chapter just completed; A Look Ahead section briefly previews the chapter to follow.

- Chapter Review Exercises cover all the concepts presented within the entire chapter.

- While the student can easily complete this textbook without the aid of a calculator, references in Chapter 10 are made to the use of the calculator.

- An appendix summarizes all constructions, postulates, and theorems and corollaries.

- End sheets serve as references for formulas, variables, abbreviations, and symbols.

Ancillaries for the Student

A *Student's Solutions Manual* (SSM) is available. It is divided into three parts:

1. "How to Study Geometry," which provides suggestions for success in the study of geometry.

2. A list of objectives for each chapter; section by section.

3. Study Notes, which provide hints, section by section.

4. The complete solution for every other odd-numbered exercise in each chapter and all the solutions for the chapter review exercises. The solutions include constructions as well as computational and proof problems.

5. Abbreviations for terms used in geometry.

6. Common variables used in geometry.

7. Formulas of geometry.

8. Symbols for geometry.

The intent of the SSM is to provide guidance without completing the student's assignments. The textbook and the SSM, used together, provide the student with a "teacher away from the classroom."

Ancillaries for the Instructor

The *Instructor's Manual* (IM) to accompany *Elementary Geometry for College Students* includes the following features:

1. Some possible course configurations

2. A list of objectives for each chapter, section by section

3. Approximately 60 reproducible transparency masters carefully selected from each chapter of the text

The *Test Item File/Quiz Item File* is a printed version of the computerized *TestPak* and *QuizPak,* which allow you to choose test items based on chapter, section, or objective. Multiple-choice, true/false, and open-ended questions are provided for each section of the text (1500 test items, 500 quiz items).

WCB TestPak 3.0, our computerized testing services, provides you with a mail-in/call-in testing program and the complete test item file on diskette for use with IBM PC, Apple, or Macintosh computers. Tests can be generated randomly, by selecting specific test items or objectives. In addition, new test items can be added and existing test items can be edited.

WCB QuizPak, a part of *TestPak 3.0,* gives students true/false, multiple-choice, and matching questions from the *Quiz Item File* for each chapter in the text. Using this portion of the program will help your students prepare for examinations. Also included with the *WCB QuizPak* is an on-line testing option that allows professor to prepare tests for students to take using the computer. The computer will automatically grade the test and update the gradebook file.

WCB GradePak, another part of *TestPak 3.0,* is a computerized Grade Management System for instructors. This program tracks student performance on examinations and assignments. It will compute each student's percentage and corresponding letter grade as well as the class average. Printouts can be made using both text and graphics.

Acknowledgments

People who are directly responsible for this finished product include the editorial staff of W. C. Brown, especially Earl McPeek, mathematics, editor; Theresa Grutz, developmental editor; Beth Meadows Dahlke; and all those working behind the scenes at Wm. C. Brown Publishing Company.

We wish to thank Linda Crowley, who typed the Solutions Manual, and Don Manning, who provided photographs for Chapters 9 and 10 of the textbook.

We also wish to thank all the reviewers for their advice in developing the manuscript: Jane C. Beattie, University of South Carolina—Aiken; Steven Blasberg, West Valley College; Dr. Patricia Clark, Indiana State University; George L. Holloway, Los Angeles Valley College; Tracey Hoy, College of Lake County; Josephine G. Lane, Eastern Kentucky University; James R. McKinney, Cal Poly—Pomona; Maurice Ngo, Chabot College; Ellen L. Rebold, Brookdale Community College; and Karen R. Swick, Palm Beach Atlantic College.

Last but not least, we thank our teachers and friends whose work, inspiration, and encouragement in some way influenced the writing of this book.

Daniel C. Alexander
Geralyn M. Koeberlein

PRELIMINARY CONCEPTS

What happens when you follow the steps in the figure? Do your powers of reasoning tell you that something is wrong? This chapter introduces some of the types of reasoning that are used in the study of geometry. Several concepts and methods used in algebra are also presented. It is possible for you to go on at once to the study of geometry in Chapter 2 and to refer back to Chapter 1 as needed.

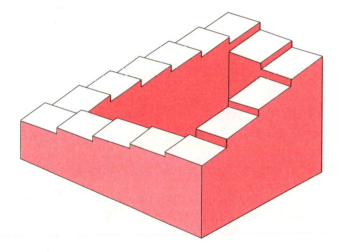

CHAPTER OUTLINE

1.1 Reasoning in Mathematics

1.2 Statements in Mathematics

1.3 Algebraic Expressions

1.4 Formulas and Equations

1.5 Inequalities

A Look Beyond The Nature of Proof

1.1 REASONING IN MATHEMATICS

The learning process in geometry requires time, careful development of the relevant vocabulary, attention to detail and order, and a great deal of thinking. The following types of thinking (or reasoning) are involved:

1. Intuition
2. Induction
3. Deduction

With time, you will become familiar with each type of reasoning and gain proficiency in its use.

In **intuition,** a sudden insight allows you to draw a conclusion without applying any formal reasoning. Your own intuitive discoveries may be prefaced by the phrase, "It occurs to me that. . . ." In cartoons, the character with the "bright idea" is using his or her intuitive powers.

As an example of intuition, consider the situation of Bill, a 14-year-old, who would like to see and hear a tape of his favorite rock group in stereo. It dawns on Bill to connect his videocassette recorder to two different television sets simultaneously. Much to his pleasure and surprise, his idea works!

Many of you have dealt in an earlier mathematics class with the notions of "line" and "line segment" (the part of a line between two points of the line). You may also have thought of a rectangle as a four-sided figure with "square" corners; the sides, of course, are line segments. These concepts are needed for Example 1.

EXAMPLE 1   What does intuition tell you about the lengths of line segments *AC* and *BD* for the rectangle *ABCD* in figure 1.1?

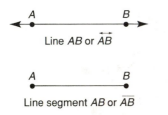
Line *AB* or $\overleftrightarrow{AB}$

Line segment *AB* or $\overline{AB}$

Rectangle *ABCD*

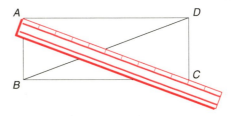

Figure 1.1

Conclusion   The lengths of the two line segments are the same.   ▲

Note: Line segments *AC* and *BD* are termed "diagonals" of the rectangle. (Later we will prove that the diagonals of a rectangle have equal lengths.)

The role intuition has played in the development of mathematics is significant. But to have an idea is not enough! If an idea or theory holds up after being tested, it moves one step closer to becoming a law of mathematics. The testing, observation, and experimentation stage of development involves inductive reasoning.

Induction is the form of reasoning in which specific observations lead to a general conclusion. As you would expect, this observation-experimentation process is quite common in laboratory and clinical settings. Chemists, physicists, doctors, psychologists, weather forecasters, and many others use specific data as a basis for drawing conclusions. In everyday life, we use induction in reaching decisions and in drawing conclusions.

EXAMPLE 2   While at the dairy case of the grocery store, you examine several 8-oz cartons of yogurt. Although the flavors and brands differ, each carton is priced at 75 cents. What do you conclude?

Conclusion   Every 8-oz container of yogurt in the store is priced at 75 cents.   ▲

Induction is also used in geometry. The rectangle you saw earlier is a special type of closed geometric figure known as a **polygon.** Each polygon has three or more straight sides (line segments with common endpoints). Any polygon with three sides, such as *ABC* in the figure, is a **triangle.**

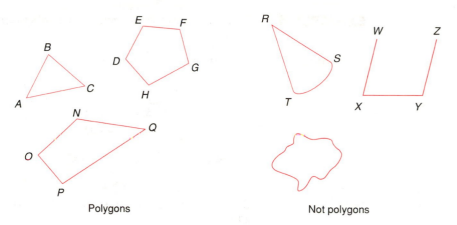

Polygons   Not polygons

EXAMPLE 3   In a geometry class, you have been asked to measure the three interior angles of each of the triangles shown in figure 1.2. You discover that two angles are equal in measure in each of triangles I, II, and IV. What may you conclude?

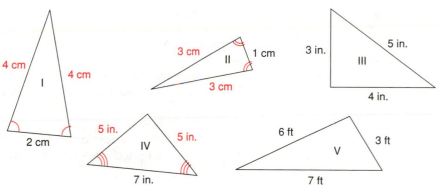

Figure 1.2

Conclusion   The triangles that have two sides of equal length also have two angles of equal measure.   ▲

Note: The device used to measure angles is a *protractor.* (See figure 1.3.) Angles are measured in degrees, denoted by °.

If too few observations are made, induction can lead to false conclusions. This fact is illustrated by Example 4, in which the phrase "prime number" is used. Perhaps enumerating such prime numbers as 2, 3, 5, 7, 11, 13, and 17 will help you recall the following definition:

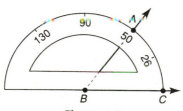

Figure 1.3

> **DEFINITION:**   A **prime number** is a positive integer greater than 1 for which the only factors (divisors) are the number itself and 1.

EXAMPLE 4   When n is a positive integer, is the number represented by $(n^2 + n + 17)$ always prime?

Solution   By examining the expression, we find that

$$n = 1 \rightarrow 1^2 + 1 + 17 = 19 \qquad \text{a prime}$$
$$n = 2 \rightarrow 2^2 + 2 + 17 = 23 \qquad \text{a prime}$$
$$n = 3 \rightarrow 3^2 + 3 + 17 = 29 \qquad \text{a prime}$$
$$n = 4 \rightarrow 4^2 + 4 + 17 = 37 \qquad \text{a prime}$$

$$.$$
$$.$$
$$.$$

But if we continue this procedure until we reach $n = 17$, the expression becomes 323:

$$17^2 + 17 + 17 = 17(17 + 1 + 1) = 17 \cdot 19 = 323$$

And since 19 is also a factor of 323, the number is not prime. Thus the number represented by $(n^2 + n + 17)$ is not always prime.   ▲

In Example 4, the observation made when $n = 17$ disproved the theory that $n^2 + n + 17$ is always prime. An observation that disproves a theory is a **counterexample.** As you can see, a counterexample might not be discovered through the inductive process. Consequently, most mathematical principles are established by **deduction** rather than by induction. When correctly applied, the deductive form of reasoning leads to certainty in conclusions.

> **DEFINITION:**   **Deduction** is the type of reasoning in which knowledge of a general principle leads to a specific conclusion.

It is convenient to illustrate the deductive type of reasoning by a form we call an **argument.** The argument consists of at least two statements that are to be accepted as true, followed by a conclusion. If the conclusion follows logically from these statements, the argument is **valid.**

EXAMPLE 5   If you accept the following statements 1 and 2 as true, what must you conclude?

1. If a student plays on the Rockville High School boys' varsity basketball team, then he is a talented athlete.

2. Todd plays on the Rockville High School boys' varsity basketball team.

Conclusion   Todd is a talented athlete.   ▲

To make it easier to see clearly the pattern for drawing conclusions deductively, we can use letters to represent statements. For instance, if

$P =$ *"It is raining," and $Q =$ "Tim will stay in the house,"*

then the statement "If P, then Q" represents

"If it is raining, then Tim will stay in the house."

DEDUCTIVE REASONING

Let P and Q represent simple statements, and assume that statements 1 and 2 are true. Then the form of a valid argument having conclusion C is given by

1. If P, then Q
2. P

C. ∴ Q

Note: The symbol ∴ means "therefore."

In the preceding form, the statement "If P, then Q" is often expressed as "P implies Q." That is, when P is known to be true, Q must follow.

EXAMPLE 6   Is the following argument valid? Assume that statements 1 and 2 are true.

1. If it is raining, then Tim will stay in the house.
2. It is raining.

C. ∴ Tim will stay in the house.

Conclusion   The argument is valid because the form of the argument is

1. If P, then Q
2. P

C. ∴ Q

with $P =$ "It is raining," and $Q =$ "Tim will stay in the house."   ▲

EXAMPLE 7   Is the following argument valid? Assume that statements 1 and 2 are true.

1. If a man lives in London, then he lives in England.
2. William lives in England.

C. ∴ William lives in London.

Conclusion The argument is not valid. Here, P = "A man lives in London," and Q = "A man lives in England." Thus the form of this argument is

1. If P, then Q
2. Q

C. $\therefore P$

But we have not set up our deductive reasoning to handle the question, "If Q, then what?" Therefore, being given Q in statement 2 does not enable us to make any valid conclusion about P. Of course, if William lives in England, he *might* live in London; but he might instead live in Liverpool, Manchester, Coventry, or any of countless other places in England. Each of these possibilities is a counterexample disproving the validity of the argument. Remember that deductive reasoning is concerned with reaching conclusions that *must be true*, given the truth of the antecedent statements. ▲

▲ *Warning: In the box, the argument on the left is valid and patterned after Example 6. The argument on the right is invalid; this form was given in Example 7.*

VALID ARGUMENT	INVALID ARGUMENT
1. If P, then Q 2. P C. $\therefore Q$	1. If P, then Q 2. Q C. $\therefore P$

We use deductive reasoning throughout our work in geometry. Suppose that you know these two facts:

1. If an angle is a right angle, then it measures 90°.
2. Angle A is a right angle.

Then you may conclude

C. Angle A measures 90°.

1.1 EXERCISES

In Exercises 1 to 8, name the type of reasoning (if any) used.

1. While participating in an Easter egg hunt, Sarah notices that each of the seven eggs she has found are numbered. Sarah concludes that all eggs used for the hunt are numbered.
2. You walk into your geometry class, look at the teacher, and conclude that you will have a quiz today.
3. Albert knows the rule that "If a number is added to each side of an equation, then the new equation has the same solution set as the given equation." Given the equation $x - 5 = 7$, Albert concludes that $x = 12$.
4. You believe that "Anyone who plays major league baseball is a talented athlete." Knowing that Duane Gibson has just been called up to the major leagues, you conclude that Duane Gibson is a talented athlete.
5. As the handcuffed man is brought into the police station, you glance at him and say to your friend, "That fellow looks guilty to me."
6. While judging a science fair project, Mr. Cange finds that each of the first 5 projects are outstanding and concludes that all 10 will be outstanding.
7. You know the rule that "If a person lives in the Santa Rosa Junior College district, then he or she will receive a tuition break at Santa Rosa." Candace tells you that she has received a tuition break. You conclude that she resides in the Santa Rosa Junior College district.

8. As Mrs. Gibson enters the doctor's waiting room, she concludes that it will be a long wait.

In Exercises 9 to 12, use intuition to state a conclusion.

9. You are told that the opposite angles formed when two lines cross are **vertical angles.** In the accompanying figure angles 1 and 2 are vertical angles. Conclusion?

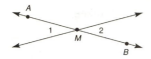

10. In the drawing for Exercise 9, point *M* is the **midpoint** of line segment *AB*. Conclusion?

11. The two triangles shown in the accompanying drawing are **similar** to each other. Conclusion?

12. Observe (but do not measure) the angles in the accompanying figure. Conclusion?

In Exercises 13 to 18, use induction to state a conclusion.

13. Several movies directed by Lawrence Garrison have won Academy Awards, while many others have received nominations. His latest work, *A Prisoner of Society,* is to be released next week. Conclusion?

14. On Monday, Matt says to you, "Andy hit his little sister at school today." On Tuesday, Matt informs you, "Andy threw his math book into the wastebasket during class." On Wednesday, Matt tells you, "Because Andy was throwing peas in the school cafeteria, he was sent to the principal's office." Conclusion?

15. While searching for a classroom, Tom stopped at an instructor's office to ask directions. On the office bookshelves are books titled *Intermediate Algebra, Calculus, Modern Geometry, Linear Algebra,* and *Differential Equations.* Conclusion?

16. At a friend's house, you see several food items, including apples, pears, grapes, oranges, and bananas. Conclusion?

17. While her daughter is having her annual dental examination, Brenda reads magazines in the waiting room. During the wait, a number of children, each carrying a small toy, return from their examinations. Conclusion?

18. On the chalkboard of the classroom you have listed the following numbers: 2, 3, 5, 7, 11, 13, 17, 19, ____ . Conclusion?

In Exercises 19 to 28, use deduction to state a conclusion, if possible.

19. If the sum of the measures of two angles is 90°, then these angles are termed "complementary." Angle 1 measures 27° and angle 2 measures 63°. Conclusion?

20. If a person attends college, then he or she will be a success in life. Kathy Jones attends Dade County Community College. Conclusion?

21. All mathematics teachers have a strange sense of humor. Alex is a mathematics teacher. Conclusion?

22. All mathematics teachers have a strange sense of humor. Alex has a strange sense of humor. Conclusion?

23. If Stewart Powers is elected president, then every family will have an automobile. Every family has an automobile. Conclusion?

24. If Tabby is meowing, then she is hungry. Tabby is hungry. Conclusion?

25. If a person is involved in politics, then that person will be in the public eye. June Jesse has been elected to the Missouri state senate. Conclusion?

26. If a student is enrolled in a literature course, then he or she will work very hard. Bram Spiegel digs ditches by hand 6 days a week. Conclusion?

27. If a person is rich and famous, then he or she is happy. Marilyn is wealthy as well as well-known. Conclusion?

28. If you study hard and hire a tutor, then you will make an A in this course. You study hard and make an A in this course. Conclusion?

In Exercises 29 to 34, either accept the claim or disprove it by stating or showing a counterexample.

29. Every prime number is odd. (*Note:* An **odd** number is a whole number that is not divisible by 2.)

30. When *n* is a positive integer, the number represented by $(n^2 + n + 19)$ is always prime.

31. A four-sided polygon, all of whose sides have the same length, must be a square.

32. The sum of a positive odd number and a positive even number is always prime. (*Note:* See Exercise 29; an **even** number is a whole number that is divisible by 2.)

33. When *n* is a positive integer, the number whose value is $(n^2 + n)$ is always divisible by 2.

34. When *n* is a positive integer, the number whose value is $(n^3 + 2n)$ is always divisible by 3. (*Note:* $n^3 = n \cdot n \cdot n$.)

STATEMENTS IN MATHEMATICS

In algebra and geometry, a **statement** is a collection of words and symbols that can be classified collectively as true or false. Of the following statements, 1 and 2 are true statements, 3 and 4 are false statements, and 5 is not a statement at all:

1. $4 + 3 = 7$

2. 17 is a prime number.

3. Ulysses S. Grant played shortstop for the Yankees.

4. $7 < 3$ (read as "7 is less than 3")

5. Look out!

Note: Some textbooks refer to "$x + 5 = 6$" as an "open statement"; it can be classified as true or false depending on the replacement value for x.

The **negation** of a given statement is a statement that makes the opposite claim to that of the given statement. If the given statement is true, its negation must be false. If the given statement is false, its negation must be true.

EXAMPLE 1

Give the negation of each statement:
(a) $4 + 3 = 7$
(b) All fish can swim.
(c) Mary is *not* a seamstress.

Solution

(a) $4 + 3 \neq 7$ ($\neq$ means "is not equal to")
(b) To negate the statement "All fish can swim," you need only assert that at least one fish cannot swim. Therefore, the negation is: Some fish cannot swim.
(c) Mary is a seamstress.   ▲

Statements composed of two or more simple statements are **compound statements.** For instance, "If P, then Q" is a type of compound statement known as a **conditional statement.** Here is one example of a conditional statement in geometry:

> *"If two sides of a triangle are equal in length, then two angles of the triangle are equal in measure." (See figure 1.4.)*

In the conditional statement, "If P, then Q," P is the **hypothesis** and Q is the **conclusion.** For the preceding claim, we have

Hypothesis: Two sides of a triangle are equal in length.

Conclusion: Two angles of the triangle are equal in measure.

Figure 1.4

EXAMPLE 2   Give the hypothesis and the conclusion of each conditional statement:
(**a**) If it rains, then Tim will stay inside.
(**b**) The rain will turn to snow if it gets very cold.
● (**c**) All right angles are congruent.

Solution   (**a**) *Hypothesis:* It rains.

Conclusion: Tim will stay inside.

(**b**) Restatement of claim: If it gets very cold, then the rain will turn to snow.

Hypothesis: It gets very cold.

Conclusion: The rain will turn to snow.

(**c**) Restatement of claim: If angles are right angles, then they are congruent.

Hypothesis: Angles are right angles.

Conclusion: They are congruent.   ▲

Three statements are related to the conditional statement: its converse, its inverse, and its contrapositive. In Table 1.1, "not *P*" means the negation of *P*, and "not *Q*" means the negation of *Q*.

Table 1.1   **Statements Related to the Conditional**

Conditional: ⊤	If *P*, then *Q*.	
Converse: F	If *Q*, then *P*. (Interchanges *P* and *Q*.)	
Inverse: F	If not *P*, then not *Q*. (Negates *P* and negates *Q*, but does not interchange their positions.)	
Contrapositive: ⊤	If not *Q*, then not *P*. (Interchanges *P* and *Q*, and negates each.)	

EXAMPLE 3   State the converse, the inverse, and the contrapositive of the following conditional statement:

If Fido eats Tastibits, then Fido is a happy puppy.

Solution   *Converse:*   If Fido is a happy puppy, then Fido eats Tastibits.

Inverse: If Fido does not eat Tastibits, then Fido is not a happy puppy.

Contrapositive: If Fido is not a happy puppy, then Fido does not eat Tastibits.
                                                                                            ▲

Whenever a conditional statement is true, its contrapositive is also true; likewise, whenever a conditional is false, its contrapositive is false. While the converse and the inverse do not necessarily have the same truth or falsity as the related conditional, they must either both be true or both be false.

EXAMPLE 4   State the converse, the inverse, and the contrapositive of the following conditional, and then determine the truth or falsity of each:

"If a man lives in Kalamazoo, then he lives in Michigan." (True)

Solution   *Converse:*   If a man lives in Michigan, then he lives in Kalamazoo. (False)

Inverse: If a man does not live in Kalamazoo, then he does not live in Michigan. (False)

Contrapositive: If a man does not live in Michigan, then he does not live in Kalamazoo. (True)   ▲

It is possible for the four statements—conditional, converse, inverse, and contrapositive—all to be true or all to be false. For instance, all four are true for the following statement:

"If an integer is even, then it is divisible by 2."

Another type of compound statement is the **conjunction,** which has the form "P and Q" and is true only when both P is true and Q is true. The accompanying table clarifies this point.

P	Q	P and Q
T	T	T
T	F	F
F	T	F
F	F	F

EXAMPLE 5   Classify the following as true or false:
(a) $2 + 3 = 5$, and $-4 < 7$.
(b) $2 + 3 = 5$, and Babe Ruth was a U.S. president.
(c) $5x + 2 < 12$, and Babe Ruth was a baseball player.

Solution   (a) True, because "$2 + 3 = 5$" is true and "$-4 < 7$" is also true.
(b) False, because "Babe Ruth was a U.S. president" is false.
(c) *Cannot* classify because $5x + 2 < 12$ is an open statement.   ▲

P	Q	P or Q
T	T	T
T	F	T
F	T	T
F	F	F

A statement of the form "P or Q" is a **disjunction** and is true when P is true, when Q is true, or when both P and Q are true. In other words, the disjunction is false only when both P and Q are false. The accompanying table clarifies this description.

Suppose that you can join the Mathematics Club "If you have an A average or if you are enrolled in a mathematics course." This means that a student may join under any of the following circumstances:

1. The student has an A average but is not enrolled in a math class.

2. The student does not have an A average but is enrolled in a math class.

3. The student has an A average and is also enrolled in a math class.

EXAMPLE 6   Classify the following as true or false:
(a) Babe Ruth was a baseball player, or $2 + 3 = 6$.
(b) Babe Ruth was a baseball player, or $2 + 3 = 5$.
(c) Either not P or Q (where P is true and Q is false).

Solution   (a) True, because the statement takes the form "*P* or *Q*" and *P* is true.
(b) True, because both *P* and *Q* are true.
(c) False, because not *P* is false and *Q* is also false.   ▲

Every definition must have a true converse. That is, every definition must be reversible. Thus, for example, the definition of "prime number" really states two claims:

1. If a number is prime, then it is an integer greater than 1 for which the only factors are the number itself and 1.

2. If an integer greater than 1 has only the number itself and 1 for factors, then it is a prime number.

EXAMPLE 7   Use the definition of "prime number" and deduction to draw a conclusion, given that:
(a) 17 is prime.
(b) 37 has only the factors 1 and 37.

Solution   (a) 17 has only the factors 1 and 17.
(b) 37 is prime.   ▲

Note: Sections 1.3 to 1.5 are devoted to the algebra needed to do geometry problems. Some of the material may be review for you. A Look Beyond is a preview of the nature of proof.

1.2 EXERCISES

In Exercises 1 and 2, which sentences are statements? If a sentence is a statement, classify it as true or false.

1. (a) Where do you live?
   (b) $4 + 7 \neq 5$.
   (c) Washington was the first U.S. president.
   (d) $x + 3 = 7$ when $x = 5$.

2. (a) Chicago is in the state of Illinois.
   (b) Get out of here!
   (c) $x < 6$ (read as "x is less than 6") when $x = 10$.
   (d) Babe Ruth is remembered as a great football player.

In Exercises 3 and 4, give the negation of each statement.

3. (a) Christopher Columbus crossed the Atlantic Ocean.
   (b) All jokes are funny.

4. (a) No one likes me.
   (b) Angle 1 is a right angle.

In Exercises 5 to 10, classify each statement as simple, conditional, a conjunction, or a disjunction.

5. If Alice plays, the volleyball team will win.

6. Alice played and the team won.

7. The first-place trophy is beautiful.

8. An integer is odd or it is even.

9. Matthew is playing shortstop.

10. You will be in trouble if you don't change your ways.

In Exercises 11 to 18, state the hypothesis and the conclusion of each statement.

11. If you go to the game, then you will have a great time.

12. If two chords of a circle have equal lengths, then the arcs of the chords are congruent.

13. If the diagonals of a parallelogram are perpendicular, then the parallelogram is a rhombus.

14. If $\dfrac{a}{b} = \dfrac{c}{d}$, where $b \neq 0$ and $d \neq 0$, then $a \cdot d = b \cdot c$.

15. Corresponding angles are congruent if two parallel lines are cut by a transversal.

16. Vertical angles are congruent when two lines intersect.

17. All squares are rectangles.

18. Base angles of an isosceles triangle are congruent.

In Exercises 19 to 22, give the converse, the inverse, and the contrapositive of each conditional statement. Then classify each of the four forms of the statement as true or false.

19. If your team wins the first-round game, then your team will win the eight-team tournament.

20. If a person lives in Sacramento, then he or she lives in California.

21. If a number is larger than 0, then it is positive.

22. If an integer is prime, then it is larger than 1.

In Exercises 23 to 28, classify each statement as true or false.

23. If a number is divisible by 6, then it is divisible by 3.

24. Rain is wet and snow is cold.

25. Rain is wet or snow is cold.

26. If Jim lives in Idaho, then he lives in Boise.

27. Triangles are round or circles are square.

28. Triangles are square or circles are round.

1.3 ALGEBRAIC EXPRESSIONS

Algebra and geometry are both examples of mathematical systems. A **mathematical system** is characterized by (1) acceptance of undefined terms on which definitions are based and (2) a set of assumed principles from which other principles can be deduced. The assumed principles of algebra are usually described as **axioms,** while those of geometry are called **postulates.** In either mathematical system, the statements that are deduced (or proved) are known as **theorems.**

FOUR PARTS OF A MATHEMATICAL SYSTEM
1. Undefined terms ⎫ vocabulary   **2.** Defined terms ⎭   **3.** Axioms or postulates ⎫ principles   **4.** Theorems ⎭

In algebra, we generally accept but do not define terms such as "addition," "multiplication," "number," "positive," and "equality." For convenience, a real number is one that has a position on the number line, as in figure 1.5.

Figure 1.5

Any real number positioned to the right of another real number is larger than the other number. For example, 4 is larger than -2; equivalently, -2 is less than 4 (smaller numbers are to the left).

Intuitively, we accept that two numerical expressions are **equal** if and only if they have the same value. For example, $2 + 3 = 5$. Some axioms of equality are listed in the following box.

AXIOMS OF EQUALITY

Reflexive ($a = a$): Any number equals itself.
Symmetric (if $a = b$, then $b = a$): Two equal numbers are equal in either order.
Transitive (if $a = b$ and $b = c$, then $a = c$): If a first number equals a second number, and if the second number equals a third number, then the first number equals the third number.
Substitution: If one numerical expression equals a second, then it may replace the second.

EXAMPLE 1   Name the axiom of each equality illustrated:
(a) If AB is the numerical length of the line segment $\overline{AB}$, then $AB = AB$.
(b) If $17 = 2x - 3$, then $2x - 3 = 17$.
(c) Given that $2x + 3x = 5x$, the statement $2x + 3x = 30$ can be replaced by $5x = 30$.

Solution   (a) Reflexive
(b) Symmetric
(c) Substitution ▲

To add two real numbers, think of the numbers as gains if positive and as losses if negative. For instance, $13 + (-5)$ represents the result of combining a gain of \$13 with a loss (or debt) of \$5. Therefore

$$13 + (-5) = 8$$

The result in addition is the **sum.** Three other examples are

$$13 + 5 = 18 \qquad (-13) + 5 = -8 \quad \text{and} \quad (-13) + (-5) = -18$$

If you multiply two real numbers, the **product** (answer) will be positive if the two numbers have the same sign, negative if the two numbers have different signs, and 0 if either or both numbers are 0.

EXAMPLE 2   Simplify each expression:
(a) $5 + (-4)$ (b) $5(-4)$ (c) $(-7)(-6)$ (d) $[5 + (-4)] + 8$

Solution   (a) $5 + (-4) = 1$
(b) $5(-4) = -20$
(c) $(-7)(-6) = 42$
(d) $[5 + (-4)] + 8 = 1 + 8 = 9$ ▲

What happens when you change the order in an addition problem? $(-3) + 9 = 6$ and $9 + (-3) = 6$. That the sums are equal when the order of the numbers added is reversed is often expressed by writing $a + b = b + a$. This property of real numbers is known as the Commutative Axiom for Addition. There is also a Commutative Axiom for Multiplication, which is illustrated by the fact that $(6)(-4) = (-4)(6)$; both products are -24.

Grouping symbols indicate which operation should be done first in a numerical expression. It is easily established that $[5 + (-4)] + 8$ equals $5 + [(-4) + 8]$, since $1 + 8$ equals $5 + 4$. In general, the fact that $(a + b) + c$ equals $a + (b + c)$ is known as the Associative Axiom for Addition. There is also an Associative Axiom for Multiplication, which is illustrated by the fact that

$$(3 \cdot 5)(-2) = 3[5(-2)]$$
$$(15)(-2) = 3(-10)$$
$$-30 = -30$$

SELECTED AXIOMS OF REAL NUMBERS

Commutative Axiom for Addition: $a + b = b + a$
Commutative Axiom for Multiplication: $a \cdot b = b \cdot a$
Associative Axiom for Addition: $a + (b + c) = (a + b) + c$
Associative Axiom for Multiplication: $a \cdot (b \cdot c) = (a \cdot b) \cdot c$

To subtract b from a (that is, to find $a - b$), change the subtraction problem to an addition problem.

DEFINITION OF SUBTRACTION

$$a - b = a + (-b)$$

where $-b$ is the additive inverse of b.

For $b = 5$, we have $-b = -5$ and for $b = -2$, we have $-b = 2$. For the subtraction $a - (b + c)$, we use the additive inverse of $b + c$, which is $(-b) + (-c)$. That is,

$$a - (b + c) = a + [(-b) + (-c)]$$

EXAMPLE 3   Simplify each expression:
(a) $5 - (-2)$ (b) $(-7) - (-3)$ (c) $12 - [3 + (-2)]$

Solution   (a) $5 - (-2) = 5 + 2 = 7$
(b) $(-7) - (-3) = (-7) + 3 = -4$
(c) $12 - [3 + (-2)] = 12 + [(-3) + 2] = 12 + (-1) = 11$ ▲

Division can be replaced by multiplication just as subtraction was replaced by addition. We cannot divide by 0!

DEFINITION OF DIVISION

For $b \neq 0$,

$$a \div b = a \cdot \frac{1}{b}$$

where $\frac{1}{b}$ is the multiplicative inverse of b.

Note: $a \div b$ is also indicated by a/b or $\frac{a}{b}$.

For $b = 5 \left(\text{that is, } b = \frac{5}{1} \right)$, we have $\frac{1}{b} = \frac{1}{5}$ and for $b = \frac{-2}{3}$, we have $\frac{1}{b} = \frac{-3}{2}$.

EXAMPLE 4   Simplify each expression:

(a) $12 \div 2$ (b) $(-5) \div \left(\frac{-2}{3} \right)$

Solution   (a) $12 \div 2 = 12 \div \frac{2}{1}$

$$= \frac{12}{1} \cdot \frac{1}{2}$$ product of two positive numbers

$$= 6$$ is a positive number

(b) $(-5) \div \left(\frac{-2}{3} \right) = \left(\frac{-5}{1} \right) \div \left(\frac{-2}{3} \right)$

$$= \left(\frac{-5}{1} \right) \cdot \left(\frac{3}{-2} \right)$$

$$= \frac{15}{2}$$ product of two negative numbers
      is a positive number   ▲

EXAMPLE 5   Mason works at the grocery store for 3 hours on Friday after school and then works for 8 hours on Saturday. If he is paid $4 per hour, how much will he be paid in all?

Solution   *Method I:*   Total the number of hours and multiply by 4.

$$4(3 + 8) = 4 \cdot 11 = \$44$$

Method II: Figure the daily wages and add them.

$$(4 \cdot 3) + (4 \cdot 8) = 12 + 32 = \$44$$

Friday's   Saturday's
wages   wages

Note: We see that $4(3 + 8) = 4 \cdot 3 + 4 \cdot 8$, where the multiplications on the right are performed before the addition is.   ▲

The Distributive Axiom was illustrated in Example 5. Because multiplications are performed before additions, we may write

$$a(b + c) = a \cdot b + a \cdot c$$
$$2(3 + 4) = 2 \cdot 3 + 2 \cdot 4$$
$$2(7) = 6 + 8$$

The "symmetric" form of the Distributive Axiom is

$$a \cdot b + a \cdot c = a(b + c)$$

This form can be used to combine *like terms* (expressions containing the same variable factors) in algebra. A **variable** is a letter that represents a **number**.

$4x + 5x = x \cdot 4 + x \cdot 5$	Commutative Axiom for Multiplication
$= x(4 + 5)$	Symmetric form of Distributive Axiom
$= x(9)$	Substitution
$= 9x$	Commutative Axiom for Multiplication
$\therefore 4x + 5x = 9x$	

The Distributive Axiom also distributes multiplication over subtraction.

FORMS OF THE DISTRIBUTIVE AXIOM

$$a(b + c) = a \cdot b + a \cdot c$$
$$a \cdot b + a \cdot c = a(b + c)$$
$$a(b - c) = a \cdot b - a \cdot c$$
$$a \cdot b - a \cdot c = a(b - c)$$

EXAMPLE 6

Combine like terms:
(a) $7x + 3x$ (b) $7x - 3x$ (c) $3x^2y + 4x^2y + 6x^2y$
(d) $3x^2y + 4xy^2 + 6xy^2$ (e) $7x + 5y$

Solution

(a) $7x + 3x = 10x$
(b) $7x - 3x = 4x$
(c) $3x^2y + 4x^2y + 6x^2y = (3x^2y + 4x^2y) + 6x^2y = 7x^2y + 6x^2y = 13x^2y$
(d) $3x^2y + 4xy^2 + 6xy^2 = 3x^2y + (4xy^2 + 6xy^2) = 3x^2y + 10xy^2$
(e) $7x + 5y;$ cannot combine unlike terms

Note: In (d), $3x^2y$ and $10xy^2$ are not like terms because $x^2y \neq xy^2$.   ▲

The statement $4x + 5x = 9x$ says that "the sum of 4 times a number and 5 times the same number equals 9 times the same number." Thus when $x = 3$, we are saying that "4 times 3 plus 5 times 3 equals 9 times 3." Because x can be any real number, we may also write, for example,

$$4\pi + 5\pi = 9\pi$$

in which π is the real number that equals approximately 3.14. Similarly,

$$4\sqrt{3} + 5\sqrt{3} = 9\sqrt{3}$$

in which $\sqrt{3}$ (read as "the positive square root of 3") is equal to approximately 1.73.

You may recall the "order of operations" from a previous class; this order is to be used when simplifying more complicated expressions.

ORDER OF OPERATIONS

1. Simplify expressions within symbols such as parentheses ( ) or brackets [ ], beginning with the innermost symbols of inclusion first.

Note: The presence of a fraction bar ——— requires that you simplify a numerator or denominator before dividing.

2. Perform all work with exponents.

3. Do all multiplications and/or divisions, in order, from left to right.

4. Finally, do all additions and/or subtractions, in order, from left to right.

EXAMPLE 7   Simplify each numerical expression:
(a) $3^2 + 4^2$ (b) $4 \cdot 7 \div 2$ (c) $2 \cdot 3 \cdot 5^2$
(d) $\dfrac{8 - 6 \div (-3)}{4 + 3(2 + 5)}$ (e) $2 + [3 + 4(5 - 1)]$

Solution   (a) $3^2 + 4^2 = 9 + 16 = 25$
(b) $4 \cdot 7 \div 2 = 28 \div 2 = 14$
(c) $2 \cdot 3 \cdot 5^2 = 2 \cdot 3 \cdot 25$
$\qquad\qquad = (2 \cdot 3) \cdot 25 = 6 \cdot 25 = 150$
(d) $\dfrac{8 - [6 \div (-3)]}{4 + 3(2 + 5)} = \dfrac{8 - (-2)}{4 + 3(7)}$

$$= \frac{10}{4 + 21} = \frac{10}{25} = \frac{2}{5}$$

(e) $2 + [3 + 4(5 - 1)] = 2 + [3 + 4(4)]$
$\qquad\qquad\qquad\qquad = 2 + (3 + 16)$
$\qquad\qquad\qquad\qquad = 2 + 19 = 21$ ▲

An expression like $(2 + 5)(6 + 4)$ can be simplified by two different methods. By following the rules of order, we have $(7)(10)$ or 70. An alternate method is described as the FOIL method: First, Outside, Inside, and Last terms are multiplied and then added. This is how it works:

$$(2 + 5)(6 + 4) = 2 \cdot 6 + 2 \cdot 4 + 5 \cdot 6 + 5 \cdot 4$$
$$= 12 + 8 + 30 + 20$$
$$= 70$$

FOIL is the Distributive Axiom in disguise. While we would not generally use FOIL to find the product of $(2 + 5)$ and $(6 + 4)$, we must use it to find the product in Example 8.

EXAMPLE 8   Use the FOIL method to find the products.
(a) $(3x + 4)(2x - 3)$ (b) $(5x + 2y)(6x - 5y)$

Solution   (a) $(3x + 4)(2x - 3) = 3x \cdot 2x + 3x(-3) + 4(2x) + 4(-3)$
$$= 6x^2 + (-9x) + 8x + (-12)$$
$$= 6x^2 - 1x - 12$$
$$= 6x^2 - x - 12$$
(b) $(5x + 2y)(6x - 5y) = 5x \cdot 6x + 5x(-5y) + 2y(6x) + 2y(-5y)$
$$= 30x^2 + (-25xy) + 12xy + (-10y^2)$$
$$= 30x^2 - 13xy - 10y^2$$ ▲

EXAMPLE 9   A rectangular garden plot has been subdivided as shown in figure 1.6 into smaller rectangles. Find the area of:
(a) rectangle I
(b) rectangle II
(c) rectangle III
(d) rectangle IV
(e) the large plot composed of I, II, III, and IV [*Hint:* Add parts (a) to (d)]
(f) the large plot determined by multiplying the dimensions
$(a + b)(c + d)$

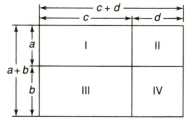

Figure 1.6

Solution   (a) $a \cdot c$ (b) $a \cdot d$ (c) $b \cdot c$ (d) $b \cdot d$
(e) $a \cdot c + a \cdot d + b \cdot c + b \cdot d$
(f) $a \cdot c + a \cdot d + b \cdot c + b \cdot d$
Note: Results in parts (e) and (f) are necessarily identical.   ▲

1.3 EXERCISES

1. Name the four parts of a mathematical system.
2. Name two examples of mathematical systems.
3. Which axiom of equality is illustrated in each of the following?
   (a) $5 = 5$

(b) If $\dfrac{1}{2} = 0.5$ and $0.5 = 50$ percent, then $\dfrac{1}{2} = 50$ percent.
(c) Because $2 + 3 = 5$, we may replace $x + (2 + 3)$ by $x + 5$.
(d) If $7 = 2x - 3$, then $2x - 3 = 7$.

4. Give an example to illustrate each axiom of equality:

   (a) Reflexive   (c) Transitive
   (b) Symmetric   (d) Substitution

5. Find each sum:

   (a) $5 + 7$ (c) $(-5) + 7$
   (b) $5 + (-7)$ (d) $(-5) + (-7)$

6. Find each sum:

   (a) $(-7) + 15$ (c) $(-7) + (-15)$
   (b) $7 + (-15)$ (d) $(-7) + [(-7) + 15]$

7. Find each product:

   (a) $5 \cdot 7$ (c) $(-5)7$
   (b) $5(-7)$ (d) $(-5)(-7)$

8. Find each product:

   (a) $(-7)(12)$ (c) $(-7)[(3)(4)]$
   (b) $(-7)(-12)$ (d) $(-7)[(3)(-4)]$

9. The area (the number of squares) of the rectangle in the accompanying drawing can be determined by multiplying the measures of the two dimensions. Will the order of multiplication change the answer? Which axiom is illustrated?

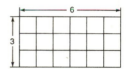

10. Identify the axiom of real numbers illustrated. Give a complete answer, such as Commutative Axiom for Multiplication.

    (a) $7(5) = 5(7)$
    (b) $(3 + 4) + 5 = 3 + (4 + 5)$
    (c) $(-2) + 3 = 3 + (-2)$
    (d) $(2 \cdot 3) \cdot 5 = 2 \cdot (3 \cdot 5)$

11. Perform each subtraction:

    (a) $7 - (-2)$ (c) $10 - 2$
    (b) $(-7) - (+2)$ (d) $(-10) - (-2)$

12. The temperature changes from $-3°F$ at 2 A.M. to $7°F$ at 7 A.M. Which expression represents the difference in temperatures from 2 A.M. to 7 A.M., $7 - (-3)$ or $(-3) - 7$?

13. Complete each division:

    (a) $12 \div (-3)$ (c) $(-12) \div \left(\dfrac{-2}{3}\right)$

    (b) $12 \div \left(\dfrac{-1}{3}\right)$ (d) $\left(\dfrac{-1}{12}\right) \div \left(\dfrac{1}{3}\right)$

14. Nine pegs are evenly spaced on a board so that the distance from each end to a peg equals the distance between any two pegs. If the board is 5 feet long, how far apart are the pegs?

15. The four owners of a shop realize a loss of $240 in February. If the loss is shared equally, what number represents the profit for each owner for that month?

16. Bill works at a weekend convention by selling copies of a book. He receives a $2 commission for each copy sold. If he sells 25 copies on Saturday and 30 copies on Sunday, what is Bill's total commission?

17. Use the Distributive Axiom to simplify each expression:

    (a) $5(6 + 7)$ (c) $\dfrac{1}{2}(7 + 11)$

    (b) $4(7 - 3)$ (d) $5x + 3x$

18. Use the Distributive Axiom to simplify each expression:

    (a) $6(9 - 4)$ (c) $7y - 2y$

    (b) $\left(\dfrac{1}{2}\right) \cdot 6(4 + 8)$ (d) $16x + 8x$

19. Simplify each expression:

    (a) $6\pi + 4\pi$ (c) $16x^2y - 9x^2y$
    (b) $8\sqrt{2} + 3\sqrt{2}$ (d) $9\sqrt{3} - 2\sqrt{3}$

20. Simplify each expression:

    (a) $\pi r^2 + 2\pi r^2$ (c) $7x^2y + 3xy^2$
    (b) $7xy + 3xy$ (d) $x + x + y$

21. Simplify each expression:

    (a) $2 + 3 \cdot 4$ (c) $2 + 3 \cdot 2^2$
    (b) $(2 + 3) \cdot 4$ (d) $2 + (3 \cdot 2)^2$

22. Simplify each expression:

    (a) $3^2 + 4^2$ (c) $3^2 + (8 - 2) \div 3$
    (b) $(3 + 4)^2$ (d) $[3^2 + (8 - 2)] \div 3$

23. Simplify each expression:

    (a) $\dfrac{8 - 2}{2 - 8}$ (c) $\dfrac{5 \cdot 2 - 6 \cdot 3}{7 - (-2)}$

    (b) $\dfrac{8 - 2 \cdot 3}{(8 - 2) \cdot 3}$ (d) $\dfrac{5 - 2 \cdot 6 + (-3)}{(-2)^2 + 4^2}$

24. Use the FOIL method to complete each multiplication:

    (a) $(2 + 3)(4 + 5)$ (b) $(7 - 2)(6 + 1)$

25. Use the FOIL method to complete each multiplication:

    (a) $(3 - 1)(5 - 2)$ (b) $(3x + 2)(4x - 5)$

26. Use the FOIL method to complete each multiplication:

    (a) $(5x + 3)(2x - 7)$ (b) $(2x + y)(3x - 5y)$

27. Using x and y, find an expression for the length of the pegged board shown in the accompanying figure.

28. The cardboard used in the construction of the box shown in the accompanying figure has an area of $xy + yz + xz + xz + yz + xy$. Simplify this expression for the total area of the cardboard.

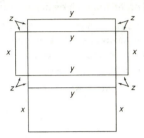

29. A large star is to be constructed, with lengths as shown in the accompanying figure. Give an expression for the total length of the wood strips used in construction.

30. The area of an enclosed plot of ground that a farmer has subdivided can be found by multiplying $(x + y)$ and $(y + z)$. Use FOIL to complete the multiplication. How does this product compare with the total of the areas of the four smaller plots?

31. The degree measures of the angles of a triangle are $3x$, $5x$, and $2x$. Find an expression for the sum of the measures of those angles in terms of x.

32. The right circular cylinder shown in the accompanying figure has circular bases that have areas of 9π square units. The side has an area of 48π square units. Find an expression for the total surface area shown.

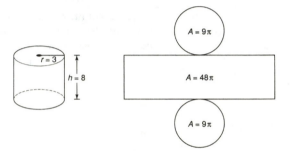

FORMULAS AND EQUATIONS

1.4

As you saw in the previous section, a **variable** is a letter used to represent an unknown number. The Greek letter π is known as a **constant** because it always equals the same number (approximately 3.14). While we will often use as variables the letters x, y, and z, it is convenient to choose r for radius, h for height, b for base, and so on.

EXAMPLE 1

For figure 1.7, combine like terms to find the perimeter (sum of the lengths of all sides) of the figure.

Solution

$$(x - 1) + (x + 2) + 2x + (2x + 1)$$
$$= x + (-1) + x + 2 + 2x + 2x + 1$$
$$= 1x + 1x + 2x + 2x + (-1) + 2 + 1$$
$$= 6x + 2 \qquad \blacktriangle$$

When the FOIL method is used with variable expressions, we can often combine like terms in the simplification.

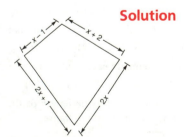

Figure 1.7

EXAMPLE 2   Find a simplified expression for the area of a rectangle with length $x + 5$ and width $x + 2$ by multiplying $(x + 5)$ times $(x + 2)$.

Solution

$$(x + 5)(x + 2)$$

$$= x \cdot x + 2 \cdot x + 5 \cdot x + 10$$
$$= x^2 + 7x + 10 \qquad \blacktriangle$$

In Example 2, we multiplied by the FOIL method before adding like terms in accordance with rules of order. When evaluating a variable expression, we must also follow that order. For instance, the value of $a^2 + b^2$ when $a = 3$ and $b = 4$ is given by

$$3^2 + 4^2 \qquad \text{or} \qquad 9 + 16 \qquad \text{or} \qquad 25$$

because exponential expressions must be simplified before addition occurs.

EXAMPLE 3   Find the value of the following expressions.
(a) $\pi r^2 h$, if $r = 3$ and $h = 4$ (leave π in answer)

(b) $\left(\dfrac{1}{2}\right) h(b + B)$, if $h = 10$, $b = 7$, and $B = 13$

Solution   (a) $\pi r^2 h = \pi \cdot 3^2 \cdot 4$
$$= \pi \cdot 9 \cdot 4 = \pi(36) = 36\pi$$

(b) $\left(\dfrac{1}{2}\right) h(b + B) = \left(\dfrac{1}{2}\right) \cdot 10(7 + 13)$

$$= \left(\dfrac{1}{2}\right) \cdot 10(20)$$

$$= \left(\dfrac{1}{2} \cdot 10\right)(20)$$

$$= 5 \cdot (20) = 100 \qquad \blacktriangle$$

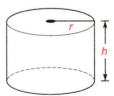

Figure 1.8

Many of the variable expressions that you will encounter are found in formulas. A **formula** is an equation that expresses a rule. For example, $V = \pi r^2 h$ is a formula for the volume V of a right circular cylinder whose height has length h and for which the circular base has a radius of length r. (See figure 1.8.)

EXAMPLE 4   Given the formula $P = 2\ell + 2w$, find the value of P when $\ell = 7$ and $w = 3$.

Solution   By substitution, $P = 2\ell + 2w$ becomes

$$P = (2 \cdot 7) + (2 \cdot 3)$$
$$= 14 + 6$$
$$= 20 \qquad \blacktriangle$$

An **equation** is an open statement that two expressions are equal. While formulas are special types of equations, most equations are not formulas. Consider the following four examples of equations:

$$x + (x + 1) = 7$$
$$2(x + 1) = 8 - 2x$$
$$x^2 - 6x + 8 = 0$$
$$P = 2\ell + 2w$$

The phrase "solving an equation" means finding the values of the variable that make the equation true when the variable is replaced by those values. These values are known as **solutions** for the equation. For example, 3 is a solution (in fact, the only solution) for the equation $x + (x + 1) = 7$ because $3 + (3 + 1) = 7$ is true.

When each side of an equation is transformed (changed) without having its solutions changed, we say that an equivalent equation is produced. Some of the properties that are used for equation solving are listed in the following box. Since these properties can be proved, they could be called theorems. However, we prefer to call an algebraic theorem by the name "property" in this book.

▲ *WARNING: We cannot multiply by 0 in solving an equation because the equation—say, $2x - 1 = 7$—collapses to $0 = 0$. Division by 0 is likewise excluded.*

PROPERTIES FOR EQUATION SOLVING

Addition Property of Equality (if $a = b$, then $a + c = b + c$): An equivalent equation results when the same number is added to each side of an equation.

Subtraction Property of Equality (if $a = b$, then $a - c = b - c$): An equivalent equation results when the same number is subtracted from each side of an equation.

Multiplication Property of Equality (if $a = b$, then $a \cdot c = b \cdot c$ for $c \neq 0$): An equivalent equation results when each side of an equation is multiplied by the same nonzero number.

Division Property of Equality $\left(\text{if } a = b, \text{ then } \dfrac{a}{c} = \dfrac{b}{c} \text{ for } c \neq 0\right)$: An equivalent equation results when each side of an equation is divided by the same nonzero number.

Addition and subtraction are **inverse operations** as are multiplication and division. In problems that involve equation solving, we will utilize inverse operations:

Add	to eliminate	a subtraction.
Subtract	to eliminate	an addition.
Multiply	to eliminate	a division.
Divide	to eliminate	a multiplication.

EXAMPLE 5 Solve the equation $2x - 3 = 7$.

Solution First add 3 (to eliminate the subtraction of 3 from $2x$):

$$2x - 3 + 3 = 7 + 3$$
$$2x = 10 \qquad \text{simplifying}$$

Now divide by 2 (to eliminate the multiplication of 2 with x):

$$\frac{2x}{2} = \frac{10}{2}$$
$$x = 5 \qquad \text{simplifying} \qquad \blacktriangle$$

Thus, 5 is the solution for the original equation. Replacing x with 5, we can confirm this:

$$2x - 3 = 7$$
$$2(5) - 3 = 7$$
$$10 - 3 = 7$$

Some steps shown in this particular solution will be eliminated in the future. We may simplify as operations are performed so that the work has this appearance:

$$2x - 3 = 7$$
$$2x = 10 \qquad \text{by addition}$$
$$x = 5 \qquad \text{by division}$$

An equation that can be written in the form $ax + b = c$ for constants a, b, and c is a **linear equation.** Our plan for solving such an equation involves getting variable terms together on one side of the equation and numerical terms together on the other side.

SOLVING A LINEAR EQUATION
1. Simplify each side of the equation.
2. Eliminate additions and/or subtractions.
3. Eliminate multiplications and/or divisions.

EXAMPLE 6 Solve the equation $2(x - 3) + 5 = 13$.

Solution

$$2(x - 3) + 5 = 13$$
$$2x - 6 + 5 = 13 \qquad \text{Distributive Axiom}$$
$$2x - 1 = 13 \qquad \text{substitution}$$
$$2x = 14 \qquad \text{addition}$$
$$x = 7 \qquad \text{division} \qquad \blacktriangle$$

Note: To check the solution, 7, we have

$$2(7 - 3) + 5 = 2(4) + 5 = 8 + 5 = 13$$

Some equations involve fractions. To avoid some of the difficulties that fractions bring, we often multiply each side of such equations by the **least common denominator (LCD)** of the fractions involved.

EXAMPLE 7   Solve the equation $\frac{x}{3} + \frac{x}{4} = 14$.

Solution   For the denominators 3 and 4, the LCD is 12. We must therefore multiply each side by 12 and use the Distributive Axiom on the left side.

$$12\left(\frac{x}{3} + \frac{x}{4}\right) = 12 \cdot 14$$

$$\frac{12}{1} \cdot \frac{x}{3} + \frac{12}{1} \cdot \frac{x}{4} = 168$$

$$4x + 3x = 168$$

$$7x = 168$$

$$x = 24 \qquad \blacktriangle$$

To check this result, we have

$$\frac{24}{3} + \frac{24}{4} = 14$$

$$8 + 6 = 14$$

It may happen that the variable appears in the denominator of one side of the equation. In such cases, our method does not change! We continue to clear an equation of fractions through multiplication by the LCD.

EXAMPLE 8   Solve the following equation for n:

$$\frac{360}{n} + 120 = 180$$

Solution   Multiplying by n (the LCD), we have

$$n\left(\frac{360}{n} + 120\right) = 180 \cdot n$$

$$\frac{n}{1} \cdot \frac{360}{n} + 120 \cdot n = 180n$$

$$360 + 120n = 180n$$

$$360 = 60n$$

$$6 = n \qquad \blacktriangle$$

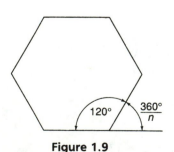

Figure 1.9

Note: n represents the number of sides possessed by the polygon in figure 1.9. $\frac{360}{n}$ and 120 represent the measures of angles in the figure.

Our final example combines many of the ideas introduced in this and the previous section. Example 9 is based on the formula for the area of a trapezoid.

EXAMPLE 9   In figure 1.10, for the formula $A = \left(\dfrac{1}{2}\right) \cdot h \cdot (b + B)$, suppose that $A = 77$, $b = 4$, and $B = 7$. Find the value of h.

Solution   Substitution leads to the equation

$$77 = \left(\frac{1}{2}\right) \cdot h \cdot (4 + 7)$$

$$77 = \left(\frac{1}{2}\right) \cdot h \cdot 11$$

$$2(77) = 2 \cdot \left(\frac{1}{2}\right) \cdot h \cdot 11 \qquad \text{multiplying by 2}$$

$$154 = 11h \qquad \text{simplifying}$$

$$14 = h \qquad \text{dividing by 11} \qquad \blacktriangle$$

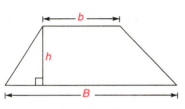

Figure 1.10

In closing, we recall that the Addition Property of Equality (if $a = b$, then $a + c = b + c$) is really a theorem. We can show that $a + c = b + c$ must be true whenever $a = b$. This proof is sketched as follows.

Because $a + c$ is the sum of two real numbers, it is a real number also.

$$\therefore a + c = a + c \qquad \text{Reflexive Axiom}$$

But

$$a = b \qquad \text{hypothesis}$$

$$\therefore a + c = b + c \qquad \text{substitution of } b \text{ for } a$$

A more generalized form of the Addition Property of Equality can be stated as follows:

If $a = b$ and $c = d$, then $a + c = b + d$.

In this restatement, d is substituted for c on the right side of the earlier equation $a + c = b + c$ to obtain $a + c = b + d$.

1.4 EXERCISES

In Exercises 1 to 6, simplify by combining similar terms.

1. $(2x + 3) + (3x + 5)$

2. $(2x + 3) - (3x - 5)$

3. $x + (3x + 2) - (2x + 4)$

4. $(3x + 2) + (2x - 3) - (x + 1)$

5. $2(x + 1) + 3(x + 2)$ (*Hint:* Multiply before adding.)

6. $3(2x + 5) - 2(3x - 1)$

In Exercises 7 to 12, simplify by using the FOIL method of multiplication.

7. $(x + 3)(x + 4)$

8. $(x - 5)(x - 7)$

9. $(2x + 5)(3x - 2)$

10. $(3x + 7)(2x + 3)$

11. $(a + b)^2 + (a - b)^2$

12. $(x + 2)^2 - (x - 2)^2$

In Exercises 13 to 16, evaluate each expression.

13. $\ell \cdot w \cdot h$, if $\ell = 4$, $w = 3$, and $h = 5$

14. $a^2 + b^2$, if $a = 5$ and $b = 7$

15. $2 \cdot \ell + 2 \cdot w$, if $\ell = 13$ and $w = 7$

16. $a \cdot b \div c$, if $a = 6$, $b = 16$, and $c = 4$

In Exercises 17 to 20, find the value of the variable named in each formula. Leave π in the answers for Exercises 19 and 20.

17. S, if $S = 2 \cdot \ell \cdot w + 2 \cdot w \cdot h + 2 \cdot \ell \cdot h$, $\ell = 6$, $w = 4$, and $h = 5$

18. A, if $A = \left(\dfrac{1}{2}\right) a(b + c + d)$, $a = 2$, $b = 6$, $c = 8$, and $d = 10$

19. V, if $V = \left(\dfrac{1}{3}\right) \pi \cdot r^2 \cdot h$, $r = 3$, and $h = 4$

20. S, if $S = 4\pi r^2$ and $r = 2$

In Exercises 21 to 32, solve each equation.

21. $2x + 3 = 17$

22. $3x - 3 = -6$

23. $\dfrac{y}{-3} + 2 = 6$

24. $3y = -21 - 4y$

25. $a + (a + 2) = 26$

26. $b = 27 - \dfrac{b}{2}$

27. $2(x + 1) = 30 - 6(x - 2)$

28. $2(x + 1) + 3(x + 2) = 22 + 4(10 - x)$

29. $\dfrac{x}{3} - \dfrac{x}{2} = -5$

30. $\dfrac{x}{2} + \dfrac{x}{3} + \dfrac{x}{4} = 26$

31. $\dfrac{360}{n} + 135 = 180$

32. $\dfrac{(n - 2) \cdot 180}{n} = 150$

In Exercises 33 to 36, find the value of the indicated variable for each given formula.

33. w if $S = 2 \cdot \ell \cdot w + 2 \cdot w \cdot h + 2 \cdot \ell \cdot h$, $S = 148$, $\ell = 5$, and $h = 6$

34. b if $A = \left(\dfrac{1}{2}\right) \cdot h \cdot (b + B)$, $A = 156$, $h = 12$, and $B = 11$

35. y if $m = \left(\dfrac{1}{2}\right)(x - y)$, $m = 23$, and $x = 78$

36. Y if $m = \dfrac{Y - y}{X - x}$, $m = \dfrac{-3}{2}$, $y = 1$, $X = 2$, and $x = -2$

In Exercises 37 and 38, write a proof like the one in the final paragraph of this section. Assume that a, b, and c are real numbers.

37. If $a = b$, then $a \cdot c = b \cdot c$.

38. If $a = b$, then $a - c = b - c$.

In Exercises 39 and 40, the statements are not always true. Cite a counterexample for each claim.

39. If $a \cdot c = b \cdot c$, then $a = b$. (*Note:* Name values of a, b, and c for which $a \cdot c = b \cdot c$ but $a \neq b$.)

40. If $a^2 = b^2$, then $a = b$.

1.5 INEQUALITIES

In geometry, we will need to work with inequalities. **Inequalities** are statements that involve one of the following relationships:

$<$	means "is less than"
$>$	means "is greater than"
$\leq$	means "is less than or equal to"
$\geq$	means "is greater than or equal to"
$\neq$	means "is not equal to"

The statement $-4 < 7$ is true because negative 4 is less than positive 7. On the number line, the smaller number is always found to the left of the larger number. An equivalent claim is $7 > -4$, which means positive 7 is greater than negative 4.

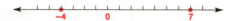

Both statements $6 \leq 6$ and $4 \leq 6$ are true because the relationship $\leq$ provides a disjunction. The statement $6 \leq 6$ could also be expressed by the statement $6 < 6$ *or* $6 = 6$, which is true because $6 = 6$ is true. Because $4 < 6$ is true, the disjunction $4 \leq 6$ is also true.

EXAMPLE 1 Give two true statements that involve the symbol $\geq$.

Solution

$\quad\quad 5 \geq 5 \quad\quad$ because $5 = 5$ is true
$\quad\quad 12 \geq 5 \quad\quad$ because $12 > 5$ is true ▲

The symbol $\neq$ is used to join any two numeric expressions that do not have the same value; for example, $2 + 3 \neq 7$.

> **DEFINITION:** *a* is less than *b* (that is, $a < b$) if and only if there is a positive number *p* for which $a + p = b$; *a* is greater than *b* (that is, $a > b$) if and only if $b < a$.

EXAMPLE 2 Find, if possible, the following:
(a) Any number *a* for which $a < a$ is true.
(b) Any numbers *a* and *b* for which $a < b$ and $b < a$ is true.

Solution
(a) There is no such number. If $a < a$, then $a + p = a$ for some positive number *p*. Subtracting *a* from each side of the equation gives $p = 0$. This statement ($p = 0$) contradicts the fact that *p* is positive.
(b) There are no such numbers. If $a < b$, then *a* is to the left of *b* on the number line. Therefore, $b < a$ is false, because this statement claims that *b* is to the left of *a*. ▲

EXAMPLE 3 What may you conclude for the numbers *x*, *y*, and *z* if $x < y$ and $x > z$?

Solution $x < y$ means that *x* is to the left of *y*, as in figure 1.11. Similarly, $x > z$ (equivalently, $z < x$) means that *z* is to the left of *x*. With *z* to the left of *x*, which is itself to the left of *y*, we clearly have *z* to the left of *y*; thus $z < y$. ▲

Figure 1.11

Example 3 suggests a transitive relationship for the inequality "is less than," and this is stated in the following property. The Transitive Property of Inequality can also be stated using $>$, $\leq$, or $\geq$.

TRANSITIVE PROPERTY OF INEQUALITY

For numbers a, b, and c, if $a < b$ and $b < c$, then $a < c$.

This property can be proved as follows:

1. $a < b$ means that $a + p_1 = b$ for some positive number p_1.
2. $b < c$ means that $b + p_2 = c$ for some positive number p_2.
3. Substituting $a + p_1$ for b (from statement 1) in the statement $b + p_2 = c$, we have $(a + p_1) + p_2 = c$.
4. Now $a + (p_1 + p_2) = c$.
5. But the sum of two positive numbers is also positive; that is, $p_1 + p_2 = p$, so statement 4 becomes $a + p = c$.
6. If $a + p = c$, then $a < c$, by the definition of "is less than."

Therefore $a < b$ and $b < c$ implies that $a < c$.   ▲

The Transitive Property of Inequality can be extended to a series of unequal expressions. When a first value is less than a second, the second is less than a third, and so on, then the first is less than the last.

EXAMPLE 4   Two angles are complementary if the sum of their measures is exactly 90°. If the measure of the first of two complementary angles is more than 27°, what must you conclude about the measure of the second angle?

Solution   The second angle must measure less than 63°. The statements needed to establish this result are as follows:

1. If x and y are the angle measures, then $x + y = 90$.
2. If $x > 27$, then $27 < x$ and $27 + p = x$ for some positive number p.
3. $x + y = 90$ becomes $(27 + p) + y = 90$.
4. Restated, this equation is $27 + (p + y) = 90$.
5. Then $p + y = 63$.
6. Therefore $y < 63$.   ▲

We now turn our attention to solving inequalities such as

$$x + (x + 1) < 7 \quad \text{and} \quad 2(x - 3) + 5 \geq 3$$

While the plan here is almost the same as the one used for equation solving, there are some very important differences.

EXAMPLE 5   For the statement $-6 < 9$, determine the statement that results when each side is changed as follows:

(a) has 4 added to it   (c) is multiplied by 3
(b) has 2 subtracted from it   (d) is divided by -3

Solution   (a) $-6 + 4 \ ? \ 9 + 4$
$$-2 \ ? \ 13 \rightarrow -2 < 13$$

(b) $-6 - 2 \ ? \ 9 - 2$
$$-8 \ ? \ 7 \rightarrow -8 < 7$$

(c) $(-6)(3) \ ? \ 9(3)$
$$-18 \ ? \ 27 \rightarrow -18 < 27$$

(d) $(-6) \div (-3) \ ? \ 9 \div (-3)$
$$2 \ ? \ -3 \rightarrow 2 > -3.$$ ▲

As Example 5 suggests, addition and subtraction preserve the inequality symbol; and multiplication and division by a *positive* number preserve the inequality symbol, but multiplication and division by a *negative* number reverse the inequality symbol.

PROPERTIES OF INEQUALITIES

Stated for $<$, these properties have counterparts involving $>$, $\leq$, and $\geq$.

Addition: If $a < b$, then $a + c < b + c$.

Subtraction: If $a < b$, then $a - c < b - c$.

Multiplication: (i) If $a < b$ and $c > 0$ (c is positive), then $a \cdot c < b \cdot c$.
                   (ii) If $a < b$ and $c < 0$ (c is negative), then $a \cdot c > b \cdot c$.

Division: (i) If $a < b$ and $c > 0$ (c is positive), then $\dfrac{a}{c} < \dfrac{b}{c}$.

                   (ii) If $a < b$ and $c < 0$ (c is negative), then $\dfrac{a}{c} > \dfrac{b}{c}$.

The plan for solving inequalities is similar to the one used for equation solving.

▲ *WARNING: Be sure to reverse the inequality symbol upon multiplying or dividing by a negative number.*

SOLVING AN INEQUALITY

1. Simplify each side of the inequality.
2. Eliminate additions and subtractions.
3. Eliminate multiplications and divisions.

EXAMPLE 6   Solve $2x - 3 \leq 7$.

Solution

$$2x - 3 + 3 \leq 7 + 3 \qquad \text{adding 3 preserves } \leq$$
$$2x \leq 10 \qquad \text{simplify}$$
$$\frac{2x}{2} \leq \frac{10}{2} \qquad \text{division by } +2 \text{ preserves } \leq$$
$$x \leq 5 \qquad \text{simplify}$$

The possible values of x are shown on a number line in figure 1.12; this picture is the **graph** of the solutions. Notice that the point above the 5 is filled solid to indicate that 5 is included as a solution.   ▲

Figure 1.12

EXAMPLE 7   Solve $x(x - 2) - (x + 1)(x + 3) < 9$.

Solution   Using the Distributive Axiom and FOIL, we simplify the left side to get

$$(x^2 - 2x) - (x^2 + 4x + 3) < 9$$

Subtraction is performed by adding the additive inverse of each term in $(x^2 + 4x + 3)$.

$$\therefore (x^2 - 2x) + (-x^2 - 4x - 3) < 9$$
$$-6x - 3 < 9 \qquad \text{simplify}$$
$$-6x < 12 \qquad \text{add 3}$$
$$\frac{-6x}{-6} > \frac{12}{-6} \qquad \begin{array}{l}\text{divide by } -6 \text{ and reverse}\\ \text{the inequality symbol}\end{array}$$
$$x > -2$$

Figure 1.13

The graph of the solution is shown in figure 1.13. Notice that the circle above the -2 is left empty to indicate that -2 is not included as a solution.   ▲

1.5 EXERCISES

1. If line segment AB and line segment CD in the accompanying drawing are drawn to scale, what does intuition tell you about the lengths of these segments?

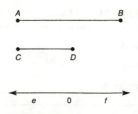

2. Using the number line shown in Exercise 1, write two statements that relate the values of e and f.

3. If angles ABC and DEF in the accompanying drawing were measured with a protractor, what does intuition tell you about the degree measures of these angles?

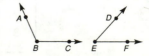

4. Consider the statement $x \geq 6$. Which of the following choice(s) of x below will make this a true statement?

$$x = -3 \qquad x = 0 \qquad x = 6 \qquad x = 9 \qquad x = 12$$

5. According to the definition of $a < b$, there is a positive number p for which $a + p = b$. Find the value of p for the statement given.

(a) $3 < 7$ (b) $-3 < 7$

6. Does the Transitive Property of Inequality hold true for four real numbers a, b, c, and d? That is, is the following statement true?

If $a < b$, $b < c$, and $c < d$, then $a < d$.

7. Of several line segments, $AB > CD$ (the length of segment AB is greater than that of segment CD), $CD > EF$, $EF > GH$, and $GH > IJ$. What does the Transitive Property of Inequality allow you to conclude regarding IJ and AB?

8. Of several angles, the degree measures are related in this way: $m\angle JKL > m\angle GHI$ (the measure of angle JKL is greater than that of angle GHI), $m\angle GHI > m\angle DEF$, and $m\angle DEF > m\angle ABC$. What does the Transitive Property of Inequality allow you to conclude regarding $m\angle ABC$ and $m\angle JKL$?

9. Classify as true or false.

(a) $5 \leq 4$ (c) $5 \leq 5$
(b) $4 \leq 5$ (d) $5 < 5$

10. Classify as true or false.

(a) $-5 \leq 4$ (c) $-5 \leq -5$
(b) $5 \leq -4$ (d) $5 \leq -5$

11. Two angles are supplementary if the sum of their measures is $180°$. If the measure of the first of two supplementary angles is less than $32°$, what must you conclude about the measure of the second angle?

12. Two trim boards need to be used together to cover a 12-ft length along one wall. If Jim recalls that one board is more than 7 ft long, what length must the second board be to span the 12-ft length?

13. Consider the inequality $-3 \leq 5$. Write the statement that results when

(a) each side is multiplied by 4
(b) -7 is added to each side
(c) each side is multiplied by (-6)
(d) each side is divided by (-1)

14. Consider the inequality $-6 > -9$. Write the statement that results when

(a) 8 is added to each side
(b) each side is multiplied by -2
(c) each side is multiplied by 2
(d) each side is divided by -3

15. Suppose that you are solving an inequality. Complete this chart for your work by indicating whether the inequality symbol should be reversed or kept by writing "change" or "no change."

	Positive	Negative
Add		
Subtract		
Multiply		
Divide		

In Exercises 16 to 26, first solve each inequality. Then draw a number line graph of the solutions.

16. $5x - 1 \leq 29$

17. $2x + 3 \leq 17$

18. $5 + 4x > 25$

19. $5 - 4x > 25$

20. $5(2 - x) \leq 30$

21. $2x + 3x < 200 - 5x$

22. $5(x + 2) < 6(9 - x)$

23. $\dfrac{x}{3} - \dfrac{x}{2} \leq 4$

24. $\dfrac{2x - 3}{-5} > 7$

25. $x^2 + 4x \leq x(x - 5) - 18$

26. $x(x + 2) < x(2 - x) + 2x^2$

In Exercises 27 to 30, the claims made are not always true. Cite a counterexample to show why each claim fails.

27. If $a < b$, then $a \cdot c < b \cdot c$.

28. If $a < b$, then $a \cdot c \neq b \cdot c$.

29. If $a < b$, then $a^2 < b^2$.

30. If $a \neq b$ and $b \neq c$, then $a \neq c$.

In Exercises 31 and 32, the statements are true and can be called theorems. Use the definition of $a < b$ to prove each statement.

31. If $a < b$ and $c < d$, then $a + c < b + d$.

32. If $a < b$, then $c - a > c - b$.

▲ A Look Beyond: The Nature of Proof

In the early stages of development, mathematical systems generally have some theorems that seem obvious. Suppose that you measure each of the right angles in figure 1.14, using a protractor. Because the measure of each angle is 90°, you readily accept the theorem, "All right angles are equal in measure." Not only is the theorem easy to accept, we will see later that it is also easy to prove!

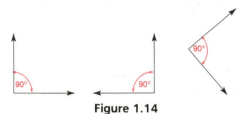

Figure 1.14

By building a sufficiently large foundation (vocabulary, postulates, and proved theorems), you gain the power to prove that less apparent statements are true. For example, it can be proved that the products of certain lengths are equal in figure 1.15.

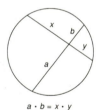

$a \cdot b = x \cdot y$

Figure 1.15

With the proper foundation, you will not only be able to prove that less obvious theorems must be true, you can also develop and justify a new theory (and thus a theorem). In fact, this describes the method by which our body of mathematical knowledge expands!

Before you can prove theorems, you must prepare for the undertaking. This is somewhat like playing a game in which you are a lawyer trying to prove your case. To convince the jury, you must be able to cite existing laws (like definitions and postulates) or earlier cases (like theorems). In addition, you must present the details of your case in an orderly manner so that a logical flow of details results.

What mathematical tools are suggested by the preceding comparison? Initially, there are the principles of logic that dictate the order of the claims made. These claims must be supported by a precise language (using terms that are understood and other terms that are defined by reference to the undefined terms), by assumptions (accepted statements

commonly called "axioms" in algebra and "postulates" in geometry), and by the previously proved theorems.

A good definition has four characteristics:

1. It names the term being defined.

2. It places the term into a set or category.

3. It distinguishes this term from other terms in the same category (without providing unnecessary additional facts).

4. It is reversible.

EXAMPLE

Explain why the following definition satisfies the four characteristics of a good definition:

DEFINITION: An **isosceles triangle** is a triangle that has two congruent sides.

SOLUTION

1. The term to be defined, "isosceles triangle," is named.

2. The term "isosceles triangle" is placed into the category "triangles."

3. The distinguishing characteristic is "two congruent sides." *Note:* "Congruent" means that the two sides have the same length.

4. Reversibility is demonstrated by the following restatements:

> *"If a triangle is isosceles, then it has two congruent sides."*
> *"If a triangle has two congruent sides, then it is isosceles."*   ▲

Let's take a closer look at the meaning of these two restatements in step 4. The reversibility of the definition allows us to draw two conclusions:

1. If we are given that a triangle is isosceles, then we can conclude that two sides are congruent.

Given: Triangle *ABC* is isosceles.

Conclude: Side $\overline{AC}$ is congruent to side $\overline{AB}$.

2. If we are given that two sides are congruent, then we can conclude that the triangle is isosceles.

Given: Side $\overline{AC}$ is congruent to side $\overline{AB}$.

Conclude: Triangle *ABC* is isosceles.

To get at the nature of proof, let us suppose that you know these facts:

1. **Undefined terms** are number and addition.

2. **Defined terms** are

Integer	Any number in the set $I = \{\ldots, -3, -2, -1, 0, 1, 2, 3, \ldots\}$
Even integer	Any number that is the product of 2 and an integer
Odd integer	Any number that is 1 larger than an even integer
Product	The result (answer) in multiplication

3. **Axioms of algebra** include such assumptions as the Distributive Axiom.

Now consider this theorem:

"The product of two odd integers is also odd."

In our plan for the proof, we first need to let *j* and *k* represent two different integers. Since we wish to represent two odd integers, these may be indicated by $(2j + 1)$ and $(2k + 1)$. We do not choose $(2j + 1)$ and $(2j + 1)$ because we would then be squaring an odd number, which might bias the results; we must avoid using a special case to prove a general relationship.

Proof

Let the two odd integers be represented by $(2j + 1)$ and $(2k + 1)$. (These were given in the hypothesis, and we know how to represent them by our definition!)

The product is indicated by

$$(2j + 1)(2k + 1)$$

Using the FOIL method, the product is

$$4jk + 2j + 2k + 1$$

The key is to recognize that this expression is 1 larger than an even integer, as shown:

$$
\begin{aligned}
&4jk + 2j + 2k + 1 \\
= \;&(4jk + 2j + 2k) + 1 \\
= \;&2(2jk + j + k) + 1 \qquad \text{Distributive Axiom}
\end{aligned}
$$

Because *j* and *k* are integers, *jk* and *2jk* are also integers. It follows that the sum $2jk + j + k$ must be an integer as well. Thus, $2(2jk + j + k) + 1$ is 1 more than twice an integer or is 1 more than an even integer.

Thus the product (by definition) is an odd integer, and the theorem is proved.

Generally, the symbol ▲ indicates that the proof is completed.

▲ SUMMARY

A LOOK BACK AT CHAPTER 1

One purpose of this chapter has been to introduce the student to the types of statements and the types of reasoning needed in the study of geometry. The three types of reasoning used are intuition, induction, and deduction. The second important aim of this chapter has been to provide the student with an introduction to (or review of, if the student has already completed an algebra course) the algebraic skills needed for the development and application of geometry. Because algebra and geometry are both examples of mathematical systems, the development of each depends on the need to construct proofs of theorems; for that reason, several illustrations of proof were provided.

A LOOK AHEAD TO CHAPTER 2

In the next chapter, we begin investigating geometry as a mathematical system. Much of the chapter is devoted to developing the basic terminology of geometry. While a few of the initial postulates and theorems of geometry are encountered in Chapter 2, we will only begin to scratch the surface of the total body of geometric knowledge covered in this textbook. In turn, what is found in this textbook is only a part of all known geometry.

IMPORTANT CONCEPTS OF CHAPTER 1

1.1 Intuition, Induction, Deduction
Line, Line Segment
Triangle, Rectangle, Polygon
Counterexample
Argument—Valid and Invalid

1.2 Statement—Simple and Compound
Negation
Conditional Statement
Hypothesis, Conclusion
Converse, Inverse, Contrapositive
Conjunction, Disjunction

1.3 Mathematical Systems
    Undefined Terms
    Definitions
    Postulates (Axioms)
    Theorems
Axioms of Equality
    Reflexive
    Symmetric
    Transitive
    Substitution

Properties of Real Numbers
    Commutative
    Associative
    Distributive
Definitions of Subtraction and Division
Evaluating an Expression
Order of Operations
FOIL

1.4 Variable, Constant
Equation
Formula
Solving an Equation
Equivalent Equations
Properties of Equation Solving
    Addition
    Subtraction
    Multiplication
    Division
Linear Equation

1.5 Inequality
Transitive Property of Inequality
Properties of Inequality
    Addition
    Subtraction
    Multiplication
    Division

A Look Beyond The Nature of Proof

▲ REVIEW EXERCISES

1. What are the four components of a mathematical system?

2. Name three types of reasoning.

In Review Exercises 3 to 5, name the type of reasoning illustrated.

3. While watching the pitcher warm up, Phillip thinks, "I'll be able to hit against him."

4. Laura is away at camp. On the first day, her mother brings her additional clothing. On the second day, her mother brings her another pair of shoes. On the third day, her mother brings her cookies. Laura concludes that her mother misses her.

5. Sarah knows the rule, "A number (not 0) divided by itself equals 1." The teacher asks Sarah, "What is 5 divided by 5?" Sarah says, "The answer is 1."

6. Either accept each claim or disprove it by citing a counterexample.

   (a) If $a > b$, then $\dfrac{1}{a} > \dfrac{1}{b}$ for nonzero a and b.

   (b) When n is a positive integer, the number whose value is $(n^2 + 3n + 2)$ is always divisible by 2.

7. State the hypothesis and conclusion for each statement:

   (a) If the diagonals of a trapezoid are equal in length, then the trapezoid is isosceles.

   (b) Diagonals of a parallelogram are congruent if the parallelogram is a rectangle.

8. Determine whether the conjunction and the disjunction are true or false for each line in the following chart:

P	Q	Conjunction P and Q	Disjunction P or Q
T	T		
T	F		
F	T		
F	F		

9. Classify each statement as simple, conditional, a conjunction, or a disjunction. Then classify the statement as true or false.

   (a) Ronald Reagan was a U.S. president, and Lassie was a horse.

   (b) Circles are round.

   (c) Ronald Reagan was a U.S. president, or Lassie was a horse.

   (d) If a person lives in Indiana, then she lives in Indianapolis.

In Review Exercises 10 to 12, draw a valid conclusion, where possible.

10. 1. If a person has a good job, then that person has a college degree.
    2. Billy Fuller has a college degree.

    C. ∴ ?

11. 1. If a person has a good job, then that person has a college degree.
    2. Jody Smithers has a good job.

    C. ∴ ?

12. 1. If the measure of an angle is 90°, then that angle is a right angle.
    2. Angle A is not a right angle.

    C. ∴ ?

13. For the statement "If the Bears win, we will be happy," write the statement for the

   (a) converse   **(b)** inverse   **(c)** contrapositive

14. Which statement—(a), (b), or (c)—in Review Exercise 13 has the same truth or falsity as the given statement?

15. Name the axiom or property illustrated.

   (a) $2 + (x + y) = (2 + x) + y$

   (b) If $a = b$, then $a \cdot c = b \cdot c$

   (c) If $a < b$, then $-2a > -2b$

   (d) $2 + (x + y) = 2 + (y + x)$

16. Simplify each expression.

   (a) $3(-4)$ **(d)** $8 \div (-4)$ **(g)** $-12 - (-4)$

   (b) $3 - (-4)$ **(e)** $-12 \div (-4)$ **(h)** $(-12)(-4)$

   (c) $8 + (-4)$ **(f)** $-12 + (-4)$

17. Simplify each expression.

   (a) $7 \cdot 3 - 6 \cdot 5$

   (b) $7 \cdot [3 + (-4)] \cdot (-5)$

   (c) $7 \cdot 3 + (-4)^3$

   (d) $\dfrac{6 \cdot (3^2 + 4^2)}{-10} + \dfrac{7 + 5}{6 \div 2}$

18. Simplify each expression.

   (a) $(3x - 2) + (4x - 8)$

   (b) $3(x - 3) + 4(2x + 1)$

   (c) $12\pi + 5\pi$

   (d) $9\sqrt{2} - 5\sqrt{2}$

19. Solve each equation.

   (a) $2(x + 3) + 3(x - 1) = 28$

   (b) $2(x + 3) - 3(x - 1) = 1$

   (c) $(x + 1)(x + 4) = x^2 + 2x + 13$

   (d) $\dfrac{x}{3} + \dfrac{2x}{5} = 22$

20. Solve each inequality, and graph the solution set on a number line.

   (a) $(2x + 3) - (3x - 2) \le 12$

   (b) $2(2x + 3) - 3(x - 2) > 12$

21. Find the value of the variable named in each formula.

   (a) S if $S = (n - 2)180$ and $n = 7$.

   (b) r if $V = \dfrac{4}{3}\pi r^3$ and $V = 36\pi$.

22. **(a)** Write an expression for the perimeter of the triangle shown in the accompanying figure.

   (b) If the perimeter is 32 centimeters, find the value of x.

   (c) Find the length of each side of the triangle.

23. The sum of the measures of all three angles of the triangle in Review Exercise 22 is 180°. If the sum of the measures of angles 1 and 2 is more than 130°, what can you conclude about the measure of angle 3?

24. Susan wants to have a 4-ft board with some pegs on it. She wants to leave 6 in. on each end and 4 in. between each peg. How many pegs will fit on the board? [*Hint:* If n represents the number of pegs, then $(n - 1)$ represents the number of equal spaces.]

2 LINE AND ANGLE RELATIONSHIPS

The sides of angles 1 and 2 meet to form 90° angles and are said to be **perpendicular.** How do you think angles 1 and 2 are related?

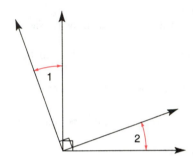

2.1 SOME UNDEFINED TERMS AND INITIAL DEFINITIONS

Recall that geometry, like algebra, is an example of a *mathematical system.*

FOUR PARTS OF A MATHEMATICAL SYSTEM
1. Undefined terms ⎫ 2. Defined terms ⎭ vocabulary 3. Axioms or postulates ⎫ 4. Theorems ⎭ principles

Many terms are classified as undefined because they do not fit into a set or category that has been determined. Terms that *are* defined, however, should be described with precision. As stated in Chapter 1, a good definition possesses four characteristics:

1. It names the term being defined.
2. It places the term into a set or category.

CHAPTER OUTLINE

2.1 Some Undefined Terms and Initial Definitions

2.2 Angles and Their Relationships

2.3 Introduction to Geometry Proofs

2.4 The Formal Proof of a Theorem

A Look Beyond Historical Sketch of Euclid

3. It distinguishes this term from other terms in the same category (without providing unnecessary additional facts).

4. It is reversible.

In geometry, the terms "point," "line," and "plane" are often used without being defined. Instead, they are simply described and serve as the building blocks for later terminology.

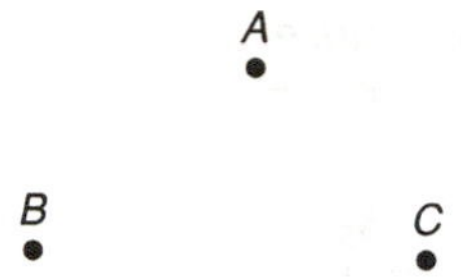

A **point,** which is represented by a dot, has location but not size; that is, a point has no dimensions. An uppercase italic letter is used to name a point. Figure 2.1 shows points *A*, *B*, and *C*. ("Point" may be abbreviated "pt." for convenience.)

The second undefined term is **line.** Lines have a quality of "straightness" that is not defined but assumed. If several points on a line are shown, these points form a straight path. Whereas a point has no dimensions, a line is one-dimensional; that is, the distance between any two points on a given line can be measured. Line *AB*, represented symbolically by $\overleftrightarrow{AB}$, extends infinitely far in opposite directions, as suggested by the arrows on the line. A line may also be represented by a single lowercase letter. Figure 2.2a and b show the lines *AB* and *m*. Often a lowercase letter is used, in context, to mean a line even though the line symbol is omitted.

A
•

B   *C*
•   •

Figure 2.1

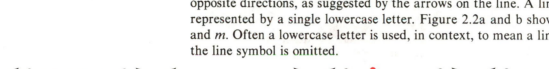

| (a) | (b) | (c) | (d) |

Figure 2.2

Note the position of point *X* on $\overleftrightarrow{AB}$ in figure 2.2c. When three points such as *A*, *X*, and *B* are on the same line, they are said to be **collinear.** In the order shown, which you may symbolize *A-X-B*, point *X* is said to be *between A and B.*

When no drawing is available, the notation *A-B-C* means that these points are collinear, with *B* between *A* and *C*. When a drawing is provided, we assume that all points in the drawing that look collinear are to be accepted as collinear, *unless otherwise stated.* Figure 2.2d shows that *A*, *B*, and *C* are collinear, with *B* between *A* and *C*.

Just as the term **axiom** is used in algebra to state an assumed truth, the term **postulate** names an assumption in geometry.

POSTULATE 1   Through two distinct points, there is exactly one line.

In this text, more important terms are defined in boxes, while less important but related terms are shown in the running text in **boldface type.**

DEFINITION: A **line segment** is the part of a line that consists of two points, known as "endpoints," and all points between them.

Postulate 1 implies that a unique line segment exists between any two points.

The symbol for line segment AB, named by its endpoints, is $\overline{AB}$. Omission of the segment symbol, as in AB, means that we are considering the *length* of the segment. These symbols are summarized in Table 2.1.

Symbol	Words for Symbol	Geometric Figure
$\overleftrightarrow{AB}$	Line AB	A ●————————————● B
$\overline{AB}$	Line segment AB	A ●————————————● B
AB	Length of segment AB	A number

Table 2.1

A ruler can of course be used to measure the length of any line segment. This length may be represented by AB or BA (for segment AB), and it must be a positive number.

POSTULATE 2   (Ruler Postulate)   The measure of any line segment is a unique positive number.

It should be clear that a length relationship exists between the line segments determined in figure 2.3. This relationship is stated in the third postulate.

A ●————————X————————● B

Figure 2.3

POSTULATE 3   (Segment-Addition Postulate)   If X is a point on $\overline{AB}$ and A-X-B, then $AX + XB = AB$.

DEFINITION: **Congruent** ($\cong$) segments are two segments that have the same length.

In general, geometric figures that can be made to coincide (fit perfectly one on top of the other) are said to be **congruent.** The symbol $\cong$ is a combination of the symbol $\sim$, which means that the figures have the same shape, and $=$, which means that the corresponding parts of the figure have the same measure. In figure 2.4, $\overline{AB} \cong \overline{CD}$, but $\overline{AB} \not\cong \overline{EF}$. Does it follow that $\overline{CD} \cong \overline{EF}$?

In figure 2.5, if A, M, and B are collinear and if $\overline{AM} \cong \overline{MB}$, then M is the **midpoint** of $\overline{AB}$. Equivalently, M is the midpoint of $\overline{AB}$ if $AM = MB$. Also, if $\overline{AM} \cong \overline{MB}$, then $\overline{CD}$ is described as a **bisector** of $\overline{AB}$. Under what condition would $\overline{AB}$ be a bisector of $\overline{CD}$?

A ●————————● B

C ●————————● D

E ●————————————————● F

Figure 2.4

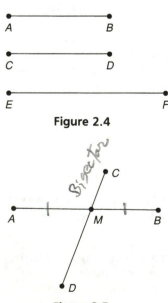

Figure 2.5

DEFINITION: **Ray AB,** denoted by $\overrightarrow{AB}$, is the union of $\overline{AB}$ and all points X on $\overleftrightarrow{AB}$ such that B is between A and X.

In figure 2.6, $\overleftrightarrow{AB}$, $\overrightarrow{AB}$, and $\overrightarrow{BA}$ are shown; note that $\overrightarrow{AB}$ and $\overrightarrow{BA}$ are not the same.

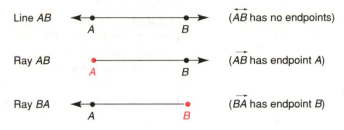

Line AB A   B   ($\overleftrightarrow{AB}$ has no endpoints)

Ray AB A   B   ($\overrightarrow{AB}$ has endpoint A)

Ray BA A   B   ($\overrightarrow{BA}$ has endpoint B)

Figure 2.6

Opposite rays are two rays that share a common endpoint and form a straight line when combined. In figure 2.7a, $\overrightarrow{BA}$ and $\overrightarrow{BC}$ are opposite rays. When two lines have a single point in common, they are said to **intersect.** Lines ℓ and m intersect at point P in figure 2.7b.

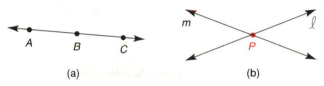

(a)   (b)

Figure 2.7

Another undefined term in geometry is **plane.** A plane is two-dimensional; that is, it has infinite length and infinite width, but no thickness. Except for its limited size, a flat surface such as the top of a table could be used as an example of a plane.

> **DEFINITION:** **Parallel lines** are lines in the same plane that do not intersect.
>
> $\ell \parallel M$

In figure 2.8, ℓ and n are parallel; in symbols, $\ell \parallel n$. However, ℓ and m are not parallel; in symbols, $\ell \nparallel m$.

Returning to the undefined term "plane," an uppercase letter will be used to name each plane. Because a plane (like a line) is infinite, we can only show a portion of the plane or planes, as in figure 2.9.

Figure 2.8

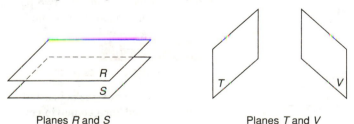

Planes R and S Planes T and V

Figure 2.9

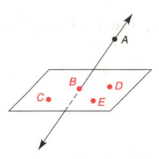

Figure 2.10

Because a plane is two-dimensional, it consists of an infinite number of points and contains an infinite number of lines. Two distinct points may determine (or "fix") a line; likewise, exactly three noncollinear points determine a plane. Just as collinear points lie on the same line, **coplanar points** lie in the same plane. In figure 2.10, points B, C, D, and E are coplanar, while A, B, C, and D are noncoplanar.

In this book, points shown in figures are assumed to be coplanar unless otherwise stated. For instance, points A, B, C, D, and E are coplanar in figure 2.11a, as are points F, G, H, J, and K in figure 2.11b.

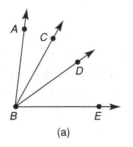
(a)

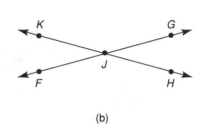
(b)

Figure 2.11

POSTULATE 4   Through three noncollinear points, there is exactly one plane.

Space is the set of all possible points. It is three-dimensional, having qualities of length, width, and depth. When two planes intersect in space, their intersection is a line. An opened greeting card suggests this relationship, as does figure 2.12a. This notion gives rise to our fifth postulate.

POSTULATE 5   If two distinct planes intersect, then their intersection is a line.

The intersection of two infinite planes is also infinite, as a line is. If two planes do not intersect, then they are **parallel.** The parallel **vertical** planes shown in figure 2.12b may remind you of the opposite walls of your classroom. The parallel **horizontal** planes in figure 2.12c suggest the relationship between ceiling and floor.

Imagine a plane and two points of that plane, points A and B. Now think of the line containing the two points and its relationship to the plane. Perhaps your conclusion can be summed up as follows:

POSTULATE 6   Given two distinct points in a plane, the line containing these points also lies in the plane.

The ancient Greeks insisted that only two tools (a compass and a straightedge) be used for **geometric constructions,** which were idealized drawings assuming a perfection in the use of these tools. The compass was assumed to allow for creating perfect circles and for marking off segments of equal length. The

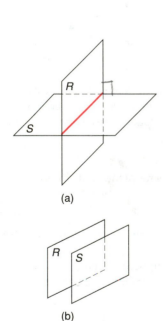

(a)

(b)

(c)

Figure 2.12

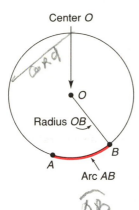

Center O

Radius OB

Arc AB

straightedge could be used to pass a line through two designated points. But with those instruments and assumptions, can any geometric figure be constructed to perfection?

A **circle** is the set of all points in a plane that are at a given distance from a particular point (known as the "center" of the circle). The part of a circle between any two of its points is known as an **arc.** Any line segment joining the center to a point on the circle is a **radius** (plural, "radii") of the circle.

CONSTRUCTION 1   To construct a segment congruent to a given segment.

Given: $\overline{AB}$ as shown in figure 2.13a

Construct: $\overline{CD}$ on line m so that $\overline{CD} \simeq \overline{AB}$ (or $CD = AB$)

Construction: With your compass open to the length of $\overline{AB}$, place the stationary point of the compass at C and mark off a length equal to AB, as shown in figure 2.13b.

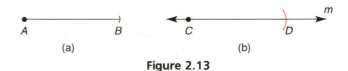

(a)   (b)

Figure 2.13   ▲

CONSTRUCTION 2   To construct the midpoint M of a given line segment AB.

Given: $\overline{AB}$ as shown in figure 2.14a

Construct: M so that $\overline{AM} \simeq \overline{MB}$

Construction: On a given line, construct a line segment congruent to $\overline{AB}$ as shown in figure 2.14. Then open the compass to a length greater than one-half of AB. Using A as the center of the arc, mark off arcs both above and below $\overline{AB}$. Now, without changing the radius and with B as the center, mark off arcs both above and below $\overline{AB}$. Denote by letter C the intersection of the arcs above $\overline{AB}$; use D for the intersection of the arcs below $\overline{AB}$, as shown in figure 2.14b.

Next draw the line or segment between points C (above) and D (below) to intersect $\overline{AB}$. Designate this point as M, the desired midpoint. *Note:* While this is simply a construction method, the uniqueness of the midpoint is proved in Theorem 2.1.1.   ▲

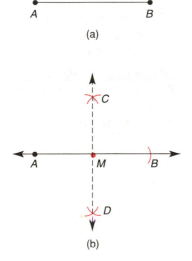

A   B

(a)

C

A   M   B

D

(b)

Figure 2.14

Geometry principles are often used along with algebra in problem solving.

EXAMPLE 1   *Given:* M is the midpoint of $\overline{CD}$; $CM = 3x + 9$ and $MD = x + 17$

Find: x and CM

Solution　Since M is the midpoint of $\overline{CD}$, $CM = MD$. Then

$$3x + 9 = x + 17$$
$$2x + 9 = 17$$
$$2x = 8$$
$$x = 4$$

By substitution, $CM = 3(4) + 9 = 12 + 9 = 21$.　　▲

We wish to call attention to the term "unique" and the general notion of uniqueness. Postulate 2 implies the following:

1. There exists a number measure for the segment.

2. Only *one* measure is permissible.

Both qualities are necessary for uniqueness! Other phrases that may replace the term "unique" include the following:

One and only one

Exactly one

One and no more than one

Construction 2 asks for *the* midpoint of a segment, thus implying its uniqueness. Because the uniqueness of the midpoint can be justified, we call the following statement a theorem.

THEOREM 2.1.1　The midpoint of a line segment is unique.

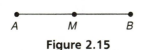

Figure 2.15

The uniqueness of the midpoint follows logically from the definition of a midpoint, the Segment-Addition Postulate, and the Ruler Postulate. Consider figure 2.15, in which M is the midpoint of $\overline{AB}$, and study the line of reasoning.

$AM = MB$	M is the midpoint, so this is true by definition
$AM + MB = AB$	M is between A and B on $\overline{AB}$; Segment-Addition Postulate
$AM + AM = AB$	AM replaces MB in the preceding claim; Substitution Axiom of Equality
$2(AM) = AB$	algebraic simplification
$AM = \dfrac{AB}{2}$	Division Property of Equality

Now M is *the* point located on $\overline{AB}$ at a distance equal to $\dfrac{AB}{2}$ units from A (Ruler Postulate).

2.1 EXERCISES

Use the drawings provided in answering the following questions.

1. In the figure for Exercises 1 and 2, name three points that appear to be

(a) collinear **(b)** noncollinear

2. How many lines can be drawn through

  (a) point A **(c)** points A, B, and C
  (b) points A and B **(d)** points A, B, and D

3. Give the meaning of $\overleftrightarrow{CD}$, $\overline{CD}$, CD, and $\overrightarrow{CD}$.

4. Explain the difference, if any, between

  (a) $\overleftrightarrow{CD}$ and $\overleftrightarrow{DC}$ **(c)** CD and DC
  (b) $\overline{CD}$ and $\overline{DC}$ **(d)** $\overrightarrow{CD}$ and $\overrightarrow{DC}$

5. In the figure for Exercises 5 through 9, name two lines that appear to be

  (a) parallel   **(b)** nonparallel

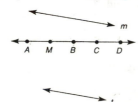

6. Classify as true or false:

  (a) $AB + BC = AD$
  (b) $AD - CD = AB$
  (c) $AD - CD = AC$
  (d) $AB + BC + CD = AD$
  (e) $AB = BC$

7. *Given:* M is the midpoint of $\overline{AB}$
     $AM = 2x + 1$ and $MB = 3x - 2$
  Find: x and AM

8. *Given:* M is the midpoint of $\overline{AB}$
     $AM = 2(x + 1)$ and $MB = 3(x - 2)$
  Find: x and AB

9. *Given:* $AM = 2x + 1$, $MB = 3x + 2$, and
     $AB = 6x - 4$
  Find: x and AB

10. Can a segment bisect a line? a segment? Can a line bisect a segment? a line?

11. In the accompanying figure, name

  (a) two opposite rays
  (b) two rays that are not opposite

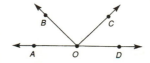

12. Suppose that (a) point C lies in plane X and (b) point D lies in plane X. What may you conclude regarding $\overleftrightarrow{CD}$?

13. Draw a sketch (*not* a construction) of

  (a) two intersecting planes
  (b) two parallel planes
  (c) two parallel planes intersected by a third plane that is not parallel to the first or second plane

14. Suppose that (a) planes M and N intersect, (b) point A lies in both planes M and N, and (c) point B lies in both planes M and N. What may you conclude regarding $\overleftrightarrow{AB}$?

15. Suppose that (a) points A, B, and C are collinear and (b) $AB > AC$. Which point can you conclude *cannot* lie between the other two?

16. In the accompanying figure, points A, B, C, and D are coplanar; B, C, and D are collinear; point E is not in plane M. How many planes contain

  (a) points A, B, and C **(c)** points A, B, C, and D
  (b) points B, C, and D **(d)** points A, B, C, and E

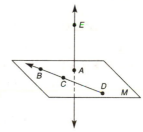

17. Using the number line provided, name the point that

  (a) is the midpoint of $\overline{AE}$
  (b) is the endpoint of a segment of length 4, if the other endpoint is point G
  (c) has a distance from B equal to $3(AC)$

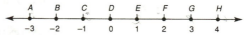

18. Suppose that (a) A-B-C-D is on $\overline{AD}$, (b) B is the midpoint of $\overline{AC}$, and (c) C is the midpoint of $\overline{BD}$. What may you conclude about the lengths of

  (a) $\overline{AB}$ and $\overline{CD}$ **(c)** $\overline{AC}$ and $\overline{CD}$
  (b) $\overline{AC}$ and $\overline{BD}$

In Exercises 19 to 22, use only a compass and a straightedge to complete each construction. Use the following figure for Exercises 19 and 20.

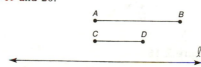

19. *Given:* $\overline{AB}$ and $\overline{CD}$ $(AB > CD)$
Construct: $\overline{MN}$ on line ℓ so that $MN = AB + CD$

20. *Given:* $\overline{AB}$ and $\overline{CD}$ $(AB > CD)$
Construct: $\overline{EF}$ so that $EF = AB - CD$

21. *Given:* $\overline{AB}$ as shown in the figure below Exercise 22
Construct: $\overline{PQ}$ on line n so that $PQ = 3(AB)$

22. *Given:* $\overline{AB}$ as shown below

Construct: $\overline{TV}$ on line n so that $TV = \dfrac{1}{2}(AB)$

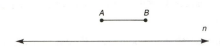

23. Can you use the construction for the midpoint of a segment to divide a line segment into

(a) three congruent parts   (c) six congruent parts
(b) four congruent parts   (d) eight congruent parts

24. Generalize your findings in Exercise 23.

25. In the accompanying figure, line ℓ is parallel to plane P (that is, it will not intersect P even if extended). Line m intersects line ℓ. What may you conclude about m and P?

*26. In the figure shown, $\overleftrightarrow{AB}$ and $\overleftrightarrow{EF}$ are said to be **skew** lines because they neither intersect nor are parallel. How many planes are determined by

(a) parallel lines AB and DC
(b) intersecting lines AB and BC
(c) skew lines AB and EF
(d) lines AB, BC, and DC
(e) points A, B, and F
(f) points A, C, and H
(g) points A, C, F, and H

ANGLES AND THEIR RELATIONSHIPS

This section introduces you to the language of angles.

> **DEFINITION:** An **angle** is the union of two rays that share a common endpoint.

In figure 2.16a, the angle is symbolized by $\angle ABC$ or $\angle CBA$. The rays BA and BC are known as the **sides** of the angle. B, the common endpoint of these rays, is known as the **vertex** of the angle. When three letters are used to name an angle, the vertex is always named in the middle. In many instances, a numeral is used to name the angle. Figure 2.16b shows $\angle 1$; $\angle 1$ may also be described as $\angle MNQ$ or simply as $\angle N$ (after the vertex).

You cannot measure an angle with a ruler! The instrument shown in figure 2.17 (and used in the measurement of angles) is a **protractor.** For example, you would express the measure of $\angle RST$ by writing $m\angle RST = 50°$. When a lowercase m is used before the angle symbol $\angle$, this means that the measure is in degrees. Measuring the angles in figure 2.16 with a protractor, we would find that $m\angle B = 62°$ and $m\angle 1 = 90°$. Postulate 7 states the limitations upon an angle's measure.

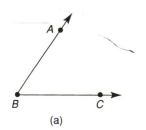

(a)

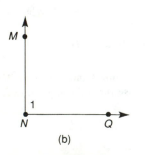

(b)

Figure 2.16

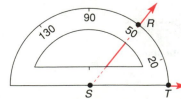

Figure 2.17

(Protractor Postulate) The measure of an angle is a unique positive number.

Note: In Chapters 1 to 9, the measures of angles are between 0° and 180°, including 180°. Angles with measures greater than 180° are discussed in Chapter 10.

An angle whose measure is less than 90° is an **acute angle.** If the angle's measure is exactly 90°, the angle is a **right angle.** If the angle's measure is between 90° and 180°, the angle is **obtuse.** Finally, an angle whose measure is exactly 180° is a **straight angle;** alternatively, a straight angle is one whose sides form opposite rays (a straight line) (see Table 2.2).

Table 2.2   **Angles**

Angle	Measure
Acute (1)	m∠1 = 23°
Right (2)	m∠2 = 90°
Obtuse (3)	m∠3 = 112°
Straight (4)	m∠4 = 180°

In figure 2.18a, ∠ABC contains the noncollinear points A, B, and C. These three points, in turn, determine a plane. The plane containing ∠ABC is separated into three subsets by the angle:

Points like D are in the *interior* of ∠ABC.

Points like E are said to be *on* ∠ABC.

Points like F are in the *exterior* of ∠ABC.

With this description, it is possible to state the counterpart of the Segment-Addition Postulate!

(Angle-Addition Postulate) If a point D lies in the interior of an angle ABC, then m∠ABD + m∠DBC = m∠ABC.

Figure 2.18b illustrates the Angle-Addition Postulate. In figure 2.18b, ∠ABD and ∠DBC are also said to be adjacent. Two angles are **adjacent** if they share a common side and a common vertex but have no interior points in common. In figure 2.18b, ∠ABC and ∠ABD are not adjacent because they have interior points in common.

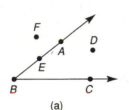

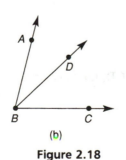

Figure 2.18

> **DEFINITION:** **Congruent angles** (≅ ∠s) are two angles with the same measure.

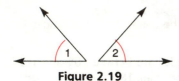

Figure 2.19

Congruent angles must coincide when one is placed over the other. (Do not consider that the sides appear to have different lengths; remember that rays are infinite in length!) In symbols, $\angle 1 \cong \angle 2$ if $m\angle 1 = m\angle 2$. In figure 2.19, similar markings indicate that $\angle 1 \cong \angle 2$.

EXAMPLE 1

Given: $\angle 1 \cong \angle 2$
$m\angle 1 = 2x + 15$
$m\angle 2 = 3x - 2$

Find: x

Solution $\angle 1 \cong \angle 2$ means $m\angle 1 = m\angle 2$. Therefore

$$2x + 15 = 3x - 2$$
$$17 = x \qquad \text{or} \qquad x = 17$$

(Note that $m\angle 1 = m\angle 2 = 49°$.)   ▲

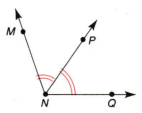

Figure 2.20

When P is located in the interior of $\angle MNQ$ so that $\angle MNP \cong \angle PNQ$, $\overrightarrow{NP}$ is said to **bisect** $\angle MNQ$. Equivalently, $\overrightarrow{NP}$ is the **bisector** of $\angle MNQ$. Thus, the angle is bisected when $m\angle MNP = m\angle PNQ$ (see figure 2.20).

Many angle relationships involve a pair of angles. For instance, two angles whose measures add up to 90° are **complementary,** and each angle is the **complement** of the other. Similarly, two angles whose measures add up to 180° are known as **supplementary,** and each angle is the **supplement** of the other.

EXAMPLE 2

Given: $\angle P$ and $\angle Q$ are complementary

$$m\angle P = \frac{x + 20}{2} \qquad \text{and} \qquad m\angle Q = \frac{5x - 20}{3}$$

Find: x

Solution
$$m\angle P + m\angle Q = 90$$
$$\frac{x + 20}{2} + \frac{5x - 20}{3} = 90$$

Multiplying by 6 (the least common denominator, or LCD, for 2 and 3), we get

$$6 \cdot \frac{x + 20}{2} + 6 \cdot \frac{5x - 20}{3} = 6 \cdot (90)$$
$$3(x + 20) + 2(5x - 20) = 540$$
$$3x + 60 + 10x - 40 = 540$$
$$13x + 20 = 540$$
$$13x = 520$$
$$x = 40$$

(For $x = 40$, $m\angle P = 30°$ and $m\angle Q = 60°$, so their sum is exactly 90°.)   ▲

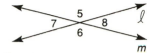

Figure 2.21

When two straight lines intersect, the pairs of nonadjacent angles formed are each known as **vertical angles.** In figure 2.21, $\angle 5$ and $\angle 6$ are vertical, as are $\angle 7$ and $\angle 8$. In addition, however, $\angle 5$ and $\angle 7$ can be described as adjacent and supplementary angles, as can $\angle 5$ and $\angle 8$. If $m\angle 7 = 300$, then what is $m\angle 5$? What is $m\angle 8$? It is true in general that vertical angles are congruent, and we will use this in Example 3.

Recall now the Addition and Subtraction Properties of Equality: If $a = b$ and $c = d$, then $a \pm b = c \pm d$. These principles can be used in solving a system of equations such as the following one:

$$\begin{array}{rl} x + y = & 5 \\ 2x - y = & 7 \\ \hline 3x \quad\;\; = & 12 \qquad \text{left and right sides are added} \\ x = & 4 \end{array}$$

Now we can substitute 4 for x in either of our original equations, to solve for y:

$$\begin{array}{rl} x + y = 5 & \\ 4 + y = 5 & \qquad \text{by substitution} \\ y = 1 & \end{array}$$

EXAMPLE 3

Given: In figure 2.21, ℓ and m intersect so that

$$m\angle 5 = 2x + 2y$$
$$m\angle 8 = 2x - y$$
$$m\angle 6 = 4x - 2y$$

Find: x and y

Solution

$\angle 5$ and $\angle 8$ are supplementary (adjacent and together form a straight angle). Therefore $m\angle 5 + m\angle 8 = 180$. $\angle 5$ and $\angle 6$ are congruent (vertical). Therefore $m\angle 5 = m\angle 6$. Consequently, we have

$$\begin{array}{rl} (2x + 2y) + (2x - y) = 180 & \qquad \text{supplementary } \angle\text{s } 5 \text{ and } 8 \\ 2x + 2y = 4x - 2y & \qquad \cong \angle\text{s } 5 \text{ and } 6 \end{array}$$

Simplifying,
$$\begin{array}{l} 4x + y = 180 \\ 2x - 4y = 0 \end{array}$$

Using the Multiplication Property of Equality, we multiply the first equation by 4, so the equivalent system is

$$\begin{array}{rl} 16x + 4y = & 720 \\ 2x - 4y = & 0 \\ \hline 18x \quad\;\; = & 720 \qquad \text{adding left, right sides} \\ x = & 40 \end{array}$$

From the equation $4x + y = 180$, it follows that

$$\begin{array}{l} 4(40) + y = 180 \\ 160 + y = 180 \\ y = 20 \end{array}$$

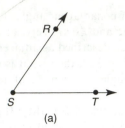

(a)

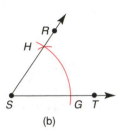

(b)

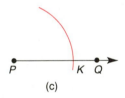

(c)

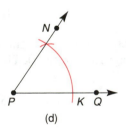

(d)

Figure 2.22

In summary, $x = 40$ and $y = 20$.

Note: m$\angle 5 = 120°$, m$\angle 8 = 60°$, and m$\angle 6 = 120°$.   ▲

In Section 2.2, you considered some basic constructions with line segments. Now consider these constructions again, using angle concepts. Later, it will become clear why these methods are valid!

CONSTRUCTION 3   To construct an angle congruent to a given angle.

Given: $\angle RST$, as shown in figure 2.22a

Construct: With $\overrightarrow{PQ}$ as one side, $\angle NPQ \cong \angle RST$

Construction: With a compass, mark an arc to intersect both sides of $\angle RST$ (at points G and H, respectively). (See figure 2.22b.) Without changing the radius, mark an arc to intersect $\overrightarrow{PQ}$ at K and the "would-be" second side of $\angle NPQ$. (See figure 2.22c.)

Now mark an arc to measure the distance from G to H. Using the same arc, mark an arc with K as center to intersect the would-be second side of the desired angle. Now draw the ray from P through the point of intersection of the two arcs. (See figure 2.22d.)

The resulting angle is the one desired, as we prove in Section 4.2, Exercise 25.   ▲

Just as a line segment can be bisected, so can an angle. This takes you to a fourth construction method.

CONSTRUCTION 4   To construct the angle bisector of a given angle.

Given: $\angle PRT$, as shown in figure 2.23a

Construct: $\overrightarrow{RS}$ so that $\angle PRS \cong \angle SRT$

Construction: Using a compass, mark an arc to intersect the sides of $\angle PRT$ at points M and N. (See figure 2.23b.) Now, with M and N as centers, mark off two arcs with equal radii to intersect at point S in the interior of $\angle PRT$, as shown. Now draw ray RS, the desired angle bisector. (See figure 2.23c.)   ▲

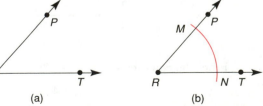

(a)   (b)   (c)

Figure 2.23

Reasoning from the definition of an angle bisector, the Angle-Addition Postulate, and the Protractor Postulate, we can justify the following theorem.

> **THEOREM 2.2.1** There is one and only one angle bisector for a given angle.

2.2 EXERCISES

Use drawings as needed to answer each of the following questions.

1. Must two rays with a common endpoint be coplanar? Must three rays with a common endpoint be coplanar? box

2. Suppose that $\overrightarrow{AB}$, $\overrightarrow{AC}$, $\overrightarrow{AD}$, $\overrightarrow{AE}$, and $\overrightarrow{AF}$ are coplanar as shown in the figure for Exercises 2 to 5. Classify the following as true or false:
   (a) $m\angle BAC + m\angle CAD = m\angle BAD$
   (b) $\angle BAC \cong \angle CAD$
   (c) $m\angle BAE - m\angle DAE = m\angle BAC$
   (d) $\angle BAC$ and $\angle DAE$ are adjacent
   (e) $m\angle BAC + m\angle CAD + m\angle DAE = m\angle BAE$

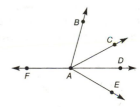

3. Without using a protractor, name the type of angle represented by
   (a) $\angle BAE$ (b) $\angle FAD$ (c) $\angle BAC$ (d) $\angle FAE$

4. Consider the figure for Exercises 2 to 5. What, if anything, is wrong with the claim $m\angle FAB + m\angle BAE = m\angle FAE$?

5. Suppose that (a) $\angle FAC$ and $\angle CAD$ are adjacent and (b) $\overrightarrow{AF}$ and $\overrightarrow{AD}$ are opposite rays. What may you conclude about $\angle FAC$ and $\angle CAD$?

In Exercises 6 to 8, use the accompanying drawing.

6. Given: $m\angle RST = 2x + 9$
   $m\angle TSV = 3x - 2$
   $m\angle RSV = 67°$
   Find: x

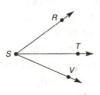

7. Given: $m\angle RST = 5(x + 1) - 3$
   $m\angle TSV = 4(x - 2) + 3$
   $m\angle RSV = 4(2x + 3) - 7$
   Find: x

8. Given: $\overrightarrow{ST}$ bisects $\angle RSV$
   $m\angle RST = 5x + y$
   $m\angle TSV = 4x + 3y + 1$
   $m\angle RSV = 14x - 8y$
   Find: x and y

9. Given: $\overleftrightarrow{AB}$ and $\overleftrightarrow{AC}$ in plane P as shown
   $\overleftrightarrow{AD}$ intersects P at point A
   $\angle CAB \cong \angle DAC$
   $\angle DAC \cong \angle DAB$
   What may you conclude?

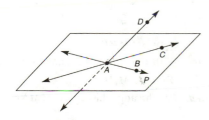

10. Two angles are complementary. One angle is 12° larger than the other. Using two variables x and y, find the size of each angle by solving a system of equations.

11. Two angles are supplementary. One angle is 24° more than twice the other. Using two variables x and y, find the measure of each angle.

12. For two complementary angles, find an expression for the measure of the second angle if the measure of the first is
    (a) $x°$
    (b) $(3x - 12)°$
    (c) $(2x + 5y)°$

13. Suppose that the two angles in Exercise 12 are supplementary. Find expressions for the supplements, using the expressions provided in Exercise 12, parts (a) to (c).

14. On the protractor shown in the figure for Exercises 14 and 15, $\overrightarrow{NP}$ bisects $\angle MNQ$. Find x.

15. On the same protractor, $\angle MNP$ and $\angle PNQ$ are complementary. Find x.

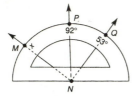

16. Classify as true or false:
    (a) If points P and Q lie in the interior of $\angle ABC$, then $\overline{PQ}$ lies in the interior of $\angle ABC$.
    (b) If points P and Q lie in the interior of $\angle ABC$, then $\overleftrightarrow{PQ}$ lies in the interior of $\angle ABC$.
    (c) If points P and Q lie in the interior of $\angle ABC$, then $\overrightarrow{PQ}$ lies in the interior of $\angle ABC$.

In Exercises 17 to 22, use only a compass and a straightedge to perform the indicated constructions. Use the figure below for Exercises 17 to 19.

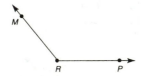

17. *Given:*   Obtuse $\angle MRP$
    Construct: With $\overrightarrow{OA}$ as one side, an angle $\cong \angle MRP$

18. *Given:*   Obtuse $\angle MRP$
    Construct: $\overrightarrow{RS}$, the angle bisector of $\angle MRP$

19. *Given:*   Obtuse $\angle MRP$
    Construct: Rays $\overrightarrow{RS}$, $\overrightarrow{RT}$, and $\overrightarrow{RU}$ so that $\angle MRP$ is divided into four $\cong$ angles

20. *Given:*   Straight $\angle DEF$ as shown
    Construct: A right angle with vertex at E
    (*Hint:* Use Construction 4.)

21. Draw a triangle with three acute angles. Construct angle bisectors for each of the three angles. What seems to be true, based on the appearance of your construction?

22. *Given:*   Acute $\angle 1$ as shown
    Construct: Triangle ABC with $\angle A \cong \angle 1$, $\angle B \cong \angle 1$, and base $\overline{AB}$

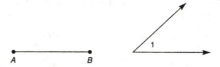

23. What seems to be true of the sides in the triangle you constructed in Exercise 22?

24. *Given:*   Straight $\angle ABC$ as shown
    Construct: Bisectors of $\angle ABD$ and $\angle DBC$
    What type of angle is formed by the bisectors of the two angles?

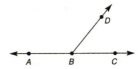

2.3
INTRODUCTION TO GEOMETRY PROOFS

This section introduces some guidelines for proving geometric properties. Several examples are offered to help you develop your own proofs. In the beginning, the form of proof will be a two-column proof, with Statements in the left column and Reasons in the right column. But where do the statements and reasons come from?

To deal with this question, you must ask What is it that is known (Given) and why should the conclusion (Prove) follow from this information? Understanding the why may mean dealing with several related conclusions and thus several intermediate whys. In correctly piecing together a proof, you will usually scratch out several conclusions and reorder these. Of course, each conclusion must be justified by citing the Given (hypothesis), a previously stated definition or postulate, or a previously proven theorem.

The typical format for a geometry proof problem is as follows:

Given: ————[Drawing]

Prove: ————

Consider this problem:

Given: A-P-B on $\overline{AB}$

Prove: $AP = AB - PB$

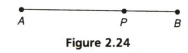

Figure 2.24

First consider the Drawing (figure 2.24), and relate it to any additional information described by the Given. Then consider the Prove. Do you understand the claim, and does it seem reasonable? If it seems reasonable, the intermediate claims must be ordered and supported to form the contents of the proof. Since a proof must begin with the Given and conclude with the Prove, the proof of the preceding problem has this form:

PROOF

Statements	Reasons
1. A-P-B on $\overline{AB}$	1. Given
2. ?	2. ?
.	.
.	.
.	.
?. $AP = AB - PB$	?. ?

To construct the proof, you must glean from the Drawing and the Given that

$$AP + PB = AB$$

You can then deduce (through subtraction) that $AP = AB - PB$. Thus the complete proof problem will have this appearance:

Given: A-P-B on $\overline{AB}$

Prove: $AP = AB - PB$

PROOF

Statements	Reasons
1. A-P-B on $\overline{AB}$	1. Given
2. $AP + PB = AB$	2. Segment-Addition Postulate
3. $AP = AB - PB$	3. Subtraction Property of Equality

Now consider this problem:

Given: $MN > PQ$

Prove: $MP > NQ$

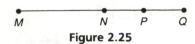

Figure 2.25

First study the Drawing (figure 2.25) and the related Given, to understand the situation. Then read the Prove with reference to the Drawing. Constructing the proof requires you to begin with the Given and end with the Prove. What may be confusing here is that the Given involves *MN* and *PQ*, while the Prove involves *MP* and *NQ*. However, this is easily remedied through the addition of *NP* to each side of the inequality $MN > PQ$.

Given: $MN > PQ$

Prove: $MP > NQ$

PROOF

Statements	Reasons
1. $MN > PQ$	1. Given
2. $MN + NP > NP + PQ$	2. Addition Property of Inequality
3. But $MN + NP = MP$ and $NP + PQ = NQ$	3. Segment-Addition Postulate
4. $MP > NQ$	4. Substitution Axiom of Equality

Note: The final reason may come as a surprise! However, the Substitution Axiom of Equality allows you to replace a quantity with its equal in *any* statement—including an inequality.

Now that you have a better idea of the development of proofs, consider an example that involves an angle bisector.

EXAMPLE 1   Study this proof, noting the order of the statements and reasons.

Given: $\overrightarrow{ST}$ bisects $\angle RSU$
$\overrightarrow{SV}$ bisects $\angle USW$

Prove: $m\angle RST + m\angle VSW = m\angle TSV$

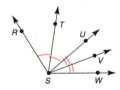

PROOF

Statements	Reasons
1. $\overrightarrow{ST}$ bisects $\angle RSU$	1. Given
2. $m\angle RST = m\angle TSU$	2. If an angle is bisected, then the measures of the resulting angles are equal
3. $\overrightarrow{SV}$ bisects $\angle USW$	3. Same as no. 1
4. $m\angle VSW = m\angle USV$	4. Same as no. 2
5. $m\angle RST + m\angle VSW = m\angle TSU + m\angle USV$	5. Addition Property of Equality (use the equations from statements 2 and 4)
6. $m\angle TSU + m\angle USV = m\angle TSV$	6. Angle-Addition Postulate
7. $m\angle RST + m\angle VSW = m\angle TSV$	7. Substitution Axiom of Equality ▲

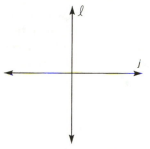

Figure 2.26

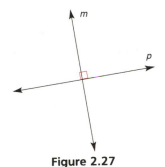

Figure 2.27

In the proof of Example 1, the Given information was split between statements 1 and 3; this was done for clarity, not out of necessity. In the proof, reason 2 was the definition of angle bisector. This should remind you of the importance of the role of definitions in a mathematical system.

Informally, a **vertical** line is one that extends up and down, like a flagpole. On the other hand, a line that extends left to right is **horizontal.** In figure 2.26, ℓ is vertical and j is horizontal. Where lines ℓ and j intersect, they appear to form angles of equal measure.

> **DEFINITION:**   **Perpendicular** lines are two lines that meet to form congruent adjacent angles.

Perpendicular lines do not have to be vertical and horizontal. In figure 2.27, the slanted lines m and p are perpendicular ($m \perp p$). Often a small square is placed in the opening of an angle to signify that it is a right angle.

The purpose of Example 2, which follows, is to establish the relationship between perpendicular lines and right angles. It is not a formal proof because those require a statement of a theorem. Study this proof, noting the order of the statements and reasons.

EXAMPLE 2

Given: $\overleftrightarrow{AB} \perp \overleftrightarrow{CD}$, intersecting at E
Prove: $\angle AEC$ is a right angle

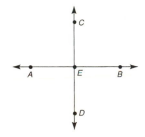

PROOF

Statements	Reasons
1. $\overleftrightarrow{AB} \perp \overleftrightarrow{CD}$, intersecting at E	1. Given
2. $\angle AEC \cong \angle CEB$	2. Perpendicular lines meet to form congruent adjacent angles
3. $m\angle AEC = m\angle CEB$	3. If two angles are congruent, their measures are equal
4. $\angle AEB$ is a straight angle and $m\angle AEB = 180°$	4. Measure of a straight angle equals 180°
5. $m\angle AEC + m\angle CEB = m\angle AEB$	5. Angle-Addition Postulate
6. $m\angle AEC + m\angle CEB = 180°$	6. Substitution Axiom of Equality
7. $m\angle AEC + m\angle AEC = 180°$ or $2 \cdot m\angle AEC = 180°$	7. Substitution Axiom of Equality
8. $m\angle AEC = 90°$	8. Division Property of Equality
9. $\angle AEC$ is a right angle	9. If the measure of an angle is 90° then the angle is a right angle ▲

Now we take up the notion of equivalence relation—a concept that has some significance in geometry but is not important in a beginning- or intermediate-level algebra course. The fact that the equality of real numbers is an equivalence relation leads to the fact that the congruence of angles is also an equivalence relation. Congruence of angles (or of line segments) is closely tied to equality of angle measures (or segment measures), by definition of congruence.

The following are some properties of the congruence of angles:

Reflexive: $\angle 1 \cong \angle 1$; an angle is congruent to itself.

Symmetric: If $\angle 1 \cong \angle 2$, then $\angle 2 \cong \angle 1$.

Transitive: If $\angle 1 \cong \angle 2$ and $\angle 2 \cong \angle 3$, then $\angle 1 \cong \angle 3$.

Like the equality of real numbers, the congruence of angles is known as an equivalence relation. An **equivalence relation** is a relation R for which elements *a, b,* and *c* possess these properties:

Reflexive: $a \, R \, a$.

Symmetric: If $a \, R \, b$, then $b \, R \, a$.

Transitive: If $a \, R \, b$ and $b \, R \, c$, then $a \, R \, c$.

In later chapters you will see that congruence of triangles and similarity of triangles are also equivalence relations.

Returning to the formulation of a proof, the final example in this section is based on the fact that vertical angles are congruent when two lines intersect. Because there are two pairs of congruent angles, the Prove could be stated:

Prove: $\angle 1 \cong \angle 3$ and $\angle 2 \cong \angle 4$

Such a conclusion is a conjunction and would be proved if both congruences were established. For simplicity, the Prove of Example 3 is stated:

Prove: $\angle 2 \cong \angle 4$

Study this proof, noting the order of the statements and reasons.

EXAMPLE 3   *Given:* $\overleftrightarrow{AC}$ intersects $\overleftrightarrow{BD}$ at O
Prove: $\angle 2 \cong \angle 4$

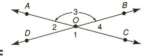

PROOF

Statements	Reasons
1. $\overleftrightarrow{AC}$ intersects $\overleftrightarrow{BD}$ at O	1. Given
2. $\angle$s *AOC* and *DOB* are straight $\angle$s, with m$\angle AOC = 180$ and m$\angle DOB = 180$	2. The measure of a straight angle is 180°
3. m$\angle AOC = $ m$\angle DOB$	3. Substitution Axiom of Equality
4. m$\angle 1 + $ m$\angle 4 = $ m$\angle DOB$ and m$\angle 1 + $ m$\angle 2 = $ m$\angle AOC$	4. Angle-Addition Postulate
5. m$\angle 1 + $ m$\angle 4 = $ m$\angle 1 + $ m$\angle 2$	5. Substitution Axiom of Equality

6.	$m\angle 4 = m\angle 2$	6. Subtraction Property of Equality
7.	$\angle 4 \cong \angle 2$	7. If two angles are equal in measure, the angles are congruent
8.	$\angle 2 \cong \angle 4$	8. Symmetric Axiom of Congruency of Angles ▲

In the preceding proof, the degree symbol (°) has been omitted from the statements, and this will continue to be done in future proofs. Moreover, there is no need to reorder the congruent angles from statement 7 to statement 8 since equivalence relations have a symmetric property; thus in the later work, statement 7 will be written to match the Prove even if the previous line does not have the right order. The same type of thinking applies to proving lines perpendicular or parallel: the order is simply not important!

Construction 2 in Section 2.1 not only determined the midpoint of $\overline{AB}$ but also that of the **perpendicular bisector** of $\overline{AB}$. In many instances, you will need the perpendicular line at a point other than the midpoint of a segment.

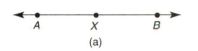
(a)

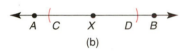
(b)

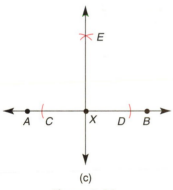
(c)

Figure 2.28

CONSTRUCTION 5 To construct the line perpendicular to a given line at a specified point on the given line.

Given: $\overleftrightarrow{AB}$ with point X, as shown in figure 2.28a

Construct: A line $\overleftrightarrow{EX}$, so that $\overleftrightarrow{EX} \perp \overleftrightarrow{AB}$

Construction: Using X as the center, mark off arcs of equal radii on each side of X to intersect $\overleftrightarrow{AB}$ at C and D. (See figure 2.28b.) Now, using C and D as centers, mark off arcs of equal radii with radius of length greater than XD so that these arcs intersect either above or below $\overleftrightarrow{AB}$.

Calling the point of intersection E, draw $\overleftrightarrow{EX}$, which is the desired perpendicular. (See figure 2.28c.)   ▲

The theorem Construction 5 is based on is a consequence of the Protractor Postulate, and we state it without proof.

> **THEOREM 2.3.1** There is exactly one line perpendicular to a given line at any point on the line.

Construction 2, which was used to locate the midpoint of a line segment, is also the method for constructing the perpendicular bisector of a line segment. In figure 2.29, $\overleftrightarrow{XY}$ is the perpendicular bisector of $\overline{RS}$. The following theorem can be proved by methods developed later in this book.

> **THEOREM 2.3.2** The perpendicular bisector of a line segment is unique.

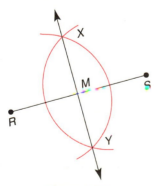

Figure 2.29

2.3 EXERCISES

In Exercises 1 and 2, supply reasons.

1. *Given:* $\angle 1 \cong \angle 3$
   Prove: $\angle MOP \cong \angle NOQ$

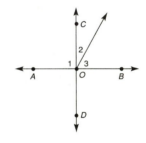

PROOF

Statements	Reasons
1. $\angle 1 \cong \angle 3$	1. ?
2. $m\angle 1 = m\angle 3$	2. ?
3. $m\angle 1 + m\angle 2 = m\angle MOP$ and $m\angle 2 + m\angle 3 = m\angle NOQ$	3. ?
4. $m\angle 1 + m\angle 2 = m\angle 3 + m\angle 2$	4. ?
5. $m\angle MOP = m\angle NOQ$	5. ?
6. $\angle MOP \cong \angle NOQ$	6. ?

2. *Given:* $\overleftrightarrow{AB}$ intersects $\overleftrightarrow{CD}$ at O so that $\angle 1$ is a right $\angle$
   Prove: $\angle 2$ and $\angle 3$ are complementary

PROOF

Statements	Reasons
1. $\overleftrightarrow{AB}$ intersects $\overleftrightarrow{CD}$ at O	1. ?
2. $\angle AOB$ is a straight $\angle$, so $m\angle AOB = 180$	2. ?
3. $m\angle AOC + m\angle COB = m\angle AOB$	3. ?
4. $m\angle 1 + m\angle COB = 180$	4. ?
5. $\angle 1$ is a right angle	5. ?
6. $m\angle 1 = 90$	6. ?
7. $90 + m\angle COB = 180$	7. ?
8. $m\angle COB = 90$	8. ?
9. $m\angle 2 + m\angle 3 = m\angle COB$	9. ?
10. $m\angle 2 + m\angle 3 = 90$	10. ?
11. $\angle 2$ and $\angle 3$ are complementary	11. ?

In Exercises 3 and 4, supply statements.

3. *Given:* $\angle 1 \cong \angle 2$ and $\angle 2 \cong \angle 3$
   Prove: $\angle 1 \cong \angle 3$

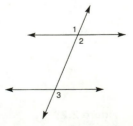

PROOF

Statements	Reasons
1. ?	1. Given
2. ?	2. Transitive Property of Congruence

4. *Given:* m∠AOB = m∠1
m∠BOC = m∠1
Prove: $\overrightarrow{OB}$ bisects ∠AOC

PROOF

Statements	Reasons
1. ?	1. Given
2. ?	2. Substitution
3. ?	3. Angles with equal measures are congruent
4. ?	4. If a ray divides an angle into two congruent angles, then the ray bisects the angle

In Exercises 5 to 7, use a compass and a straightedge to complete these constructions.

5. *Given:*   Point N on line s
     Construct: Line m through
           N so that m ⊥ s

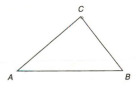

6. *Given:* $\overrightarrow{OA}$
     Construct: Right angle BOA
     (*Hint:* Use the straightedge to extend $\overrightarrow{OA}$ to the left.)

7. *Given:*   Triangle ABC
     Construct: The perpendicular bisectors of sides $\overline{AB}$, $\overline{AC}$, and $\overline{BC}$

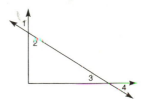

8. Draw a conclusion based on the results of Exercise 7.

In Exercises 9 to 12, complete the proofs by supplying ordered lists of statements and reasons. Use the following figure for Exercises 9 and 10.

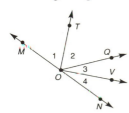

9. *Given:* ∠1 and ∠3 are complementary
     ∠2 and ∠3 are complementary
     Prove: ∠1 ≅ ∠2

10. *Given:* ∠1 ≅ ∠2
     ∠3 ≅ ∠4
     ∠2 and ∠3 are complementary
     Prove: ∠1 and ∠4 are complementary

11. *Given:* $\overleftrightarrow{AB}$, $\overleftrightarrow{DE}$, and $\overleftrightarrow{CF}$ as shown
     ∠BCF and ∠CFE are supplementary
     $\overrightarrow{CG}$ bisects ∠BCF, while $\overrightarrow{FG}$ bisects ∠CFE
     Prove: ∠2 and ∠3 are complementary

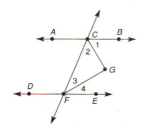

12. *Given:* Rays OM, OT, OQ, OV, and ON
     ∠TOV is a right angle
     $\overrightarrow{OT}$ bisects ∠MOQ
     $\overrightarrow{OV}$ bisects ∠QON
     Prove: ∠MON is a straight angle

13. Does the relation "is perpendicular to" have a reflexive property (consider line ℓ)? a symmetric property (consider lines ℓ and m)? a transitive property (consider lines ℓ, m, and n)? Is this relation an equivalence relation?

14. Does the relation "is greater than" have a reflexive property (consider real number *a*)? a symmetric property (consider real numbers *a* and *b*)? a transitive property (consider real numbers *a*, *b*, and *c*)? Is this relation an equivalence relation?

15. Does the relation "is a brother of" have a reflexive property (consider one male)? a symmetric property (consider two males)? a transitive property (consider three males)? Is this relation an equivalence relation?

16. Does the relation "is in love with" have a reflexive property (consider one person)? a symmetric property (consider two people)? a transitive property (consider three people)? Is this relation an equivalence relation?

17. By this time, the text has used numerous symbols and abbreviations. In this exercise, provide the *word* repre-

sented or abbreviated by each of the following:

(a) $\perp$ (f) adj.
(b) $\angle$s   (g) comp.
(c) supp.   (h) $\overrightarrow{AB}$
(d) rt.   (i) $\cong$
(e) m$\angle$1   (j) vert.

18. If there were no understood restriction to lines in a plane in Theorem 2.3.1, the theorem would be false. Explain why the following statement is false: "In space, there is exactly one line perpendicular to a given line at any point on the line."

19. If there were no understood restriction to lines in a plane in Theorem 2.3.2, the theorem would be false. Explain why the following statement is false: "In space, the perpendicular bisector of a line segment is unique."

*20. In the accompanying proof, reference numbers to earlier statements are provided at the left, to relate the claims made to earlier statements so that you will see how the new statement is deduced. Provide the missing reasons in the proof.

Given: $\angle$1 and $\angle$2 are complementary
       $\angle$1 is acute

Prove: $\angle$2 is also acute

PROOF

Reference	Statements	Reasons
	1. $\angle$1 and $\angle$2 are complementary	**1.** ?
(1)	**2.** m$\angle$1 + m$\angle$2 = 90	**2.** ?
	3. $\angle$1 is acute	**3.** ?
(3)	**4.** Where m$\angle$1 = x, 0 < x < 90	**4.** ?
(2)	**5.** x + m$\angle$2 = 90	**5.** ?
(5)	**6.** m$\angle$2 = 90 − x	**6.** ?
(4)	**7.** −x < 0 < 90 − x	**7.** ?
(7)	**8.** 90 − x < 90 < 180 − x	**8.** ?
(7, 8)	**9.** 0 < 90 − x < 90	**9.** ?
(6, 9)	**10.** 0 < m$\angle$2 < 90	**10.** ?
(10)	**11.** $\angle$2 is acute	**11.** ?

2.4

THE FORMAL PROOF OF A THEOREM

Statements that follow logically from known undefined terms, definitions, and postulates are called **theorems.** In other words, a theorem is a statement that can be proven. The formal proof of a theorem has several parts. To begin to understand how these are related, you need to consider carefully the terms "hypothesis" and "conclusion." The hypothesis of a statement describes the given situation (Given), while the conclusion describes what you need to establish (Prove). When a statement has the form "If H, then C," the hypothesis is the H statement and the conclusion is the C statement. Some theorems must be reworded to fit into "If . . . , then . . ." form so that the hypothesis and conclusion are easy to recognize.

EXAMPLE 1   Give the hypothesis H and conclusion C for each of these statements:
(a) If two lines intersect, then the vertical angles formed are congruent.
(b) All right angles are congruent.

(c) Parallel lines do not intersect.

(d) Lines are perpendicular when they meet to form congruent adjacent angles.

Solution

(a) As is   H: Two lines intersect.
              C: The vertical angles formed are congruent.

(b) Reworded If two angles are right angles, then these angles are congruent.
              H: Two angles are right angles.
              C: The angles are congruent.

(c) Reworded If two lines are parallel, then these lines do not intersect.
              H: Two lines are parallel.
              C: The lines do not intersect.

(d) Reordered When (if) two lines meet to form congruent adjacent angles, these lines are perpendicular.
              H: Two lines meet to form congruent adjacent angles.
              C: The lines are perpendicular.   ▲

Why do we need to distinguish between the hypothesis and the conclusion? For a theorem, the hypothesis determines the Drawing and the Given, providing a description of the Drawing's known characteristics. The conclusion determines what you wish to establish (the Prove) concerning the Drawing.

The five necessary parts of a formal proof are listed in the accompanying box in the order in which they should be developed.

ESSENTIAL PARTS OF THE FORMAL PROOF OF A THEOREM

1. *Statement:* States the theorem to be proved.
2. *Drawing:* Represents the hypothesis of the theorem.
3. *Given:* Describes the Drawing according to the information found in the hypothesis of the theorem.
4. *Prove:* Describes the Drawing according to the claim made in the conclusion of the theorem.
5. *Proof:* Orders a list of claims (Statements) and justifications (Reasons), beginning with the Given and ending with the Prove; there must be a logical flow in this Proof.

The most difficult part of a formal proof is the thinking process that must take place between parts 4 and 5. This game plan or analysis involves deducing and ordering conclusions based on the given situation. One must be something of a lawyer—selecting the claims that help prove the case, while discarding those that are superfluous. In the process of ordering the statements, it may be possible to think in reverse order, like so:

The Prove statement would be true if what else were true?

This technique can be quite beneficial! In any event, the final proof must be arranged in an order that allows one to reason from an earlier statement to a later claim in accordance with this logical pattern (perhaps several times):

H: hypothesis
P: principle
∴ C: conclusion

THEOREM 2.4.1   If two lines are perpendicular, then they meet to form right angles.

EXAMPLE 2

Solution

Write a formal proof of Theorem 2.4.1.

1. State the theorem.

    If two lines are perpendicular, then they meet to form right angles.

2. The hypothesis is H: Two lines are perpendicular.
    Make a Drawing to fit this description. (See figure 2.30.)

3. State the Given, using the Drawing and based on H: Two lines are ⊥.

    Given: $\overleftrightarrow{AB} \perp \overleftrightarrow{CD}$ intersecting at E

4. State the Prove, using the Drawing and based on C: They meet to form right angles.

    Prove: $\angle AEC$ is a right angle

5. Construct the Proof. This proof is found in Example 2, Section 2.3. A formal proof would differ from the one in that example only by being preceded by the statement of the theorem!   ▲

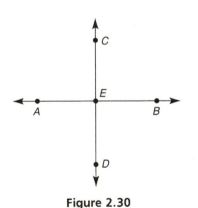

Figure 2.30

Sometimes the converse of a statement is also true. In fact, Example 3 presents the formal proof of a theorem that is the converse of the theorem in Example 2. Remember that the **converse** of "If P, then Q is the statement if Q, then P."

Once a theorem has been proved, it may be cited thereafter as a reason in future proofs. Thus, any theorem found in this section can be used for justification in later sections.

The proof that follows is nearly complete! It is difficult to provide a complete formal proof that explains the "how to" and simultaneously indicates the polished form. Thus, Example 2 aims at the how to, while Example 3 illustrates the polished form. What you do not see with Example 3 are the considerable thought and used piece(s) of scratch paper needed to piece together this puzzle.

The proof of a theorem is not unique! From the start, the Drawings need not match, although the same relationships should be indicated. Certainly different letters are likely to be chosen in illustrating the hypothesis.

▲ *WARNING: You should not make a drawing that embeds qualities beyond those described in the hypothesis; neither should your drawing indicate fewer qualities than the hypothesis prescribes!*

THEOREM 2.4.2 If two lines meet to form a right angle, then these lines are perpendicular.

EXAMPLE 3   Give a formal proof for Theorem 2.4.2.

If two lines meet to form a right angle, then these lines are perpendicular.

Given: $\overleftrightarrow{AB}$ and $\overleftrightarrow{CD}$ intersect at E so that $\angle ABC$ is a right angle
Prove: $\overleftrightarrow{AB} \perp \overleftrightarrow{CD}$

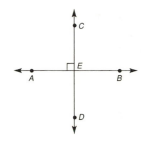

PROOF

Statements	Reasons
1. $\overleftrightarrow{AB}$ and $\overleftrightarrow{CD}$ intersect so that $\angle AEC$ is a right angle	1. Given
2. $m\angle AEC = 90$	2. If an $\angle$ is a right $\angle$, its measure is 90
3. $\angle AEB$ is a straight $\angle$, so $m\angle AEB = 180$	3. If an $\angle$ is a straight $\angle$, its measure is 180
4. $m\angle AEC + m\angle CEB = m\angle AEB$	4. Angle-Addition Postulate
5. $90 + m\angle CEB = 180$	5. Substitution
6. $m\angle CEB = 90$	6. Subtraction Property of Equality
7. $m\angle AEC = m\angle CEB$	7. Substitution
8. $\angle AEC \cong \angle CEB$	8. If two $\angle$s have = measures, the $\angle$s are $\cong$
9. $\overleftrightarrow{AB} \perp \overleftrightarrow{CD}$	9. If two lines form $\cong$ adjacent $\angle$s, these lines are $\perp$ ▲

Several other theorems are now stated, most of which are left as exercises. This list contains theorems that are quite useful when cited as reasons in later proofs. Proof of Theorem 2.4.8 is provided.

THEOREM 2.4.3 If two angles are complementary to the same angle (or to congruent angles), then these angles are congruent.

THEOREM 2.4.4 If two angles are supplementary to the same angle (or to congruent angles), then these angles are congruent.

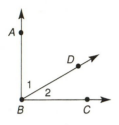

∠s 1 and 2 are
complementary

Figure 2.31

THEOREM 2.4.5 If two lines intersect, then the vertical angles formed are congruent.

THEOREM 2.4.6 Any two right angles are congruent.

THEOREM 2.4.7 If the exterior sides of two adjacent angles form perpendicular rays, then these angles are complementary.

For the Drawing of Theorem 2.4.7, see figure 2.31; $\overrightarrow{BA} \perp \overrightarrow{BC}$.

THEOREM 2.4.8 If the exterior sides of two adjacent angles form a straight line, then these angles are supplementary.

Given: ∠3 and ∠4 and $\overleftrightarrow{EG}$

Prove: ∠3 and ∠4 are supplementary

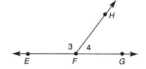

PROOF

Statements	Reasons
1. ∠3 and ∠4 and $\overleftrightarrow{EG}$	1. Given
2. m∠3 + m∠4 = m∠EFG	2. Angle-Addition Postulate
3. ∠EFG is a straight angle	3. If the sides of an ∠ are opposite rays, it is a straight ∠
4. m∠EFG = 180	4. The measure of a straight ∠ is 180
5. m∠3 + m∠4 = 180	5. Substitution Axiom of Equality
6. ∠3 and ∠4 are supplementary	6. If the sum of measures of two ∠s is 180, the ∠s are supplementary

▲

The final two theorems in this section are stated for convenience. They will make later proofs easier to prove.

THEOREM 2.4.9 If two segments are congruent, then their midpoints separate these segments into four congruent segments.

> **THEOREM 2.4.10** If two angles are congruent, then their bisectors separate these angles into four congruent angles.

2.4 EXERCISES

In Exercises 1 to 6, state the hypothesis H and conclusion C for each statement.

1. If a line segment is bisected, then each of the equal segments has half the length of the original segment.

2. If two sides of a triangle are congruent, then the triangle is isosceles.

3. All squares are quadrilaterals.

4. Every regular polygon has congruent interior angles.

5. Two angles are congruent if each is a right angle.

6. The lengths of corresponding sides of similar polygons are proportional.

7. Name, in order, the five parts of the formal proof of a theorem.

8. Which part (hypothesis or conclusion) of a theorem determines the
   (a) Drawing   (b) Given   (c) Prove

In Exercises 9 to 17, write a formal proof of each theorem. Be sure to include the statement of the theorem.

9. If two angles are complementary to the same angle (or to congruent angles), then these angles are congruent.

10. If two angles are supplementary to the same angle (or to congruent angles), then these angles are congruent.

11. If two lines intersect, then the vertical angles formed are congruent.

12. Any two right angles are congruent.

13. If the exterior sides of two adjacent angles form perpendicular rays, then these angles are complementary.

14. If two segments are congruent, then their midpoints separate these segments into four congruent segments.

15. If two angles are congruent, then their bisectors separate these angles into four congruent angles.

*16. The bisectors of two adjacent supplementary angles form a right angle.

*17. The supplement of an acute angle is obtuse.

▲ A LOOK BEYOND: HISTORICAL SKETCH OF EUCLID

Names often associated with the early development of Greek mathematics, beginning in approximately 600 B.C., include Thales, Pythagoras, Archimedes, Appolonius, Diophantus, Eratosthenes, and Heron. However, the name most often associated with traditional geometry is that of Euclid, who lived around 300 B.C.

Although Euclid was Greek, he was asked to head the mathematics department at the University of Alexandria (in Egypt), which was the center of Greek learning. It is believed that Euclid told Ptolemy (the local ruler) that "There is no royal road to geometry," in response to Ptolemy's request for a quick and easy knowledge of the subject.

Euclid's best known work is the *Elements,* a systematic treatment of geometry with some algebra and number theory. That work, which consists of 13 volumes, has dominated the study of geometry for more than 2000 years. Most secondary-level geometry courses, even today, are based on Euclid's *Elements* and in particular on these volumes:

Book I: Triangles and congruence, parallels, quadrilat-

erals, the Pythagorean theorem; and area relationships

Book III: Circles, chords, secants, tangents, and angle measurement

Book IV: Constructions and regular polygons

Book VI: Similar triangles, proportions, and the Angle Bisector theorem

Book XI: Lines and planes in space, and parallelepipeds

One of Euclid's theorems was a forerunner of the property of trigonometry known as the Law of Cosines. Although it is difficult to understand now, it will make sense to you later. As stated by Euclid, "In an obtuse-angled triangle, the square of the side opposite the obtuse angle equals the sum of squares of the other two sides and the product of one side and the projection of the other upon it."

While it is believed that Euclid was a great teacher, he is also recognized as a great mathematician and as the first author of an elaborate textbook. In Chapter 3, Euclid's Parallel Postulate is central to the study of plane geometry.

▲ SUMMARY

A LOOK BACK AT CHAPTER 2

Our goal in this chapter has been to deal with a mathematical system known as geometry. We encountered the four elements of a mathematical system: undefined terms, definitions, postulates, and theorems. The undefined terms were needed to lay the foundation for defining new terms. The postulates were needed to lay the foundation for the theorems we proved here and for the theorems that lie ahead. Constructions were also introduced in this chapter as a means of establishing the uniqueness of a midpoint of a segment, the bisector of an angle, and the perpendicular to a line from a point on the line.

A LOOK AHEAD TO CHAPTER 3

In the next chapter, the theorems we will prove are based on a postulate known as the Parallel Postulate. A new method of proof, called indirect proof, will be introduced and used throughout this chapter. While many of the theorems in Chapter 3 deal with parallel lines, other theorems in the chapter deal with the angles of a polygon.

IMPORTANT TERMS AND CONCEPTS OF CHAPTER 2

2.1 Point
   Line
   Line Segment
   Ruler Postulate
   Collinear Points
   Segment-Addition Postulate
   Congruent Segments
   Midpoint of a Segment
   Bisector of a Segment
   Ray
   Opposite Rays
   Parallel Lines
   Plane
   Space
   Circle
2.2 Angle
   Protractor Postulate
   Acute, Right, Obtuse, and Straight Angles
   Angle-Addition Postulate
   Adjacent Angles
   Congruent Angles
   Bisector of an Angle
   Complementary and Supplementary Angles
   Vertical Angles
2.3 Perpendicular Lines
   Equivalence Relation (Reflexive, Symmetric, Transitive)
   Perpendicular Bisector
2.4 Hypothesis, Conclusion
   Converse
A Look Beyond Historical Sketch of Euclid

▲ REVIEW EXERCISES

Sketch and label the figures described for Review Exercises 1 to 3.

1. Points A, B, C, and D are coplanar. A, B, and C are the only three of these points that are collinear.

2. Line ℓ intersects plane X at point P.

3. Plane M contains intersecting lines j and k.

Use the following figure for Review Exercises 4 to 6.

4. Name the coordinate of the midpoint of $\overline{AD}$.

5. Complete: $AB + BC =$ _____ .

6. Name a ray opposite ray $\overrightarrow{BC}$.

7. A, B, and C are three points on a line. $AC = 8$, $BC = 4$, and $AB = 12$. Which point must be between the two other points?

8. *Given:* $\overrightarrow{BD}$ bisects $\angle ABC$
   $m\angle ABD = 5x - 11$
   $m\angle DBC = x + 21$
   Find: $m\angle ABC$

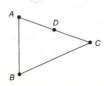

9. *Given:* D is the midpoint of $\overline{AC}$
   $\overline{AC} \cong \overline{BC}$
   $CD = 2x + 5$
   $BC = x + 28$
   Find: AC

10. *Given:* m∠3 = 7x − 21
m∠4 = 3x + 7
Find: m∠FMH

11. *Given:* m∠FMH = 4x + 1
m∠4 = x + 4
Find: m∠4

12. *Given:* ∠EFG is a right angle
m∠HFG = 2x − 6
m∠EFH = 3 · m∠HFG
Find: m∠EFH

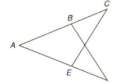

13. Two angles are supplementary. One angle is 40° more than four times the other. Find the measures of the two angles.

State whether the statements in Review Exercises 14 to 18 are always true (A), sometimes true (S), or never true (N).

14. If AM = MB, then A, M, and B are collinear.

15. If two angles are congruent, then they are right angles.

16. The bisectors of vertical angles are opposite rays.

17. Complementary angles are congruent.

18. The supplement of an obtuse angle is another obtuse angle.

Write two-column proofs for Exercises 19 to 25.

19. *Given:* $\overline{AB} \cong \overline{AE}$
$\overline{BC} \cong \overline{DE}$
Prove: $\overline{AC} \cong \overline{AD}$

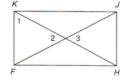

20. *Given:* $\overline{KF} \perp \overline{FH}$
∠JHF is a right ∠
Prove: ∠KFH ≅ ∠JHF

21. *Given:* ∠1 ≅ ∠3
Prove: ∠1 ≅ ∠2

22. *Given:* ∠1 is comp. to ∠M
∠2 is comp. to ∠M
Prove: ∠1 ≅ ∠2

23. *Given:* ∠MOP ≅ ∠MPO
$\overrightarrow{OR}$ bisects ∠MOP
$\overrightarrow{PR}$ bisects ∠MPO
Prove: ∠1 ≅ ∠2

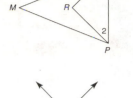

24. *Given:* ∠3 is supp. to ∠5
∠4 is supp. to ∠6
Prove: ∠3 ≅ ∠6

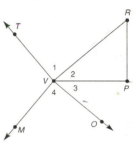

25. *Given:* ∠1 ≅ ∠P
∠4 ≅ ∠P
$\overrightarrow{VP}$ bisects ∠RVO
Prove: ∠TVP ≅ ∠MVP

26. *Given:* $\overline{VP}$
Construct: $\overline{VW}$ such that
VW = 4 · VP

27. Construct a 135° angle.

28. *Given:*   Triangle PQR
Construct: The three angle bisec-
tors
What did you discover about the three angle bisectors of this triangle?

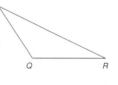

29. *Given:* $\overline{AB}$, $\overline{BC}$, and ∠B, as shown
Construct: Triangle ABC

30. *Given:*   m∠B = 50°
Construct: An angle whose measure is 20°

3

PARALLEL LINES

CHAPTER OUTLINE

3.1 The Parallel Postulate, Transversals, and Special Angles

3.2 Indirect Proof

3.3 Proving Lines Parallel

3.4 Applications of Parallels to Triangles and Other Polygons

A Look Beyond Non-Euclidean Geometries

The sides of angles 1 and 2 are **parallel.** How are these angles related? The sides of angles 2 and 3 are parallel, too. How are angles 2 and 3 related?

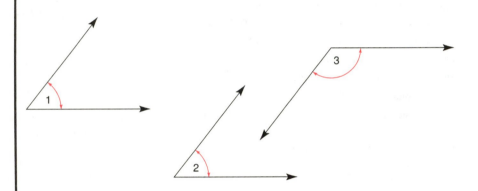

3.1 THE PARALLEL POSTULATE, TRANSVERSALS, AND SPECIAL ANGLES

By definition, two lines (or segments or rays) are **perpendicular** if they meet to form congruent adjacent angles. Using this definition, we proved the theorem stating that "perpendicular lines meet to form right angles." It is worth noting that two rays or line segments are perpendicular if they are adjacent or intersecting parts of two lines that are perpendicular. We now consider another method for constructing a perpendicular.

CONSTRUCTION 6 To construct the line that is perpendicular to a given line from a point not on the given line.

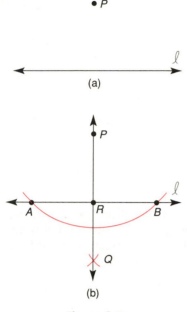

(a)

(b)

Figure 3.1

Given: In figure 3.1a, line ℓ and point P not on ℓ

Construct: $\overleftrightarrow{PQ} \perp \ell$

Construction: With P as the center, open the compass to a length large enough to intersect ℓ in two points (A and B), as shown in figure 3.1b.

With A and B as centers, mark off arcs of equal radii (not changing the compass opening) to intersect at a point Q, as shown.

Draw $\overleftrightarrow{PQ}$ to complete the desired perpendicular.   ▲

In this construction, $\angle PRA$ and $\angle PRB$ are right angles. The arcs drawn from A and B to intersect below line ℓ provide greater accuracy than would arcs intersecting above ℓ.

Construction 6 suggests a uniqueness relationship that can be proved.

THEOREM 3.1.1   Exactly one line is perpendicular to a given line from a point not on that line.

The term "perpendicular" is extended to include line-plane and plane-plane relationships. The drawings in figure 3.2 indicate two perpendicular lines, a line perpendicular to a plane, and two perpendicular planes.

Similarly, the word "parallel" is used to describe possible relationships among lines and planes. However, the lines must be in the same plane, as the following definition emphasizes.

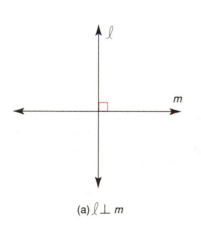

(a) $\ell \perp m$

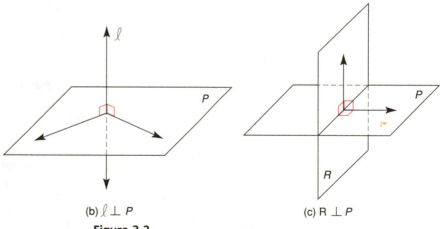

(b) $\ell \perp P$

(c) $R \perp P$

Figure 3.2

DEFINITION: **Parallel lines** are lines in the same plane that do not intersect.

More generally, two lines in a plane, a line and a plane, or two planes are parallel if they do not intersect. Segments or rays are parallel if they are parts

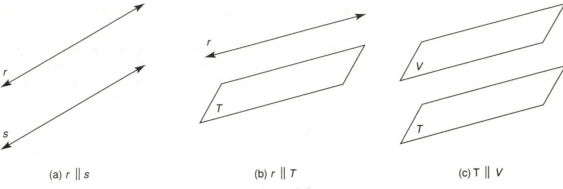

(a) r ∥ s   (b) r ∥ T   (c) T ∥ V

Figure 3.3

of two parallel lines. Figure 3.3 shows drawings that indicate possible applications of the word "parallel." In figure 3.4, two parallel planes M and N are intersected by a third plane G. How must the lines of intersection, a and b, be related?

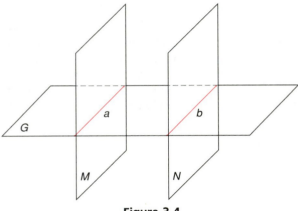

Figure 3.4

The type of geometry you are studying is known as Euclidean geometry. In this geometry, a plane is a flat, two-dimensional surface in which the line segment joining any two points of the plane lies entirely within the plane; informally, we say the line segment is "straight." While the postulate that follows characterizes Euclidean geometry, the Look Beyond section of this chapter will consider alternative geometries. Postulate 9, the Euclidean Parallel Postulate, is easy to accept intuitively because of the way we perceive a plane.

POSTULATE 9   **(Parallel Postulate)**   Through a point not on a line, exactly one line is parallel to the given line.

In particular, consider figure 3.5, in which line m and point P (with P not on m) both lie in plane R. It seems reasonable that exactly one line can be drawn through P parallel to line m. The method of construction for the unique line through P parallel to m is provided in the Look Beyond section of this chapter.

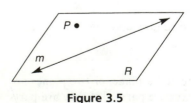

Figure 3.5

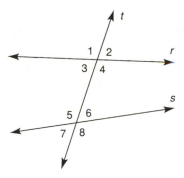

Figure 3.6

A **transversal** is a line that intersects two (or more) other lines at distinct points; all of the lines lie in the same plane. In figure 3.6, t is a transversal for lines r and s. Angles that are formed between r and s are **interior angles;** those outside r and s are **exterior angles.**

Interior angles: ∠3, ∠4, ∠5, ∠6
Exterior angles: ∠1, ∠2, ∠7, ∠8

Two angles that lie in the same relative positions when two lines are cut by a transversal are **corresponding angles.** In figure 3.6, ∠1 and ∠5 are corresponding angles; each angle is above the line and to the left of the transversal that helps form the angle.

Corresponding angles: ∠1 and ∠5   above left
(must be in pairs)   ∠3 and ∠7   below left
∠2 and ∠6   above right
∠4 and ∠8   below right

Two interior angles that have different vertices (plural of "vertex") and lie on opposite sides of the transversal are **alternate interior angles.** Two exterior angles that have different vertices and lie on opposite sides of the transversal are **alternate exterior angles.** Both types of alternate angles must occur in pairs. (See figure 3.6.)

Alternate interior angles: ∠3 and ∠6
∠4 and ∠5
Alternate exterior angles: ∠1 and ∠8
∠2 and ∠7

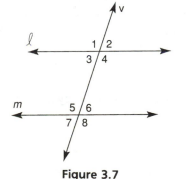

Figure 3.7

Now suppose that parallel lines ℓ and m in figure 3.7 are cut by transversal v. If a protractor were used to measure ∠1 and ∠5, these angles would have equal measures. That is, they are congruent. Similarly, any other pair of corresponding angles will be congruent, as long as $\ell \parallel m$.

POSTULATE 10 If two parallel lines are cut by a transversal, then the corresponding angles are congruent.

EXAMPLE 1 In figure 3.7, if $\ell \parallel m$ and m∠1 = 117, find:
(a) m∠2   (c) m∠4
(b) m∠5   (d) m∠8

Solution (a) m∠2 = 63   supplementary to ∠1
(b) m∠5 = 117   corresponding to ∠1
(c) m∠4 = 117   vertical ∠ to ∠1
(d) m∠8 = 117   corresponding to ∠4 (found in part (c))   ▲

Several theorems follow from these postulates; for all these theorems, formal proofs are provided. Study the proofs and be able to state all the theorems. Later, you can cite the theorems as reasons in subsequent proofs.

THEOREM 3.1.2   If two parallel lines are cut by a transversal, then the alternate interior angles are congruent.

Given: In figure 3.8, $a \parallel b$
         transversal k

Prove: $\angle 3 \cong \angle 6$

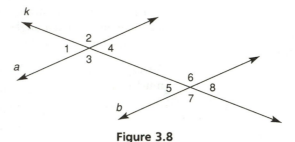

Figure 3.8

PROOF

Statements	Reasons
1. $a \parallel b$; transversal k	1. Given
2. $\angle 2 \cong \angle 6$	2. If two $\parallel$ lines are cut by a transversal, corresponding $\angle$s are $\cong$
3. $\angle 3 \cong \angle 2$	3. If two lines intersect, vertical $\angle$s formed are $\cong$
4. $\angle 3 \cong \angle 6$	4. Transitive (of $\cong$) ▲

It is easy to prove that $\angle 4$ and $\angle 5$ are congruent, because they are supplements to $\angle 3$ and $\angle 6$. Another theorem that is easy to prove is now stated, but the proof is left as an exercise.

THEOREM 3.1.3   If two parallel lines are cut by a transversal, then the alternate exterior angles are congruent.

Whenever two parallel lines are cut by a transversal, an interesting relationship exists between the two interior angles on the same side of the transversal. The following proof establishes that these interior angles are supplementary; a similar claim can be made for the pair of exterior angles on the same side of the transversal.

THEOREM 3.1.4   If two parallel lines are cut by a transversal, then the interior angles on the same side of the transversal are supplementary.

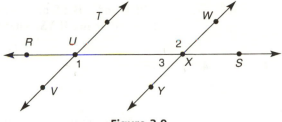

Figure 3.9

Given: In figure 3.9, $\overleftrightarrow{TV} \parallel \overleftrightarrow{WY}$, with transversal $\overleftrightarrow{RS}$

Prove: ∠1 and ∠3 are supplementary

PROOF

Statements	Reasons
1. $\overleftrightarrow{TV} \parallel \overleftrightarrow{WY}$; transversal $\overleftrightarrow{RS}$	1. Given
2. ∠1 ≅ ∠2	2. If two ∥ lines are cut by a transversal, alternate interior ∠s are ≅
3. m∠1 = m∠2	3. If two ∠s are ≅, their measures are =
4. ∠WXY is a straight ∠, so m∠WXY = 180	4. If an ∠ is a straight ∠, its measure is 180
5. m∠2 + m∠3 = m∠WXY	5. Angle-Addition Postulate
6. m∠2 + m∠3 = 180	6. Substitution
7. m∠1 + m∠3 = 180	7. Substitution
8. ∠1 and ∠3 are supplementary	8. If the sum of measures of two ∠s is 180, the ∠s are supplementary

▲

The proof of the following theorem is left as an exercise.

THEOREM 3.1.5   If two parallel lines are cut by a transversal, then the exterior angles on the same side of the transversal are supplementary.

The final examples in this section make use of methods from algebra and deal with the angles formed when two parallel lines are cut by a transversal.

EXAMPLE 2   *Given:* In figure 3.9, $\overleftrightarrow{TV} \parallel \overleftrightarrow{WY}$, with transversal $\overleftrightarrow{RS}$
$\quad\quad$ m∠RUV = $(x + 4)(x - 3)$
$\quad\quad$ m∠WXS = $x^2 - 3$

Find: x

Solution $\angle RUV$ and $\angle WXS$ are alternate exterior angles, so they are congruent. Then $m\angle RUV = m\angle WXS$. Therefore

$$(x + 4)(x - 3) = x^2 - 3$$
$$x^2 + x - 12 = x^2 - 3$$
$$x - 12 = -3$$
$$x = 9$$

▲

Notice that both angles measure 78° when $x = 9$.

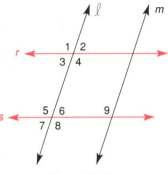

Figure 3.10

In figure 3.10, lines r and s are parallel. However, ℓ and m are not known to be parallel. Which angle, $\angle 1$ or $\angle 9$, must be congruent to $\angle 5$? To answer the question, it might be helpful to use a second color to draw in the lines that are known to be parallel. Now decide which transversal to use, ℓ or m. In this case, ℓ is the transversal for r and s that forms corresponding angles 5 and 1. Therefore, $\angle 1$ must be congruent to $\angle 5$.

EXAMPLE 3   *Given:* In figure 3.10, $r \parallel s$ and transversal ℓ
    $m\angle 3 = 4x + y$
    $m\angle 5 = 6x + 5y$
    $m\angle 6 = 5x - 2y$

Find: x and y

Solution $\angle 3$ and $\angle 6$ are congruent alternate interior angles, while $\angle 3$ and $\angle 5$ are supplementary angles, by Theorem 3.1.4. These facts lead us to the following system of equations:

$$4x + y = 5x - 2y$$
$$(4x + y) + (6x + 5y) = 180$$

These equations can be simplified to

$$x - 3y = 0$$
$$10x + 6y = 180$$

After we divide each term of the second equation by 2, the system becomes

$$x - 3y = 0$$
$$5x + 3y = 90$$

Addition leads to the equation $6x = 90$, so $x = 15$. Substituting this into $x - 3y = 0$, we have

$$15 - 3y = 0$$
$$3y = 15$$
$$y = 5$$

Our solution, $x = 15$ and $y = 5$, yields the following angle measures:

$$m\angle 3 = 65$$
$$m\angle 5 = 115$$
$$m\angle 6 = 65$$

▲

Notice that these angle measures are consistent with figure 3.10 and the required relationships for the angles named. For instance, $m\angle 3 + m\angle 5 = 180$, and we see that interior angles on the same side of the transversal are supplementary.

3.1 EXERCISES

In Exercises 1 to 9, use drawings, as needed, to answer each question.

1. Does the relation "is parallel to" have a
   (a) reflexive property (consider a line m)
   (b) symmetric property (consider lines m and n in a plane)
   (c) transitive property (consider coplanar lines m, n, and q)

   Is this relation an equivalence relation?

2. In a plane, $\ell \perp m$ and $t \perp m$. By appearance, how are ℓ and t related?

3. In the accompanying drawing, $r \parallel s$. Each interior angle on the right side of the transversal t has been bisected. Using your intuition, what appears to be true of $\angle 9$ formed by the bisectors?

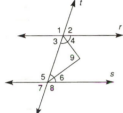

4. Make a sketch to represent two planes that are
   (a) parallel
   (b) perpendicular

5. In the drawing, r is parallel to s, and $m\angle 2 = 87$. Find
   (a) $m\angle 3$
   (b) $m\angle 6$
   (c) $m\angle 1$
   (d) $m\angle 7$

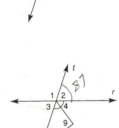

6. In Euclidean geometry, how many lines can be drawn through a point P not on a line ℓ that are
   (a) parallel to line ℓ
   (b) perpendicular to line ℓ

7. In the drawing, lines r and s are cut by transversal t. Which angle
   (a) corresponds to $\angle 1$
   (b) is the alternate interior $\angle$ for $\angle 4$
   (c) is the alternate exterior $\angle$ for $\angle 1$
   (d) is the other interior angle on the same side of transversal t as $\angle 3$

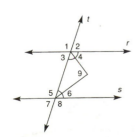

8. In the accompanying figure, $\overline{AD} \parallel \overline{BC}$, $\overline{AB} \parallel \overline{DC}$, and $m\angle A = 92°$. Find
   (a) $m\angle B$
   (b) $m\angle C$
   (c) $m\angle D$

9. In the drawing, $\ell \parallel m$, with transversal t, and $\overrightarrow{OQ}$ bisects $\angle MON$. If $m\angle 1 = 112°$, find the following:
   (a) $m\angle 2$
   (b) $m\angle 4$
   (c) $m\angle 5$
   (d) $m\angle MOQ$

10. *Given:* $\ell \parallel m$
    transversal t
    $m\angle 1 = 4x + 2$
    $m\angle 6 = 4x - 2$
    Find: x and $m\angle 5$

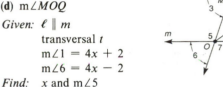

Use the accompanying figure for Exercises 11 to 13.

11. *Given:* $m \parallel n$
    transversal k
    $m\angle 3 = x^2 - 3x$
    $m\angle 6 = (x + 4)(x - 5)$
    Find: x and $m\angle 4$

12. *Given:* $m \parallel n$
    transversal k
    $m\angle 1 = 5x + y$
    $m\angle 2 = 3x + y$
    $m\angle 8 = 3x + 5y$
    Find: x, y, and $m\angle 8$

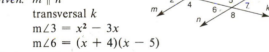

13. *Given:* $m \parallel n$
    transversal k
    $m\angle 3 = 6x + y$
    $m\angle 5 = 8x + 2y$
    $m\angle 6 = 4x + 7y$
    Find: x, y, and $m\angle 7$

14. In the three-dimensional figure, $\overline{CA} \perp \overleftrightarrow{AB}$ and $\overleftrightarrow{BE} \perp \overline{AB}$. Are $\overline{CA}$ and $\overline{BE}$ parallel to each other? (Compare to Exercise 2.)

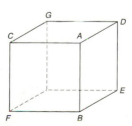

In Exercises 15 to 17, use the accompanying figure.

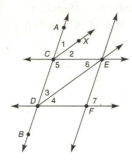

15. *Given:* $\overleftrightarrow{CE} \parallel \overleftrightarrow{DF}$
transversal $\overleftrightarrow{AB}$
$\overrightarrow{CX}$ bisects $\angle ACE$
$\overrightarrow{DE}$ bisects $\angle CDF$
    Prove: $\angle 1 \cong \angle 3$

16. *Given:* $\overleftrightarrow{CE} \parallel \overleftrightarrow{DF}$
transversal $\overleftrightarrow{AB}$
$\overrightarrow{DE}$ bisects $\angle CDF$
    Prove: $\angle 3 \cong \angle 6$

17. *Given:* $\overleftrightarrow{CE} \parallel \overleftrightarrow{DF}$ and $\overleftrightarrow{CD} \parallel \overleftrightarrow{EF}$
    Prove: $\angle 5$ and $\angle 7$ are supplementary

18. *Given:* $r \parallel s$
transversal t
$\angle 1$ is a right $\angle$
    Prove: $\angle 2$ is a right $\angle$

19. In triangle *ABC*, line *t* is
drawn through vertex *A* so
that $\overleftrightarrow{t} \parallel \overline{BC}$.

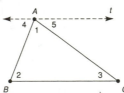

(a) Which pairs of $\angle$s are $\cong$?
(b) What is the sum of m$\angle 1$, m$\angle 4$, and m$\angle 5$?
(c) What is the sum of measures of $\angle$s of $\triangle ABC$?

In Exercises 20 to 22, write a formal proof of each theorem.

20. If two parallel lines are cut by a transversal, then the alternate exterior angles are congruent.

21. If two parallel lines are cut by a transversal, then the exterior angles on the same side of the transversal are supplementary.

22. If a transversal is perpendicular to one of two parallel lines, then it is also perpendicular to the other line.

23. Suppose that two lines are cut by a transversal in such a way that corresponding angles are not congruent. Can those two lines be parallel?

24. *Given:*   Line ℓ and point P not on ℓ
    Construct: $\overleftrightarrow{PQ} \perp \ell$

25. *Given:*   Triangle *ABC* with three acute angles
    Construct: $\overline{BD} \perp \overline{AC}$

26. *Given:*   Triangle *ABC* with obtuse $\angle ACB$
    Construct: $\overline{CE} \perp \overline{AB}$

27. *Given:*   Triangle *MNQ* with obtuse $\angle MNQ$
    Construct: $\overline{MR} \perp \overline{NQ}$
    (*Hint:* Extend $\overline{NQ}$.)

28. *Given:* A line m and a point T not on m
    Suppose that you do the following:
    (i) Construct a perpendicular line r from T to m.
    (ii) Construct a line s perpendicular to line r at point T.
    What relationship holds between s and m?

3.2

INDIRECT PROOF

Let $P \rightarrow Q$ represent the statement "If P, then Q." The following statements are related to this conditional statement (note that $\sim P$ represents the negation of P):

Conditional:	$P \rightarrow Q$
Converse:	$Q \rightarrow P$
Inverse:	$\sim P \rightarrow \sim Q$
Contrapositive:	$\sim Q \rightarrow \sim P$

For example, consider the following conditional statement:

If Tom lives in San Diego, then he lives in California.

This true statement has these related statements:

Converse: If Tom lives in California, then he lives in San Diego. (false)

Inverse: If Tom does not live in San Diego, then he does not live in California. (false)

Contrapositive: If Tom does not live in California, then he does not live in San Diego. (true)

In general, the conditional statement and its contrapositive are either both true or both false! In advanced courses, a statement of the form "If *P*, then *Q*" may be established by proving that its contrapositive is true, if that seems the easier proof to complete. Similarly, the converse and the inverse are also either both true or both false.

EXAMPLE 1   For the conditional statement that follows, give the converse, the inverse, and the contrapositive. Then classify each as true or false.

"If two angles are vertical angles, then they are congruent angles."

Solution   *Converse:*   If two angles are congruent angles, then they are vertical angles. (false)

Inverse: If two angles are not vertical angles, then they are not congruent angles. (false)

Contrapositive: If two angles are not congruent angles, then they are not vertical angles. (true)   ▲

Consider the following circumstances, and accept each premise as true:

1. If Matt cleans his room, then he will go to the movie.

2. Matt does not get to go to the movie.

What may you conclude? You should have deduced that Matt did not clean his room; if he had, he would have gone to the movie. This "backdoor" reasoning is based on the following principle:

PRINCIPLE OF NEGATIVE INFERENCE
$P \rightarrow Q$
$\dfrac{\sim Q}{\therefore \quad \sim P}$

The situation just described is closely related to a method known as **indirect proof.**

You will need to know when to use the indirect method of proof. Generally, look for a statement of the form $P \rightarrow Q$ in which Q contains a negation and denies some claim. For instance, an indirect proof might be best if Q reads in one of these ways:

c is *not* equal to *d*

ℓ is *not* perpendicular to $\hat{m}$

However, we will see in Example 4 of this section that the indirect method can be used to prove that line ℓ is parallel to line *m*. Another instance in which indirect proof is used is for proving existence and uniqueness theorems.

The method of indirect proof is illustrated in Example 2. All indirect proofs in this book are given in paragraph form (as are many of the direct proofs).

In any paragraph proof, each statement must still be justified (or at least justifiable). Because of the need to order your statements properly, this type of proof may have a positive impact on the essays you write for your other classes.

EXAMPLE 2

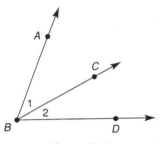

Figure 3.11

Given: In figure 3.11, $\overrightarrow{BA}$ is *not* perpendicular to $\overrightarrow{BD}$

Prove: ∠1 and ∠2 are *not* complementary

Proof: Suppose that ∠1 and ∠2 are complementary. Then m∠1 + m∠2 = 90 because the sum of the measures of two complementary ∠s is 90. We also know that m∠1 + m∠2 = m∠ABD, by the Angle-Addition Postulate. In turn, m∠ABD = 90, by substitution. Then ∠ABD is a right angle. In turn, $\overrightarrow{BA} \perp \overrightarrow{BD}$. But this contradicts the given hypothesis; therefore the supposition must be false, and it follows that ∠1 and ∠2 are *not* complementary.   ▲

In Example 2 and in all indirect proofs, the first statement takes the form

"Suppose that . . ."   or   *"Assume that . . ."*

By its very nature, this statement cannot be supported, while every other statement in the proof must be justified; thus, when a contradiction is reached, the finger of blame points to the supposition. At that stage of the proof, we may say that the claim involving ~Q has failed and is false; thus, our only recourse is to conclude that Q is true. Following is an outline of this technique.

METHOD OF INDIRECT PROOF

To prove the statement P → Q or to complete the proof problem of the form

Given: P

Prove: Q

where Q is often a negation, use the following steps:

1. Suppose that ~Q is true.
2. Reason from the supposition until you reach a contradiction.
3. Note that the supposition claiming that ~Q is true must be false, and that Q must therefore be true.

Step 3 completes the proof.

EXAMPLE 3   Complete a formal proof of the following theorem:

If two lines are cut by a transversal so that corresponding angles are not congruent, then the two lines are not parallel.

Given: In figure 3.12, ℓ and m are cut by transversal t
$\angle 1 \not\cong \angle 5$

Prove: $\ell \not\parallel m$

Proof: Assume that $\ell \parallel m$. When these lines are cut by transversal t, the corresponding angles (including $\angle 1$ and $\angle 5$) are congruent. But $\angle 1 \not\cong \angle 5$, by hypothesis. Thus the assumed statement, which claims that $\ell \parallel m$, must be false. It follows that $\ell \not\parallel m$.   ▲

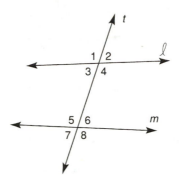

Figure 3.12

The versatility of the indirect method is shown in the final examples of this section. The indirect proofs preceding Example 4 all contain a negative in the conclusion (Prove); the proofs in the final illustrations use the indirect method to achieve a positive conclusion.

EXAMPLE 4   *Given:* In figure 3.13, parallel planes P and Q are intersected by plane T in lines ℓ and m

Prove: $\ell \parallel m$

Proof: Assume that ℓ is not parallel to m. Then ℓ and m intersect at some point A. But if so, point A must be on both planes P and Q, which means that planes P and Q intersect; but P and Q are parallel by hypothesis. Therefore, the assumption that ℓ and m are not parallel must be false, and it follows that $\ell \parallel m$.   ▲

Indirect proofs are also used to establish uniqueness theorems, as Example 5 illustrates.

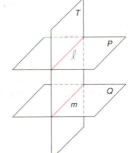

Figure 3.13

EXAMPLE 5   Prove the statement, "The angle bisector of an angle is unique."

Given: In figure 3.14a, $\overrightarrow{BD}$ bisects $\angle ABC$

Prove: $\overrightarrow{BD}$ is the only angle bisector for $\angle ABC$

Proof: $\overrightarrow{BD}$ bisects $\angle ABC$, so $m\angle ABD = \frac{1}{2}m\angle ABC$. Suppose that $\overrightarrow{BE}$ (as shown in figure 3.14b) is also a bisector of $\angle ABC$ and that $m\angle ABE = \frac{1}{2}m\angle ABC$.

By the Angle-Addition Postulate, $m\angle ABD = m\angle ABE + m\angle EBD$. By substitution, $\frac{1}{2}m\angle ABC = \frac{1}{2}m\angle ABC + m\angle EBD$; but then $m\angle EBD = 0$ by subtraction. An angle with a measure of 0 contradicts the Protractor Postulate, which states that the measure of an angle is a unique positive number. Therefore the assumed statement must be false, and it follows that the angle bisector of an angle is unique.   ▲

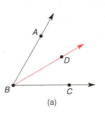

(a)

(b)

Figure 3.14

3.2 EXERCISES

In Exercises 1 to 4, write the converse, the inverse, and the contrapositive of each statement. When possible, classify the statement as true or false.

1. If Juan wins the state lottery, then he will be rich.

2. If $x > 2$, then $x \neq 0$.

3. Two angles are complementary if the sum of their measures is 90°.

4. In a plane, if two lines are not perpendicular to the same line, then these lines are not parallel.

In Exercises 5 to 8, draw a conclusion where possible.

5. **a.** If two triangles are congruent, then the triangles are similar.
   b. Triangles ABC and DEF are not congruent.
   c. ?

6. **a.** If two triangles are congruent, then the triangles are similar.
   b. Triangles ABC and DEF are not similar.
   c. ?

7. **a.** If $x > 3$ and $x \neq 7$, then $x = 5$.
   b. $x > 3$ and $x \neq 7$
   c. ?

8. **a.** If $x > 3$ and $x \neq 7$, then $x = 5$.
   b. $x \neq 5$
   c. ?

9. Which of the following statements would you prove by the indirect method?
   (a) In triangle ABC, if $m\angle A > m\angle B$, then $AC \neq BC$.
   (b) If alternate exterior $\angle 1 \not\equiv$ alternate exterior $\angle 8$, then ℓ is not parallel to m.
   (c) If $(x + 2) \cdot (x - 3) = 0$, then $x = -2$ or $x = 3$.
   (d) If two sides of a triangle are congruent, then the two angles opposite these sides are also congruent.
   (e) The perpendicular bisector of a line segment is unique.

10. For each statement in Exercise 9 that can be proved by the indirect method, give the first statement in each proof.

In Exercises 11 to 22, give the indirect proof for each problem or statement.

11. *Given:* $\angle AOD \not\equiv \angle AFE$
    Prove: $\overleftrightarrow{DC} \not\parallel \overleftrightarrow{EG}$

12. *Given:* $\angle 1 \not\equiv \angle 2$
    Prove: $\overrightarrow{OB}$ does not bisect $\angle AOC$

13. *Given:* $AO > HF$
             $OH = FI$
    Prove: H is not the midpoint of $\overline{AI}$

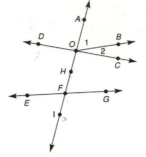

14. *Given:* Obtuse $\angle MQS$
             $\overrightarrow{QN}$ bisects $\angle MQP$
             $\overrightarrow{QR}$ bisects $\angle PQS$
    Prove: $\angle NQR$ is not a right angle

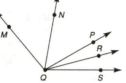

15. If two angles are not congruent, then these angles are not vertical angles.

16. If $x^2 \neq 25$, then $x \neq 5$.

17. If alternate interior angles are not congruent when two lines are cut by a transversal, then the lines so cut are not parallel.

18. If a and b are positive numbers, then
    $\sqrt{a^2 + b^2} \neq a + b$.

19. The midpoint of a line segment is unique.

20. There is exactly one line perpendicular to a given line at a point on the line.

*21. In a plane, if two lines are parallel to a third line, then the two lines are parallel to each other.

*22. In a plane, if two lines are intersected by a transversal so that the corresponding angles are congruent, then the lines are parallel.

3.3 PROVING LINES PARALLEL

In Section 3.1, several methods for proving angles congruent or supplementary were developed, using parallel lines. Here is a quick review of the relevant postulate and theorems.

POSTULATE 10 If two parallel lines are cut by a transversal, then the corresponding angles are congruent.

THEOREM 3.1.2   If two parallel lines are cut by a transversal, then the alternate interior angles are congruent.

THEOREM 3.1.3   If two parallel lines are cut by a transversal, then the alternate exterior angles are congruent.

THEOREM 3.1.4   If two parallel lines are cut by a transversal, then the interior angles on the same side of the transversal are supplementary.

THEOREM 3.1.5   If two parallel lines are cut by a transversal, then the exterior angles on the same side of the transversal are supplementary.

Suppose that we now wish to prove that lines are parallel rather than to establish an angle relationship (as the previous statements do). At present, the only method we have of proving lines parallel is based on the definition of parallel lines. Establishing the conditions of the definition (that coplanar lines do *not* intersect) is virtually impossible! Thus we begin to develop methods for proving that lines in a plane are parallel by proving Theorem 3.3.1 by the indirect method.

THEOREM 3.3.1   If two lines are cut by a transversal so that the corresponding angles are congruent, then these lines are parallel.

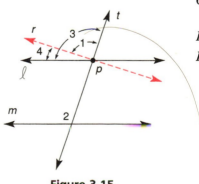

Figure 3.15

Given: ℓ and m cut by transversal t
 $\angle 1 \cong \angle 2$, as in figure 3.15

Prove: $\ell \parallel m$

Proof: Suppose that $\ell \nparallel m$. Then a line r can be drawn through point P so that it is parallel to m; this follows from the Parallel Postulate. If $r \parallel m$, then $\angle 3 \cong \angle 2$ since these angles correspond. But $\angle 1 \cong \angle 2$ by hypothesis. Now $\angle 3 \cong \angle 1$, by the Transitive Property of Congruence; therefore m$\angle 3$ = m$\angle 1$. But m$\angle 1$ + m$\angle 4$ = m$\angle 3$. (See figure 3.15.) Therefore m$\angle 1$ + m$\angle 4$ = m$\angle 1$; and by subtraction, m$\angle 4$ = 0. This contradicts the Protractor Postulate, which states that the measure of any angle must be a positive number. Consequently, r and ℓ must coincide, and it follows that $\ell \parallel m$. ▲

Once proved, Theorem 3.3.1 opens the doors to a host of direct techniques for proving that lines are parallel. In all such claims, the lines are assumed to be coplanar. Each claim is the converse of its counterpart in Section 3.1.

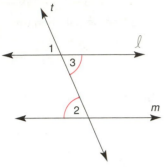

Figure 3.16

> **THEOREM 3.3.2** If two lines are cut by a transversal so that the alternate interior angles are congruent, then these lines are parallel.

Given: Lines ℓ and m and transversal t
$\angle 2 \cong \angle 3$, as in figure 3.16

Prove: $\ell \parallel m$

Proof: ℓ and m are cut by transversal t so that $\angle 2 \cong \angle 3$. Because $\angle 1$ and $\angle 3$ are vertical angles, $\angle 1 \cong \angle 3$. Then $\angle 1 \cong \angle 2$; that is, corresponding angles are congruent. It follows that ℓ and m are parallel, by Theorem 3.3.1. ▲

The remaining converses of the Section 3.1 theorems are left as exercises. Each can be proved by the direct method.

EXAMPLE 1

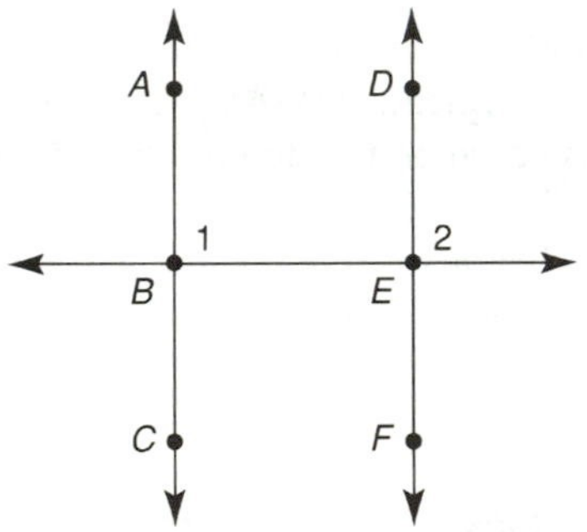

Figure 3.17

Complete the proof.

Given: $\overleftrightarrow{AC} \perp \overleftrightarrow{BE}$ and $\overleftrightarrow{DF} \perp \overleftrightarrow{BE}$, as in figure 3.17

Prove: $\overleftrightarrow{AC} \parallel \overleftrightarrow{DF}$

Proof: Because $\overleftrightarrow{AC} \perp \overleftrightarrow{BE}$, $\angle 1$ is a right angle. Similarly, since $\overleftrightarrow{DF} \perp \overleftrightarrow{BE}$, $\angle 2$ is a right angle. Then $\angle 1 \cong \angle 2$, since all right angles are congruent. And since $\angle 1$ and $\angle 2$ are congruent corresponding angles, it follows that $\overleftrightarrow{AC} \parallel \overleftrightarrow{DF}$. ▲

In a more involved drawing, it may be difficult to decide which lines are parallel because of congruent angles. Consider figure 3.18. Suppose that $\angle 1 \cong \angle 3$. Which lines must be parallel? The resulting confusion (since it appears that a may be parallel to b *and* c may be parallel to d) can be overcome by asking, "Which lines help form $\angle 1$ and $\angle 3$?" In this case, $a \parallel b$ while c acts as a transversal. If you were given instead that $\angle 3 \cong \angle 8$, then c would be parallel to d, with b as the transversal.

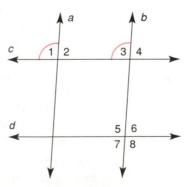

Figure 3.18

EXAMPLE 2   Complete the proof.

Given: $a \parallel b$
          $\angle 2 \cong \angle 7$, as in figure 3.18

Prove: $c \parallel d$

Proof: If $a \parallel b$, then $\angle 2 \cong \angle 4$, because these are corresponding angles. By hypothesis, $\angle 2 \cong \angle 7$, and it follows that $\angle 4 \cong \angle 7$. But the vertical $\angle$s are $\cong$, so $\angle 7 \cong \angle 6$. By transitivity, it follows that $\angle 4$ and $\angle 6$ are congruent, and they too are corresponding angles. Therefore $c \parallel d$.   ▲

A minor difficulty occurs when segments, rather than lines, are provided in a drawing, as in figure 3.19. This is easily remedied by extending the segments.

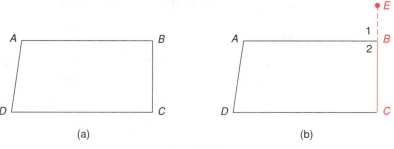

(a)   (b)

Figure 3.19

EXAMPLE 3   Complete the proof.

Given: $\angle B$ is supplementary to $\angle C$

Prove: $\overline{AB} \parallel \overline{DC}$

Proof: Extend $\overline{BC}$ to some point E. Now $\angle 2$ is supplementary to $\angle 1$. But $\angle 2$ is supplementary to $\angle C$ by hypothesis. Then $\angle 1 \cong \angle C$, because they are supplements of the same angle. Because $\angle 1$ and $\angle C$ are $\cong$ corresponding angles with respect to $\overline{AB}$ and $\overline{DC}$ (cut by transversal $\overline{BC}$), it follows that $\overline{AB} \parallel \overline{DC}$.   ▲

Suppose that you are asked to prove the following statement:

"If two angles are not congruent, then the angles are not vertical angles."

And suppose that you have already proved the following theorem:

If two angles are vertical angles, then these angles are congruent.

You may simply say that the first statement follows by contraposition; that is, the two statements are contrapositives, so both are true or both are false.

▲ *WARNING: To use the approach described, it is necessary to prove P → Q or ~Q → ~P, but not both.*

EXAMPLE 4   Suppose that you have proved the following statement:

"If two sides of a triangle are congruent, then two angles of the triangle are congruent."

(a) State the contrapositive of this statement.

(b) Describe a proof for the contrapositive stated in (a).

Solution   (a) If no two angles of a triangle are congruent, then no two sides of the triangle are congruent.

(b) This statement is the contrapositive of the claim, "If two sides of a triangle are congruent, then two angles of the triangle are congruent." Because we are to assume that the quoted theorem has already been proved, the statement is true by contraposition.   ▲

Although the proofs of this section have been given in paragraph form, you should give proofs for the Section 3.3 exercises in standard two-column form. Later, we will give all proofs in paragraph form since that style is preferred in advanced mathematics courses. Our final example is designed to illustrate once more the two-column proof.

EXAMPLE 5   Complete the proof.

Given: In figure 3.20, $\overline{RV}$ and $\overline{ST}$ intersect at point X
∠1 and ∠3 are complementary
∠2 and ∠4 are also complementary

Prove: $\overline{RS} \parallel \overline{TV}$

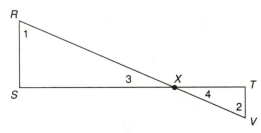

Figure 3.20

PROOF

Statements	Reasons
1. $\overline{RV}$ and $\overline{ST}$ intersect at point X ∠1 and ∠3 are complementary ∠2 and ∠4 are complementary	1. Given
2. ∠3 ≅ ∠4	2. If two lines intersect, vertical ∠s are ≅
3. ∠1 ≅ ∠2	3. Complements of ≅ ∠s are ≅
4. $\overline{RS} \parallel \overline{TV}$	4. If two lines are cut by a transversal so that alternate interior ∠s are ≅, these lines are parallel ▲

Some additional theorems (whose proofs are left as exercises) are now stated. Each will be helpful in constructing future proofs.

THEOREM 3.3.3   If two lines are cut by a transversal so that the alternate exterior angles are congruent, then these lines are parallel.

THEOREM 3.3.4   If two lines are cut by a transversal so that the interior angles on the same side of the transversal are supplementary, then these lines are parallel.

THEOREM 3.3.5   If two lines are cut by a transversal so that the exterior angles on the same side of the transversal are supplementary, then these lines are parallel.

THEOREM 3.3.6   If two lines are each parallel to a third line, then these lines are parallel to each other.

THEOREM 3.3.7   If two lines are each perpendicular to a third line, then these lines are parallel to each other.

3.3 EXERCISES

In Exercises 1 to 10, name the lines (if any) that must be parallel under the given conditions.

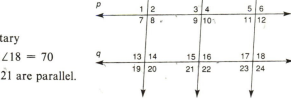

1. $\angle 1 \cong \angle 20$
2. $\angle 3 \cong \angle 10$
3. $\angle 9 \cong \angle 14$
4. $\angle 7 \cong \angle 11$
5. $\ell \perp p$ and $n \perp p$

6. $\ell \parallel m$ and $m \parallel n$
7. $\ell \perp p$ and $m \perp q$
8. $\angle 8$ and $\angle 9$ are supplementary
9. $m\angle 8 = 110$, $p \parallel q$, and $m\angle 18 = 70$
10. The bisectors of $\angle 9$ and $\angle 21$ are parallel.

In Exercises 11 to 18, construct the proof for each problem.

11. *Given:* $\angle 1$ and $\angle 2$ are complementary
    $\angle 3$ and $\angle 1$ are complementary
   Prove: $\overline{BC} \parallel \overline{DE}$

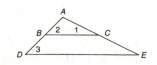

12. *Given:* $\ell \parallel m$
    $\angle 3 \cong \angle 4$
   Prove: $\ell \parallel n$

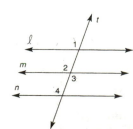

13. *Given:* $\overline{AD} \perp \overline{DC}$
$\overline{BC} \perp \overline{DC}$
Prove: $\overline{AD} \parallel \overline{BC}$

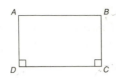

14. *Given:* $m\angle 2 + m\angle 3 = 90$
$\overrightarrow{BE}$ bisects $\angle ABC$
$\overrightarrow{CE}$ bisects $\angle BCD$
Prove: $\ell \parallel n$

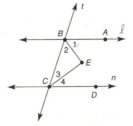

15. *Given:* $\overrightarrow{DE}$ bisects $\angle CDA$
$\angle 3 \cong \angle 1$
Prove: $\overline{ED} \parallel \overline{AB}$

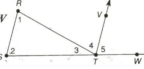

*16. *Given:* $m\angle 1 + m\angle 2 + m\angle 3 = 180$
$\angle 1 \cong \angle 2$
$\overrightarrow{TV}$ bisects $\angle RTW$
Prove: $\overline{RS} \parallel \overline{VT}$

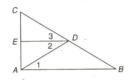

17. *Given:* $\overline{XY} \parallel \overline{WZ}$
$\angle 1 \cong \angle 2$
Prove: $\overline{MN} \parallel \overline{XY}$

18. *Given:* $\ell \parallel m$
$\angle 3 \cong \angle 4$
Prove: $\overline{AB} \parallel \overline{DF}$

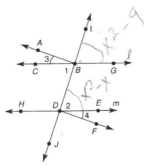

In Exercises 19 to 22, determine the value of x so that line ℓ will be parallel to line m. (Use the drawing that accompanies Exercise 18.)

19. $m\angle IBG = x^2 - 9$
$m\angle 2 = x(x - 1)$

20. $m\angle DBG = 2x^2 - 3x + 6$
$m\angle HDB = 2x(x - 1) - 2$

21. $m\angle 1 = (x + 1)(x + 4)$
$m\angle HDB = 16(x + 3) -$
$(x^2 - 2)$

22. $m\angle IBG = (x^2 - 1)(x + 1)$
$m\angle JDE = 185 - x^2(x + 1)$

In Exercises 23 to 27, give a formal proof for each theorem.

23. If two lines are cut by a transversal so that the alternate exterior angles are congruent, then these lines are parallel.

24. If two lines are cut by a transversal so that the interior angles on the same side of the transversal are supplementary, then these lines are parallel.

25. If two lines are cut by a transversal so that the exterior angles on the same side of the transversal are supplementary, then these lines are parallel.

26. If two lines are parallel to the same line, then these lines are parallel to each other.
(Although this statement is also true for parallel lines in space, *prove only for coplanar lines.*)

27. If two lines are perpendicular to the same line, then these lines are parallel to each other.

In Exercises 28 to 30, prove each statement by contraposition.

28. If the alternate interior angles are not congruent when two lines are cut by a transversal, then these lines are not parallel.

29. If the interior angles on the same side of a transversal are not supplementary when two lines are cut by that transversal, then these lines are not parallel.

30. If the exterior angles on the same side of a transversal are not supplementary when two lines are cut by that transversal, then these lines are not parallel.

31. Suppose that you are in a calculus class and have done the following:
  (i) Proved the theorem, "If an infinite series converges, then the general term of the series has a limit of zero."
  (ii) Learned the definition, "An infinite series is said to diverge when it does not converge." (In effect, "diverge" is the opposite of "converge.")
  (a) State the contrapositive of the theorem stated in (i), using the term "diverge" in the resulting statement.
  (b) Provide a proof for the statement in part (a).

3.4
APPLICATIONS OF PARALLELS TO TRIANGLES AND OTHER POLYGONS

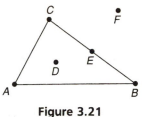

Figure 3.21

> **DEFINITION:** A **triangle** is the union of three line segments that are determined by three noncollinear points.

Symbolized by Δ, the triangle is a figure you have encountered many times. In the triangle in figure 3.21, known as ΔABC, each point A, B, and C is a **vertex** of the triangle; collectively, these three points are the **vertices** of the triangle. $\overline{AB}$, $\overline{BC}$, and $\overline{AC}$ are the **sides** of the triangle. Point D is in the interior of the triangle; point E is on the triangle; and point F is in the exterior of the triangle.

Triangles may be categorized by the lengths of their sides. Table 3.1 presents the type of triangle, the relationship among its sides, and a drawing in which congruent parts (by definition) are indicated.

Table 3.1   **Triangles Classified by Congruent Sides**

Type		Number of Congruent Sides
Scalene		None
Isosceles		At least two congruent sides
Equilateral		All three sides congruent

equilangelar

Triangles may also be classified according to their angles, as shown in Table 3.2.

Table 3.2   Triangles Classified by Angles

Type		Angle(s)	Type		Angle(s)
Acute		All angles acute	Right		One right angle
Obtuse		One obtuse angle	Equiangular		All angles congruent

EXAMPLE 1   In ΔHJK, $HJ = 4$, $JK = 4$, and m$\angle J = 90°$. Describe completely the type of triangle being represented.

Solution $\triangle HJK$ is a right isosceles $\triangle$, or $\triangle HJK$ is an isosceles right triangle.   ▲

In an earlier exercise, it was suggested that the sum of the measures of the three interior angles of a triangle is 180°. This is now stated as a theorem and proved through the use of an auxiliary (or helping) line. When an auxiliary line is added to the drawing for a proof, a justification must be given for the existence of that line, such as:

There is exactly one line through two distinct points.
An angle has exactly one bisector.
There is only one line perpendicular to another line at a point on that line.

The list of possibilities continues.

When an auxiliary line is introduced into a proof, the original drawing is sometimes redrawn for the sake of clarity. This has been done in the proof of the following important theorem.

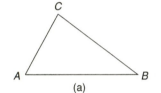

(a)

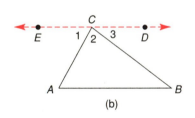

(b)

Figure 3.22

THEOREM 3.4.1   In a triangle, the sum of the measures of the interior angles is 180°.

Given: $\triangle ABC$ in figure 3.22a

Prove: $m\angle A + m\angle B + m\angle C = 180°$

PROOF

Statements	Reasons
1. Draw $\overleftrightarrow{ED}$ through C so that $\overleftrightarrow{ED} \parallel \overleftrightarrow{AB}$, as in figure 3.22a	1. Through a point outside a line, exactly one line is parallel to the given line
2. $m\angle 1 + m\angle 2 = m\angle ECB$	2. Angle-Addition Postulate
3. $m\angle ECB + m\angle 3 = m\angle ECD$	3. Angle-Addition Postulate
4. $m\angle 1 + m\angle 2 + m\angle 3 = m\angle ECD$	4. Substitution
5. $\angle ECD$ is a straight $\angle$, so $m\angle ECD = 180°$	5. If an $\angle$ is a straight $\angle$, it measures 180°
6. $m\angle 1 + m\angle 2 + m\angle 3 = 180°$	6. Substitution
7. $\angle 1 \cong \angle A$ and $\angle 3 \cong \angle B$	7. If two $\parallel$ lines are cut by a transversal, alternate interior $\angle$s are $\cong$
8. $m\angle 1 = m\angle A$ and $m\angle 3 = m\angle B$	8. If two $\angle$s are $\cong$, they have $=$ measures
9. $m\angle A + m\angle B + m\angle ACB = 180°$	9. Substitution ▲

Statements 7 and 8 are related so closely that we cannot have one without the other. At times, we will use these notions of the equality and congruence of angles interchangeably within a proof, without stating both.

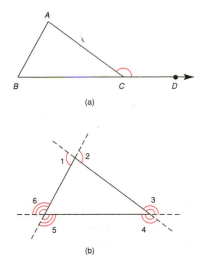

(a)

(b)

Figure 3.23

When the sides of a triangle are extended, each angle that is formed by a side and an extension of the adjacent side is an **exterior angle** of the triangle. In figure 3.23a, $\angle ACD$ is an exterior angle of $\triangle ABC$; for a triangle, there are a total of six exterior angles—two at each vertex. (See figure 3.23b.)

A theorem that follows directly from a previous theorem is known as a **corollary** of that theorem. Corollaries, like theorems, must be proved before they can be used. These proofs are often brief, but they depend on the related theorem. Here are some corollaries of Theorem 3.4.1.

> **COROLLARY 3.4.2**   Each angle of an equiangular triangle measures 60°.

> **COROLLARY 3.4.3**   The acute angles of a right triangle are complementary.

> **COROLLARY 3.4.4**   The measure of an exterior angle of a triangle equals the sum of the measures of the two nonadjacent interior angles.

> **COROLLARY 3.4.5**   If two angles of one triangle are congruent to two angles of another triangle, then the third angles are also congruent.

EXAMPLE 2   *Given:* $\angle M$ is a right $\angle$ in $\triangle NMQ$
$m\angle N = 57°$

Find: $m\angle Q$

Solution   Since the acute $\angle$s of a right triangle are complementary,

$$m\angle N + m\angle Q = 90°$$
$$\therefore 57° + m\angle Q = 90°$$
$$m\angle Q = 33°$$

▲

EXAMPLE 3   *Given:* In figure 3.24,

$$m\angle 1 = x^2 + 2x$$
$$m\angle S = x^2 - 2x$$
$$m\angle T = 3x + 10$$

Find: x

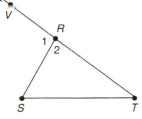

Figure 3.24

Solution   By Corollary 3.4.4,

$$m\angle 1 = m\angle S + m\angle T$$
$$x^2 + 2x = (x^2 - 2x) + (3x + 10)$$
$$2x = x + 10$$
$$x = 10$$

Note: $m\angle 1 = 120°$, $m\angle S = 80°$, and $m\angle T = 40°$, so $120 = 80 + 40$ becomes the check.   ▲

> **DEFINITION:**   A **polygon** is a closed figure whose sides are line segments that intersect only at the endpoints.

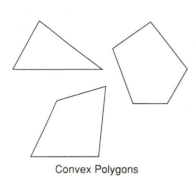

Convex Polygons
Figure 3.25

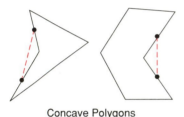

Concave Polygons
Figure 3.26

The polygons we generally consider in this textbook are **convex;** the angle measures of convex polygons are between 0° and 180°. Convex polygons are shown in figure 3.25; those in figure 3.26 are **concave.** A line segment joining two points of a concave polygon can contain points in the exterior of the polygon.

Following are some special names for polygons with fixed numbers of sides.

Polygon	No. of Sides	Polygon	No. of Sides
Triangle	3	Heptagon	7
Quadrilateral	4	Octagon	8
Pentagon	5	Nonagon	9
Hexagon	6	Decagon	10

Figure 3.27 shows heptagon *ABCDEFG* for which $\angle GAB$, $\angle B$, and $\angle BCD$ are interior angles and $\angle 1$, $\angle 2$, and $\angle 3$ are exterior angles. $\overline{AB}$, $\overline{BC}$, and $\overline{CD}$ are some of the sides of the heptagon, since these join consecutive vertices. $\overline{AC}$, $\overline{AD}$, and $\overline{AE}$ are among the many diagonals of the polygon, each of which joins nonconsecutive vertices of *ABCDEFG*.

Figure 3.28 shows polygons that are, respectively, (a) equilateral, (b) equiangular, and (c) regular (both sides and angles are congruent). Note the parts that are congruent.

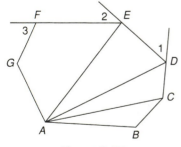

Figure 3.27

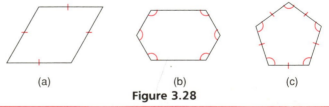

(a)   (b)   (c)
Figure 3.28

> **THEOREM 3.4.6**   The sum of the measures of the interior angles of a polygon with *n* sides is given by $(n - 2) \cdot 180°$. Note that $n > 2$ for any polygon.

The sum of the measures of the angles of a triangle is 180°. The corresponding sum for a quadrilateral is 360°. For a quadrilateral, $n = 4$, which works out to $(4 - 2) \cdot 180°$ or $2(180°)$ or 360°. Intuition will tell you that the four angles of a square are right angles, and the sum of the measures of those angles is certainly 360°.

Let us consider an informal proof of Theorem 3.4.6 for the special case of a pentagon. The proof would change for a polygon of a different number of sides, but only by the number of triangles into which the polygon can be separated.

Proof

Consider the pentagon *ABCDE* in figure 3.29, with auxiliary segments (diagonals from one vertex) as shown. Two points determine a line or segment.

With angles marked as shown, you can see that

$$
\begin{aligned}
\text{m}\angle 1 + \qquad\quad \text{m}\angle 2 + \text{m}\angle 3 &= 180° \\
\text{m}\angle 6 + \text{m}\angle 5 \qquad\quad + \text{m}\angle 4 &= 180° \\
\underline{\text{m}\angle 8 + \text{m}\angle 9 + \text{m}\angle 7 \qquad\qquad\qquad\quad} &= 180° \\
\text{m}\angle E + \text{m}\angle A + \text{m}\angle D + \text{m}\angle B + \text{m}\angle C &= 540° \qquad \text{adding}
\end{aligned}
$$

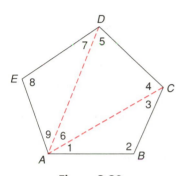

Figure 3.29

For pentagon *ABCDE*, in which $n = 5$, the sum of the measures of the interior angles is $(5 - 2) \cdot 180°$.

If you draw diagonals from one vertex of a polygon of *n* sides, you can form $(n - 2)$ triangles. The sum of the measures of the interior angles always equals $(n - 2) \cdot 180°$. ▲

EXAMPLE 4

Find the sum of the measures of the interior angles of a hexagon. Then find the measure of each interior angle of an equiangular hexagon.

Solution

For the hexagon, $n = 6$, so the sum of the measures of the interior angles is $(6 - 2) \cdot 180°$ or $4(180°)$ or 720°.

In an equiangular hexagon, each of the six interior angles measures $\dfrac{720°}{6}$ or 120°. ▲

From Example 4, you can see that the formula for the measure of each interior angle of a regular polygon is

$$\frac{(n - 2) \cdot 180}{n}$$

EXAMPLE 5

Each interior angle of a certain regular polygon has a measure of 144°. Find its number of sides, and identify the type of polygon it is.

Solution

Let *n* be the number of sides the polygon has. All *n* of the interior angles are equal in measure.

The measure of each interior angle is given by

$$\frac{(n - 2) \cdot 180}{n}$$

Then

$$\frac{(n - 2) \cdot 180}{n} = 144$$

$$(n - 2) \cdot 180 = 144n$$

$$180n - 360 = 144n$$

$$36n = 360$$

$$n = 10$$

The polygon is a regular decagon.   ▲

The following interesting corollary to Theorem 3.4.6 can be established through algebra.

> **COROLLARY 3.4.7**   The sum of the measures of the exterior angles, one at each vertex, of a polygon is 360°.

For the proof that follows, notice that each interior angle of a polygon and its adjacent exterior angle are supplementary. Therefore, the sum of the measures of each of the *n* pairs of these interior and exterior angles must be 180°.

We now consider an algebraic proof for Corollary 3.4.7,

"The sum of the measures of the exterior angles, one at each vertex, of a polygon of n sides is 360°."

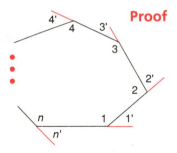

Figure 3.30

Proof A polygon of *n* sides has *n* interior angles and *n* exterior angles, if one is considered at each vertex. As shown in figure 3.30, these interior and exterior angles may be grouped into pairs of supplementary angles. Since there are *n* pairs of angles, the sum of the measures of all pairs is $180 \cdot n$ degrees.

In turn, the sum of the measures of the interior angles is $(n - 2) \cdot 180$. In words, we have

Sum of Measures of Interior Angles	+	Sum of Measures of Exterior Angles	=	Sum of Measures of All Supplementary Pairs	
$(n - 2) \cdot 180$	+	sum of exterior ∠s	=	$180n$	
$180n - 360$	+	sum of exterior ∠s	=	$180n$	Distributive Axiom
-360	+	sum of exterior ∠s	=	0	subtraction
∴		sum of exterior ∠s	=	360	addition ▲

EXAMPLE 6 Use Corollary 3.4.7 to find the number of sides of a regular polygon if each interior angle measures 144°. (Notice that we are repeating Example 5.)

Solution If each interior angle measures 144°, then each exterior angle measures 36° (they are supplementary, since exterior sides of these adjacent angles form a straight line).

Now each exterior angle has the measure given by

$$\frac{360°}{n}$$

In this case, $\dfrac{360}{n} = 36$, and it follows that $36n = 360$, so $n = 10$. The polygon (a decagon) has 10 sides. ▲

3.4 EXERCISES

In Exercises 1 and 2, make drawings as needed.

1. Suppose that for $\triangle ABC$ and $\triangle MNQ$, you know that $\angle A \cong \angle M$ and $\angle B \cong \angle N$. What is true of $\angle C$ and $\angle Q$?

2. Suppose that T is a point on side $\overline{PQ}$ of $\triangle PQR$ and that $\overrightarrow{RT}$ bisects $\angle PRQ$, while $\angle P \cong \angle Q$. What is true of the angles formed when $\overrightarrow{RT}$ intersects $\overline{PQ}$?

In Exercises 3 to 5, you are given $j \parallel k$ and $\triangle ABC$ as shown in the drawing.

3. *Given:* $m\angle 3 = 50°$, $m\angle 4 = 72°$
   Find: $m\angle 1$, $m\angle 2$, and $m\angle 5$

4. *Given:* $m\angle 3 = 55°$, $m\angle 2 = 74°$
   Find: $m\angle 1$, $m\angle 4$, and $m\angle 5$

5. *Given:* $m\angle 1 = 124°$, $m\angle 5 = 42°$
   Find: $m\angle 2$, $m\angle 3$, and $m\angle 4$

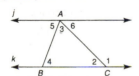

6. *Given:* $\overline{MN} \perp \overline{NQ}$ and $\angle$s as shown below
   Find: x, y, and z

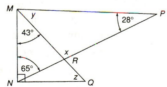

7. *Given:* $\overline{AB} \parallel \overline{DC}$, $\overline{AD} \parallel \overline{BC}$, $\overline{AE} \parallel \overline{FC}$, and $\angle$s as shown below.
   Find: x, y, and z

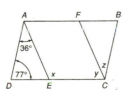

8. Find the total number of diagonals for a polygon with n sides if
   (a) $n = 4$ (b) $n = 5$ (c) $n = 6$ (d) $n = 8$

9. Find the sum of the measures of the interior angles of a polygon that has n sides if
   (a) $n = 4$ (b) $n = 5$ (c) $n = 6$ (d) $n = 8$

10. Find the measure of each interior angle and each exterior angle of a regular polygon with n sides if
    (a) $n = 4$ (b) $n = 5$ (c) $n = 6$ (d) $n = 8$

11. Find the number of sides a polygon has if the sum of its interior angles' measures is
    (a) $900°$ (b) $1260°$ (c) $1980°$ (d) $2340°$

12. Find the number of sides a regular polygon has if the measure of each interior angle is
    (a) $108°$ (b) $144°$ (c) $150°$ (d) $168°$

13. Find the number of sides a regular polygon has if the measure of each exterior angle is
    (a) $24°$ (b) $18°$ (c) $45°$ (d) $9°$

14. Lug bolts are equally spaced about the wheel to form the equal angles shown in the accompanying figure. What is the measure of each of the equal acute angles?

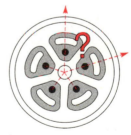

15. What is the measure of each interior angle of a stop sign?

In Exercises 16 and 17, use the accompanying figure.

16. *Given:* Right $\triangle ABC$ with right $\angle C$
    $m\angle 1 = 7x + 4$
    $m\angle 2 = 5x + 2$
    Find: x

17. *Given:* $m\angle 1 = x$
    $m\angle 2 = y$
    $m\angle 3 = 3x - \dfrac{1}{2}$
    Find: x and y

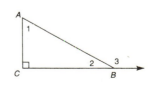

In Exercises 18 and 19, use the accompanying drawing.

18. *Given:* $m\angle 1 = 8(x + 2)$
$m\angle 3 = 5x - 3$
$m\angle 5 = 5(x + 1) - 2$

Find: x

19. *Given:* $m\angle 1 = x$
$m\angle 2 = 4y$
$m\angle 3 = 2y$
$m\angle 4 = 2x - y - 40$

Find: $x, y, m\angle 5$

20. *Given:* Equiangular $\triangle RST$
$\overrightarrow{RV}$ bisects $\angle SRT$

Prove: $\triangle RVS$ is a right $\triangle$

21. *Given:* $\overline{MN}$ and $\overline{PQ}$ intersect at K
$\angle M \cong \angle Q$

Prove: $\angle P \cong \angle N$

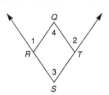

22. *Given:* Quadrilateral $RSTQ$ with exterior $\angle$s at R and T

Prove: $m\angle 1 + m\angle 2 = m\angle 3 + m\angle 4$

23. Draw, if possible, an

(a) isosceles obtuse triangle
(b) equilateral right triangle

24. A father wishes to make a home plate for his son to use in practicing baseball. Find the size of each of the equal angles if the home plate is modeled on the one in:

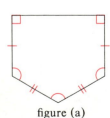

figure (a)   figure (b)

25. The adjacent interior and exterior angles of a certain polygon are supplementary, as indicated in the accompanying drawing. Assume that you know that the measure of each interior angle of a regular polygon is $\dfrac{(n - 2)180}{n}$.

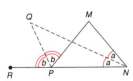

(a) Express the measure of each exterior angle as the supplement of the interior angle.
(b) Simplify the expression in part a to show that each exterior angle has a measure of $\dfrac{360}{n}$.

26. Consider any regular polygon; find and join (in order) the midpoints of the sides. What does intuition tell you about the resulting polygon?

In Exercises 27 to 30, use a formal proof to prove each corollary.

27. Each interior angle of an equiangular triangle measures 60°.

28. The acute angles of a right triangle are complementary.

29. The measure of an exterior angle of a triangle equals the sum of the measures of the two nonadjacent interior angles.

30. If two angles of one triangle are congruent to two angles of another triangle, then the third angles are also congruent.

31. Use an indirect proof to establish the following theorem: A triangle cannot have more than one right angle.

*32. *Given:* $\overrightarrow{NQ}$ bisects $\angle MNP$
$\overrightarrow{PQ}$ bisects $\angle MPR$
$m\angle Q = 42$

Find: $m\angle M$

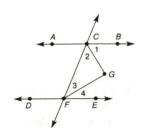

*33. *Given:* $\overleftrightarrow{AB}$, $\overleftrightarrow{DE}$, and $\overleftrightarrow{CF}$ as shown
$\overleftrightarrow{AB} \parallel \overleftrightarrow{DE}$
$\overrightarrow{CG}$ bisects $\angle BCF$
$\overrightarrow{FG}$ bisects $\angle CFE$

Prove: $\angle G$ is a right angle

▲ A Look Beyond: Non-Euclidean Geometries

The geometry we present in this book is often described as Euclidean geometry. A non-Euclidean geometry is a geometry characterized by the existence of at least one contradiction of a Euclidean geometry postulate. To appreciate this subject, you need to realize the importance of the word "plane" in the Parallel Postulate. Thus the Parallel Postulate is now restated.

PARALLEL POSTULATE   *In a plane*, through a point not on a line, exactly one line is parallel to the given line.

One of the consequences of the Parallel Postulate is also restated.

> **THEOREM 3.3.1**   *In a plane*, if two lines are cut by a transversal so that the corresponding angles are congruent, then these lines are parallel.

Suggested by the Parallel Postulate is the construction problem that follows! Notice that you are asked to construct the unique line parallel to a given line through an external point.

CONSTRUCTION 7   To construct the line parallel to a given line from a point not on that line.

Given: $\overleftrightarrow{AB}$ and point P not on $\overleftrightarrow{AB}$, as in figure 3.31a

(a)

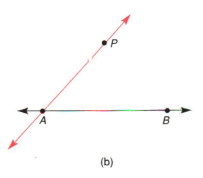

(b)

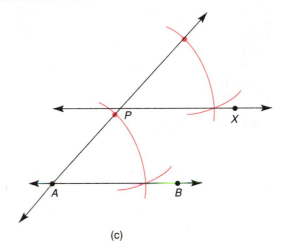

(c)

Figure 3.31

Construct: The line through point P parallel to $\overleftrightarrow{AB}$

Construction: Draw a line (to become a transversal) through point P and some point on $\overleftrightarrow{AB}$, with $\overleftrightarrow{AP}$ as shown in figure 3.31b.

Using P as the vertex, construct the angle that corresponds to $\angle PAB$ and is congruent to $\angle PAB$. The line PX shown in figure 3.31c is the desired line parallel to $\overleftrightarrow{AB}$.   ▲

The Parallel Postulate characterizes a course in plane geometry; it corresponds to the theory that the earth is flat. On a small scale (most applications aren't global), the theory works well and serves the needs of carpenters, designers, and most engineers.

To begin the move to a different geometry, consider the surface of a sphere (like the earth). (See figure 3.32.) By definition, a **sphere** is the set of all points in space that are at a fixed distance from a given point. If a line segment on the surface of the sphere is extended to form a line, it becomes a

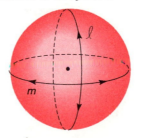

(a) ℓ and m are lines in spherical geometry

(b) These circles are *not* lines in spherical geometry

Figure 3.32

great circle (like the equator of the earth). Each line in this geometry, known as "spherical geometry," is the intersection of a plane containing the center of the sphere with the sphere.

Spherical geometry (or elliptic geometry) is actually a model of Riemannian geometry, named in honor of Georg F. B. Riemann (1826–1866), the German mathematician responsible for the next postulate. The Riemannian Postulate is not numbered in this book, because it does not characterize Euclidean geometry.

RIEMANNIAN POSTULATE   Through a point not on a line, there are no lines parallel to the given line.

To understand the Riemannian Postulate, consider a sphere (as in figure 3.33) containing line ℓ and point P not

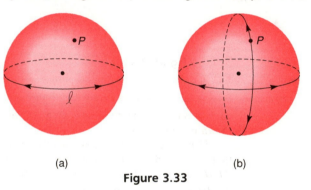

(a)   (b)

Figure 3.33

on ℓ. Any line drawn through point P must intersect ℓ in two points. To see this develop, follow the frames in figure 3.34, which depict an attempt to draw a line parallel to ℓ through point P.

Consider the natural extension to Riemannian geometry of the claim that the shortest distance between two points is a straight line. For the sake of efficiency and common sense, a person traveling from New York City to London will follow the path of a line as it is known in spherical geometry. As you might guess, this is used to chart international flights between cities. In Euclidean geometry, the claim suggests that a person tunnel under the earth's surface from one city to the other.

A second type of non-Euclidean geometry is attributed to the works of a German, Karl F. Gauss (1777–1855); a Russian, Nikolai Lobachevski (1793–1856); and a Hungarian, Johann Bolyai (1802–1862). The postulate for this system of non-Euclidean geometry is as follows.

LOBACHEVSKIAN POSTULATE   Through a point not on a line, there are infinitely many lines parallel to the given line.

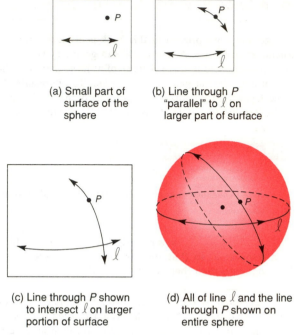

(a) Small part of surface of the sphere

(b) Line through P "parallel" to ℓ on larger part of surface

(c) Line through P shown to intersect ℓ on larger portion of surface

(d) All of line ℓ and the line through P shown on entire sphere

Figure 3.34

This form of non-Euclidean geometry is termed "hyperbolic geometry." Rather than use the plane or sphere as the surface for study, mathematicians use a saddle-like surface known as a **hyperbolic paraboloid.** (See figure 3.35.) A line ℓ is the intersection of a plane with this surface. Clearly, more than one plane can intersect this surface to form a line containing P that does not intersect ℓ. In fact, an infinite number of planes intersect the surface in an infinite number of lines parallel to ℓ and containing P. Table 3.3 compares the three types of geometry.

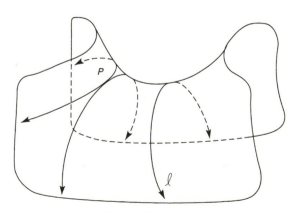

Figure 3.35

Table 3.3 Comparison of Types of Geometry

Postulate	Model	Line	Number of Lines through P Parallel to ℓ
Parallel (Euclidean)	Plane geometry	Intersection of two planes	One
Riemannian	Spherical geometry	Intersection of plane with sphere (including center point of sphere)	None
Lobachevskian	Hyperbolic geometry	Intersection of plane with hyperbolic paraboloid	Infinitely many

▲ SUMMARY

A LOOK BACK AT CHAPTER 3

The goal of this chapter has been to prove several theorems based on the postulate, "If two parallel lines are cut by a transversal, then the corresponding angles are congruent." The method of indirect proof was introduced as a basis for proving lines parallel if the corresponding angles are congruent. Several methods of proving lines parallel were then demonstrated by the direct method. The Parallel Postulate was used to prove that the sum of the measures of the interior angles of a triangle is 180°. Several corollaries naturally followed from this theorem. A sum formula was then developed for the interior angles of any polygon.

A LOOK AHEAD TO CHAPTER 4

In the next chapter, the concept of congruence will be extended to triangles, and several methods of proving triangles congruent will be developed. Several theorems dealing with the inequalities of a triangle will also be proved. Properties will be used to classify quadrilaterals.

IMPORTANT TERMS AND CONCEPTS OF CHAPTER 3

3.1 Perpendicular Lines, Perpendicular Planes
Parallel Lines, Parallel Planes

Parallel Postulate
Alternate Interior Angles, Alternate Exterior Angles, Corresponding Angles
3.2 Indirect Proof
Conditional, Converse, Inverse, Contrapositive
Principle of Negative Inference
3.3 Proving Lines Parallel
Contraposition
3.4 Triangle
Vertices, Sides
Interior and Exterior of a Triangle
Scalene Triangle, Isosceles Triangle, Equilateral Triangle
Acute Triangle, Obtuse Triangle, Right Triangle, Equiangular Triangle
Auxiliary Line
Corollary
Convex Polygons (Triangle, Quadrilateral, Pentagon, Hexagon, Heptagon, Octagon, Nonagon, Decagon)
Regular Polygon, Equilateral Polygon, Equiangular Polygon
A Look Beyond Non-Euclidean Geometries

▲ REVIEW EXERCISES

1. In the following drawings, if m∠1 = m∠2, which lines are parallel?

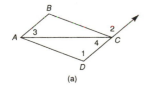

(a)

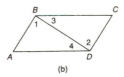

(b)

Use the accompanying figure for Review Exercises 2 and 3.

2. *Given:* m∠13 = 70°
   Find: m∠3

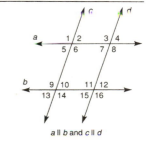

a ∥ b and c ∥ d

3. *Given:* m∠9 = 2x + 17
        m∠11 = 5x − 94
    Find: x

Use the accompanying figure for Review Exercises 4 and 5.

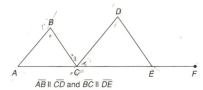

AB ∥ *CD* and *BC* ∥ *DE*

4. *Given:* m∠B = 75°, m∠DCE = 50°
    Find: m∠D and m∠DEF

5. *Given:* m∠DCA = 130°
        m∠BAC = 2x + y
        m∠BCE = 150°
        m∠DEC = 2x − y
    Find: x and y

6. *Given:* In the drawing for Review Exercises 6 to 11,
        AC ∥ *DF*
        AE ∥ *BF*
        m∠AEF = 3y
        m∠BFE = x + 45
        m∠FBC = 2x + 15
    Find: x and y

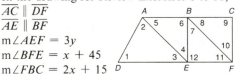

Use the given information to name the segments that must be parallel. If there are no such segments, write "none." Assume *A-B-C* and *D-E-F*.

7. ∠3 ≅ ∠11

8. ∠4 ≅ ∠5

9. ∠7 ≅ ∠10

10. ∠6 ≅ ∠9

11. ∠8 ≅ ∠5 ≅ ∠3

For Review Exercises 12 to 15, find the values of x and y.

12.

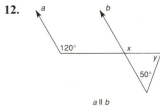

a ∥ *b*

13.

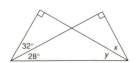

14.

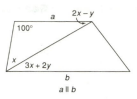

a ∥ *b*

15.

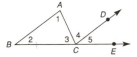

16. *Given:* m∠1 = x² − 12
        m∠4 = x(x − 2)
    Find: x so that $\overrightarrow{AB}$ ∥ $\overrightarrow{CD}$

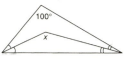

17. *Given:* $\overrightarrow{AB}$ ∥ $\overrightarrow{CD}$
        m∠2 = x² − 3x + 4
        m∠1 = 17x − x² − 5
        m∠ACE = 111
    Find: m∠3, m∠4, m∠5

18. *Given:* $\overline{DC}$ ∥ $\overline{AB}$
        ∠A ≅ ∠C
        m∠A = 3x + y
        m∠D = 5x + 10
        m∠C = 5y + 20
    Find: m∠B

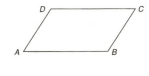

For Review Exercises 19 to 24, state whether the statements are always true (A), sometimes true (S), or never true (N).

19. An isosceles triangle is a right triangle.

20. An equilateral triangle is a right triangle.

21. A scalene triangle is an isosceles triangle.

22. An obtuse triangle is an isosceles triangle.

23. A right triangle has two congruent angles.

24. A right triangle has two complementary angles.

25. Complete the following table for regular polygons.

Number of sides	8	12	20			
Measure of each exterior ∠				24	36	
Measure of each interior ∠					157.5	178

For Review Exercises 26 to 29, sketch, if possible, the polygon described.

26. A quadrilateral that is equiangular but not equilateral

27. A quadrilateral that is equilateral but not equiangular

28. A triangle that is equilateral but not equiangular

29. A hexagon that is equilateral but not equiangular

For Review Exercises 30 and 31, write the converse, inverse, and contrapositive of each statement.

30. If two angles are right angles, then the angles are congruent.

31. If it is not raining, then I am happy.

32. *Given:* $\overline{BE} \perp \overline{DA}$
    $\overline{CD} \perp \overline{DA}$
    Prove: $\angle 1 \cong \angle 2$

33. *Given:* $\angle C \cong \angle 3$
    $\overline{BE} \perp \overline{DA}$
    Prove: $\overline{CD} \perp \overline{DA}$

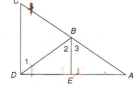

34. *Given:* $\angle 1 \cong \angle 2$
    $\overline{BD} \parallel \overline{AE}$
    Prove: $\angle A \cong \angle E$

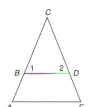

35. *Given:* $\angle A \cong \angle C$
    $\overline{DC} \parallel \overrightarrow{AB}$
    Prove: $\overline{DA} \parallel \overline{CB}$

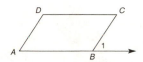

In Review Exercises 36 and 37, use the accompanying figure.

36. *Given:* $m \not\parallel n$
    Prove: $\angle 1 \not\cong \angle 2$

37. *Given:* $\angle 1 \not\cong \angle 3$
    Prove: $m \not\parallel n$

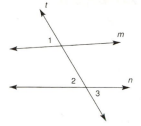

38. Construct the line through C parallel to $\overline{AB}$.

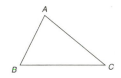

39. Construct an equilateral triangle ABC with side $\overline{AB}$.

TRIANGLES AND QUADRILATERALS

In much of this chapter, we will be using the notion of congruent triangles. A pair of triangles are congruent if one fits perfectly over the other. Do you see any such pairs of triangles in the roof truss shown?

We will also study quadrilaterals—figures that have four sides. One type of quadrilateral, known as a trapezoid, has a pair of parallel sides. Another type of quadrilateral, known as a parallelogram, has both pairs of opposite sides parallel. Can you find a trapezoid and a parallelogram in the drawing?

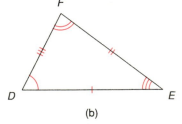

CHAPTER OUTLINE

4.1 Congruent Triangles
4.2 CPCTC and HL
4.3 Isosceles Triangles
4.4 Quadrilaterals and Parallelograms
4.5 More Quadrilaterals
4.6 Inequalities in a Triangle
A Look Beyond Historical Sketch of Archimedes

4.1 CONGRUENT TRIANGLES

Two triangles are **congruent** if one coincides (fits perfectly) with the other. In figure 4.1, we say that $\triangle ABC \cong \triangle DEF$ if these congruences hold:

$$\angle A \cong \angle D \qquad \overline{AB} \cong \overline{DE}$$
$$\angle B \cong \angle E \qquad \overline{BC} \cong \overline{EF}$$
$$\angle C \cong \angle F \qquad \overline{AC} \cong \overline{DF}$$

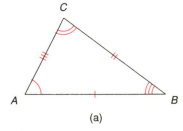

(a)

(b)

Figure 4.1

It is best to state the congruence of the triangles in an order in which corresponding vertices occur; these are the vertices at which congruent angles exist. For the triangles in figure 4.1, this correspondence of vertices is given by

$$A \leftrightarrow D \qquad B \leftrightarrow E \qquad C \leftrightarrow F$$

When it is stated that $\Delta MNQ \cong \Delta RST$, you may conclude that M corresponds to R, that N corresponds to S, that $\angle Q \cong \angle T$, and that $\overline{NQ} \cong \overline{ST}$, to mention a few of the possibilities.

> **DEFINITION:** Two triangles are **congruent** when the six parts of the first triangle are congruent to the six corresponding parts of the second triangle.

As always, any definition is reversible! If two triangles are known to be congruent, you may deduce that the corresponding parts are congruent. Moreover, if the six pairs of parts are known to be congruent, then so are the triangles! In ΔMNQ and ΔRST in figure 4.2, corresponding congruent parts have been indicated, and you may conclude that $\Delta MNQ \cong \Delta RST$.

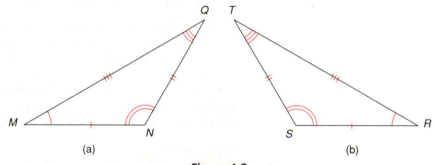

(a)   (b)

Figure 4.2

By stating the vertices in corresponding order, you will be able to keep track of congruences even when a drawing has not been provided.

Here are some facts about congruent triangles:

1. Any triangle is congruent to itself: $\Delta ABC \cong \Delta ABC$.

2. If $\Delta ABC \cong \Delta DEF$, then $\Delta DEF \cong \Delta ABC$.

3. If $\Delta ABC \cong \Delta DEF$ and $\Delta DEF \cong \Delta GHI$, then $\Delta ABC \cong \Delta GHI$.

You probably recognize these properties by the names Reflexive, Symmetric, and Transitive. It follows that the congruence of triangles is an equivalence relation.

It would be difficult and quite time-consuming to establish that triangles were congruent if six pairs of congruent parts had to be verified first. Fortunately, it is possible to prove triangles congruent with fewer than six pairs of congruences. To suggest a first method, consider the construction in Example 1.

EXAMPLE 1   Construct a triangle whose sides have the lengths of segments provided in figure 4.3a (not drawn to scale).

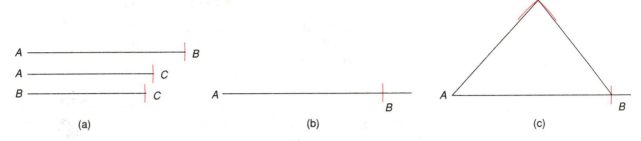

(a)   (b)   (c)

Figure 4.3

Solution   Choose one segment as the base (bottom side), and mark its length as shown in figure 4.3b.

Using the left endpoint, mark off a length equal to that of another segment. Then mark the length of the last segment from the right endpoint so that these arcs intersect, and locate the third vertex of the triangle. Joining this point to the endpoints of the base completes the desired triangle. (See figure 4.3c.) ▲

If another triangle were constructed (as in Example 1), it would be congruent to the one shown. It might be necessary to flip or rotate it to have corresponding vertices match, but that is perfectly acceptable! The point of Example 1 is that it does provide a method for establishing the congruence of triangles, using only three pairs of parts. If corresponding angles are measured in the previous triangle or in any other triangle constructed with the same lengths for sides, these pairs of angles will also be congruent!

POSTULATE 11   If the three sides of one triangle are congruent to the three sides of a second triangle, then the triangles are congruent (SSS).

The designation SSS will be cited as a reason in the proof that follows.

EXAMPLE 2   *Given:* $\overline{AB}$ and $\overline{CD}$ bisect each other at M
$\overline{AC} \cong \overline{DB}$

Prove: $\triangle AMC \cong \triangle BMD$

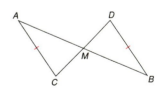

PROOF

Statements	Reasons
1. $\overline{AB}$ and $\overline{CD}$ bisect each other at M	**1.** Given
2. $\therefore \overline{AM} \cong \overline{MB}$ $\overline{CM} \cong \overline{MD}$	**2.** If a segment is bisected, the segments formed are $\cong$
3. $\overline{AC} \cong \overline{DB}$	**3.** Given
4. $\triangle AMC \cong \triangle BMD$	**4.** SSS ▲

The two sides that form an angle of a triangle are said to **include that angle** of the triangle. In $\triangle TUV$ in figure 4.4a, sides $\overline{TU}$ and $\overline{TV}$ form $\angle T$; therefore, $\overline{TU}$ and $\overline{TV}$ include $\angle T$. In turn, $\angle T$ is said to be the included angle for $\overline{TU}$ and $\overline{TV}$. Similarly, any two angles must have a common side, and these two angles are said to **include that side** of the triangle. In $\triangle TUV$, $\angle U$ and $\angle T$ share the common side $\overline{UT}$; therefore, $\angle U$ and $\angle T$ include the side $\overline{UT}$. $\overline{UT}$ is described as the side included by $\angle U$ and $\angle T$.

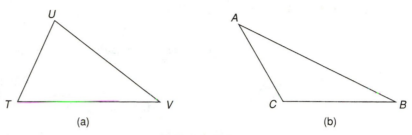

Figure 4.4

EXAMPLE 3

In $\triangle ABC$ in figure 4.4b:
(a) Which angle is included by $\overline{AC}$ and $\overline{CB}$?
(b) Which sides include $\angle B$?
(c) What is the included side for $\angle A$ and $\angle B$?
(d) Which angles include $\overline{CB}$?

Solution

(a) $\angle C$ (since it is formed by $\overline{AC}$ and $\overline{BC}$)
(b) $\overline{AB}$ and $\overline{BC}$ (since these form $\angle B$)
(c) $\overline{AB}$ (since it is the common side for $\angle A$ and $\angle B$)
(d) $\angle C$ and $\angle B$ (since $\overline{CB}$ is a side of each angle)   ▲

A second way of establishing that two triangles are congruent involves showing that two sides and the included angle of one triangle are congruent to two sides and the included angle of a second triangle. If you and another person each draw a triangle for which two of the sides measure 2 inches and 3 inches, respectively, and the included angle measures 54°, then those triangles are congruent. (See figure 4.5.)

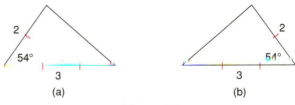

Figure 4.5

POSTULATE 12   If two sides and the included angle of one triangle are congruent to two sides and the included angle of a second triangle, then the triangles are congruent (SAS).

In Example 4, which follows, the two triangles share a common side; the statement $\overline{PN} \cong \overline{PN}$ is justified by the Reflexive Property of Congruence. In some textbooks, this reason is simply written Identity. However stated, this justification applies when triangles (or perhaps other polygons) have a part in common.

EXAMPLE 4   *Given:* $\overline{PN} \perp \overline{MQ}$
$\overline{MN} \cong \overline{NQ}$

Prove: $\triangle PNM \cong \triangle PNQ$

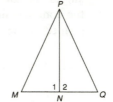

PROOF

Statements	Reasons
1. $\overline{PN} \perp \overline{MQ}$	1. Given
2. $\angle 1 \cong \angle 2$	2. If two lines are $\perp$, they meet to form $\cong$ adjacent $\angle$s
3. $\overline{MN} \cong \overline{NQ}$	3. Given
4. $\overline{PN} \cong \overline{PN}$	4. Identity (or Reflexive)
5. $\triangle PNM \cong \triangle PNQ$	5. SAS ▲

The next method for proving triangles congruent requires a combination of two angles and the included side. If two people each draw a triangle for which two of the angles measure 33° and 47° and the included side measures 5 inches, then those triangles are congruent. (See figure 4.6.)

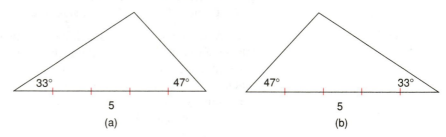

Figure 4.6

POSTULATE 13   If two angles and the included side of one triangle are congruent to two angles and the included side of a second triangle, then the triangles are congruent (ASA).

While this method is written compactly as ASA, you must not get careless as you write these abbreviations! For example, ASA refers to two angles and the included side, while SAS refers to two sides and the included angle. To use either postulate, the specific conditions described in it must be satisfied.

While SSS, SAS, and ASA are all valid methods of proving triangles congruent, SSA is *not* a method and *cannot* be used. In figure 4.7, the two triangles are marked using SSA, and yet the two triangles are *not* congruent.

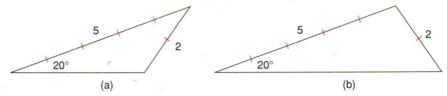

Figure 4.7

EXAMPLE 5

Given: $\overline{AB} \cong \overline{DE}$
$\overline{BC} \cong \overline{EC}$
$\angle 1 \cong \angle 2$

Prove: $\triangle ACE \cong \triangle DCB$

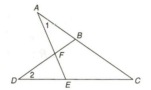

PROOF

Statements	Reasons
1. $\overline{AB} \cong \overline{DE}$ and $\overline{BC} \cong \overline{EC}$	1. Given
2. $AB = DE$ and $BC = EC$	2. If two segments are $\cong$, their measures are $=$
3. $AB + BC = DE + EC$	3. Addition Property of Equality
4. $AB + BC = AC$ $DE + EC = DC$	4. Segment-Addition Postulate
5. $AC = DC$	5. Substitution
6. $\overline{AC} \cong \overline{DC}$	6. If segments are $=$ in measure, they are $\cong$
7. $\angle 1 \cong \angle 2$	7. Given
8. $\angle C \cong \angle C$	8. Identity
9. $\triangle ACE \cong \triangle DCB$	9. ASA ▲

We now consider a theorem (proven by the ASA postulate) that is convenient as a reason in many proofs.

> **THEOREM 4.1.1**   If two angles and the nonincluded side of one triangle are congruent to two angles and the nonincluded side of a second triangle, then the two triangles are congruent (AAS).

Given: $\angle T \cong \angle K$, $\angle S \cong \angle J$, and $\overline{SR} \cong \overline{HJ}$ as in figure 4.8

Prove: $\triangle TSR \cong \triangle KJH$

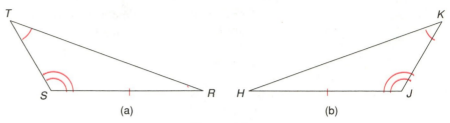

Figure 4.8

PROOF

Statements	Reasons
1. $\angle T \cong \angle K$ $\angle S \cong \angle J$	1. Given
2. $\angle R \cong \angle H$	2. If two ∠s of one △ are ≅ to two ∠s of another △, then the third ∠s are also congruent
3. $\overline{SR} \cong \overline{HJ}$	3. Given
4. $\triangle TSR \cong \triangle KJH$	4. ASA ▲

▲ **WARNING:** *Do not use the order SSA, since it is simply not valid for proving triangles congruent; with SSA, the triangles may or may not be congruent.*

In summary, you may use SSS, SAS, ASA, or AAS to prove that triangles are congruent. Identity may be used to state a self-congruence when a side or angle is common to two triangles.

4.1 EXERCISES

In Exercises 1 to 6, use drawings as needed to answer each question.

1. In the accompanying drawing, name a common angle and a common side for $\triangle ABC$ and $\triangle ABD$. If $\overline{BC} \cong \overline{BD}$, can you conclude that $\triangle ABC$ and $\triangle ABD$ are congruent? Do you believe that SSA can be used as a reason for proving triangles congruent?

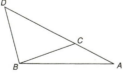

2. Given that the triangles shown in the accompanying drawing are congruent, with corresponding angles marked, find values for *a, b,* and *x.*

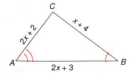

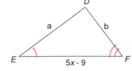

3. In a right triangle, the sides that form the right angle are **legs;** the longest side (opposite the right angle) is the **hypotenuse.** Suppose that two right triangles have congruent pairs of legs. Some textbooks say that the right triangles are congruent by the reason LL. In our work, LL is just a special case of one of the postulates in this section. Which postulate is that?

4. In the figure for Exercises 2 and 4, write a statement that the triangles are congruent, with due attention to the order of corresponding vertices.

5. In $\triangle ABC,$ the midpoints of the sides are joined. What does intuition tell you about the relationship between $\triangle AED$ and $\triangle FDE$? (We will prove this relationship later.)

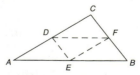

6. Suppose that you wish to prove that $\triangle RST \cong \triangle SRV.$

Using the reason Identity, name one pair of corresponding parts that are congruent.

In Exercises 7 to 10, use the accompanying figure, in which the congruent sides or angles are indicated. Where possible, state the additional information needed to prove the triangles congruent by the specified method.

7. (a) SAS **(c)** ASA
  (b) SSS **(d)** AAS

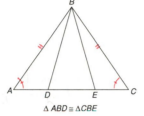

Δ*ABD* ≅ Δ*CBE*

8. (a) SAS **(c)** ASA
  (b) SSS **(d)** AAS

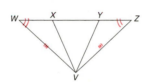

Δ*WVY* ≅ Δ*ZVX*

9. (a) SAS **(c)** ASA
  (b) SSS **(d)** AAS

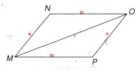

Δ*MNO* ≅ Δ*OPM*

10. (a) SAS **(c)** ASA
  (b) SSS **(d)** AAS

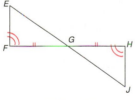

Δ*EFG* ≅ Δ*JHG*

In Exercises 11 to 18, use SSS, SAS, ASA, or AAS to prove triangles congruent.

11. *Given:* $\overline{AB} \cong \overline{CD}$ and $\overline{AD} \cong \overline{CB}$ as shown below
  Prove: Δ*ABC* ≅ Δ*CDA*

12. *Given:* $\overline{DC} \parallel \overline{AB}$ and $\overline{AD} \parallel \overline{BC}$ as shown below.
  Prove: Δ*ABC* ≅ Δ*CDA*

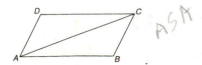

13. *Given:* $\overrightarrow{PQ}$ bisects ∠*MPN*
  $\overline{MP} \cong \overline{NP}$
  Prove: Δ*MQP* ≅ Δ*NQP*

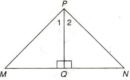

14. *Given:* $\overline{PQ} \perp \overline{MN}$ and ∠1 ≅ ∠2
  Prove: Δ*MQP* ≅ Δ*NQP*

15. *Given:* $\overline{AB} \perp \overline{BC}$ and $\overline{AB} \perp \overline{BD}$
  $\overline{BC} \cong \overline{BD}$
  Prove: Δ*ABC* ≅ Δ*ABD*

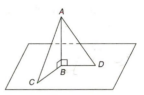

16. *Given:* $\overline{PN}$ bisects $\overline{MQ}$
  ∠*M* and ∠*Q* are right angles
  Prove: Δ*PQR* ≅ Δ*NMR*

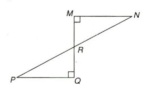

17. *Given:* ∠1 ≅ ∠2 and ∠3 ≅ ∠4
  Prove: Δ*RST* ≅ Δ*SRV*

18. *Given:* $\overline{VS} \cong \overline{TR}$ and ∠1 ≅ ∠2
  Prove: Δ*RST* ≅ Δ*SRV*

In Exercises 19 to 22, the methods to be used are SSS, SAS, ASA, and AAS.

19. Given that Δ*RST* ≅ Δ*RVU*, does it follow that Δ*RSU* is also congruent to Δ*RVT*? Name the method, if any, used in arriving at this conclusion.

20. Given that ∠*S* ≅ ∠*V* and $\overline{ST} \cong \overline{UV}$, does it follow that Δ*RST* ≅ Δ*RVU*? Which method, if any, did you use?

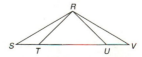

21. Given that $\angle A \cong \angle E$ and $\angle B \cong \angle D$, does it follow that $\triangle ABC \cong \triangle DEC$? If so, cite the method used in arriving at this conclusion.

22. Given that $\angle A \cong \angle E$ and $\overline{BC} \cong \overline{DC}$, does it follow that $\triangle ABC \cong \triangle DEC$? Cite the method, if any, used in reaching this conclusion.

23. In quadrilateral $ABCD$, $\overline{AC}$ and $\overline{BD}$ are perpendicular bisectors of each other. Name *all* triangles that are congruent to:

   (a) $\triangle ABE$ **(b)** $\triangle ABC$ **(c)** $\triangle ABD$

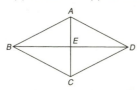

24. In $\triangle ABC$ and $\triangle DEF$, you know that $\angle A \cong \angle D$, $\angle C \cong \angle F$, and $\overline{AB} \cong \overline{DE}$. Before concluding that the triangles are congruent by ASA, you need to show that $\angle B \cong \angle E$. State the postulate or theorem that allows you to confirm this statement ($\angle B \cong \angle E$).

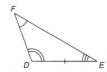

In Exercises 25 to 27, prove the triangles congruent.

25. *Given:* Planes P and Q with $\overline{AC} \perp \overline{CB}$ and $\overline{DC} \perp \overline{CB}$
        $m\angle ABC = 30°$ and $m\angle D = 60°$
   Prove: $\triangle ABC \cong \triangle DBC$

26. *Given:* $\overline{SP} \cong \overline{SQ}$ and $\overline{ST} \cong \overline{SV}$
   Prove: $\triangle SPV \cong \triangle SQT$ and $\triangle TPQ \cong \triangle VQP$

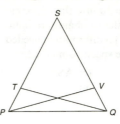

27. *Given:* Plane M
       C is the midpoint of $\overline{EB}$
       $\overline{AD} \perp \overline{BE}$ and $\overline{AB} \parallel \overline{ED}$
   Prove: $\triangle ABC \cong \triangle DEC$

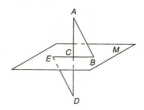

In Exercises 28 to 31, use the figure below to complete each construction.

28. Construct $\triangle ABC$ to have sides $\overline{AB}$, $\overline{AC}$, and $\overline{BC}$.

29. Construct $\triangle ABC$ to have sides $\overline{AB}$ and $\overline{AC}$ and the included angle $\angle A$.

30. Construct $\triangle ABC$ to have $\angle A$ and $\angle B$ and the included side $\overline{AB}$.

31. Construct $\triangle ABC$ to have sides $\overline{AB}$ and $\overline{BC}$ and the included angle $\angle B$.

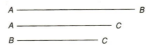

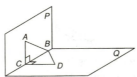

4.2

CPCTC AND HL

Recall that the definition of congruent triangles states that *all* six parts (three sides and three angles) of one triangle are congruent respectively to the six corresponding parts of the second triangle. If we have proved that $\triangle ABC \cong \triangle DEF$ by SAS (the congruent parts are marked in figure 4.9), then we can draw conclusions such as $\angle C \cong \angle F$ and $\overline{AC} \cong \overline{DF}$. The following reason is often cited for drawing such conclusions and is based on the definition of congruent triangles.

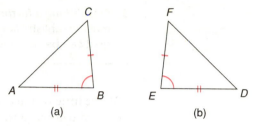

Figure 4.9

CPCTC: Corresponding parts of congruent triangles are congruent.

EXAMPLE 1

Given: $\overrightarrow{WZ}$ bisects $\angle TWV$
        $\overline{WT} \cong \overline{WV}$
Prove: $\overline{TZ} \cong \overline{VZ}$

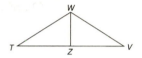

PROOF

Statements	Reasons
1. $\overrightarrow{WZ}$ bisects $\angle TWV$	1. Given
2. $\angle TWZ \cong \angle VWZ$	2. The bisector of an angle separates it into two $\cong$ $\angle$s
3. $\overline{WT} \cong \overline{WV}$	3. Given
4. $\overline{WZ} \cong \overline{WZ}$	4. Identity
5. $\triangle TWZ \cong \triangle VWZ$	5. SAS
6. $\overline{TZ} \cong \overline{VZ}$	6. CPCTC ▲

We can easily extend this proof to prove that $\overline{WZ}$ bisects $\overline{TV}$, simply by adding a seventh statement and reason:

7. $\overline{WZ}$ bisects $\overline{TV}$	7. If a line segment is divided into two $\cong$ parts, then it has been bisected

In our work with triangles, we will be establishing three types of conclusions:

1. *Proving triangles congruent,* like $\triangle TWZ \cong \triangle VWZ$

2. *Proving corresponding parts of congruent triangles congruent,* like

$\overline{TZ} \cong \overline{VZ}$ (Notice that two $\triangle$s have to be proved $\cong$ before CPCTC can be used.)

3. *Establishing a further relationship,* like $\overline{WZ}$ bisects $\overline{TV}$ (Notice that we must establish that two $\triangle$s are $\cong$ and also apply CPCTC before this goal can be reached.)

While little is stated in this book about a "plan for proof," every geometry student and teacher must have one before a proof can be completed.

EXAMPLE 2

Given: $\overline{ZW} \cong \overline{YX}$
$\overline{ZY} \cong \overline{WX}$
Prove: $\overline{ZY} \parallel \overline{WX}$

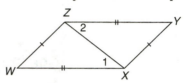

Plan for Proof

By showing that $\triangle ZWX \cong \triangle ZYX$, we can show that $\angle 1 \cong \angle 2$. But $\angle 1$ and $\angle 2$ are alternate interior angles for $\overline{ZY}$ and $\overline{WX}$.

PROOF

Statements	Reasons
1. $\overline{ZW} \cong \overline{YX}$; $\overline{ZY} \cong \overline{WX}$	1. Given
2. $\overline{ZX} \cong \overline{ZX}$	2. Identity
3. $\triangle ZWX \cong \triangle ZYX$	3. SSS
4. $\angle 1 \cong \angle 2$	4. CPCTC
5. $\overline{ZY} \parallel \overline{WX}$	5. If two lines are cut by a transversal so that the alt. int. $\angle$s are $\cong$, these lines are $\parallel$ ▲

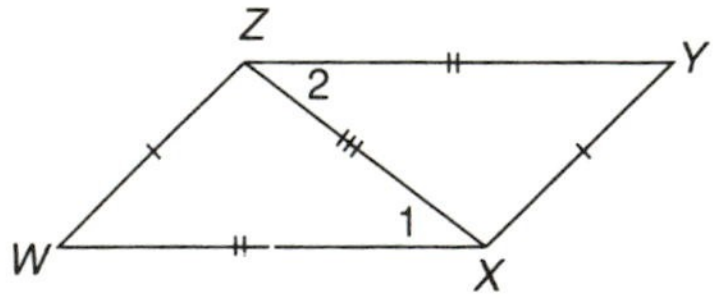

As you can see in figure 4.10, like marks are often used to indicate corresponding parts of congruent triangles. The markings generally used include the following:

A square in the opening of a right angle

The same number of marks on corresponding sides of $\cong$ $\triangle$s

The same number of arcs in openings of $\cong$ $\angle$s

Figure 4.10

It may also be convenient for you to trace in different colors the two triangles that your plan of proof has you prove congruent.

In a right triangle, the side opposite the right angle is the **hypotenuse** of the triangle, while the sides of the right angle are the **legs** of the triangle. These parts of a right triangle are illustrated in figure 4.11.

In addition to the methods discussed earlier for proving triangles congruent is the HL method, which applies exclusively to right triangles. In HL, H refers to hypotenuse and L refers to leg. The proof of this method will be delayed until Section 5.5.

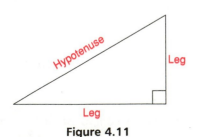

Figure 4.11

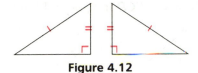

Figure 4.12

THEOREM 4.2.1   If the hypotenuse and a leg of one right triangle are congruent to the hypotenuse and a leg of a second right triangle, then the triangles are congruent (HL).

For a depiction of the relationship described in Theorem 4.2.1, see figure 4.12.

EXAMPLE 3   *Given:* $\overline{AB}$ and $\overline{CD}$ as shown in figure 4.13a

Construct: The right triangle whose hypotenuse has a length equal to AB and one of whose legs has a length equal to CD

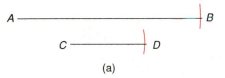

(a)

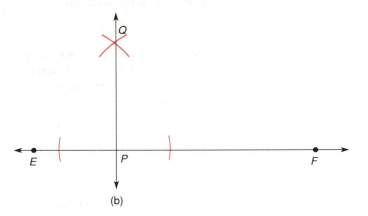

(b)

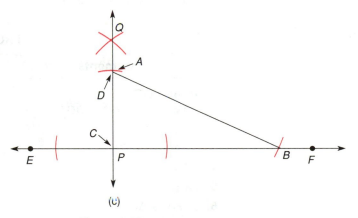

(c)

Figure 4.13

Solution   First we construct the line perpendicular to $\overleftrightarrow{EF}$ at point P. (See figure 4.13b.)
Second, mark off the length of $\overline{CD}$ on $\overleftrightarrow{PQ}$, as shown in figure 4.13c.
Finally, with point D as center, mark off a length equal to that of $\overline{AB}$, as shown in figure 4.13c. ΔABC is the desired right Δ.   ▲

EXAMPLE 4 Cite the reason why the right triangles $\triangle ABC$ and $\triangle ECD$ in figure 4.14 are congruent if:

(a) $\overline{AB} \cong \overline{EC}$ and $\overline{AC} \cong \overline{ED}$
(b) $\angle A \cong \angle E$ and C is the midpoint of $\overline{BD}$
(c) $\overline{BC} \cong \overline{CD}$ and $\angle 1 \cong \angle 2$
(d) $\overline{AB} \cong \overline{EC}$ and $\overline{EC}$ bisects $\overline{BD}$

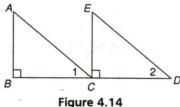

Figure 4.14

Solution (a) HL   (b) AAS   (c) ASA   (d) SAS   ▲

In Example 5, we wish to validate one of our methods of construction— that of constructing the unique line that is perpendicular to a given line from a point not on the given line.

EXAMPLE 5 *Given:* P not on ℓ, as shown in figure 4.15
    $\overline{PA} \cong \overline{PB}$ (by construction)
    $\overline{AQ} \cong \overline{BQ}$ (by construction)
Prove: $\overline{PQ} \perp \overline{AB}$

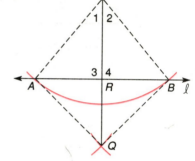

Figure 4.15

Plan for Proof We first show that $\triangle PAQ \cong \triangle PBQ$. Using corresponding parts of the first $\triangle$s, we can also show that $\triangle PRA \cong \triangle PRB$. It follows that $\angle 3 \cong \angle 4$, so $\overline{PQ} \perp \overline{AB}$.

PROOF

Statements	Reasons
1. P not on ℓ $\overline{PA} \cong \overline{PB}$ and $\overline{AQ} \cong \overline{BQ}$	1. Given
2. $\overline{PQ} \cong \overline{PQ}$	2. Reflexive
3. $\triangle PAQ \cong \triangle PQB$	3. SSS
4. $\angle 1 \cong \angle 2$	4. CPCTC
5. $\overline{PR} \cong \overline{PR}$	5. Reflexive
6. $\triangle PRA \cong \triangle PRB$	6. SAS
7. $\angle 3 \cong \angle 4$	7. CPCTC
8. $\overline{PQ} \perp \overline{AB}$	8. If two lines meet to form $\cong$ adjacent $\angle$s, these lines are $\perp$ ▲

Note: While we generally do not provide a plan for proof, you must formulate such a plan before you can write any proof in a logical order.

4.2 EXERCISES

In Exercises 1 to 8, plan the proof and name the method that allows you to prove triangles congruent.

1. *Given:* ∠1 and ∠2 are right ∠s as shown below.
$\overline{CA} \cong \overline{DA}$
Prove: $\triangle ABC \cong \triangle ABD$

2. *Given:* ∠1 and ∠2 are right ∠s
$\overrightarrow{AB}$ bisects ∠CAD
Prove: $\triangle ABC \cong \triangle ABD$

ASA

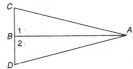

3. *Given:* P is the midpoint of both $\overline{MR}$ and $\overline{NQ}$ as shown below.
Prove: $\triangle MNP \cong \triangle RQP$

4. *Given:* $\overline{MN} \parallel \overline{QR}$ and $\overline{MN} \cong \overline{QR}$
Prove: $\triangle MNP \cong \triangle RQP$

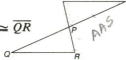

AAS

In Exercises 5 to 8, use the drawing below.

5. *Given:* ∠R and ∠V are right ∠s
∠1 ≅ ∠2
Prove: $\triangle RST \cong \triangle VST$

6. *Given:* ∠1 ≅ ∠2 and ∠3 ≅ ∠4
Prove: $\triangle RST \cong \triangle VST$

7. *Given:* $\overline{SR} \cong \overline{SV}$ and $\overline{RT} \cong \overline{VT}$
Prove: $\triangle RST \cong \triangle VST$

8. *Given:* ∠R and ∠V are right ∠s
$\overline{RT} \cong \overline{VT}$
Prove: $\triangle RST \cong \triangle VST$

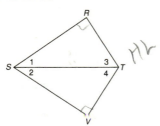

HL

In Exercises 9 and 10, refer to the drawing and find the size of each angle named.

9. *Given:* $\overline{UW} \parallel \overline{XZ}, \overline{VY} \perp \overline{UW}$, and
$\overline{VY} \perp \overline{XZ}$
m∠1 = m∠4 = 42°
Find: m∠2, m∠3, m∠5, and m∠6

10. *Given:* $\overline{UW} \parallel \overline{XZ}, \overline{VY} \perp \overline{UW}$, and
$\overline{VY} \perp \overline{XZ}$
m∠1 = m∠4 = 4x + 3
m∠2 = 6x − 3
Find: m∠1, m∠2, m∠3, m∠4, m∠5, and m∠6

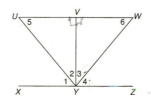

In Exercises 11 to 16, prove that the indicated triangles in the figure below are congruent.

11. *Given:* $\overline{HJ} \perp \overline{KL}$ and $\overline{HK} \cong \overline{HL}$
Prove: $\triangle HJK \cong \triangle HJL$

12. *Given:* $\overrightarrow{HJ}$ bisects ∠KHL
$\overline{HJ} \perp \overline{KL}$
Prove: $\triangle HJK \cong \triangle HJL$

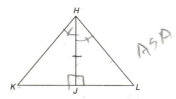

ASA

13. *Given:* ∠P and ∠R are right ∠s
M is the midpoint of $\overline{PR}$
Prove: $\triangle NPM \cong \triangle QRM$

14. *Given:* M is the midpoint of $\overline{NQ}$
$\overline{NP} \parallel \overline{RQ}$ with transversals $\overline{PR}$ and $\overline{NQ}$
Prove: $\triangle NPM \cong \triangle QRM$

15. *Given:* ∠1 and ∠2 are right ∠s
H is the midpoint of $\overline{FK}$
$\overline{FG} \parallel \overline{HJ}$
Prove: $\triangle FHG \cong \triangle HKJ$

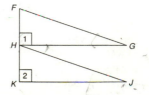

16. *Given:* $\overline{DE} \perp \overline{EF}$ and $\overline{CB} \perp \overline{AB}$
$\overline{AB} \parallel \overline{FE}$
$\overline{AD} \cong \overline{FC}$
Prove: $\triangle ABC \cong \triangle FED$

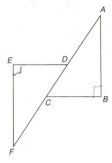

In Exercises 17 to 24, prove the indicated relationship.

17. *Given:* $\overline{DF} \cong \overline{DG}$ and $\overline{FE} \cong \overline{EG}$
Prove: $\angle F \cong \angle G$

18. *Given:* $\overrightarrow{DE}$ bisects $\angle FDG$
$\angle F \cong \angle G$
Prove: $\overline{DF} \cong \overline{DG}$

19. *Given:* E is the midpoint of $\overline{FG}$
$\overline{DF} \cong \overline{DG}$
Prove: $\overline{DE} \perp \overline{FG}$

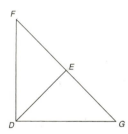

20. *Given:* $\angle MQP$ and $\angle NPQ$ are rt. $\angle$s
$\overline{MQ} \cong \overline{NP}$
Prove: $\overline{MP} \cong \overline{NQ}$

21. *Given:* $\angle 1 \cong \angle 2$ and $\overline{MN} \cong \overline{QP}$
Prove: $\overline{MQ} \parallel \overline{NP}$

22. *Given:* $\overline{MN} \parallel \overline{QP}$ and $\overline{MQ} \parallel \overline{NP}$
Prove: $\overline{MQ} \cong \overline{NP}$

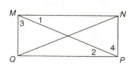

23. *Given:* $\overrightarrow{RW}$ bisects $\angle SRU$
$\overline{RS} \cong \overline{RU}$
Prove: $\angle T \cong \angle V$ (*Hint:* First show
that $\triangle RSW \cong \triangle RUW$.)

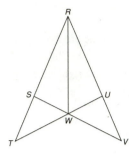

24. *Given:* $\overline{DB} \perp \overline{BC}$ and $\overline{CE} \perp \overline{DE}$
$\overline{AB} \cong \overline{AE}$
Prove: $\angle BDC \cong \angle ECD$ (*Hint:*
First show that $\triangle ACE \cong \triangle ADB$.)

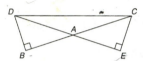

In Exercises 25 to 27, establish the construction method described.

25. Angle congruent to a given angle
Given: $\angle ABC$
$\overline{BD} \cong \overline{BE} \cong \overline{ST} \cong \overline{SR}$ (by construction)
$\overline{DE} \cong \overline{TR}$ (by construction)
Prove: $\angle B \cong \angle S$

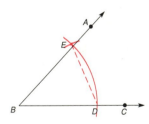
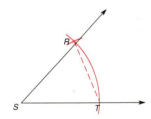

26. Bisector of a given angle
Given: $\angle XYZ$
$\overline{YM} \cong \overline{YN}$ (by construction)
$\overline{MW} \cong \overline{NW}$ (by construction)
Prove: $\overrightarrow{YW}$ bisects $\angle XYZ$

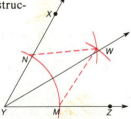

27. Perpendicular to a line at a point on the line

Given: Line *m*, with point *P* on *m*

$\overline{PQ} \cong \overline{PR}$ (by construction)

$\overline{QS} \cong \overline{RS}$ (by construction)

Prove: $\overleftrightarrow{SP} \perp m$

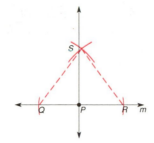

4.3 ISOSCELES TRIANGLES

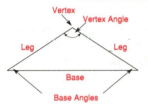

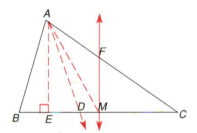

Figure 4.16

In an isosceles triangle, the two sides of equal length are **legs** and the third side is the **base**. The point at which the two legs meet is the **vertex** of the triangle, so the angle formed by the legs (and opposite the base) is the **vertex angle**. The two remaining angles are **base angles**.

In **any** triangle, a number of segments, rays, or lines are related to the triangle. (See figure 4.16.) Each angle of a triangle has a unique **angle bisector,** and this may be indicated by a ray or segment from the vertex of the bisected angle. Just as an angle bisector begins at the vertex of an angle, so does the **median,** which joins a vertex to the midpoint of the opposite side. Generally, the median from a vertex of a triangle is not the same as the angle bisector from that vertex. An **altitude** is a line segment drawn from a vertex to the opposite side such that it is perpendicular to the opposite side. Finally, the **perpendicular bisector** of a side of a triangle is shown as a line in figure 4.16. A segment or ray could also perpendicularly bisect a side of the triangle. In figure 4.16, $\overrightarrow{AD}$ is the angle bisector of $\angle BAC$; $\overline{AE}$ is the altitude from *A* to $\overline{BC}$; *M* is the midpoint of $\overline{BC}$; $\overline{AM}$ is the median from *A* to $\overline{BC}$; and $\overleftrightarrow{FM}$ is the perpendicular bisector of $\overline{BC}$. (Notice that *M* is the midpoint of $\overline{BC}$.)

An altitude can actually lie in the exterior of a triangle. In figure 4.17, which shows the obtuse triangle $\triangle RST$, the altitude from *R* must be drawn to an extension of side $\overline{ST}$. Later we will use the length of the altitude $\overline{RH}$ in place of *h* in the following standard formula for the area of a triangle:

$$A = \frac{1}{2} \cdot b \cdot h$$

The angle bisector and the median necessarily lie in the interior of the triangle.

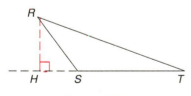

Figure 4.17

Each triangle has three altitudes—one from each vertex. As these are shown for $\triangle ABC$ in figure 4.18, do the three altitudes seem to meet at a common point?

We now consider a statement that involves the altitudes of congruent triangles.

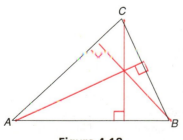

Figure 4.18

> **THEOREM 4.3.1** Corresponding altitudes of congruent triangles are congruent. *Heights*

Given: $\triangle ABC \cong \triangle RST$
altitudes $\overline{CD}$ to $\overline{AB}$ and $\overline{TV}$ to $\overline{RS}$

Prove: $\overline{CD} \cong \overline{TV}$

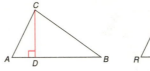 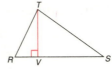

PROOF

Statements	Reasons
1. $\triangle ABC \cong \triangle RST$ altitudes $\overline{CD}$ to $\overline{AB}$ and $\overline{TV}$ to $\overline{RS}$	1. Given
2. $\overline{CD} \perp \overline{AB}$ and $\overline{TV} \perp \overline{RS}$	2. An altitude of a $\triangle$ is the line segment from one vertex drawn $\perp$ to the opposite side
3. $\angle CDA$ and $\angle TVR$ are right $\angle$s	3. If two lines are $\perp$, they form right $\angle$s
4. $\angle CDA \cong \angle TVR$	4. All right angles are $\cong$
5. $\overline{AC} \cong \overline{RT}$ and $\angle A \cong \angle R$	5. CPCTC
6. $\triangle CDA \cong \triangle TVR$	6. AAS
7. $\overline{CD} \cong \overline{TV}$	7. CPCTC ▲

Each triangle has three medians—one from each vertex to the midpoint of the opposite side. As the medians are drawn for $\triangle DEF$ in figure 4.19a, does it appear that the three medians intersect at a point?

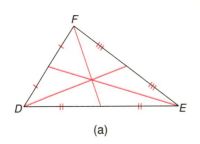

(a)

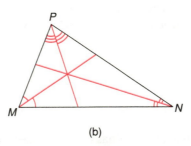

(b)

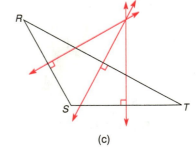

(c)

Figure 4.19

Each triangle has three angle bisectors—one for each of the three angles. As these are shown for $\triangle MNP$ in figure 4.19b, does it appear that the three angle bisectors have a point in common?

Each triangle has three perpendicular bisectors for its sides; these are shown for $\triangle RST$ in figure 4.19c. Like the altitudes, medians, and angle bisectors, the perpendicular bisectors of the sides also meet at a single point. These points of intersection will be given more attention in Chapter 6.

EXAMPLE 1 Give a formal proof of Theorem 4.3.2.

> **THEOREM 4.3.2** The bisector of the vertex angle of an isosceles triangle separates the triangle into two congruent triangles.

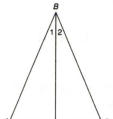

Given: Isosceles triangle $\triangle ABC$, with $\overline{AB} \cong \overline{BC}$
$\overrightarrow{BD}$ bisects $\angle ABC$

Prove: $\triangle ABD \cong \triangle CBD$

PROOF

Statements	Reasons
1. Isosceles $\triangle ABC$ with $\overline{AB} \cong \overline{BC}$	1. Given
2. $\overrightarrow{BD}$ bisects $\angle ABC$	2. Given
3. $\angle 1 \cong \angle 2$	3. The bisector of an $\angle$ separates it into two $\cong \angle$s
4. $\overline{BD} \cong \overline{BD}$	4. Identity
5. $\triangle ABD \cong \triangle CBD$	5. SAS ▲

In many proofs, **auxiliary** (or helping) lines, segments, or rays are needed in constructing a proof. You must be careful to account for the unique line, segment, or ray as it is introduced into the existing drawing. That is, each auxiliary figure must be **determined,** but it must not be **underdetermined** or **overdetermined.** A figure is underdetermined when there is more than one possible figure described. On the other extreme, a figure is overdetermined when it is impossible for *all* conditions described to be satisfied.

Consider figure 4.20 and the following three descriptions, which are coded D for determined, U for underdetermined, and O for overdetermined:

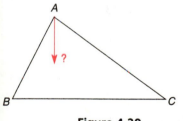

Figure 4.20

D: Draw a segment from A perpendicular to $\overline{BC}$ so that the terminal point is on $\overline{BC}$.

U: Draw a segment from A to $\overline{BC}$ so that the terminal point is on $\overline{BC}$.

O: Draw a segment through A perpendicular to $\overline{BC}$ so that it bisects $\overline{BC}$; let the terminal point of the segment lie on $\overline{BC}$.

In Example 2, an auxiliary segment is needed. As you study the proof, note the uniqueness of the segment and its justification in the proof.

EXAMPLE 2 Give a formal proof of Theorem 4.3.3.

> **THEOREM 4.3.3** If two sides of a triangle are congruent, then the angles opposite these sides are also congruent.

Given: Isosceles $\triangle MNP$
with $\overline{MP} \cong \overline{NP}$

Prove: $\angle M \cong \angle N$

(*Note:* Figure 4.21b shows the auxiliary segment.)

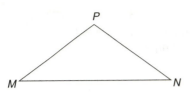

(a)

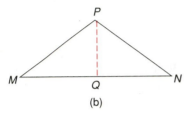

(b)

Figure 4.21

PROOF

Statements	Reasons
1. Isosceles $\triangle MNP$ with $\overline{MP} \cong \overline{NP}$	**1.** Given
2. Draw $\angle$ bisector $\overrightarrow{PQ}$ from P to $\overline{MN}$	**2.** Every angle has one and only one bisector
3. $\triangle MPQ \cong \triangle NPQ$	**3.** The bisector of the vertex angle of an isosceles $\triangle$ separates it into two $\cong$ $\triangle$s
4. $\angle M \cong \angle N$	**4.** CPCTC ▲

Theorem 4.3.3 is sometimes stated as, "Base angles of an isosceles triangle are congruent." We use this concept in Example 3.

EXAMPLE 3   Find the size of each angle of the isosceles triangle shown in figure 4.22 if:
(**a**) $m\angle 1 = 36°$
(**b**) Measure of each base $\angle$ is 5° less than twice the measure of the vertex angle

Solution   (**a**) $m\angle 1 + m\angle 2 + m\angle 3 = 180$. Since $m\angle 1 = 36$ and $\angle 2$ and $\angle 3$ are $\cong$, we have

Figure 4.22

$$36 + 2(m\angle 2) = 180$$
$$2(m\angle 2) = 144$$
$$m\angle 2 = 72$$

Now $m\angle 1 = 36$, while $m\angle 2 = m\angle 3 = 72$.
(**b**) Let the vertex angle measure be given by x. Then the size of each base angle is $2x - 5$. Because the sum of measures is 180°,

$$x + (2x - 5) + (2x - 5) = 180$$
$$5x - 10 = 180$$
$$5x = 190$$
$$x = 38$$
$$2x - 5 = 71$$

Therefore $m\angle 1 = 38$ and $m\angle 2 = m\angle 3 = 71$.   ▲

In some instances, a carpenter may want to get a quick, accurate measurement without having to go get his or her tools. Suppose that the carpenter's square shown in figure 4.23 is handy but that a miter box is not nearby. If two marks are made at lengths of 4 inches from the corner of the square and these are then joined, what size angle is determined?

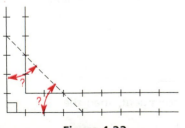

Figure 4.23

Example 4 shows us that the converse of the theorem, "Base angles of an isosceles Δ are congruent," is also true. Be careful, however, not to assume that the converse of an "If, then" statement always is true!

EXAMPLE 4   Give a formal proof of Theorem 4.3.4.

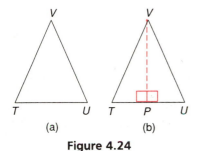

Figure 4.24

> **THEOREM 4.3.4**   If two angles of a triangle are congruent, then the sides opposite these angles are also congruent.

Given: $\triangle TUV$ with $\angle T \cong \angle U$, as in figure 4.24a
Prove: $\overline{VU} \cong \overline{VT}$

PROOF

Statements	Reasons
1. $\triangle TUV$ with $\angle T \cong \angle U$	1. Given
2. Draw $\overline{VP}$, the segment from $P \perp$ to $\overline{TU}$, as in figure 4.24b	2. There is exactly one perpendicular from a point to a line
3. $\therefore \angle VPT \cong \angle VPU$	3. $\perp$ lines meet to form $\cong$ adjacent $\angle$s
4. $\overline{VP} \cong \overline{VP}$	4. Identity
5. $\triangle TPV \cong \triangle UPV$	5. AAS
6. $\overline{VU} \cong \overline{VT}$	6. CPCTC ▲

When all three sides of a triangle are congruent, the triangle is **equilateral.** If all three angles are congruent, then the triangle is **equiangular.** Theorems 4.3.3 and 4.3.4 can be used to prove the following corollaries.

> **COROLLARY 4.3.5**   An equilateral triangle is also equiangular.

> **COROLLARY 4.3.6**   An equiangular triangle is also equilateral.

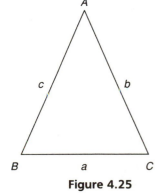

Figure 4.25

> **DEFINITION**   The **perimeter** of a triangle is the sum of the lengths of its sides. Thus, if *a, b,* and *c* are the lengths of the three sides, then the perimeter *P* is given by $P = a + b + c$. (See figure 4.25.)

EXAMPLE 5   *Given:* $\angle B \cong \angle C$
                            $AB = 5.3$ and $BC = 3.6$

               Find: The perimeter of $\triangle ABC$

Solution   If $\angle B \cong \angle C$, then $AC = AB = 5.3$. Therefore

$$P = a + b + c$$
$$= 3.6 + 5.3 + 5.3$$
$$= 14.2$$

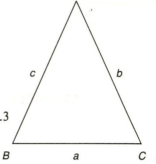

4.3 EXERCISES

In Exercises 1 to 6, describe the segment as determined, underdetermined, or overdetermined. Use the accompanying drawing for reference.

1. Draw a segment through point A.

2. Draw a segment with endpoints A and B.

3. Draw a segment $\overline{AB}$ parallel to line m.

4. Draw segment $\overline{AB}$ perpendicular to m.

5. Draw a segment through A perpendicular to m.

6. Draw $\overline{AB}$ so that line m bisects $\overline{AB}$.

7. A surveyor knows that a lot has the shape of an isosceles triangle. If the vertex angle measures 70° and each equal side is 160 feet long, what measure has each of the base angles?

8. Construct an angle whose measure is 60°. Then bisect it to form two angles of 30° each.

9. Construct an angle whose measure is 90°. Then bisect it to form two angles of 45° each.

10. In concave quadrilateral $ABCD$, the angle at A measures 40°. $\triangle ABD$ is isosceles, and $\overrightarrow{BC}$ bisects $\angle ABD$, while $\overrightarrow{DC}$ bisects $\angle ADB$. What are the measures of $\angle ABC$, $\angle ADC$, and $\angle 1$?

In Exercises 11 to 16, use arithmetic or algebra as needed to find the measures indicated. Note the use of dashes on equal sides of the given isosceles triangles.

11. Using the accompanying drawing, find $m\angle 1$ and $m\angle 2$ if $m\angle 3 = 68°$.

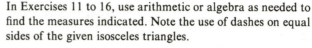

12. Using the accompanying drawing, find $m\angle 4$, the angle formed by the bisectors of $\angle 3$ and $\angle 2$, if $m\angle 3 = 68°$.

13. Using the same drawing, find the measure of $\angle 5$, which is formed by the bisectors of $\angle 1$ and $\angle 3$. Again let $m\angle 3 = 68°$.

14. Using the drawing accompanying Exercise 12, find an expression for the measure of $\angle 5$ if $m\angle 3 = 2x$ and the segments shown bisect the angles of the isosceles triangle.

15. In isosceles $\triangle ABC$ shown in the drawing on page 119, the base angles are 12° larger than the vertex angle. Find the measure of each angle.

16. In $\triangle ABC$, vertex angle A is 5° more than one-half of base angle B. Find the size of each angle.

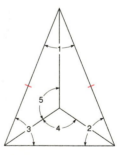

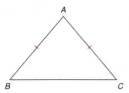

In Exercises 17 to 22, construct proofs.

17. *Given:* $\angle 3 \cong \angle 1$
Prove: $\overline{AB} \cong \overline{AC}$

18. *Given:* $\overrightarrow{AG}$ bisects $\angle FAC$
$\overrightarrow{AG} \parallel \overline{BC}$
Prove: $\triangle ABC$ is isosceles

19. *Given:* $\overline{AB} \cong \overline{AC}$
Prove: $\angle 6 \cong \angle 7$

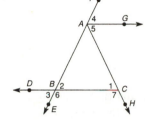

20. *Given:* $\angle 1 \cong \angle 3$
$\overline{RU} \cong \overline{VU}$
Prove: $\triangle STU$ is isosceles

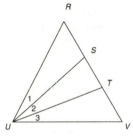

21. *Given:* isosceles $\triangle MNP$ with
vertex P
isosceles $\triangle MNQ$ with
vertex Q
Prove: $\triangle MQP \cong \triangle NQP$

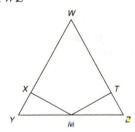

22. *Given:* $\overline{WY} \cong \overline{WZ}$
M is the midpoint of $\overline{YZ}$
$\overline{MX} \perp \overline{WY}$ and $\overline{MT} \perp \overline{WZ}$
Prove: $\overline{MX} \cong \overline{MT}$

In Exercises 23 to 26, complete formal proofs for each statement.

23. The altitude from the vertex of an isosceles triangle is also the median to the base of the triangle.

24. The bisector of the vertex angle of an isosceles triangle bisects the base.

25. The angle bisectors of the base angles of an isosceles triangle, together with the base, form an isosceles triangle.

26. Each acute angle in an isosceles right triangle measures 45°.

27. Construct an isosceles triangle $\triangle ABC$ for which $\overline{AB}$ is the base and $\overline{BC}$ is one of the two legs.

28. Construct an isosceles triangle $\triangle DEF$, given that $\overline{EF}$ is the base and $\angle E$ is one of the two congruent base angles.

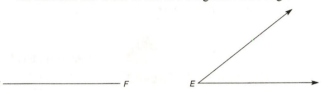

29. Construct an isosceles triangle $\triangle HJK$ for which $\overline{HK}$ is the base and $\overrightarrow{JM}$ is the bisector of the vertex angle, with point M on $\overline{HK}$.

30. Draw an acute triangle, and construct the three medians of the triangle.

31. **(a)** Draw an obtuse triangle, and construct the three altitudes of the triangle.
(b) Given a right triangle, how many altitudes must be constructed?

32. Draw an obtuse triangle, and construct the three angle bisectors of the triangle.

33. Draw an acute triangle, and construct the three perpendicular bisectors of the sides of the triangle.

34. Construct an equilateral triangle and its three altitudes. What does intuition tell you about the medians, the angle bisectors, and the perpendicular bisectors of the sides?

35. A carpenter has placed a square over an angle in such a manner (see the accompanying drawing) that $\overline{AB} \cong \overline{AC}$ and $\overline{BD} \cong \overline{CD}$. What may you conclude about the location of point D?

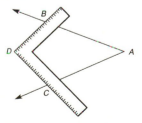

36. In $\triangle BAT$, $\overline{BR} \cong \overline{BT} \cong \overline{AR}$, while $m\angle RBT = 20°$. Find $m\angle A$.

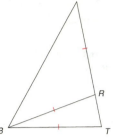

37. In $\triangle PMN$, $\overline{PM} \cong \overline{PN}$. $\overrightarrow{MB}$ bisects $\angle PMN$, and $\overrightarrow{NA}$

bisects $\angle PNM$. Name all isosceles triangles in the drawing, if $m\angle P = 36°$.

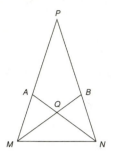

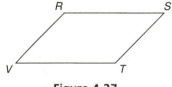

4.4

QUADRILATERALS AND PARALLELOGRAMS

A **quadrilateral** is a polygon that has four sides. Unless otherwise stated, the term "quadrilateral" refers to a figure such as *ABCD* in figure 4.26a, in which the line segment sides lie within a single plane. When the sides of the quadrilateral do not lie within one plane, as with *MNPQ* in figure 4.26b, the quadrilateral is said to be **skew.** Thus, *MNPQ* is a skew quadrilateral.

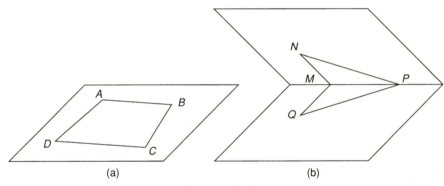

(a)   (b)

Figure 4.26

DEFINITION: A **parallelogram** is a quadrilateral in which both pairs of opposite sides are parallel.

Figure 4.27

The symbol for parallelogram is ▱. Shown in figure 4.27 is ▱*RSTV*.

EXAMPLE 1   Give a formal proof of Theorem 4.4.1.

THEOREM 4.4.1   A diagonal of a parallelogram separates it into two congruent triangles.

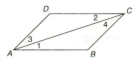

Given: ▱ABCD, with $\overline{AB} \parallel \overline{DC}$, $\overline{AD} \parallel \overline{BC}$, and diagonal $\overline{AC}$
Prove: △ACD ≅ △ACB

PROOF

Statements	Reasons
1. ▱ABCD; $\overline{AB} \parallel \overline{CD}$	1. Given
2. ∠1 ≅ ∠2	2. If two ∥ lines are cut by a transversal, the alternate interior ∠s are congruent
3. $\overline{AD} \parallel \overline{BC}$	3. Given
4. ∠3 ≅ ∠4	4. Same as reason 2
5. $\overline{AC} \cong \overline{AC}$	5. Identity
6. △ACD ≅ △ACB	6. ASA

▲

Some corollaries of Theorem 4.4.1 are now stated.

COROLLARY 4.4.2 Opposite angles of a parallelogram are congruent.

COROLLARY 4.4.3 Opposite sides of a parallelogram are congruent.

COROLLARY 4.4.4 Diagonals of a parallelogram bisect each other.

Recall Theorem 3.1.4: "If two parallel lines are cut by a transversal, then the interior angles on the same side of the transversal are supplementary." A corollary of that theorem is stated next.

COROLLARY 4.4.5 Consecutive angles of a parallelogram are supplementary.

Example 2 illustrates the fact that two parallel lines are everywhere equidistant. In general, the phrase "distance between two parallel lines" refers to the length of the perpendicular segment between the two parallel lines.

EXAMPLE 2 Given: $\overleftrightarrow{AB} \parallel \overleftrightarrow{CD}$
$\overline{AC} \perp \overleftrightarrow{CD}$ and $\overline{BD} \perp \overleftrightarrow{CD}$
Prove: $\overline{AC} \cong \overline{BD}$

PROOF

Statements	Reasons
1. $\overleftrightarrow{AB} \parallel \overleftrightarrow{CD}$	1. Given
2. $\overline{AC} \perp \overleftrightarrow{CD}$ and $\overline{BD} \perp \overleftrightarrow{CD}$	2. Given
3. $\overline{AC} \parallel \overline{BD}$	3. If two lines are $\perp$ to the same line, they are parallel
4. $ABDC$ is a $\square$	4. If both pairs of opposite sides of a quadrilateral are $\parallel$, the quadrilateral is a $\square$.
5. $\overline{AC} \cong \overline{BD}$	5. Opposite sides of a $\square$ are congruent ▲

In Example 2, we used the definition of a parallelogram to prove that a particular quadrilateral was a parallelogram, but there are other ways of establishing that a given quadrilateral is a parallelogram.

EXAMPLE 3   Give a formal proof of Theorem 4.4.2.

> **THEOREM 4.4.6**   If two sides of a quadrilateral are both congruent and parallel, then the quadrilateral is a parallelogram.

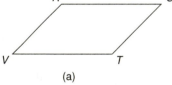

(a)

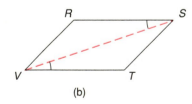

(b)

Figure 4.28

Given: In figure 4.28a, $\overline{RS} \parallel \overline{VT}$ and $\overline{RS} \cong \overline{VT}$
Prove: $RSTV$ is a $\square$

PROOF

Statements	Reasons
1. $\overline{RS} \parallel \overline{VT}$ and $\overline{RS} \cong \overline{VT}$	1. Given
2. Draw diagonal $\overline{VS}$, as in figure 4.28b	2. Exactly one line passes through two points
3. $\angle RSV \cong \angle SVT$	3. If two $\parallel$ lines are cut by a transversal, alternate interior $\angle$s are $\cong$
4. $\triangle RSV \cong \triangle SVT$	4. SAS
5. $\therefore \angle RVS \cong \angle VST$	5. CPCTC
6. $\overline{RV} \parallel \overline{ST}$	6. If two lines are cut by a transversal so that alternate interior $\angle$s are $\cong$, these lines are $\parallel$
7. $RSTV$ is a $\square$	7. If both pairs of opposite sides of a quadrilateral are $\parallel$, the quadrilateral is a parallelogram ▲

Other properties a quadrilateral may have that will enable you to conclude that it is a parallelogram are stated in the theorems that follow.

> **THEOREM 4.4.7** If both pairs of opposite sides of a quadrilateral are congruent, then it is a parallelogram.

> **THEOREM 4.4.8** If the diagonals of a quadrilateral bisect each other, then the quadrilateral is a parallelogram.

When you draw a figure that tests the preceding conditions, be sure not to include more conditions than the hypothesis states. For instance, if you drew two diagonals that not only bisected each other but were of equal lengths, as well, the quadrilateral would be a special type of parallelogram known as a **rectangle.** We will deal with rectangles more in the following section.

The next quadrilateral we will consider was not a figure of special interest in Euclid's time. Known as a kite, this quadrilateral corresponds to the child's toy pictured in figure 4.29. In the toy's construction, two pieces of wood are joined at right angles so that one piece is bisected and the other is not. This leads to the formal definition of a kite.

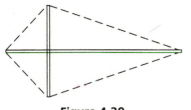

Figure 4.29

> **DEFINITION:** A **kite** is a quadrilateral in which one and only one diagonal is the perpendicular bisector of the other.

> **THEOREM 4.4.9** A kite has two pairs of congruent adjacent sides.

The proof of Theorem 4.4.9 is left as an exercise.

As you observe an old barn or shed, you may see that it has, with age, begun to lean. While a triangle is rigid in shape (figure 4.30a) and bends only

(a)

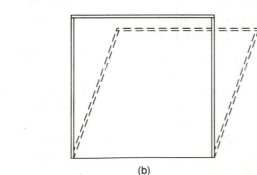

(b)

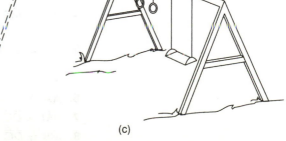

(c)

Figure 4.30

when broken, a quadrilateral (figure 4.30b) does *not* provide the same level of strength and stability. In observing the construction of a house, bridge, or building, note the use of wooden or metal triangles as braces. The brace in the swing set in figure 4.30c suggests the following theorem.

> **THEOREM 4.4.10**   The segment that joins the midpoints of two sides of a triangle is parallel to the third side and has a length equal to one-half the length of the third side.

We will prove the first part of this theorem but leave the second part as an exercise. The part we want to prove now is: "The segment that joins the midpoints of two sides of a triangle is parallel to the third side."

Given: In figure 4.31a, $\triangle ABC$, with M and N the midpoints of $\overline{AB}$ and $\overline{AC}$, respectively

Prove: $\overline{MN} \parallel \overline{BC}$

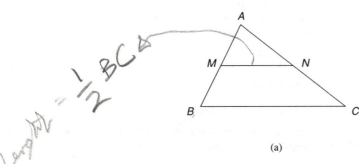

(a)   (b)

Figure 4.31

PROOF

Statements	Reasons
1. Through C, construct $\overleftrightarrow{CE} \parallel \overline{AB}$, as in figure 4.31b	1. Parallel Postulate
2. Extend $\overline{MN}$ to meet $\overleftrightarrow{CE}$ at D, as in figure 4.31b	2. Exactly one line passes through two points
3. $\triangle ABC$, with midpoints M and N of $\overline{AB}$ and $\overline{AC}$, respectively	3. Given
4. $\overline{AM} \cong \overline{MB}$ and $\overline{AN} \cong \overline{NC}$	4. The midpoint of a segment divides it into $\cong$ segments
5. $\angle 1 \cong \angle 2$ and $\angle 4 \cong \angle 3$	5. If two $\parallel$ lines are cut by a transversal, alternate interior $\angle$s are $\cong$
6. $\triangle ANM \cong \triangle CND$	6. AAS
7. $\overline{AM} \cong \overline{DC}$	7. CPCTC
8. $\overline{MB} \cong \overline{DC}$	8. Transitive (both are $\cong$ to $\overline{AM}$)

9. Quadrilateral *BMDC* is a ▱

10. $\overline{MN} \parallel \overline{BC}$

9. If two sides of a quadrilateral are both ≅ and ∥, the quadrilateral is a parallelogram

10. Opposite sides of a ▱ are ∥ ▲

Theorem 4.4.10 also asserts that the segment formed by joining the midpoints of two sides of a triangle has a length equal to one-half the length of the third side. This part of the theorem is used in Example 4.

EXAMPLE 4

Given: $\triangle ABC$, with D the midpoint of $\overline{AC}$ and E the midpoint of $\overline{BC}$
$DE = 2x + 1$; $AB = 5x - 1$

Find: x, DE, and AB

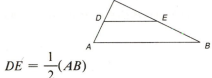

Solution   By Theorem 4.4.6,

$$DE = \frac{1}{2}(AB)$$

so

$$2x + 1 = \frac{1}{2}(5x - 1)$$

Multiplying by 2, we have

$$4x + 2 = 5x - 1$$
$$3 = x$$

Therefore $DE = 2 \cdot 3 + 1 = 7$. And similarly, $AB = 5 \cdot 3 - 1 = 14$. ▲

4.4 EXERCISES

In Exercises 1 to 4, refer to the accompanying drawing, in which *ABCD* is a parallelogram.

1. Given that $AB = 3x + 2$, $BC = 4x + 1$, and $CD = 5x - 2$, find the length of each side of ▱*ABCD*.

2. Given that $m\angle A = 2x + 3$ and $m\angle C = 3x - 27$, find the measure of each angle of ▱*ABCD*.

3. Given that $m\angle A = 2x + 3$ and $m\angle B = 3x - 23$, find the size of each angle of ▱*ABCD*.

4. Given that $m\angle A = 2x + y$, $m\angle B = 2x + 3y - 20$, and $m\angle C = 3x - y + 16$, find the measure of each angle of ▱*ABCD*.

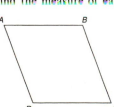

5. In quadrilateral *RSTV*, the midpoints of consecutive sides are joined in order. Try drawing other quadrilaterals and joining their midpoints. What can you conclude about the resulting quadrilateral in each case?

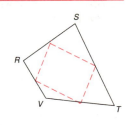

6. In quadrilateral *ABCD*, the midpoints of opposite sides are joined to form two intersecting segments. Try drawing other quadrilaterals and joining opposite midpoints on them. What can you conclude about these segments in each case?

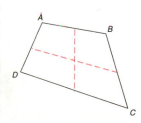

In Exercises 7 to 12, complete each proof.

7. *Given:* ∠1 ≅ ∠2 and ∠3 ≅ ∠4
Prove: $\overline{MP} \perp \overline{NQ}$

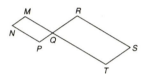

8. *Given:* M-Q-T and P-Q-R so that MNPQ and QRST are ▱s
Prove: ∠N ≅ ∠S

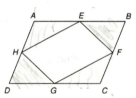

9. *Given:* ▱ABCD, with midpoints E, F, G, and H of the sides, as shown in the accompanying drawing
Prove: EFGH is a ▱

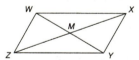

10. *Given:* ▱WXYZ with diagonals $\overline{WY}$ and $\overline{XZ}$
Prove: △WMX ≅ △YMZ

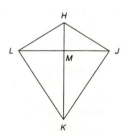

11. *Given:* Quadrilateral HJKL, with M the midpoint of $\overline{LJ}$ and $\overline{LJ} \perp \overline{HK}$
Prove: $\overrightarrow{HM}$ bisects ∠LHJ, and $\overline{LK} \cong \overline{JK}$

12. *Given:* ▱MNPQ, with T the midpoint of $\overline{MN}$ and S the midpoint of $\overline{QP}$
Prove: △QMS ≅ △NPT, and MSPT is a ▱

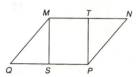

In Exercises 13 to 16, write a formal proof for each theorem or corollary.

13. The opposite angles of a parallelogram are congruent.

14. The opposite sides of a parallelogram are congruent.

15. If both pairs of opposite sides of a quadrilateral are congruent, then the quadrilateral is a parallelogram.

16. If the diagonals of a quadrilateral bisect each other, then the quadrilateral is a parallelogram.

17. A kite has two pairs of congruent adjacent sides.

In Exercises 18 to 20, use the drawings provided to deal with each problem.

18. The bisectors of two consecutive angles of a parallelogram meet as shown. What can you conclude about ∠P?

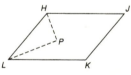

19. When the bisectors of two consecutive angles of a parallelogram meet at a point on the remaining side, as shown, what may you conclude about △DEC? About △ADE? About △BCE?

20. RSTV is a kite, with $\overline{RS} \perp \overline{ST}$ and $\overline{RV} \perp \overline{VT}$. If m∠STV = 40°, how large is the angle formed by the bisectors of ∠RST and ∠STV? The bisectors of ∠SRV and ∠RST?

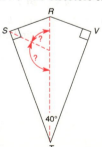

*21. Prove that the segment that joins the midpoints of two sides of a triangle has a length equal to one-half the length of the third side. (*Hint:* Use the accompanying drawing, in which $\overline{MN}$ is extended to D, a point on $\overline{CD}$, which is parallel to $\overline{AB}$.)

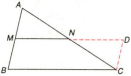

*22. Prove that, when the midpoints of consecutive sides of a quadrilateral are joined in order, the resulting quadrilateral is a parallelogram. (*Hint:* Use the result in Exercise 21.)

In Exercises 23 to 25, use the accompanying illustration, in which $\triangle RST$ has M and N for midpoints of sides $\overline{RS}$ and $\overline{RT}$, respectively.

23. *Given:* $MN = 2y - 3$
  $ST = 3y$
  Find: y, MN, and ST

24. *Given:* $MN = x^2 + 5$
  $ST = x(2x + 5)$
  Find: x, MN, and ST

25. *Given:* $RM = RN = 2x + 1$
  $ST = 5x - 3$
  $m\angle R = 60°$
  Find: x, RM, and ST

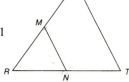

4.5

MORE QUADRILATERALS

Figure 4.32

In this section, we continue to discuss various quadrilaterals. The first of these is the rectangle, which we will now define.

> **DEFINITION:**   A **rectangle** is a parallelogram that has a right angle.

Rectangle $ABCD$ is shown in figure 4.32.

> **COROLLARY 4.5.1**   All angles of a rectangle are right angles.

EXAMPLE 1   Give a formal proof of Theorem 4.5.2.

> **THEOREM 4.5.2**   The diagonals of a rectangle are congruent.

Given: Rectangle $MNPQ$ with diagonals $\overline{MP}$ and $\overline{NQ}$
Prove: $\overline{MP} \cong \overline{NQ}$

PROOF

Statements	Reasons
1. Rectangle $MNPQ$ with diagonals $\overline{MP}$ and $\overline{NQ}$	**1.** Given
2. $MNPQ$ is a ▱	**2.** By definition, a rectangle is a ▱ with a right angle
3. $\overline{MN} \cong \overline{QP}$	**3.** Opposite sides of a ▱ are ≅

4. $\overline{MQ} \cong \overline{MQ}$	4. Identity
5. $\angle NMQ$ and $\angle PQM$ are right $\angle$s	5. By Corollary 4.5.1, the four $\angle$s of a rectangle are right $\angle$s
6. $\angle NMQ \cong \angle PQM$	6. All right $\angle$s are $\cong$
7. $\triangle NMQ \cong \triangle PQM$	7. SAS
8. $\overline{MP} \cong \overline{NQ}$	8. CPCTC ▲

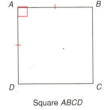

Square *ABCD*

> **DEFINITION:**   A **square** is a rectangle that has two congruent adjacent sides.

> **COROLLARY 4.5.3**   All sides of a square are congruent.

Because a square is a rectangle, it has four right angles and its diagonals are congruent. Because a square is also a parallelogram, its opposite sides are parallel.

The next type of quadrilateral we will consider is the rhombus.

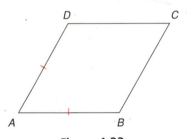

Figure 4.33

> **DEFINITION:**   A **rhombus** is a parallelogram with two congruent adjacent sides.

In figure 4.33, adjacent sides $\overline{AB}$ and $\overline{AD}$ of rhombus *ABCD* are marked congruent. Because a rhombus is a type of parallelogram, it is also necessary that $\overline{AB} \cong \overline{DC}$ and $\overline{AD} \cong \overline{BC}$.

> **COROLLARY 4.5.4**   All sides of a rhombus are congruent.

EXAMPLE 2   Construct a formal proof for Theorem 4.5.5.

> **THEOREM 4.5.5**   The diagonals of a rhombus are perpendicular.

Given: Rhombus *ABCD*, with diagonals $\overline{AC}$ and $\overline{DB}$
Prove: $\overline{AC} \perp \overline{DB}$

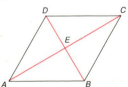

PROOF

Statements	Reasons
1. Rhombus *ABCD* with diagonals $\overline{AC}$ and $\overline{DB}$	1. Given
2. *ABCD* is a ▱	2. A rhombus is a ▱ with two ≅ adjacent sides
3. $\overline{DB}$ bisects $\overline{AC}$	3. Diagonals of a ▱ bisect each other
4. $\overline{AE} \cong \overline{EC}$	4. If a segment is bisected, it is divided into two ≅ segments
5. $\overline{AD} \cong \overline{DC}$	5. A rhombus is a ▱ with two ≅ adjacent sides
6. $\overline{DE} \cong \overline{DE}$	6. Identity
7. $\triangle ADE \cong \triangle CDE$	7. SSS
8. $\angle DEA \cong \angle DEC$	8. CPCTC
9. $\overline{AC} \perp \overline{DB}$	9. If two lines meet to form ≅ adjacent ∠s, the lines are ⊥ ▲

An alternate definition of "square" is, "A square is a rhombus whose sides form a right angle." Therefore, a further property of a square is that its diagonals are perpendicular.

> **DEFINITION:** A **trapezoid** is a quadrilateral with exactly two parallel sides.

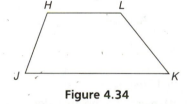

Figure 4.34

Figure 4.34 shows trapezoid *HJKL*, in which $\overline{HL} \parallel \overline{JK}$. The parallel sides $\overline{HL}$ and $\overline{JK}$ are **bases,** and the nonparallel sides $\overline{HJ}$ and $\overline{LK}$ are **legs.** Because $\angle J$ and $\angle K$ each have $\overline{JK}$ for a side, they are **base angles** of the trapezoid; $\angle H$ and $\angle L$ are also base angles since $\overline{HL}$ is a base.

When the midpoints of the two legs of a trapezoid are joined, the resulting segment is known as the **median** of the trapezoid. Given that *M* and *N* are the midpoints of the legs $\overline{HJ}$ and $\overline{LK}$ in trapezoid *HJKL*, $\overline{MN}$ is the median of the trapezoid. (See figure 4.35a.)

If the two legs of a trapezoid are congruent, the trapezoid is known

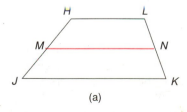

(a)

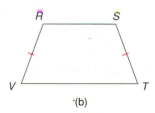

(b)

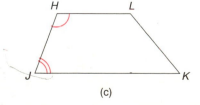

(c)

Figure 4.35

as an **isosceles trapezoid.** In figure 4.35b, *RSTV* is an isosceles trapezoid if $\overline{RV} \cong \overline{ST}$ and $\overline{RS} \parallel \overline{VT}$.

Every trapezoid contains two pairs of consecutive interior angles that are supplementary. Each of these pairs of angles is formed when parallel lines are cut by a transversal. In figure 4.35c, angles *H* and *J* are supplementary, as are angles *L* and *K*.

EXAMPLE 3   Give a proof of Theorem 4.5.6.

> **THEOREM 4.5.6**   The base angles of an isosceles trapezoid are congruent.

Given: In figure 4.36a, trapezoid *RSTV*, with $\overline{RV} \cong \overline{ST}$ and $\overline{RS} \parallel \overline{VT}$

Prove: $\angle V \cong \angle T$ and $\angle R \cong \angle S$

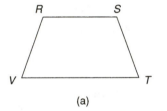 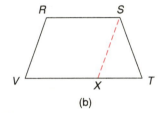

(a)   (b)

Figure 4.36

PROOF

Statements	Reasons
1. Trapezoid *RSTV* with $\overline{RV} \cong \overline{ST}$ and $\overline{RS} \parallel \overline{VT}$	1. Given
2. Draw $\overline{SX} \parallel \overline{RV}$, as in figure 4.36b	2. Parallel Postulate
3. *RSXV* is a ▱	3. If both pairs of opposite sides of a quadrilateral are ∥, it is a ▱
4. $\overline{RV} \cong \overline{SX}$	4. Opposite sides of a ▱ are ≅
5. $\overline{ST} \cong \overline{SX}$	5. Transitive
6. $\angle SXT \cong \angle T$	6. If two sides of a △ are ≅, the angles opposite those sides are also ≅
7. $\angle V \cong \angle SXT$	7. If two ∥ lines are cut by a transversal, corresponding ∠s are ≅
8. $\angle V \cong \angle T$	8. Transitive
9. $\angle V$ is supplementary to $\angle VRS$ and $\angle T$ is supplementary to $\angle TSR$	9. If two ∥ lines are cut by a transversal, interior ∠s on the same side of the transversal are supplementary
10. $\angle VRS \cong \angle TSR$	10. Supplements of ≅ ∠s are ≅ ▲

A corollary of Theorem 4.5.6, covering the diagonals of an isosceles triangle, now follows.

> **COROLLARY 4.5.7**   The diagonals of an isosceles trapezoid are congruent.

Two theorems that apply to any trapezoid can now be stated. Their proofs are left as exercises.

> **THEOREM 4.5.8**   The length of the median of a trapezoid equals one-half the sum of the lengths of the two bases.

> **THEOREM 4.5.9**   The median of a trapezoid is parallel to each base.

Table 4.1 summarizes the properties of quadrilaterals that have been expressed in various definitions and theorems. The table suggests that rectangles have nine properties (including those of the quadrilateral and the parallelogram).

Table 4.1   **Properties of Quadrilaterals**

Quadrilateral
   1. Four sides, by definition

Trapezoid
   1. Four sides, by definition
   2. Two sides parallel, by definition

Kite

   1. Four sides, by definition
   2. One diagonal is the perpendicular
      bisector of the other, by definition
   3. Two pairs of congruent adjacent sides

Isosceles Trapezoid
   1. Four sides, by definition
   2. Two sides parallel, by definition
   3. Congruent legs, by definition
   4. Congruent base angles
   5. Congruent diagonals

Table 4.1   **Properties of Quadrilaterals (*cont'd.*)**

Parallelogram
1. Four sides, by definition
2. Opposite sides parallel, by definition
3. Opposite sides congruent
4. Opposite angles congruent
5. Consecutive angles supplementary
6. Diagonals bisect each other

Rhombus
1. Four sides, by definition
2. Opposite sides parallel, by definition
3. Opposite sides congruent
4. Opposite angles congruent
5. Consecutive angles supplementary
6. Diagonals bisect each other
7. Two congruent adjacent sides, by definition
8. Four congruent sides
9. Perpendicular diagonals

Rectangle
1. Four sides, by definition
2. Opposite sides parallel, by definition
3. Opposite sides congruent
4. Opposite angles congruent
5. Consecutive angles supplementary
6. Diagonals bisect each other
7. A right angle, by definition
8. Four right angles
9. Congruent diagonals

Square
1. Four sides, by definition
2. Opposite sides parallel, by definition
3. Opposite sides congruent
4. Opposite angles congruent
5. Consecutive angles supplementary
6. Diagonals bisect each other
7. A right angle, by definition
8. Four right angles
9. Congruent diagonals
10. Two congruent adjacent sides, by definition
11. Four congruent sides
12. Perpendicular diagonals

4.5 EXERCISES

Use the drawings provided in answering Exercises 1 to 10.

1. If diagonal $\overline{DB}$ is congruent to each side of rhombus *ABCD*, what is the measure of $\angle A$? Of $\angle ABC$?

2. If the diagonals of a parallelogram are perpendicular, what may you conclude about the parallelogram? (*Hint:* Make a number of drawings in which you use only the information just suggested.)

3. If the diagonals of a parallelogram are congruent, what may you conclude about the parallelogram?

4. If the diagonals of a parallelogram are perpendicular and congruent, what may you conclude about the parallelogram?

5. If the diagonals of a trapezoid are congruent, what may you conclude about the trapezoid?

6. If the diagonals of a rhombus are congruent, what may you conclude about the rhombus?

7. Without writing a formal proof, explain why

$$MN = \frac{1}{2}(AB + DC)$$

in the accompanying drawing.

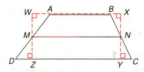

Use the result from Exercise 7 in Exercises 8 to 10, in which you are given trapezoid *ABCD*, with $\overline{AB} \parallel \overline{DC}$; *M* and *N* are midpoints of $\overline{AD}$ and $\overline{BC}$, respectively.

8. *Given:* $MN = 6.3$ and $DC = 7.5$
   Find: *AB*

9. *Given:* $AB = 6x + 3$,
   $DC = 8x - 1$
   Find: *MN*, in terms of *x*

10. *Given:* $AB = 7x + 5$,
   $DC = 4x - 2$,
   $MN = 5x + 3$
   Find: *x*

In Exercises 11 to 14, complete each proof.

11. *Given:* *ABCD* is an isosceles trapezoid
   Prove: $\triangle ABE$ is isosceles

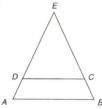

12. *Given:* Rectangle *WXYZ*
   Prove: $\overline{WY} \cong \overline{XZ}$

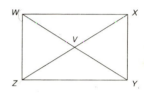

13. *Given:* Rhombus *PQST*, with midpoints *A*, *B*, *C*, and *D* of the sides, as shown
   Prove: *ABCD* is a ▱ (*Note: ABCD* is also a rectangle. This is more easily proved in a later chapter, using analytic methods.)

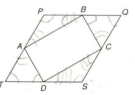

In Exercises 14 to 17, give a formal proof of each corollary.

14. All angles of a rectangle are right angles.

15. All sides of a rhombus are congruent.

16. All sides of a square are congruent.

17. The diagonals of an isosceles trapezoid are congruent.

In Exercises 18 to 26, give a formal proof of each statement.

18. The diagonals of a square are perpendicular.

19. A diagonal of a rhombus bisects two angles of the rhombus.

20. If the diagonals of a parallelogram are congruent, then the parallelogram is a rectangle.

21. If the diagonals of a parallelogram are perpendicular, then the parallelogram is a rhombus.

22. If the diagonals of a parallelogram are congruent and perpendicular, then the parallelogram is a square.

23. If the midpoints of the sides of a rectangle are joined in order, then the quadrilateral formed is a rhombus.

*24. The length of the median of a trapezoid equals one-half the sum of the lengths of the two bases.

25. The median of a trapezoid is parallel to each base.

26. If the midpoints of the sides of an isosceles trapezoid are joined in order, then the quadrilateral formed is a rhombus.

27. *Given:* $\overline{AB} \parallel \overline{DC}$
$m\angle A = m\angle B = 56°$
$\overline{CE} \parallel \overline{DA}$ and $\overrightarrow{CF}$ bisects $\angle DCB$

Find: $m\angle FCE$

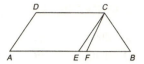

28. (a) Argue that the midpoint of the hypotenuse of a right triangle is equidistant from the three vertices of the triangle. Use the fact that the congruent diagonals of a rectangle bisect each other. Be sure to provide a drawing.

(b) Use the relationship from part a to find *CM*, the length of the median to the hypotenuse of right $\triangle ABC$, in which $m\angle C = 90°$, $AC = 6$, $BC = 8$, and $AB = 10$.

4.6
INEQUALITIES IN A TRIANGLE

Recall that $a > b$ if and only if there is a positive number p for which $a = b + p$. In several instances, geometric inequalities are obvious; some of these cases are now cited.

> **CASE 1**   If *B* is between *A* and *C* on $\overline{AC}$, then $AC > AB$ and $AC > BC$. (The measure of a segment is greater than the measure of any of its parts.)

$AC = AB + BC$ and, since $BC > 0$, it follows that $AC > AB$. Similarly, $AC > BC$. These relationships follow logically from the definition of $a > b$, and each is obvious in the preceding drawing.   ▲

> **CASE 2**   If $\overrightarrow{BD}$ separates $\angle ABC$ into two parts, then $m\angle ABC > m\angle 1$ and $m\angle ABC > m\angle 2$. (The measure of an angle is greater than the measure of any of its parts.)

You know that $m\angle ABC = m\angle 1 + m\angle 2$. Because $m\angle 2 > 0$, it follows that $m\angle ABC > m\angle 1$. It also follows that $m\angle ABC > m\angle 2$.   ▲

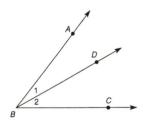

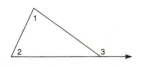

> **CASE 3**   If $\angle 3$ is an exterior angle of a triangle, then $m\angle 3 > m\angle 1$ and $m\angle 3 > m\angle 2$. (The measure of an exterior angle of a triangle is greater than the measure of either of the nonadjacent interior angles.)

$m\angle 3 = m\angle 1 + m\angle 2$, so $m\angle 3 > m\angle 1$ and $m\angle 3 > m\angle 2$.   ▲

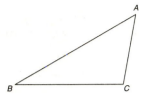

> **CASE 4**   In △*ABC*, if ∠*C* is a right angle or an obtuse angle, then m∠*C* > m∠*A* and m∠*C* > m∠*B*. (If a triangle contains a right or an obtuse angle, then the measure of this angle is greater than the measure of either of the remaining angles.)

While a proof of this could be written, the following is an informal explanation. You know that

$$m\angle A + m\angle B + m\angle C = 180°$$

With m∠*C* ≥ 90°, it follows that m∠*A* + m∠*B* ≤ 90°, and each angle (∠*A* and ∠*B*) must be acute. Thus m∠*C* > m∠*A* and m∠*C* > m∠*B*.   ▲

Sometimes it is useful to prove a particular theorem so that it can be used to help complete the proof of other theorems. Such a theorem is a **lemma.**

> **LEMMA 4.6.1 (Addition Property of Inequality)**   If *a* > *b* and *c* > *d*, then *a* + *c* > *b* + *d*.

Given: *a* > *b* and *c* > *d*

Prove: *a* + *c* > *b* + *d*

PROOF

Statements	Reasons
1. *a* > *b* and *c* > *d*	**1.** Given
2. *a* = *b* + p_1 and *c* = *d* + p_2, where p_1 and p_2 are positive	**2.** *a* > *b* if and only if there is a positive number *p* for which *a* = *b* + *p*.
3. *a* + *c* = (*b* + *d*) + (p_1 + p_2)	**3.** Addition Property of Equality
4. *a* + *c* = (*b* + *d*) + p_3, where p_1 + p_2 = p_3	**4.** Substitution
5. *a* + *c* > *b* + *d*	**5.** *a* > *b* if and only if there is a positive number *p* for which *a* = *b* + *p*. ▲

Lemma 4.6.1 is used in the proof of Example 1. Notice that the proof is less formal here, in that reasons may not always be stated in the paragraph style of proof.

EXAMPLE 1   Give a paragraph proof for the following problem.

Given: *AB* > *CD* and *BC* > *DE*

Prove: *AC* > *CE*

Proof: If $AB > CD$ and $BC > DE$, then $AB + BC > CD + DE$, by Lemma 4.6.1. But $AB + BC = AB$ and $CD + DE = CE$. By substitution, it follows that $AC > CE$.   ▲

You may prefer to state the second statement in the preceding proof as "But $AB + BC = AB$ and $CD + DE = CE$, by the Segment-Addition Postulate." Because this postulate has been used so many times by now, the reasoning behind it may be second nature. Less obvious supporting properties should always be stated. Remember that the proof must be written for the reader to follow and understand easily.

The paragraph proof in Example 1 uses the same logical and sequential flow as the earlier proofs. Its lack of formality should not be taken to suggest that each claim need not be supported. When a reason is obvious, it may not be stated, but every claim must be justifiable and ordered. Keeping these thoughts in mind may help you write improved paragraphs in an English class as well!

Example 2 illustrates the use of an informal proof for a theorem. Before supplying the proof, you must still provide the statement of theorem, a drawing, the given, and the prove.

EXAMPLE 2   Given an informal proof of Theorem 4.6.1.

THEOREM 4.6.2   If one side of a triangle is longer than a second side, then the measure of the angle opposite the first side is greater than the measure of the angle opposite the second side.

Given: $\triangle ABC$, with $AC > BC$

Prove: $m\angle B > m\angle A$

Proof: Given $\triangle ABC$ with $AC > BC$, locate point D on $\overline{AC}$ so that $\overline{CD} \cong \overline{BC}$, as in figure 4.37b. Now $m\angle 2 = m\angle 5$ in the isosceles triangle. It is obvious that $m\angle ABC > m\angle 2$; therefore, $m\angle ABC > m\angle 5$ (*). But $m\angle 5 > m\angle A$ (*) since $\angle 5$ is an exterior angle of $\triangle ADB$. Using the two starred statements, we can conclude by the Transitive Property of Inequality that $m\angle ABC > m\angle A$; that is, $m\angle B > m\angle A$ in figure 4.37a.   ▲

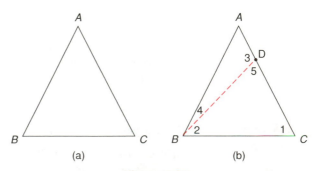

(a)   (b)

Figure 4.37

The relationship just described extends, of course, to all sides and all angles of a triangle. That is, the largest of the three angles of a triangle is opposite the longest side and the smallest angle is opposite the shortest side.

EXAMPLE 3 Given that the three sides of $\triangle ABC$ are $AB = 4$, $BC = 5$, and $AC = 6$, arrange the angles by size.

Solution Because $AC > BC > AB$, the largest angle lies opposite $\overline{AC}$. The angle intermediate in size lies opposite $\overline{BC}$, while the smallest angle lies opposite $\overline{AB}$. Thus, the order of the angles by size is

$$\text{m}\angle B > \text{m}\angle A > \text{m}\angle C \qquad \blacktriangle$$

The converse of Theorem 4.6.2 is also true. It is necessary, however, to use an indirect proof to establish the converse. Recall that with this method of proof we begin by supposing the exact opposite of what we want to show. When this assumption leads us to a contradiction, we know that the assumption must be faulty and that the desired claim is therefore true.

EXAMPLE 4 Prove Theorem 4.6.3 by using an indirect approach and the paragraph form.

> **THEOREM 4.6.3** If the measure of one angle of a triangle is greater than the measure of a second angle, then the side opposite the first angle is longer than the side opposite the second angle.

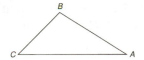

Given: $\triangle ABC$ with $\text{m}\angle B > \text{m}\angle A$

Prove: $AC > BC$

Proof: Given $\triangle ABC$ with $\text{m}\angle B > \text{m}\angle A$, assume that $AC \leq BC$. But if $AC = BC$, then $\text{m}\angle B = \text{m}\angle A$, which contradicts the hypothesis. Also, if $AC < BC$, then it follows by the previous theorem that $\text{m}\angle B < \text{m}\angle A$, which also contradicts the hypothesis. Thus the supposed statement must be false, and it follows that $AC > BC$. $\qquad \blacktriangle$

EXAMPLE 5 Given $\triangle RST$ in which $\text{m}\angle R = 90°$, $\text{m}\angle S = 60°$, and $\text{m}\angle T = 30°$, write an inequality that compares the lengths of the sides.

Solution With $\text{m}\angle R > \text{m}\angle S > \text{m}\angle T$, it follows that the sides opposite these $\angle$s are unequal in the same order. That is,

$$ST > RT > SR \qquad \blacktriangle$$

The following corollary is a consequence of Theorem 4.6.3.

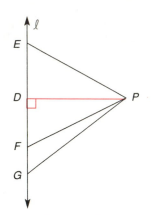

Figure 4.38

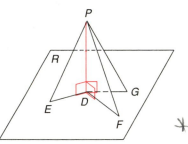

Figure 4.39

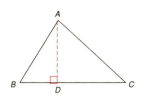

> **COROLLARY 4.6.4**   The perpendicular segment from a point to a line is the shortest segment that can be drawn from the point to the line.

In figure 4.38, $PD < PE$, $PD < PF$, and $PD < PG$. In every case, $\overline{PD}$ is opposite an acute angle of a triangle, while the second segment is always opposite a right angle (necessarily the largest angle of the triangle involved).

Corollary 4.6.4 can easily be extended to three dimensions.

> **COROLLARY 4.6.5**   The perpendicular segment from a point to a plane is the shortest segment that can be drawn from the point to the plane.

In figure 4.39, $\overline{PD}$ has a length less than that of $\overline{PE}$, $\overline{PF}$, or $\overline{PG}$.

Our final theorem shows that no side of a triangle can have a length greater than or equal to the sum of the other two sides. In the proof, the relationship is validated for only one of three possible inequalities. Theorem 4.6.6 is often called the Triangle Inequality.

> **THEOREM 4.6.6** (**Triangle Inequality**)   The sum of the lengths of any two sides of a triangle is greater than the length of the third side.

Given: $\triangle ABC$

Prove: $BA + CA > BC$

Proof: Draw $\overline{AD} \perp \overline{BC}$. Since the shortest segment from a point to a line is the perpendicular segment, $BA > BD$ and $CA > CD$. By adding the stated inequalities, $BA + CA > BD + CD$. By the Segment-Addition Postulate, the sum $BD + DC$ can be replaced by BC, to give the result $BA + CA > BC$.   ▲

EXAMPLE 6   Can a triangle have sides of the following lengths?
(**a**) 3, 4, and 5
(**b**) 3, 4, and 7
(**c**) 3, 4, and 8

Solution   (**a**) Yes, since no side has a length greater than or equal to the sum of the lengths of the other two sides
(**b**) No, because $7 = 3 + 4$
(**c**) No, since $8 > 3 + 4$ ▲

From Example 6, you can see that the length of one side cannot be greater than or equal to the sum of the lengths of the other two sides.

4.6 EXERCISES

In Exercises 1 to 10, classify each statement as true or false.
In Exercises 1 and 2, use the accompanying figure.

1. $\overline{AB}$ is the longest side of $\triangle ABC$.

2. $AB < BC$

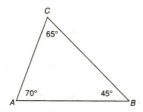

In Exercises 3 and 4, use the accompanying figure.

3. $DB > AB$

4. Because $m\angle A = m\angle B$, it follows that $DA = DC$.

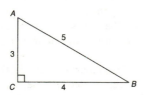

5. $m\angle A + m\angle B = m\angle C$

6. $m\angle A > m\angle B$

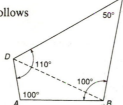

7. $DF > DE + EF$

8. If $\overrightarrow{DG}$ is the bisector of $\angle EDF$, then $DG > DE$.

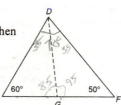

9. $DA > AC$

10. $CE = ED$

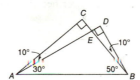

11. If possible, draw a triangle whose angles measure:
  (a) 100°, 100°, and 60°
  (b) 45°, 45°, and 90°

12. If possible, draw a triangle whose sides measure:
  (a) 7, 7, and 14

  (b) 6, 7, and 14
  (c) 6, 7, and 8

In Exercises 13 to 16, construct proofs.

13. *Given:* $m\angle ABC > m\angle DBE$
  $m\angle CBD > m\angle EBF$
  Prove: $m\angle ABD > m\angle DBF$

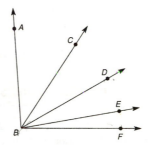

14. *Given:* Equilateral $\triangle ABC$
  Prove: $DA > AC$

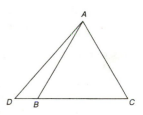

15. *Given:* Quadrilateral $RSTU$ with diagonal $\overline{US}$
  $\angle R$ and $\angle TUS$ are right $\angle$s
  Prove: $TS > UR$

16. *Given:* Isosceles trapezoid $ABCD$ with $\overline{AB} \parallel \overline{CD}$ and $m\angle A > m\angle D$
  Prove: $DC > AB$

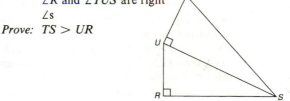

17. For $\triangle ABC$ and $\triangle DEF$, suppose that $\overline{AC} \cong \overline{DF}$ and $\overline{AB} \cong \overline{DE}$, but that $m\angle A < m\angle D$. Draw a conclusion regarding $\overline{BC}$ and $\overline{EF}$.

18. In $\triangle MNP$, $\overrightarrow{MQ}$ bisects $\angle NMP$ and $MN < MP$. Draw a conclusion about the relative lengths of $\overline{NQ}$ and $\overline{QP}$.

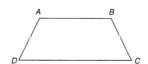

In Exercises 19 to 22, use the fact that "The sum of the lengths of two sides of a triangle is greater than the length of the third side."

19. The sides of a triangle have lengths of 4, 6, and x. Write an inequality that states the possible values of x.

20. The sides of a triangle have lengths of 7, 13, and x. As in Exercise 19, write an inequality that describes possible values of x.

21. If the lengths of two sides of a triangle are represented by $2x + 5$ and $3x + 7$ (in which x is positive), describe in terms of x the possible lengths of the third side whose length is represented by y.

22. Prove by the indirect method: "The diagonal of a square

is not equal in length to the length of any of the sides of the square."

23. Prove the following statement by the indirect method.

Given: $\triangle MPN$ is not isosceles
Prove: $PM \neq PN$

In Exercises 24 and 25, prove each theorem.

24. The length of the median of a trapezoid is less than one-half the sum of the lengths of the four sides of the trapezoid.

25. The length of an altitude of a triangle that does not contain a right angle is less than the length of either side containing the same vertex as the altitude.

▲ A Look Beyond: Historical Sketch of Archimedes

While Euclid (see *A Look Beyond,* Chapter 2) was a great teacher and wrote so that the majority might understand the principles of geometry, Archimedes wrote only for the very well-educated mathematicians and scientists of his day. Archimedes (287 B.C.–212 B.C.) wrote on such topics as the measure of the circle, the quadrature of the parabola, and spirals. In his works, Archimedes found a very good approximation of π. His other geometric works included investigations of conic sections and spirals, and he also wrote about physics. He was a great inventor and is probably remembered more for his inventions than for his writing.

Several historical events concerning the life of Archimedes have been substantiated, and one account involves his detection of a dishonest goldsmith. In that story, Archimedes was called upon to determine whether the crown that had been ordered by the king was constructed entirely of gold. By applying the principle of hydrostatics (which he had discovered), Archimedes established that the goldsmith had not constructed the crown entirely of gold. (The principle of hydrostatics states that an object placed in a fluid displaces an amount of fluid equal in weight to the amount of weight the object loses while submerged.)

One of his inventions is known as Archimedes' screw. This device allows water to flow from one level to a higher level so that, for example, holds of ships can be emptied of water. Archimedes' screw was used in Egypt to drain fields when the Nile River overflowed its banks.

When Syracuse (where Archimedes lived) came under siege by the Romans, Archimedes designed a long-range catapult that was so effective that Syracuse was able to fight off the powerful Roman army for three years before being overcome.

One report concerning the inventiveness of Archimedes has been treated as false, because his result has not been duplicated. It was said that he designed a wall of mirrors that could focus and reflect the sun's heat with such intensity as to set fire to Roman ships at sea. Because recent experiments with concave mirrors have failed to produce such intense heat, the account is difficult to believe.

Archimedes eventually died at the hands of a Roman soldier, even though the Roman army had been given orders not to harm him. After his death, the Romans honored his brilliance with a tremendous monument displaying the figure of a sphere inscribed in a right circular cylinder.

▲ Summary

A Look Back at Chapter 4

The goal of this chapter has been to develop several methods for proving triangles congruent. Inequality relationships for the sides and angles of a triangle were also developed. Quadrilaterals were named and classified according to their properties.

A Look Ahead to Chapter 5

In the next chapter, similarity will be defined for all polygons, with an emphasis on triangles. The Pythagorean Theorem will be proved, and special right triangles will be discussed.

IMPORTANT TERMS AND CONCEPTS OF CHAPTER 4

4.1 Congruent Triangles
SSS, SAS, ASA, AAS
Included Side, Included Angle
Reflexive Property of Congruence (Identity)
Hypotenuse and Legs of a Right Triangle
4.2 HL
CPCTC
4.3 Isosceles Triangle
Vertex, Legs, and Base of an Isosceles Triangle
Base Angles, Vertex Angle

4.4 Quadrilateral
Parallelogram
Kite
4.5 Rectangle
Rhombus
Trapezoid (Bases, Legs, Median, and Base Angles)
Isosceles Trapezoid
Square
4.6 Lemma
Triangle Inequality

A Look Beyond Historical Sketch of Archimedes

▲ **REVIEW EXERCISES**

1. *Given:* ∠AEB ≅ ∠DEC
   $\overline{AE} ≅ \overline{ED}$
   Prove: △AEB ≅ △DEC

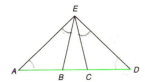

2. *Given:* $\overline{AB} = \overline{EF}$
   $\overline{AC} = \overline{DF}$
   ∠1 ≅ ∠2
   Prove: ∠B ≅ ∠E

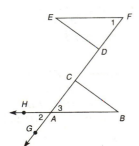

3. *Given:* $\overline{AD}$ bisects $\overline{BC}$
   $\overline{AB} ⊥ \overline{BC}$
   $\overline{DC} ⊥ \overline{BC}$
   Prove: $\overline{AE} ≅ \overline{ED}$

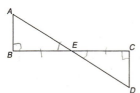

4. *Given:* ∠BAD ≅ ∠CDA
   $\overline{AB} ≅ \overline{CD}$
   Prove: $\overline{AE} ≅ \overline{ED}$

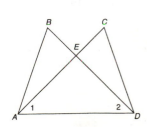

5. *Given:* $\overline{BE}$ is the altitude to $\overline{AC}$
   $\overline{AD}$ is the altitude to $\overline{CE}$
   $\overline{BC} ≅ \overline{CD}$
   Prove: $\overline{BE} ≅ \overline{AD}$

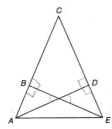

6. *Given:* $\overline{OA} ≅ \overline{OB}$
   $\overline{OC}$ is the median to $\overline{AB}$
   Prove: $\overline{OC} ⊥ \overline{AB}$

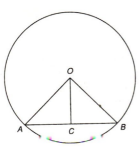

7. *Given:* $\overline{AB} ≅ \overline{DE}$
   $\overline{AB} ∥ \overline{DE}$
   $\overline{AF} ≅ \overline{DC}$
   Prove: $\overline{BC} ∥ \overline{FE}$

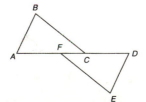

8. *Given:* *B* is the midpoint of $\overline{AC}$
$\overline{BD} \perp \overline{AC}$
Prove: $\triangle ADC$ is isosceles

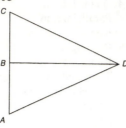

9. *Given:* $\overline{JM} \perp \overline{GM}$
$\overline{GK} \perp \overline{KJ}$
Prove: $\angle G \cong \angle J$

10. *Given:* $\overline{TN} \cong \overline{TR}$
$\overline{TO} \perp \overline{NP}$
$\overline{TS} \perp \overline{PR}$
$\overline{TO} \cong \overline{TS}$
Prove: $\angle N \cong \angle R$

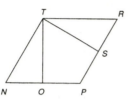

11. *Given:* $\overline{AC} \cong \overline{AE}$
$\angle CBD \cong \angle EFD$
D is the midpoint of $\overline{CE}$
Prove: $\overline{BD} \cong \overline{DF}$

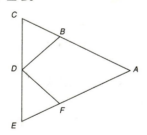

12. *Given:* $\overline{AF} \cong \overline{GC}$
$\overline{AB} \cong \overline{BC}$
D is the midpoint of $\overline{AB}$
E is the midpoint of $\overline{BC}$
Prove: $\overline{DF} \cong \overline{EG}$

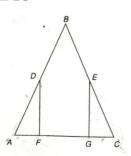

13. *Given:* $\overline{YZ}$ is the base of an isosceles triangle
$\overrightarrow{XA} \parallel \overline{YZ}$
Prove: $\angle 1 \cong \angle 2$

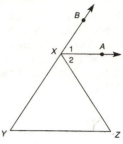

14. *Given:* $\overline{AB} \cong \overline{CD}$
$\angle BAD \cong \angle CDA$
Prove: $\triangle AED$ is isosceles

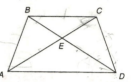

15. *Given:* $\overline{AB} \parallel \overline{DC}$
$\overline{AB} \cong \overline{DC}$
C is the midpoint of $\overline{BE}$
Prove: $\overline{AC} \parallel \overline{DE}$

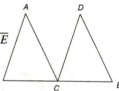

16. *Given:* $ABCD$ is a $\square$
$\overline{AF} \cong \overline{CE}$
Prove: $\overline{DF} \parallel \overline{EB}$

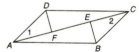

17. *Given:* $ABEF$ is a rectangle
$BCDE$ is a rectangle
$\overline{FE} \cong \overline{ED}$
Prove: $\overline{AE} \cong \overline{BD}$ and $\overline{AE} \parallel \overline{BD}$

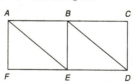

18. *Given:* $\overline{DE}$ is a median in $\triangle ADC$
$\overline{BE} \cong \overline{FD}$
$\overline{EF} \cong \overline{FD}$
Prove: $ABCF$ is a $\square$

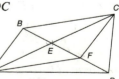

19. *Given:* $\triangle FAB \cong \triangle HCD$
$\triangle EAD \cong \triangle GCB$
Prove: $ABCD$ is a $\square$

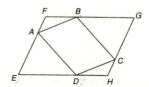

20. *Given:* △*TWX* is isosceles, with
base $\overline{WX}$
$\overline{RY} \parallel \overline{WX}$

Prove: *RWXY* is an isosceles trapezoid

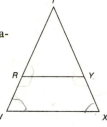

21. *Given:* $\overrightarrow{AC}$ bisects ∠*BAD*
Prove: *AD > CD*

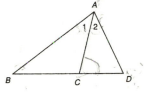

22. *Given:* *ABCD* is a parallelogram
$\overline{DC} \cong \overline{BN}$
∠3 ≅ ∠4

Prove: *ABCD* is a rhombus

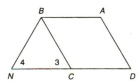

State whether the statements in Review Exercises 23 to 31 are always true (A), sometimes true (S), or never true (N).

23. A square is a rectangle.

24. If two of the angles of a trapezoid are congruent, then the trapezoid is isosceles.

25. The diagonals of a trapezoid bisect each other.

26. The diagonals of a parallelogram are perpendicular.

27. A rectangle is a rhombus.

28. Segments with lengths 10, 19, and 33 cm are sides of a triangle.

29. The diagonals of a square are perpendicular.

30. Two consecutive angles of a parallelogram are supplementary.

31. Opposite angles of a rhombus are congruent.

32. In △*PQR*, m∠*P* = 67° and m∠*Q* = 23°.

  (a) Name the shortest side.
  (b) Name the longest side.

33. In △*ABC*, m∠*A* = 40° and m∠*B* = 65°. List the sides in order of their lengths, starting with the smallest side.

34. In △*PQR*, *PQ* = 1.5, *PR* = 2, and *QR* = 2.5. List the angles in order of size, starting with the smallest angle.

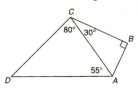

35. Name the longest segment in the accompanying figure.

36. Which of the following can be lengths of sides of a triangle?

  (a) 3, 6, 9   **(b)** 4, 5, 8   **(c)** 2, 3, 8

37. Two sides of a triangle have lengths 15 and 20. The length of the third side can be any number between _____ and _____ .

38. *Given:* $\overline{DB} \perp \overline{AC}$
$\overline{AD} \cong \overline{DC}$
m∠*C* = 70°

  Find: m∠*ADB*

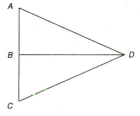

39. *Given:* $\overline{AB} \cong \overline{BC}$
∠*DAC* ≅ ∠*BCD*
m∠*B* = 50°

  Find: m∠*ADC*

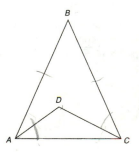

40. *Given:* ▱*ABCD*
m∠*A* = 2*x* + 6
m∠*B* = *x* + 24

  Find: m∠*C*

41. *Given:* *ABCD* is a parallelogram
m∠*A* = 4*x*
m∠*C* = 2*x* + 50

  Find: m∠*A* and m∠*D*

In Review Exercises 42 to 46, use the accompanying figure.

42. *Given:* △*ABC* is isosceles with base $\overline{AB}$
m∠2 = 3*x* + 10
m∠4 = $\frac{5}{2}x$ + 18

  Find: m∠*C*

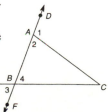

43. *Given:* $\triangle ABC$ with perimeter 40
    $AB = 10$
    $BC = x + 6$
    $AC = 2x - 3$
    Find: Whether $\triangle ABC$ is scalene, isosceles, or equilateral

44. *Given:* $\triangle ABC$ is isosceles
    $AB = y + 7$
    $BC = 3y + 5$
    $AC = 9 - y$
    Find: Whether $\triangle ABC$ is also equilateral

45. *Given:* $\triangle ABC$ is isosceles, and its perimeter is less than 44
    $AB = 10$
    $AC = x + 7$
    $BC = 2x - 8$
    Find: Which side of $\triangle ABC$ is the base

46. *Given:* $\overline{AC}$ and $\overline{BC}$ are the legs of isosceles $\triangle ABC$
    $m\angle 1 = 5x$
    $m\angle 3 = 2x + 12$
    Find: $m\angle 2$

In Review Exercises 47 to 49, M and N in the accompanying figure are the midpoints of $\overline{FJ}$ and $\overline{FH}$, respectively.

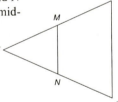

47. *Given:* Isosceles $\triangle FJH$ with
    $\overline{FJ} \cong \overline{FH}$
    $FM = 2y + 3$
    $NH = 5y - 9$
    $JH = 2y$
    Find: The perimeter of $\triangle FMN$

48. *Given:* $JH = 12$
    $m\angle J = 80$
    $m\angle F = 60$
    Find: MN, $m\angle FMN$, $m\angle FNM$

49. *Given:* $MN = x^2 + 6$
    $JH = 2x(x + 2)$
    Find: x, MN, JH

50. Construct another isosceles triangle, in which the base angles are half as large as the given base angles.

51. Construct a rhombus, given these lengths for the diagonals.

5 SIMILAR TRIANGLES

The picture on the right is an enlargement of the one on the left. The proportionality of the person's features tells us of the relationship between the two portraits. Likewise the rectangular frame on the right is an enlargement of the other frame. In geometry, we say that the two rectangles (shapes of the frames) are similar. In this chapter, we study similar figures while emphasizing the similarity of triangles.

The early work of this chapter may be a review of algebra for some students. If you have studied this material elsewhere, you may be able to omit Section 5.1 in your study of this chapter.

CHAPTER OUTLINE

5.1 Quadratic Equations and Solutions

5.2 Ratio and Proportion

5.3 Similar Polygons and Triangles

5.4 Segments Divided Proportionally

5.5 Similar Right Triangles and the Pythagorean Theorem

5.6 Special Right Triangles

A Look Beyond An Unusual Application of Similar Triangles

 5.1 QUADRATIC EQUATIONS AND SOLUTIONS

An equation that can be written in the form

$$ax^2 + bx + c = 0 \qquad a \neq 0$$

is a **quadratic equation.** For example, the following three equations are quadratic:

$$x^2 - 7x + 12 = 0 \qquad 6x^2 = 7x + 3 \qquad 4x^2 = 9$$

Many quadratic equations can be solved by a factoring method that depends on the Zero Product Property.

ZERO PRODUCT PROPERTY   If $a \cdot b = 0$, then $a = 0$ or $b = 0$.

When this property is stated in words, it reads: "If the product of two expressions equals 0, then at least one of the factors must equal 0."

EXAMPLE 1   Solve $x^2 - 7x + 12 = 0$.

Solution   First you must factor the polynomial; then check the factors by using the FOIL method of multiplication.

$$(x - 3)(x - 4) = 0 \qquad \text{factoring}$$
$$x - 3 = 0 \quad \text{or} \quad x - 4 = 0 \qquad \text{Zero Product Property}$$
$$x = 3 \quad \text{or} \quad x = 4 \qquad \text{Addition Property}$$

To check $x = 3$, substitute into the given equation:

$$3^2 - (7 \cdot 3) + 12 = 9 - 21 + 12 = 0$$

Similarly, to check when $x = 4$, substitute again:

$$4^2 - (7 \cdot 4) + 12 = 16 - 28 + 12 = 0$$

Checks for subsequent problems generally will not be provided. The solutions are usually expressed as a set, $\{3, 4\}$.   ▲

If you were asked to solve either quadratic equation

$$6x^2 = 7x + 3 \qquad \text{or} \qquad 4x^2 = 9$$

it would be necessary to change the equation so that one side would be equal to 0. The form $ax^2 + bx + c = 0$ is the **standard form** of a quadratic equation.

EXAMPLE 2   Solve $6x^2 = 7x + 3$.

Solution   First changing to standard form, you have

$$6x^2 - 7x - 3 = 0 \qquad \text{standard form}$$
$$(2x - 3)(3x + 1) = 0 \qquad \text{factoring}$$
$$2x - 3 = 0 \quad \text{or} \quad 3x + 1 = 0 \qquad \text{Zero Product Property}$$
$$2x = 3 \quad \text{or} \quad 3x = -1 \qquad \text{Addition-Subtraction Property}$$
$$x = \frac{3}{2} \quad \text{or} \quad x = \frac{-1}{3} \qquad \text{division}$$

Therefore $\left\{ \dfrac{3}{2}, \dfrac{-1}{3} \right\}$ is the solution set.   ▲

In some instances, a common factor can be extracted from each term in the factoring step. In the example that follows, the common factor is the constant 6, which cannot be set equal to 0 (that is, $6 \neq 0$). The factoring is generally easier when the common factor has been removed because there are fewer "guesses" with the smaller numbers that remain.

EXAMPLE 3   Solve $6x^2 + 30x = 144$.

Solution
$$6x^2 + 30x - 144 = 0$$
$$6(x^2 + 5x - 24) = 0$$
$$6(x + 8)(x - 3) = 0$$

$$x + 8 = 0 \qquad \text{or} \qquad x - 3 = 0$$
$$x = -8 \qquad \text{or} \qquad x = 3$$

The solution set is $\{-8, 3\}$.   ▲

Equations such as $4x^2 = 9$ and $4x^2 - 12x = 0$ are incomplete quadratic equations because one term is missing from the standard form; the linear term (having exponent 1) is missing from the first equation, and the constant term is omitted in the second. Either equation can, however, be solved by factoring; in particular, the factoring is given by

$$4x^2 - 9 = (2x + 3)(2x - 3)$$
while
$$4x^2 - 12x = 4x(x - 3)$$

When solutions to $ax^2 + bx + c = 0$ cannot be found by factoring, they may be determined by the following formula:

QUADRATIC FORMULA $x = \dfrac{-b \pm \sqrt{b^2 - 4ac}}{2a}$ are solutions for

$ax^2 + bx + c = 0$, where $a \neq 0$.

Although the formula may provide two solutions for the equation, an application problem in geometry may have a single positive solution representing a segment (or angle) measure. Recall that for $a > 0$, $\sqrt{a}$ represents the principal square root of a. That is,

$$\sqrt{a} = \text{the positive number for which } (\sqrt{a})^2 = a$$

EXAMPLE 4   Simplify each expression, if possible:
(a) $\sqrt{16}$ (b) $\sqrt{0}$ (c) $\sqrt{7}$ (d) $\sqrt{400}$ (e) $\sqrt{-4}$

Solution
(a) $\sqrt{16} = 4$ because $4^2 = 16$
(b) $\sqrt{0} = 0$ because $0^2 = 0$
(c) $\sqrt{7}$ cannot be simplified
(d) $\sqrt{400} = 20$ because $20^2 = 400$
(e) $\sqrt{-4}$ is not a real number   ▲

In expressions such as $\sqrt{9 + 16}$ and $\sqrt{4 + 9}$, you must simplify within the radical symbol first and then extract the square root. Thus

$$\sqrt{9 + 16} = 5 \quad \text{and} \quad \sqrt{4 + 9} = \sqrt{13}$$

Looking back to part (d), Example 4, you may not find it obvious that the square root of 400 is 20. When a larger number appears in the position of radicand (number under the square root sign), you may be able to use the Product Property of Square Roots to simplify.

> **PRODUCT PROPERTY OF SQUARE ROOTS**   For $a \geq 0$ and $b \geq 0$, $\sqrt{a \cdot b} = \sqrt{a} \cdot \sqrt{b}$.

Since $\sqrt{400} = \sqrt{4 \cdot 100}$, it follows by the Product Property of Square Roots that

$$\sqrt{400} = \sqrt{4 \cdot 100} = \sqrt{4} \cdot \sqrt{100} = 2 \cdot 10 = 20$$

Not only will this property allow you to simplify integer results, it will also allow you to reduce the size of the radicand (much like a fraction reduced to lowest terms). To accomplish this, choose a factor that is the largest possible perfect square. That is, choose the largest factor from the list 4, 9, 16, 25, 36, 49, 64, 81, 100, 121, . . . As an example,

$$\begin{aligned} \sqrt{45} &= \sqrt{9 \cdot 5} \\ &= \sqrt{9} \cdot \sqrt{5} \\ &= 3\sqrt{5} \end{aligned}$$

The radicand has now been reduced from 45 to 5.

EXAMPLE 5   Simplify the following radicals:
(a) $\sqrt{27}$ (b) $\sqrt{125}$ (c) $\sqrt{196}$

Solution   (a) 9 is the largest perfect square factor of 27. Therefore

$$\sqrt{27} = \sqrt{9 \cdot 3} = \sqrt{9} \cdot \sqrt{3} = 3\sqrt{3}$$

(b) 25 is a perfect square factor of 125. Therefore

$$\sqrt{125} = \sqrt{25 \cdot 5} = \sqrt{25} \cdot \sqrt{5} = 5\sqrt{5}$$

(c) 4 is a perfect square factor of 196. In fact, $196 = 4 \cdot 49$. Thus

$$\sqrt{196} = \sqrt{4 \cdot 49} = \sqrt{4} \cdot \sqrt{49} = 2 \cdot 7 = 14 \qquad \blacktriangle$$

In some instances, the symmetric form of the Product Property of Square Roots can be used to simplify products of two or more radicals. While this is possible, it is not the main focus of the property. The symmetric form looks like this:

$$\sqrt{a} \cdot \sqrt{b} = \sqrt{a \cdot b} \qquad \text{where } a \geq 0 \text{ and } b \geq 0$$

You can replace the expression $\sqrt{2} \cdot \sqrt{3}$ with $\sqrt{2 \cdot 3}$ or with $\sqrt{6}$ by the preceding property. Even if a calculator is available as you work with radicals, it is still better to simplify first since there is less chance of an error when you perform fewer key-ins.

EXAMPLE 6 Simplify the following products:
(a) $\sqrt{5} \cdot \sqrt{3}$ (b) $\sqrt{2} \cdot \sqrt{8}$ (c) $\sqrt{2} \cdot \sqrt{3} \cdot \sqrt{6}$

Solution (a) $\sqrt{5} \cdot \sqrt{3} = \sqrt{5 \cdot 3} = \sqrt{15}$
(b) $\sqrt{2} \cdot \sqrt{8} = \sqrt{2 \cdot 8} = \sqrt{16} = 4$
(c) $\sqrt{2} \cdot \sqrt{3} \cdot \sqrt{6} = (\sqrt{2} \cdot \sqrt{3})\sqrt{6} = \sqrt{6} \cdot \sqrt{6} = \sqrt{36} = 6$ ▲

When solutions to quadratic equations cannot be found by factoring, they are often found by using the Quadratic Formula. Recall that the equation must be in standard form:

$$ax^2 + bx + c = 0$$

Values for a, b, and c are then substituted into the formula. The procedure is illustrated in Example 7. Notice that the radical has been simplified.

EXAMPLE 7 Solve $x^2 - 4x - 6 = 0$.

Solution Values are $a = 1$ (understood), $b = -4$, and $c = -6$. Therefore

$$x = \frac{-b \pm \sqrt{b^2 - 4ac}}{2a}$$

$$= \frac{4 \pm \sqrt{4^2 - [4 \cdot 1(-6)]}}{2(1)}$$

$$= \frac{4 \pm \sqrt{16 + 24}}{2}$$

$$= \frac{4 \pm \sqrt{40}}{2}$$

$$= \frac{4 \pm \sqrt{4 \cdot 10}}{2}$$

$$= \frac{4 \pm 2\sqrt{10}}{2}$$

$$= \frac{2(2 \pm \sqrt{10})}{2}$$

$$= 2 \pm \sqrt{10}$$

Thus $\{2 + \sqrt{10}, 2 - \sqrt{10}\}$ is the solution set.   ▲

The Product Property of Square Roots has a counterpart for quotients, which we will use in Examples 8 and 9.

> **QUOTIENT PROPERTY OF SQUARE ROOTS**   For $a \geq 0$ and $b > 0$,
> $$\sqrt{\frac{a}{b}} = \frac{\sqrt{a}}{\sqrt{b}}.$$

EXAMPLE 8   Simplify the following quotients:

(a) $\sqrt{\dfrac{16}{9}}$ (b) $\sqrt{\dfrac{3}{4}}$ (c) $\sqrt{\dfrac{3}{5}}$

Solution   (a) $\sqrt{\dfrac{16}{9}} = \dfrac{\sqrt{16}}{\sqrt{9}} = \dfrac{4}{3}$

(b) $\sqrt{\dfrac{3}{4}} = \dfrac{\sqrt{3}}{\sqrt{4}} = \dfrac{\sqrt{3}}{2}$

(c) $\sqrt{\dfrac{3}{5}} = \dfrac{\sqrt{3}}{\sqrt{5}}$

However, a radical is not generally left in the denominator. This can be remedied as follows:

$$\frac{\sqrt{3}}{\sqrt{5}} = \frac{\sqrt{3}}{\sqrt{5}} \cdot \frac{\sqrt{5}}{\sqrt{5}} = \frac{\sqrt{3 \cdot 5}}{\sqrt{5 \cdot 5}} = \frac{\sqrt{15}}{\sqrt{25}} = \frac{\sqrt{15}}{5}$$ ▲

In Example 8, the computation $\sqrt{5} \cdot \sqrt{5} = (\sqrt{5})^2 = 5$ illustrates that squaring and extracting the square root of a nonnegative number are inverse operations. Thus $(\sqrt{7})^2 = 7$ and $\sqrt{5^2} = 5$.

In some instances, it is convenient to solve a quadratic equation by the method of square roots:

> **SQUARE ROOTS PROPERTY**   If $x^2 = p$ where $p \geq 0$, then $x = \pm \sqrt{p}$.

This technique is convenient for solving a quadratic equation like $4x^2 = 9$. This equation is easily changed to $x^2 = \dfrac{9}{4}$, so

$$x = \pm \sqrt{\frac{9}{4}}$$
$$= \pm \frac{\sqrt{9}}{\sqrt{4}}$$
$$= \pm \frac{3}{2}$$

EXAMPLE 9   Solve $5r^2 - 12 = 0$.

Solution

$$5r^2 = 12$$

$$r^2 = \frac{12}{5}$$

$$r = \pm\frac{\sqrt{12}}{\sqrt{5}}$$

$$= \pm\frac{\sqrt{4} \cdot \sqrt{3}}{\sqrt{5}}$$

$$= \pm\frac{2\sqrt{3}}{\sqrt{5}}$$

$$= \pm\frac{2\sqrt{3}}{\sqrt{5}} \cdot \frac{\sqrt{5}}{\sqrt{5}}$$

$$= \pm\frac{2\sqrt{15}}{5}$$

Thus the solution set is $\left\{ \pm\dfrac{2\sqrt{15}}{5} \right\}$. ▲

In summary, quadratic equations have the form $ax^2 + bx + c = 0$ and are solved by one of the following methods:

1. Factoring, when the left member is easily factored
2. The Quadratic Formula,

$$x = \frac{-b \pm \sqrt{b^2 - 4ac}}{2a}$$

   when $ax^2 + bx + c = 0$ is not easily factored
3. The Square Roots Property, when $b = 0$.

Integers beneath radicals should be made as small as possible! Denominators should not contain radicals! The simplification of radicals can be achieved by using these properties:

For $a \geq 0, b \geq 0$ $\quad \sqrt{a \cdot b} = \sqrt{a} \cdot \sqrt{b}$ $\qquad$ Product Property

For $a \geq 0, b > 0$ $\quad \sqrt{\dfrac{a}{b}} = \dfrac{\sqrt{a}}{\sqrt{b}}$ $\qquad$ Quotient Property

---

In using these properties to simplify square root expressions, you should always do the following:

1. Leave the smallest possible integer under the square root symbol.
2. Leave no fractions with square roots in the denominator.
3. Leave no fractions under a square root symbol.

5.1 EXERCISES

1. Which equations are quadratic?

   (a) $2x^2 - 5x + 3 = 0$ (d) $\dfrac{1}{2}x^2 - \dfrac{1}{4}x - \dfrac{1}{8} = 0$

   (b) $x^2 = x^2 + 4$ (e) $\sqrt{2x - 1} = 3$

   (c) $x^2 = 4$ (f) $(x + 1)(x - 1) = 15$

2. Which equations are incomplete quadratic equations?

   (a) $x^2 - 4 = 0$ (d) $2x^2 - 4 = 2x^2 + 8x$

   (b) $x^2 - 4x = 0$ (e) $x^2 = \dfrac{9}{4}$

   (c) $3x^2 = 2x$ (f) $x^2 - 2x - 3 = 0$

3. Simplify each expression by using the Product Property of Square Roots:

   (a) $\sqrt{8}$ (d) $\sqrt{900}$

   (b) $\sqrt{45}$ (e) $(\sqrt{3})^2$

   (c) $\sqrt{3} \cdot \sqrt{27}$ (f) $\sqrt{3} \cdot \sqrt{6}$

4. Simplify each expression by using the Product Property of Square Roots:

   (a) $\sqrt{28}$ (d) $\sqrt{200}$

   (b) $\sqrt{32}$ (e) $\sqrt{5} \cdot \sqrt{7}$

   (c) $\sqrt{54}$ (f) $\sqrt{2} \cdot \sqrt{8}$

5. Simplify each expression by using the Quotient Property of Square Roots:

   (a) $\sqrt{\dfrac{9}{16}}$ (d) $\sqrt{\dfrac{6}{9}}$

   (b) $\sqrt{\dfrac{25}{49}}$ (e) $\sqrt{\dfrac{2}{3}}$

   (c) $\sqrt{\dfrac{7}{16}}$ (f) $\sqrt{\dfrac{5}{8}}$

6. Simplify each square root expression:

   (a) $\dfrac{10}{\sqrt{2}}$ (d) $\dfrac{\sqrt{6}}{\sqrt{3}}$

   (b) $\sqrt{\dfrac{2}{5}}$ (e) $\dfrac{\sqrt{27}}{\sqrt{3}}$

   (c) $\dfrac{10}{\sqrt{5}}$ (f) $\dfrac{\sqrt{2}}{\sqrt{10}}$

In Exercises 7 to 14, solve each quadratic equation by factoring.

7. $x^2 - 6x + 8 = 0$

8. $x^2 + 4x = 21$

9. $3x^2 - 51x + 180 = 0$ (*Hint:* There is a common factor.)

10. $2x^2 + x - 6 = 0$

11. $3x^2 = 10x + 8$

12. $8x^2 + 40x - 112 = 0$

13. $6x^2 = 5x - 1$

14. $12x^2 + 10x = 12$

In Exercises 15 to 22, solve each equation by using the Quadratic Formula. Simplify solutions as much as possible.

15. $x^2 - 7x + 10 = 0$

16. $x^2 + 7x + 12 = 0$

17. $x^2 + 9 = 7x$

18. $2x^2 + 3x = 6$

19. $x^2 - 4x - 8 = 0$

20. $x^2 - 6x - 2 = 0$

21. $5x^2 = 3x + 7$

22. $2x^2 = 8x - 1$

In Exercises 23 to 28, solve each incomplete quadratic equation. Use the Square Roots Property as needed.

23. $2x^2 = 14$

24. $2x^2 = 14x$

25. $4x^2 - 25 = 0$

26. $4x^2 - 25x = 0$

27. $ax^2 - bx = 0$

28. $ax^2 - b = 0$

29. The length of a rectangle is 3 more than its width. If the area of the rectangle is 40, the dimensions x and $x + 3$ can be found by solving the equation $x(x + 3) = 40$. Find these dimensions.

30. To find the lengths of $\overline{CP}$ (which is x), $\overline{PD}$, $\overline{AP}$, and $\overline{PB}$, one must solve the equation

$$x \cdot (x + 5) = (x + 1) \cdot 4$$

Find the length of $\overline{CP}$.

5.2 RATIO AND PROPORTION

A **ratio** is the quotient $\dfrac{a}{b}$ (where $b \neq 0$) that provides a comparison between numbers a and b. Since every fraction indicates a division, every fraction represents a ratio. Read "a to b," the ratio is sometimes written in the form $a:b$.

It is generally preferable to provide the ratio in simplest form, so the ratio 6 to 8 would be reduced (in fraction form) from $\dfrac{6}{8}$ to $\dfrac{3}{4}$. If units of measure are involved, these units must be **commensurable,** which means that they must be convertible to the same unit of measure. When simplifying the ratio of two quantities that are expressed in the same unit, you eliminate the common unit in the process. If two quantities cannot be compared because no common unit of measure is possible, the quantities are **incommensurable.**

EXAMPLE 1   Find the best form of each ratio:

(a) 12 to 20   (d) 5 lb to 20 oz
(b) 12 in. to 24 in.   (e) 5 lb to 2 ft
(c) 12 in. to 3 ft   (f) 4 m to 30 cm

Solution   (a) $\dfrac{12}{20} = \dfrac{3}{5}$ (b) $\dfrac{12 \text{ in.}}{24 \text{ in.}} = \dfrac{12}{24} = \dfrac{1}{2}$

(c) $\dfrac{12 \text{ in.}}{3 \text{ ft}} = \dfrac{12 \text{ in.}}{3(12 \text{ in.})} = \dfrac{12 \text{ in.}}{36 \text{ in.}} = \dfrac{1}{3}$

(d) $\dfrac{5 \text{ lb}}{20 \text{ oz}} = \dfrac{5(16 \text{ oz})}{20 \text{ oz}} = \dfrac{80 \text{ oz}}{20 \text{ oz}} = \dfrac{4}{1}$

(e) $\dfrac{5 \text{ lb}}{2 \text{ ft}}$ is incommensurable!

(f) $\dfrac{4 \text{ m}}{30 \text{ cm}} = \dfrac{4(100 \text{ cm})}{30 \text{ cm}} = \dfrac{400 \text{ cm}}{30 \text{ cm}} = \dfrac{40}{3}$ ▲

A **proportion** is a statement that two ratios are equal. Thus $\dfrac{a}{b} = \dfrac{c}{d}$ is a proportion and may be read as "*a* is to *b* as *c* is to *d*." In the order read, *a* is the first term of the proportion, *b* is the second term, *c* is the third term, and *d* is the fourth term. The first and last terms (*a* and *d*) of the proportion are the **extremes,** while the second and third terms (*b* and *c*) are the **means.**

As you will see in Example 3, the units involved in the ratios of a proportion are not necessarily commensurable. The following property is convenient for solving many proportions:

> **PROPERTY 1 (Means-Extremes Property)**   In a proportion, the product of the means equals the product of the extremes; that is, if $\dfrac{a}{b} = \dfrac{c}{d}$ (where $b \neq 0$ and $d \neq 0$), then $a \cdot d = b \cdot c$.

In the false proportion $\dfrac{9}{12} = \dfrac{2}{3}$, it is obvious that $9 \cdot 3 \neq 12 \cdot 2$; on the other hand, the truth of the statement $\dfrac{9}{12} = \dfrac{3}{4}$ is evident from the fact that

$9 \cdot 4 = 12 \cdot 3$. Any future proportion given in this text is intended to be a true proportion.

EXAMPLE 2   Use the Means-Extremes Property to solve each proportion:

(a) $\dfrac{x}{8} = \dfrac{5}{12}$

(b) $\dfrac{x + 1}{9} = \dfrac{x - 3}{3}$

(c) $\dfrac{3}{x} = \dfrac{x}{2}$

(d) $\dfrac{x + 3}{3} = \dfrac{9}{x - 3}$

(e) $\dfrac{x + 2}{5} = \dfrac{4}{x - 1}$

Solution   (a) $x \cdot 12 = 8 \cdot 5$

$12x = 40$

$x = \dfrac{40}{12} = \dfrac{10}{3}$

(b) $3(x + 1) = 9(x - 3)$

$3x + 3 = 9x - 27$

$30 = 6x$

$x = 5$

(c) $3 \cdot 2 = x \cdot x$

$x^2 = 6$

$x = \pm \sqrt{6}$

(d) $(x + 3)(x - 3) = 3 \cdot 9$

$x^2 - 9 = 27$

$x^2 - 36 = 0$

$(x + 6)(x - 6) = 0$

$x + 6 = 0 \qquad \text{or} \qquad x - 6 = 0$

$x = -6 \qquad \text{or} \qquad x = 6$

(e) $(x + 2)(x - 1) = 5 \cdot 4$

$x^2 + x - 2 = 20$

$x^2 + x - 22 = 0$

$x = \dfrac{-b \pm \sqrt{b^2 - 4ac}}{2a}$

$= \dfrac{-1 \pm \sqrt{(1)^2 - 4(1)(-22)}}{2(1)}$

$= \dfrac{-1 \pm \sqrt{1 + 88}}{2}$

$= \dfrac{-1 \pm \sqrt{89}}{2}$

▲

In application problems involving proportions, it is essential to keep together quantities that are related. The following example illustrates the care that must be taken in forming the proportion for an application.

EXAMPLE 3 If an automobile can travel 90 mi on 4 gal of gasoline, how far can it travel on 6 gal of gasoline?

Solution By form,

$$\frac{\text{No. miles first trip}}{\text{No. gallons first trip}} = \frac{\text{no. miles second trip}}{\text{no. gallons second trip}}$$

Where x represents the number of miles on the second trip, we have

$$\frac{90}{4} = \frac{x}{6}$$
$$4x = 540$$
$$x = 135$$

Thus the car can travel 135 mi on 6 gal of gasoline. ▲

In a proportion like $\frac{a}{b} = \frac{b}{c}$, in which the second and third terms are identical, the positive value of b is the **geometric mean** of a and c. For example, 6 is the geometric mean of 4 and 9 because $\frac{4}{6} = \frac{6}{9}$; the geometric mean is also called a "mean proportional."

EXAMPLE 4 In figure 5.1, AD is the geometric mean of BD and DC. If $BC = 10$ and $BD = 4$, determine AD.

Solution $\frac{BD}{AD} = \frac{AD}{DC}$. Because $DC = BC - BD$, we know that $DC = 10 - 4 = 6$. Therefore

$$\frac{4}{x} = \frac{x}{6}$$

in which x is the length of $\overline{AD}$. Applying the Means-Extremes Property, we get

$$x^2 = 24$$
$$x = \pm\sqrt{24} = \pm\sqrt{4 \cdot 6} = \pm 2\sqrt{6}$$

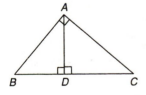

Figure 5.1

To have a permissable length for $\overline{AD}$, the geometric mean is the positive solution. Thus $AD = 2\sqrt{6}$. ▲

An **extended ratio** compares more than two quantities and must be expressed in a form such as $a:b:c$ or $d:e:f:g$. If you know that the angles of a triangle are 90°, 60°, and 30°, then the ratio that compares these measures is 90:60:30 or 3:2:1 (since 90, 60, and 30 have a greatest common factor of 30).

Unknown quantities in the ratio $a:b:c:d:\ldots$ are generally represented by variable expressions such as $ax, bx, cx, dx, \ldots$

EXAMPLE 5 Suppose that the sum of the measures of the sides of a quadrilateral is 70 and that the sides are in the ratio of 2:3:4:5. Find the measure of each side.

Solution Let the lengths of the sides be represented by $2x$, $3x$, $4x$, and $5x$. Then

$$2x + 3x + 4x + 5x = 70$$
$$14x = 70$$
$$x = 5$$

The lengths of sides are

$$2x = 10 \qquad 3x = 15 \qquad 4x = 20 \qquad 5x = 25 \qquad \blacktriangle$$

It is possible to solve certain problems in more ways than one, as is illustrated in the next example. However, the solution is unique and cannot be affected by your choice of a particular method.

EXAMPLE 6 Two complementary angles are in the ratio 2 to 3. Find the measure of each angle.

Solution Let the first of the complementary angles have measure x; then the second has the measure $90 - x$. Thus we have

$$\frac{x}{90 - x} = \frac{2}{3}$$

Using the Means-Extremes Property, we have

$$3x = 2(90 - x)$$
$$3x = 180 - 2x$$
$$5x = 180$$
$$x = 36$$
$$90 - x = 54$$

The angles have measures of 36° and 54°. $\blacktriangle$

Alternative solution: Let the measures of the two angles be $2x$ and $3x$ since they are in the ratio 2:3. Then

$$2x + 3x = 90$$
$$5x = 90$$
$$x = 18$$

Now $2x = 36$ and $3x = 54$, so the measures of the two angles are 36° and 54°. $\blacktriangle$

Some other properties of proportions are now stated without titles.

PROPERTY 2 In a proportion, the means or the extremes (or both the means and the extremes) may be interchanged; that is, if $\dfrac{a}{b} = \dfrac{c}{d}$ (where a, b, c, and d are nonzero), then $\dfrac{a}{c} = \dfrac{b}{d}$, $\dfrac{d}{b} = \dfrac{c}{a}$, $\dfrac{d}{c} = \dfrac{b}{a}$.

Note: The last proportion is the inverted form of the given proportion. This property can easily be remembered by considering that $a \cdot d = b \cdot c$ in all proportions.

PROPERTY 3 If $\dfrac{a}{b} = \dfrac{c}{d}$ (where $b \neq 0$ and $d \neq 0$), then $\dfrac{a + b}{b} = \dfrac{c + d}{d}$.

Note: When each denominator is added to each numerator on each side of an equation, another true proportion is obtained.

Proof of Property 3

$$\frac{a}{b} = \frac{c}{d} \qquad \text{hypothesis}$$

$$\frac{a}{b} + 1 = \frac{c}{d} + 1 \qquad \text{equals added to equals are equal}$$

$$1 = \frac{b}{b} = \frac{d}{d} \qquad \text{identity}$$

$$\frac{a}{b} + \frac{b}{b} = \frac{c}{d} + \frac{d}{d}$$

$$\frac{a + b}{b} = \frac{c + d}{d} \qquad \text{simplifying terms} \qquad \blacktriangle$$

Just as there are extended ratios, there are also extended proportions, such as

$$\frac{a}{b} = \frac{c}{d} = \frac{e}{f} = \cdots$$

Suggested by different numbers of servings of a particular recipe, the statement below is an extended proportion comparing numbers of eggs to numbers of cups of milk:

$$\frac{2 \text{ eggs}}{3 \text{ cups}} = \frac{4 \text{ eggs}}{6 \text{ cups}} = \frac{6 \text{ eggs}}{9 \text{ cups}}$$

EXAMPLE 7   In the triangles shown in figure 5.2,

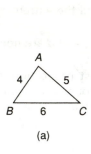

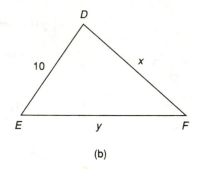

(a)   (b)

Figure 5.2

$$\frac{AB}{DE} = \frac{AC}{DF} = \frac{BC}{EF}$$

Find the lengths of $\overline{DF}$ and $\overline{EF}$.

Solution   Substituting, we have

$$\frac{4}{10} = \frac{5}{x} = \frac{6}{y}$$

From the equation

$$\frac{4}{10} = \frac{5}{x}$$

it follows that $4x = 50$ and that $x = DF = 12.5$. Using the equation

$$\frac{4}{10} = \frac{6}{y}$$

we find that $4y = 60$, so $y = EF = 15$.   ▲

5.2 EXERCISES

In Exercises 1 to 4, give the ratios in simplified form.

1. (a) 12 to 15
   (b) 12 in. to 15 in.
   (c) 1 ft to 18 in.
   (d) 1 ft to 18 oz

2. (a) 20 to 36
   (b) 24 oz to 52 oz
   (c) 20 oz to 2 lb
   (d) 2 lb to 20 oz

3. (a) 15:24
   (b) 2 ft:2 yd
   (c) 2 m:150 cm
   (d) 2 m:1 lb

4. (a) 24:32
   (b) 12 in.:2 yd
   (c) 150 cm:2 m
   (d) 1 gal:24 mi

In Exercises 5 to 14, find the values of x in each proportion.

5. (a) $\dfrac{x}{4} = \dfrac{9}{12}$

   (b) $\dfrac{7}{x} = \dfrac{21}{24}$

6. (a) $\dfrac{x - 1}{10} = \dfrac{3}{5}$

   (b) $\dfrac{x + 1}{6} = \dfrac{10}{12}$

7. (a) $\dfrac{x - 3}{8} = \dfrac{x + 3}{24}$

   (b) $\dfrac{x + 1}{6} = \dfrac{4x - 1}{18}$

8. (a) $\dfrac{9}{x} = \dfrac{x}{16}$

   (b) $\dfrac{32}{x} = \dfrac{x}{2}$

9. (a) $\dfrac{x}{4} = \dfrac{7}{x}$ (b) $\dfrac{x}{6} = \dfrac{3}{x}$

10. (a) $\dfrac{x+1}{3} = \dfrac{10}{x+2}$

(b) $\dfrac{x-2}{5} = \dfrac{12}{x+2}$

11. (a) $\dfrac{x+1}{x} = \dfrac{10}{2x}$

(b) $\dfrac{2x+1}{x+1} = \dfrac{14}{3x-1}$

12. (a) $\dfrac{x+1}{2} = \dfrac{7}{x-1}$

(b) $\dfrac{x+1}{3} = \dfrac{5}{x-2}$

13. (a) $\dfrac{x+1}{x} = \dfrac{2x}{3}$

(b) $\dfrac{x+1}{x-1} = \dfrac{2x}{5}$

14. (a) $\dfrac{x+1}{x} = \dfrac{x}{x-1}$

(b) $\dfrac{x+2}{x} = \dfrac{2x}{x-2}$

In Exercises 15 to 24, use proportions to solve each problem.

15. A recipe calls for 4 eggs and 3 cups of milk. To prepare for a larger number of guests, a cook uses 14 eggs. How many cups of milk are needed?

16. If a school secretary copies 168 worksheets for a class of 28 students, how many must be prepared for a class of 32 students?

17. An electrician installs 20 electrical outlets in a new six-room house. Assuming proportionality, how many outlets should be installed in a new construction having seven rooms? (Give answer to the nearest integer.)

18. The secretarial pool (15 secretaries in all) on one floor of a corporate complex has access to 11 telephones. If on a different floor, there are 23 secretaries, approximately what number of telephones should be available? (Assume a proportionality.)

19. Assume that AD in the illustration below is the geometric mean of BD and DC in $\triangle ABC$.

(a) Find AD if $BD = 6$ and $DC = 8$.
(b) Find BD if $AD = 6$ and $DC = 8$.

20. Using the same drawing as for Exercise 19, assume that AB is the geometric mean of BD and BC.

(a) Find AB if $BD = 6$ and $DC = 10$.
(b) Find DC if $AB = 10$ and $BC = 15$.

21. The ratio among the salaries of a secretary, a sales-person, and a vice president for a retail sales company are in the ratio 2:3:5. If their combined annual salaries amount to $92,500, what is the annual salary of each?

22. If the angles of a quadrilateral are in the ratio of 2:3:4:6, find the measure of each angle.

23. Two complementary angles are in the ratio 4:5. Find the measure of each angle, using the two methods shown in Example 6.

24. Two supplementary angles are in the ratio of 2:7. Find the measure of each angle, using the two methods of Example 6.

25. *Prove:* If $\dfrac{a}{b} = \dfrac{c}{d}$ (where a, b, c, and d are nonzero), then $\dfrac{b}{a} = \dfrac{d}{c}$.

26. *Prove:* If $\dfrac{a}{b} = \dfrac{c}{d}$ (where $b \neq 0$ and $d \neq 0$), then $\dfrac{a-b}{b} = \dfrac{c-d}{d}$.

27. After finding that $x = 12.5$ in Example 7, we could have found the value of y by solving the equation $\dfrac{5}{x} = \dfrac{6}{y}$. Cite two reasons why this method is not recommended.

28. Given that $18 \cdot 16 = 4 \cdot 72$, name a pair of fractions that must be equal.

29. Two numbers a and b are in the ratio 3:4. If the first number is decreased by 2 and the second is decreased by 1, they are in the ratio 2:3. Find a and b.

30. If the ratio of the measure of the complement of an angle to the measure of its supplement is 1:4, what is the measure of the angle?

SIMILAR POLYGONS AND TRIANGLES

When two geometric figures have the same shape, they are **similar;** the symbol for "is similar to" is $\sim$. When two figures have the same shape ($\sim$) and all parts have equal ($=$) measures, the two figures are **congruent** ($\cong$). Notice that the symbol for congruence combines the symbols for similarity and equality.

While two-dimensional figures can be similar, like $\triangle ABC$ and $\triangle DEF$ illustrated in figure 5.3, it is also possible for three-dimensional figures to be sim-

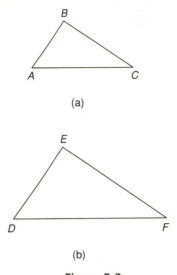

(a)

(b)

Figure 5.3

ilar; similar orange juice containers are shown in figure 5.4a and b. Informally, two figures are termed "similar" if they have the same size or if one is an enlargement of the other. Thus, a tuna fish can and an orange juice can are *not* similar even if both are right-circular cylinders. (See figure 5.4b and c.)

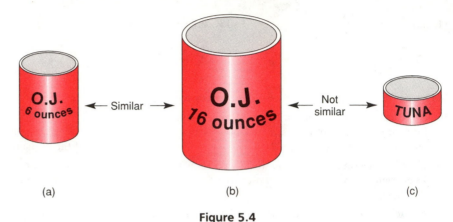

(a)   (b)   (c)

Figure 5.4

Generally our discussion of similarity will be limited to two-dimensional figures such as polygons.

For two polygons to be similar, each angle of one polygon must be congruent to the corresponding angle of the other. While this congruence of angles is necessary for similarity, it alone is not sufficient to establish similarity. The vertices of the congruent angles are **corresponding vertices** of the similar polygons. If $\angle A$ in one polygon is congruent to $\angle M$ in the second polygon, then vertex A corresponds to vertex M, and this is symbolized $A \leftrightarrow M$; we can indicate that $\angle A$ corresponds to $\angle M$ by $\angle A \leftrightarrow \angle M$. A pair of angles like $\angle A$ and $\angle M$ are **corresponding angles,** and the sides determined by consecutive and corresponding vertices are **corresponding sides** of the similar polygons.

EXAMPLE 1   Given that quadrilaterals $ABCD$ and $HJKL$ are similar, with congruent angles indicated in figure 5.5, name the vertices, angles, and sides that correspond to each other.

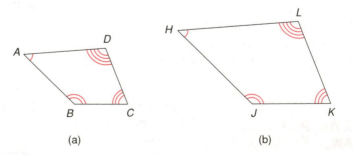

(a)   (b)

Figure 5.5

Solution   Since $\angle A \cong \angle H$, it follows that

$$A \leftrightarrow H \qquad \text{and} \qquad \angle A \leftrightarrow \angle H$$

Similarly,

$B \leftrightarrow J$	and	$\angle B \leftrightarrow \angle J$
$C \leftrightarrow K$	and	$\angle C \leftrightarrow \angle K$
$D \leftrightarrow L$	and	$\angle D \leftrightarrow \angle L$

By associating pairs of consecutive and corresponding vertices, the corresponding sides are included between the corresponding angles.

$$\overline{AB} \leftrightarrow \overline{HJ}, \quad \overline{BC} \leftrightarrow \overline{JK}, \quad \overline{CD} \leftrightarrow \overline{KL}, \quad \overline{AD} \leftrightarrow \overline{HL} \qquad \blacktriangle$$

> **DEFINITION:** Two polygons are **similar** if and only if two conditions are satisfied:
>
> 1. All pairs of corresponding angles are congruent.
> 2. All pairs of corresponding sides are proportional.

The second condition for similarity requires that the following extended proportion exist for the sides of the similar quadrilaterals of Example 1:

$$\frac{AB}{HJ} = \frac{BC}{JK} = \frac{CD}{KL} = \frac{AD}{HL}$$

Notice that *both* conditions for similarity are necessary! While condition 1 is satisfied for square *EFGH* and rectangle *RSTU* (see figure 5.6a and b), the figures are not similar—that is, one is not an enlargement of the other—because the extended proportion is not true. On the other hand, condition 2 is satisfied for square *EFGH* and rhombus *WXYZ* (see figure 5.6a and c), but the figures are not similar because the angles are not congruent.

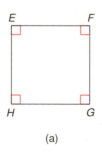
(a)

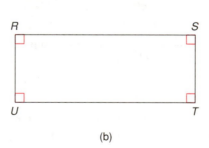
(b)

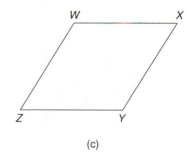
(c)

Figure 5.6

The practice of naming corresponding vertices in consecutive order for the two polygons is most convenient! For instance, if pentagon *ABCDE* is similar to pentagon *MNPQR*, then you know that $A \leftrightarrow M$, $B \leftrightarrow N$, $C \leftrightarrow P$, $D \leftrightarrow Q$, $E \leftrightarrow R$, $\angle A \cong \angle M$, $\angle B \cong \angle N$, $\angle C \cong \angle P$, $\angle D \cong \angle Q$, and $\angle E \cong \angle R$. Because the vertices correspond, you also know that

$$\frac{AB}{MN} = \frac{BC}{NP} = \frac{CD}{PQ} = \frac{DE}{QR} = \frac{EA}{RM}$$

EXAMPLE 2   If $\triangle ABC \sim \triangle DEF$ in figure 5.7, use the indicated measures to find the measures of the remaining parts of each of the triangles.

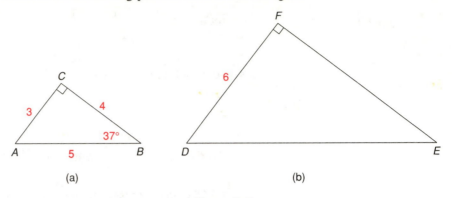

(a)   (b)

Figure 5.7

Solution   Because the sum of the measures of the angles of a triangle is 180°,

$$m\angle A = 180 - (90 + 37) = 53°$$

Due to the similarity and the correspondence of vertices,

$$m\angle D = 53°, \qquad m\angle E = 37°, \qquad m\angle F = 90°$$

The proportion that relates the lengths of the sides is

$$\frac{AC}{DF} = \frac{CB}{FE} = \frac{AB}{DE} \qquad \text{so} \qquad \frac{3}{6} = \frac{4}{FE} = \frac{5}{DE}$$

From $\dfrac{3}{6} = \dfrac{4}{FE}$, you can see that

$$3 \cdot FE = 6 \cdot 4 = 24$$
$$FE = 8$$

From $\dfrac{3}{6} = \dfrac{5}{DE}$, you can see that

$$3 \cdot DE = 6 \cdot 5 = 30$$
$$DE = 10 \qquad \qquad \blacktriangle$$

In a proportion, the ratios can *all* be inverted; thus Example 2 could have been solved by using the ratio

$$\frac{DF}{AC} = \frac{FE}{CB} = \frac{DE}{AB}$$

In an extended proportion, the equal ratios must all be equal to the same constant value. By designating this number (often called the "constant of proportionality") by k, we see that

$$\frac{DF}{AC} = k \qquad \frac{FE}{CB} = k \qquad \frac{DE}{AB} = k$$

It follows that $DF = k \cdot AC$, $FE = k \cdot CB$, and $DE = k \cdot AB$. In Example 2, this constant of proportionality had the value $k = 2$, which means that each side of the larger triangle was double the length of each side of the smaller triangle.

Our definition of similar polygons (and therefore of similar triangles) is almost impossible to use as a method of proof. Fortunately, some easier methods are available for proving triangles similar. If two triangles are carefully sketched or constructed so that their angles are congruent, they will appear to be similar, as shown in figure 5.8.

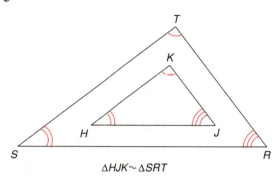

$\triangle HJK \sim \triangle SRT$

Figure 5.8

POSTULATE 14   If the three angles of one triangle are congruent to the three angles of a second triangle, then the triangles are similar (AAA).

COROLLARY 5.3.1   If two angles of one triangle are congruent to two angles of another triangle, then the triangles are similar (AA).

Corollary 5.3.1, abbreviated AA, follows from knowing that if two angles of one triangle are congruent to two angles of another triangle, then the third angles *must* also be congruent.

EXAMPLE 3

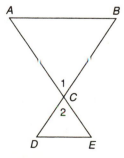

Figure 5.9

Provide an informal proof of the following problem.

Given: $\overline{AB} \parallel \overline{DE}$ in figure 5.9.

Prove: $\triangle ABC \sim \triangle EDC$

Proof: With $\overline{AB} \parallel \overline{DE}$, $\angle A$ and $\angle E$ are congruent alternate interior angles. In addition, $\angle 1 \cong \angle 2$ (vertical angles). Therefore, $\triangle ABC \sim \triangle EDC$ by AA.   ▲

In many instances, we wish to prove something beyond the similarity of triangles. This is possible through use of the definition of "similarity," which states that corresponding sides of similar triangles (or more generally of similar polygons) are proportional. This part of the definition is often cited as a reason in a proof, as shown in Example 4.

EXAMPLE 4   Complete the following informal proof.

Given: $\angle ADE \cong \angle B$ in figure 5.10.

Prove: $\dfrac{DE}{BC} = \dfrac{AE}{AC}$

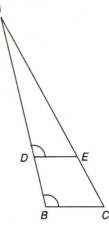

Figure 5.10

Proof: By hypothesis, $\angle ADE \cong \angle B$. In addition, $\angle A \cong \angle A$ by the Reflexive Property. It follows by AA that $\triangle ADE \sim \triangle ABC$. Using the definition of similar triangles, we can conclude that $\dfrac{DE}{BC} = \dfrac{AE}{AC}$.   ▲

In this proof, DE appears above BC because the sides with these names lie opposite $\angle A$ in the two similar triangles. AE and AC are the lengths of sides opposite the congruent and corresponding angles $\angle ADE$ and $\angle B$. That is, corresponding sides of similar triangles always lie opposite corresponding angles.

EXAMPLE 5 $\angle ADE \cong \angle B$ in figure 5.11. If $DE = 3$, $AC = 16$, and $EC = BC$, find the length BC.

Solution   From the similar triangles, we proved (in Example 4) that $\dfrac{DE}{BC} = \dfrac{AE}{AC}$. With $AC = AE + EC$, and letting the lengths of the congruent segments ($\overline{EC}$ and $\overline{BC}$) be denoted by x, we have

$$16 = AE + x \qquad \text{so} \qquad AE = 16 - x$$

Substituting into the proportion, we have

$$\frac{3}{x} = \frac{16 - x}{16}$$

Thus it follows that

$$x(16 - x) = 3 \cdot 16$$
$$16x - x^2 = 48$$
$$x^2 - 16x + 48 = 0$$
$$(x - 4)(x - 12) = 0$$

Now x (or BC) equals 4 or 12. Each length is acceptable, but the drawings differ, as illustrated in figure 5.12.   ▲

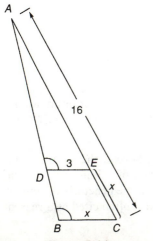

Figure 5.11

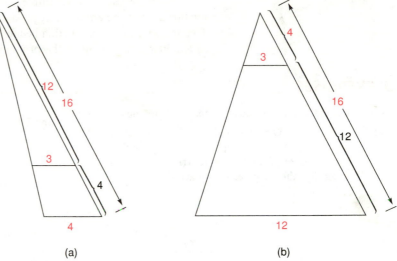

(a)   (b)

Figure 5.12

In Example 6, you are asked to prove that the product of lengths of two segments equals the product of lengths of two other segments. Here is a plan for establishing such a relationship:

1. Show that two triangles are similar, like $\triangle ABC \sim \triangle DEF$.

2. Form a proportion by choosing the ratios from

$$\frac{AB}{DE} = \frac{BC}{EF} = \frac{AC}{DF}$$

3. Use the Means-Extremes Property to obtain equal products, such as

$$AB \cdot EF = DE \cdot BC$$

when $\dfrac{AB}{DE} = \dfrac{BC}{EF}$ is the proportion used.

EXAMPLE 6

Figure 5.13

Use an informal proof for this problem.

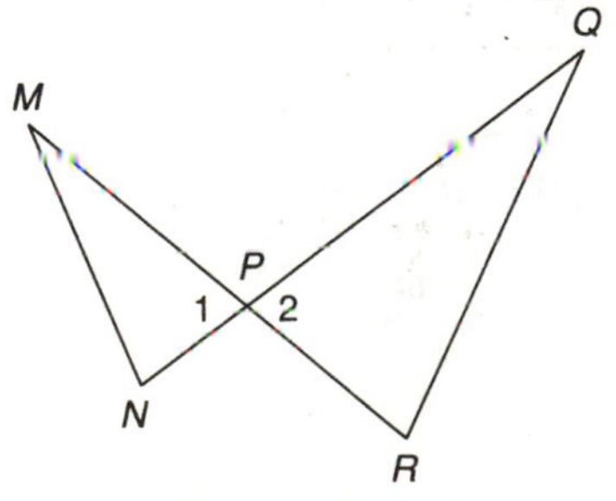

Given: $\angle M \cong \angle Q$ in figure 5.13.

Prove: $NP \cdot QR = RP \cdot MN$

Proof: By hypothesis, $\angle M \cong \angle Q$. Also, $\angle 1 \cong \angle 2$ by the fact that vertical angles are congruent. Now $\triangle MPN \sim \triangle QPR$ by AA. Then

$$\frac{NP}{RP} = \frac{MN}{QR}$$

because corresponding sides of similar triangles are proportional. Using the Means-Extremes Property, it follows that

$$NP \cdot QR = RP \cdot MN$$

▲

It was necessary in Example 6 to be selective of the sides used in the proportion so that the appropriate products could be formed. We included *NP*, *QR*, *RP*, and *MN* in the proportion because they were the lengths of segments found in the Prove statement of the problem.

5.3 EXERCISES

In Exercises 1 and 2, refer to the accompanying drawing.

1. (a) Given that $A \leftrightarrow X$, $B \leftrightarrow T$, and $C \leftrightarrow N$, write a statement claiming that the triangles shown are similar.
   (b) If $A \leftrightarrow N$, $C \leftrightarrow X$, and $B \leftrightarrow T$, write a statement claiming that the triangles shown are similar.

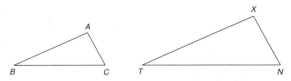

2. (a) If $\triangle ABC \sim \triangle XTN$, which angle of $\triangle ABC$ corresponds to $\angle N$ of $\triangle XTN$?
   (b) If $\triangle ABC \sim \triangle TXN$, which side of $\triangle TXN$ corresponds to side $\overline{AC}$ of $\triangle ABC$?
3. A **sphere** is the three-dimensional surface that contains all points in space lying at a fixed distance from a point known as the center of the sphere. Consider the two spheres shown. Are these two spheres similar? Are any two spheres similar? Explain.

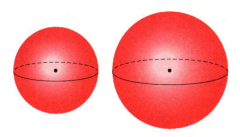

4. Given that rectangle *ABCE* is similar to rectangle *MNPR* and that $\triangle CDE \sim \triangle PQR$, what can you conclude regarding pentagon *ABCDE* and pentagon *MNPQR*?

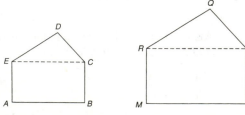

5. In $\triangle RST$ and $\triangle UVW$, $WU = \dfrac{3}{2} \cdot TR$, $WV = \dfrac{3}{2} \cdot TS$, and $UV = \dfrac{3}{2} \cdot RS$. Use intuition to

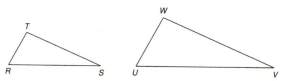

draw a conclusion about the two triangles. If your conclusion includes a similarity relationship, write an extended proportion relating the sides of the two triangles.

6. In $\triangle DGH$ and $\triangle DEF$, $DE = 3 \cdot DG$ and $DF = 3 \cdot DH$. Use intuition to form a conclusion about the triangle relationship. If the conclusion involves a similarity, state an extended proportion relating the sides of the two triangles.

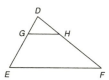

7. Given: $\triangle MNP \sim \triangle QRS$, m$\angle M = 56°$, m$\angle R = 82°$, $MN = 9$, $QR = 6$, $RS = 7$, $MP = 12$
   Find: (a) m$\angle N$ (c) NP
         (b) m$\angle P$ (d) QS

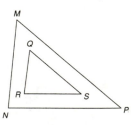

8. Given: $\triangle ABC \sim \triangle PRC$, m$\angle A = 67°$, $PC = 5$, $CR = 12$, $PR = 13$, $AB = 26$
   Find: (a) m$\angle B$ (c) AC
         (b) m$\angle RPC$ (d) CB

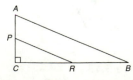

9. **(a)** Does the similarity relationship have a **reflexive** property for triangles (and polygons in general)?

(b) Is there a **symmetric** property for the similarity of triangles (and polygons)?

(c) Is there a **transitive** property for the similarity of triangles (and polygons)?

(d) Is the similarity of triangles (and polygons) an equivalence relation?

10. Using the names of properties from Exercise 9, identify the property illustrated by each statement:

(a) If $\triangle 1 \sim \triangle 2$, then $\triangle 2 \sim \triangle 1$

(b) If $\triangle 1 \sim \triangle 2$, $\triangle 2 \sim \triangle 3$, and $\triangle 3 \sim \triangle 4$, then $\triangle 1 \sim \triangle 4$

(c) $\triangle 1 \sim \triangle 1$

In Exercises 11 to 22, construct informal proofs.

In Exercises 11 and 12, use the accompanying drawing.

11. *Given:* $\overline{MN} \perp \overline{NP}, \overline{QR} \perp \overline{RP}$
Prove: $\triangle MNP \sim \triangle QRP$

12. *Given:* $\overline{MN} \parallel \overline{QR}$
Prove: $\triangle MNP \sim \triangle QRP$

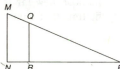

Use the accompanying figure for Exercises 13 and 14.

13. *Given:* $\angle H \cong \angle F$
Prove: $\triangle HJK \sim \triangle FGK$

14. *Given:* $\overline{HJ} \perp \overline{JF}, \overline{HG} \perp \overline{FG}$
Prove: $\triangle HJK \sim \triangle FGK$

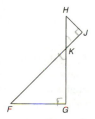

In Exercises 15 and 16, use the accompanying figure.

15. *Given:* $\overline{AB} \parallel \overline{DF}, \overline{BD} \parallel \overline{FG}$
Prove: $\triangle ABC \sim \triangle EFG$

16. *Given:* $\overline{AB} \cong \overline{BC}, \overline{CD} \cong \overline{DE}$,
and $\overline{EF} \cong \overline{FG}$
Prove: $\triangle ABC \sim \triangle EFG$

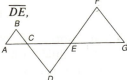

17. *Given:* $\angle RVU \cong \angle S, \angle RUV \cong \angle Q$
Prove: $\triangle PQR \sim \triangle STR$

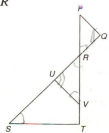

18. *Given:* $\overline{RS} \perp \overline{AB}, \overline{CB} \perp \overline{AC}$
Prove: $\triangle BSR \sim \triangle BCA$

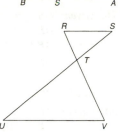

19. *Given:* $\overline{RS} \parallel \overline{UV}$
Prove: $\dfrac{RT}{VT} = \dfrac{RS}{VU}$

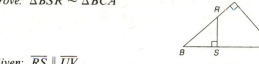

20. *Given:* $\overline{AB} \parallel \overline{DC}, \overline{AC} \parallel \overline{DE}$
Prove: $\dfrac{AB}{DC} = \dfrac{BC}{CE}$

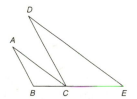

In Exercises 21 and 22, use the accompanying drawing.

21. *Given:* $\overline{RS} \parallel \overline{YZ}, \overline{RU} \parallel \overline{XZ}$
Prove: $RS \cdot ZX = ZY \cdot RT$

22. *Given:* $\triangle RST \sim \triangle UYT$
$\overline{RU} \parallel \overline{XZ}$
Prove: $\triangle RST \sim \triangle ZYX$

$$\frac{RS}{YZ} = \frac{RT}{XZ}$$

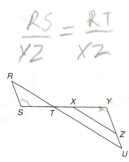

In Exercises 23 to 26, $\triangle ABC \sim \triangle DBE$ in the accompanying drawing.

23. *Given:* $AC = 8$, $DE = 6$, $CB = 6$
Find: EB (*Hint:* Let $EB = x$, and solve an equation.)

24. *Given:* $AC = 10$, $CB = 12$
E the midpoint of $\overline{CB}$
Find: DE

25. *Given:* $AC = 10$, $DE = 8$,
$AD = 4$
Find: DB

26. *Given:* $CB = 12$, $CE = 4$,
$AD = 5$
Find: DB

In Exercises 27 to 30, $\triangle ADE \sim \triangle ABC$ in the accompanying drawing.

27. *Given:* $DE = 4$, $AE = 6$,
    $EC = BC$
    Find: BC

28. *Given:* $DE = 5$, $AD = 8$,
    $DB = BC$
    Find: AB (*Hint:* Find DB first.)

29. *Given:* $DE = 4$, $AC = 20$,
    $EC = BC$
    Find: BC

30. *Given:* $AD = 4$, $AC = 18$,
    $DB = AE$
    Find: AE

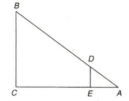

31. A person who is walking away from a 10-ft lamppost casts a shadow 6 ft long. If the person is at a distance of 10 ft from the lamppost at that moment, what is the person's height?

32. With 100 ft of string out, a kite is 64 ft above ground level. When the girl flying the kite pulls in 40 ft of string, the angle formed by the string and the ground does not

change. What is the height of the kite above the ground after the 40 ft of string have been taken in?

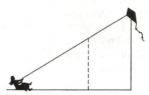

33. While admiring a rather tall tree, Fred notes that the shadow of his 6-ft frame has a length of 3 paces. On the level ground, he walks off 37 paces in the complete shadow of the tree. How tall is the tree?

34. As a garage door closes, light is cast 6 ft beyond the base of the door (as shown in the accompanying drawing) by a light fixture that is set in the garage ceiling 10 ft back from the door. If the ceiling of the garage is 10 ft above the floor, how far is the garage door above the floor at the time that light is cast 6 ft beyond the door?

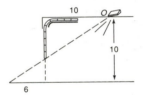

35. Prove that the altitude drawn to the hypotenuse of a right triangle separates the right triangle into two right triangles which are similar to each other and to the original right triangle.

36. Prove that the line segment joining the midpoints of two sides of a triangle forms a triangle which is similar to the original triangle.

5.4

SEGMENTS DIVIDED PROPORTIONALLY

In this section, we begin with an informal description of the phrase "divided proportionally." Suppose that three children have been provided with a joint savings account by their parents. Equal monthly deposits have been made to the account for each child since birth. If the ages of the children are 2, 4, and 6 (assume exactness for the sake of simplicity) and the total in the account is $7200, then the amount that each child should receive can be found by solving the equation

$$2x + 4x + 6x = 7200$$

Solving leads to the solution $1200 for the 2-year-old, $2400 for the 4-year-old, and $3600 for the 6-year-old. We may say that the amount has been divided proportionally. Expressed as a proportion, this is

$$\frac{1200}{2} = \frac{2400}{4} = \frac{3600}{6}$$

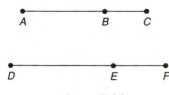

Figure 5.14

In figure 5.14, $\overline{AC}$ and $\overline{DF}$ are divided proportionally by points B and E if

$$\frac{AB}{DE} = \frac{BC}{EF}.$$

Of course, a pair of segments may be divided proportionally by several points, as shown in figure 5.15. In this case, $\overline{RW}$ and $\overline{HM}$ are divided proportionally when

$$\frac{RS}{HJ} = \frac{ST}{JK} = \frac{TV}{KL} = \frac{VW}{LM}$$

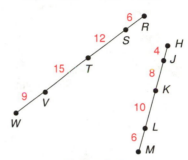

Figure 5.15

EXAMPLE 1

In figure 5.16, points D and E divide $\overline{AB}$ and $\overline{AC}$ proportionally. If $AD = 4$, $DB = 7$, and $EC = 6$, find AE.

Solution

$\dfrac{AD}{AE} = \dfrac{DB}{EC}$, so $\dfrac{4}{x} = \dfrac{7}{6}$, where $x = AE$. Then $7x = 24$, so $x = AE = \dfrac{24}{7} = 3\dfrac{3}{7}$. ▲

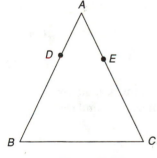

Figure 5.16

A property that will be proved in Exercise 24 of this section is

$$\text{``If } \frac{a}{b} = \frac{c}{d}, \text{ then } \frac{a+c}{b+d} = \frac{a}{b} = \frac{c}{d}.\text{''}$$

In words, we may restate this property as follows:

"The fraction whose numerator and denominator are determined, respectively, by adding numerators and denominators of equal fractions is equal to each of those equal fractions."

Here is a numerical example of this claim:

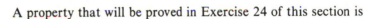

$$\text{``If } \frac{2}{3} = \frac{4}{6}, \text{ then } \frac{2+4}{3+6} = \frac{2}{3} = \frac{4}{6}.\text{''}$$

In Example 2, the preceding property is necessary as a reason.

EXAMPLE 2

Given: $\overline{RW}$ and $\overline{HM}$ are divided proportionally at the points shown in figure 5.17.

Prove: $\dfrac{RT}{HK} = \dfrac{TW}{KM}$

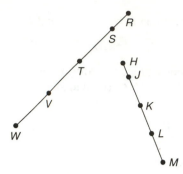

Figure 5.17

Proof: $\overline{RW}$ and $\overline{HM}$ are divided proportionally so that

$$\frac{RS}{HJ} = \frac{ST}{JK} = \frac{TV}{KL} = \frac{VW}{LM}$$

With all ratios being equal, we have

$$\frac{RS}{HJ} + \frac{ST}{JK} = \frac{TV}{KL} + \frac{VW}{LM} \qquad \text{Addition Property of Equality}$$

Using the property that if $\dfrac{a}{b} = \dfrac{c}{d}$, then $\dfrac{a+c}{b+d} = \dfrac{a}{b} = \dfrac{c}{d}$, we have

$$\frac{RS + ST}{HJ + JK} = \frac{TV + VW}{KL + LM}$$

Simplifying,

$$\frac{RT}{HK} = \frac{TW}{KM} \qquad\qquad\qquad \blacktriangle$$

Two properties that were proved earlier (Property 3 and Exercise 26, Section 5.2) are now combined as follows:

$$If \frac{a}{b} = \frac{c}{d}, \ then \ \frac{a \pm b}{b} = \frac{c \pm d}{d}.$$

The subtraction part of the property is needed for the proof of Theorem 5.4.1.

THEOREM 5.4.1   If a line is parallel to one side of a triangle and intersects the other two sides, then it divides these sides proportionally.

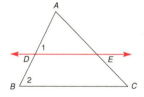

Given: $\triangle ABC$ with $\overleftrightarrow{DE} \parallel \overline{BC}$ and with $\overleftrightarrow{DE}$ intersecting $\overline{AB}$ at D and $\overline{AC}$ at E

Prove: $\dfrac{AD}{DB} = \dfrac{AE}{EC}$

Proof: Since $\overleftrightarrow{DE} \parallel \overline{BC}$, $\angle 1 \cong \angle 2$. With $\angle A$ as a common angle for $\triangle ADE$ and $\triangle ABC$, it follows by AA that these triangles are similar. Now

$$\frac{AB}{AD} = \frac{AC}{AE},$$

since these are corresponding sides of similar triangles. It follows from our fraction property (Exercise 26, Section 5.2) that

$$\frac{AB - AD}{AD} = \frac{AC - AE}{AE}$$

Because $AB - AD = DB$ and $AC - AE = EC$, the proportion becomes

$$\frac{DB}{AD} = \frac{EC}{AE}$$

Inverting both fractions gives the desired conclusion:

$$\frac{AD}{DB} = \frac{AE}{EC}$$

▲

COROLLARY 5.4.2   When three (or more) parallel lines are cut by a pair of transversals, the transversals are divided proportionally by the parallel lines.

Corollary 5.4.2 is now illustrated, and its proof is briefly described. This corollary claims, for example, that $\frac{AB}{BC} = \frac{DE}{EF}$ as shown in figure 5.18, where $p_1 \parallel p_2 \parallel p_3$. The proof requires the use of an auxiliary segment $\overline{AF}$. Then

$$\frac{AB}{BC} = \frac{AG}{GF} \qquad \text{and} \qquad \frac{AG}{GF} = \frac{DE}{EF}$$

Consequently,

$$\frac{AB}{BC} = \frac{DE}{EF} \qquad \text{by the Transitive Property of Equality}$$

Note: By interchanging the means, we could have written this proportion as $\frac{AB}{DE} = \frac{BC}{EF}$.

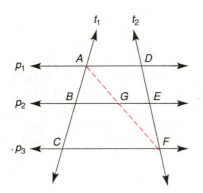

Figure 5.18

EXAMPLE 3   Given parallel lines p_1, p_2, p_3, and p_4 cut by t_1 and t_2 so that $AB = 4$, $EF = 3$, $BC = 2$, $GH = 5$, find FG and CD. (See figure 5.19.)

Solution   Because the transversals are divided proportionally,

$$\frac{AB}{EF} = \frac{BC}{FG} = \frac{CD}{GH}$$

so

$$\frac{4}{3} = \frac{2}{FG} = \frac{CD}{5}$$

Then

$$4 \cdot FG = 6 \qquad \text{and} \qquad 3 \cdot CD = 20$$

$$FG = \frac{3}{2} = 1\frac{1}{2} \qquad \text{and} \qquad CD = \frac{20}{3} = 6\frac{2}{3}$$

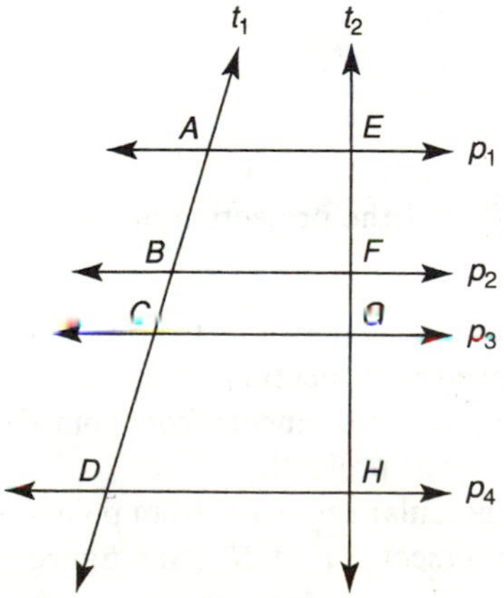

▲

Figure 5.19

The **projection of a point on a line** is the point that results from dropping a perpendicular segment from a point to a line. If the point lies on the line, it is its own projection. In figure 5.20, P' (read as "P prime") is the projection of point P on line ℓ.

The **projection of a line segment on a line** is the geometric figure that results

(a)   (b)

Figure 5.20

when all its points are projected on the line. Generally (as shown in figure 5.21a and b), the projection of a segment on a line is a segment. In figure 5.21a, the projection of $\overline{AB}$ on ℓ is $\overline{A'B'}$; in figure 5.21b, which shows $\triangle ABC$, the projection of side $\overline{AB}$ on side $\overline{AC}$ is $\overline{AB'}$. When the segment being projected is perpendicular to the line (as shown in figure 5.21c, the projection of the segment ($\overline{AB}$) is a point (A').

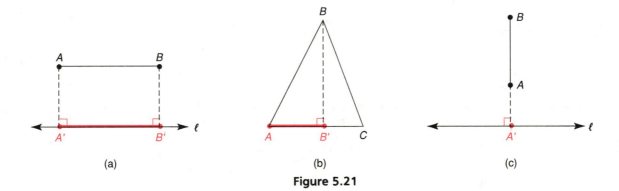

(a)   (b)   (c)

Figure 5.21

EXAMPLE 4

In figure 5.22, given right $\triangle RST$ with $\overline{SX} \perp \overline{RT}$, find the projection of:
(a) $\overline{RS}$ on $\overline{RT}$ (b) $\overline{ST}$ on $\overline{RT}$ (c) $\overline{RS}$ on $\overline{ST}$

Solution

(a) To find the projection of $\overline{RS}$ on $\overline{RT}$, drop perpendicular segments from points on $\overline{RS}$ to $\overline{RT}$. (See figure 5.22b.) $\overline{RX}$ is the desired projection.
(b) To find the projection of $\overline{ST}$ on $\overline{RT}$, drop perpendicular segments from points on $\overline{ST}$ to $\overline{RT}$. (See figure 5.22c.) $\overline{XT}$ is the desired projection.
(c) To find the projection of $\overline{RS}$ on $\overline{ST}$, drop perpendicular segments from points on $\overline{RS}$ to $\overline{ST}$. All perpendicular segments intersect $\overline{ST}$ at S. (See figure 5.22d.) Therefore point S is the desired projection.   ▲

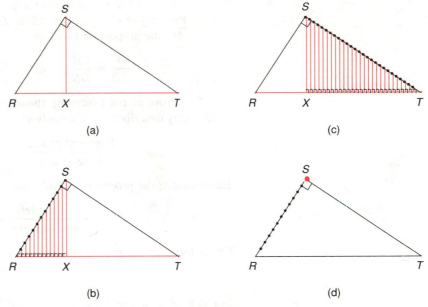

Figure 5.22

A somewhat surprising result is now stated as the final theorem of this section. Its proof follows from Theorem 5.4.1, even though the claims seem to differ vastly in content.

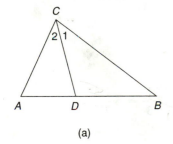

(a)

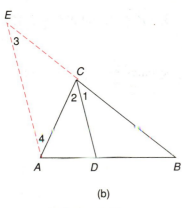

(b)

Figure 5.23

> **THEOREM 5.4.3**   If a ray bisects one angle of a triangle, then it divides the opposite side into segments that are proportional to the two sides which form that angle.

Given: $\triangle ABC$ in figure 5.23, in which $\overrightarrow{CD}$ bisects $\angle ACB$

Prove: $\dfrac{AD}{AC} = \dfrac{DB}{CB}$

Proof: We begin by extending $\overline{BC}$ beyond C (there is only one line through B and C) to meet the line through A drawn parallel to $\overline{DC}$. (See figure 5.23b.) Let E be the point of intersection (these lines must intersect; otherwise $\overline{AE}$ would have two parallels, $\overline{BC}$ and $\overline{CD}$, through point C). Since $\overline{CD} \parallel \overline{EA}$ we have

$$\frac{EC}{AD} = \frac{CB}{DB}$$

by Theorem 5.4.1. Now $\angle 1 \cong \angle 2$ (due to the angle bisector), while $\angle 1 \cong \angle 3$ (since they are corresponding angles for parallel lines) and

$\angle 2 \cong \angle 4$ (alternate interior angles for parallel lines). By the Transitive Property of Congruence, $\angle 3 \cong \angle 4$, so $\triangle ACE$ is isosceles. With $\overline{EC} \cong \overline{AC}$, the proportion becomes

$$\frac{AC}{AD} = \frac{CB}{DB} \quad \text{or} \quad \frac{AD}{AC} = \frac{DB}{CB} \quad \text{by inversion} \quad \blacktriangle$$

The Prove of the preceding theorem indicates that one form of the proportionality described is informally given by

$$\frac{\text{Segment at left}}{\text{Side at left}} = \frac{\text{segment at right}}{\text{side at right}}$$

Equivalently, the proportion could state

$$\frac{\text{Segment at left}}{\text{Segment at right}} = \frac{\text{side at left}}{\text{side at right}}$$

Other forms of the proportionality are also possible!

EXAMPLE 5   For $\triangle XYZ$ in figure 5.24, $XY = 3$ and $YZ = 5$. If $\overrightarrow{YW}$ bisects $\angle XYZ$ and $XW = 2$, find XZ.

Solution   Let $WZ = x$. We know that $\dfrac{YX}{XW} = \dfrac{YZ}{WZ}$, so $\dfrac{3}{2} = \dfrac{5}{x}$. Therefore

$$3x = 10$$

$$x = \frac{10}{3} = 3\frac{1}{3}$$

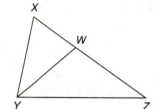

Figure 5.24

Because $XZ = XW + WZ$, we have $XZ = 2 + 3\dfrac{1}{3} = 5\dfrac{1}{3}$. $\blacktriangle$

EXAMPLE 6   In figure 5.24, $\triangle XYZ$ has sides of lengths $XY = 3$, $YZ = 4$, and $XZ = 5$. If $\overrightarrow{YW}$ bisects $\angle XYZ$, find XW and WZ.

Solution   Let $XW = y$; then $WZ = 5 - y$, and $\dfrac{XY}{YZ} = \dfrac{XW}{WZ}$ becomes $\dfrac{3}{4} = \dfrac{y}{5 - y}$. From this we can find y.

$$3(5 - y) = 4y$$
$$15 - 3y = 4y$$
$$15 = 7y$$
$$y = \frac{15}{7}$$

Then $XZ = \dfrac{15}{7} = 2\dfrac{1}{7}$, while $WZ = 5 - 2\dfrac{1}{7} = 2\dfrac{6}{7}$. $\blacktriangle$

5.4 EXERCISES

1. In preparing a recipe, 5 oz of ingredient A, 4 oz of ingredient B, and 6 oz of ingredient C are used. If 90 oz of this dish are needed, how many ounces of each ingredient should be used?

2. In a chemical mixture, 2 g of chemical A are used for each gram of chemical B, and 3 g of chemical C are needed for each gram of B. If 72 g of the mixture are prepared, what amount (in grams) of each chemical is needed?

3. Given that $\frac{AB}{EF} = \frac{BC}{FG} = \frac{CD}{GH}$ in the illustration, do the following proportions hold?

    (a) $\frac{AC}{EG} = \frac{CD}{GH}$

    (b) $\frac{AB}{EF} = \frac{BD}{FH}$

4. Given that $\overleftrightarrow{XY} \parallel \overline{TS}$, do the following proportions hold?

    (a) $\frac{TX}{XR} = \frac{RY}{YS}$

    (b) $\frac{TR}{XR} = \frac{SR}{YR}$

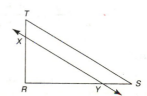

In Exercises 5 and 6, use the accompanying drawing.

5. Given: $\ell_1 \parallel \ell_2 \parallel \ell_3 \parallel \ell_4$, $AB = 5$,
    $BC = 4$, $CD = 3$,
    $EH = 10$
    Find: EF, FG, and GH

6. Given: $\ell_1 \parallel \ell_2 \parallel \ell_3 \parallel \ell_4$, $AB = 7$,
    $BC = 5$, $CD = 4$, $EF = 6$
    Find: FG, GH, EH

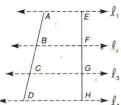

Use the accompanying figure for Exercises 7 and 8.

7. Given: $\ell_1 \parallel \ell_2 \parallel \ell_3$, $AB = 4$,
    $BC = 5$, $DE = x$,
    $EF = 12 - x$

Find: x, DE, EF

8. Given: $\ell_1 \parallel \ell_2 \parallel \ell_3$, $AB = 5$,
    $BC = x$, $DE = x - 2$,
    $EF = 7$
    Find: x, BC, DE

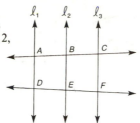

In Exercises 9 to 12, use the drawing shown.

9. Given: $\overleftrightarrow{DE} \parallel \overline{BC}$, $AD = 5$,
    $DB = 12$, $AE = 7$
    Find: EC

10. Given: $\overleftrightarrow{DE} \parallel \overline{BC}$, $AD = 6$,
    $DB = 10$, $AC = 20$
    Find: EC

11. Given: $\overleftrightarrow{DE} \parallel \overline{BC}$,
    $AD = a - 1$,
    $DB = 2a + 2$,
    $AE = a$,
    $EC = 4a - 5$
    Find: a and AD

12. Given: $\overleftrightarrow{DE} \parallel \overline{BC}$, $AD = 5$,
    $DB = a + 3$,
    $AE = a + 1$,
    $EC = 3(a - 1)$
    Find: a and EC

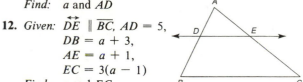

Use the figure shown for Exercises 13 and 14.

13. Given: $\overrightarrow{RW}$ bisects $\angle SRT$
    Do the following equalities hold?
    (a) $SW = WT$
    (b) $\frac{RS}{RT} = \frac{SW}{WT}$

14. Given: $\overrightarrow{RW}$ bisects $\angle SRT$
    Do the following equalities hold?
    (a) $\frac{RS}{SW} = \frac{RT}{WT}$
    (b) $\angle S = \angle T$

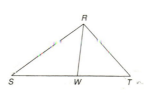

For Exercises 15 and 16, use the figure shown.

15. Given: $\overrightarrow{UT}$ bisects $\angle WUV$,
    $WU = 8$, $UV = 12$,
    $WT = 6$
    Find: TV

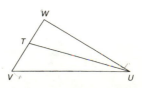

16. *Given:* $\overrightarrow{UT}$ bisects ∠*WUV*,
        $WU = 9, UV = 12,$
        $WV = 9$
   Find: *WT*

Use the drawing shown for Exercises 17 and 18.

17. *Given:* $\overrightarrow{NQ}$ bisects ∠*MNP*,
        $NP = MQ, QP = 8,$
        $MN = 12$
   Find: *NP*

18. *Given:* $\overrightarrow{NQ}$ bisects ∠*MNP*,
        $MN = 6, QP = 2,$
        $NP - MQ = 1$
   Find: *MQ*

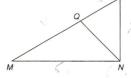

19. Refer to the accompanying figure in doing this exercise.

   (a) Name the projection of point *P* on line ℓ.
   (b) Which segment, among $\overline{PQ}, \overline{PR},$ and $\overline{PS}$, has the smallest length? Why?

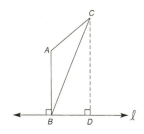

20. Name the projection on line ℓ of
   (a) $\overline{AB}$ **(b)** $\overline{AC}$ **(c)** $\overline{BC}$

21. **(a)** Name the projection of $\overline{AC}$ on $\overline{AB}$.
   (b) Name the projection of $\overline{AB}$ on $\overline{BC}$.
   (c) Name the projection of $\overline{BC}$ on $\overline{AC}$.

22. *Given:* *AC* is the geometric mean between *AD* and *AB*,
        $AD = 4, DB = 6$
   Find: *AC*

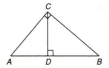

23. *Given:* $\overrightarrow{RV}$ bisects ∠*SRT*, $RS = x - 6$, $SV = 3$,
        $RT = 2 - x, VT = x + 2$
   Find: *x*

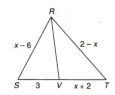

24. Complete the proof of this property:

$$\text{If } \frac{a}{b} = \frac{c}{d} \text{, then } \frac{a+c}{b+d} = \frac{a}{b} \text{ and } \frac{a+c}{b+d} = \frac{c}{d}$$

PROOF

Statements	Reasons
1. $\dfrac{a}{b} = \dfrac{c}{d}$	**1.** ?
2. $b \cdot c = a \cdot d$	**2.** ?
3. $ab + bc = ab + ad$	**3.** ?
4. $b(a + c) = a(b + d)$	**4.** ?
5. $\dfrac{a+c}{b+d} = \dfrac{a}{b}$	**5.** Means-Extremes Property (symmetric form)
6. $\dfrac{a+c}{b+d} = \dfrac{c}{d}$	**6.** ?

25. *Given:* △*RST*, with $\overleftrightarrow{XY} \parallel \overline{RT}, \overleftrightarrow{YZ} \parallel \overline{RS}$
   Prove: $\dfrac{RX}{XS} = \dfrac{ZT}{RZ}$

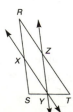

SIMILAR RIGHT TRIANGLES AND THE PYTHAGOREAN THEOREM

The following theorem was proved in Exercise 35, Section 5.3.

THEOREM 5.5.1 The altitude drawn to the hypotenuse of a right triangle separates the right triangle into two right triangles that are similar to each other and to the original right triangle.

Theorem 5.5.1 is illustrated by figure 5.25, in which the right triangle $\triangle ABC$ has its right angle at vertex C so that $\overline{CD}$ is the altitude to hypotenuse $\overline{AB}$. The smaller triangles are shown in figure 5.25b and c, and the original triangle is shown in figure 5.25d. Notice the matched arcs indicating congruent angles.

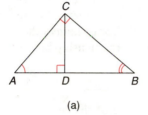
(a)

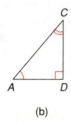
(b)

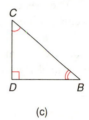
(c)

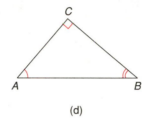
(d)

Figure 5.25

Two less obvious claims are now stated as theorems, and proofs of these are provided. In figure 5.25a, $\overline{AD}$ and $\overline{DB}$ are known as segments (parts) of the hypotenuse $\overline{AB}$.

THEOREM 5.5.2 The length of the altitude to the hypotenuse of a right triangle is the geometric mean of the lengths of the segments of the hypotenuse.

Given: $\triangle ABC$, with right $\angle ACB$, $\overline{CD} \perp \overline{AB}$

Prove: $\dfrac{AD}{CD} = \dfrac{CD}{DB}$

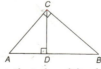

Proof: With right $\triangle ABC$ and $\overline{CD} \perp \overline{AB}$, we can use the fact that the altitude drawn to the hypotenuse of a right triangle separates it into two right triangles that are similar to each other. Because $\triangle ADC \sim \triangle CDB$, it follows that

$$\frac{AD}{CD} = \frac{CD}{DB}$$

because corresponding sides of similar triangles are proportional. ▲

THEOREM 5.5.3 The length of each leg of a right triangle is the geometric mean of the length of the hypotenuse and the length of the projection of the leg on the hypotenuse.

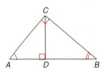

Given: $\triangle ABC$, with right $\angle ACB$, $\overline{CD} \perp \overline{AB}$

Prove: $\dfrac{AB}{AC} = \dfrac{AC}{AD}$ (Note that $\overline{AD}$ is the projection of $\overline{AC}$ on $\overline{AB}$.)

Proof: With right $\triangle ABC$ and $\overline{CD} \perp \overline{AB}$, we know that $\triangle ADC \sim \triangle ACB$ because the altitude drawn to the hypotenuse of a right triangle forms two right triangles that are similar to the original right triangle. In turn, $\dfrac{AB}{AC} = \dfrac{AC}{AD}$. $\left(\text{Similarly, it can be shown that } \dfrac{AB}{CB} = \dfrac{CB}{DB}, \text{ in which } \overline{DB}\right.$

is the projection of $\overline{CB}$ on $\overline{AB}.\Big)$ ▲

The preceding theorem opens the doors to a proof of the famous Pythagorean Theorem, one of the most often applied relationships in geometry. While the theorem's title gives credit to the Greek geometer Pythagoras, many other proofs are known, and the ancient Chinese were aware of the relationship before the time of Pythagoras.

> **THEOREM 5.5.4 (Pythagorean Theorem)** The square of the length of the hypotenuse of a right triangle is equal to the sum of the squares of the lengths of the legs.

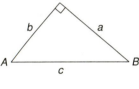

(a)

Thus, where c is the length of the hypotenuse, and a and b are the lengths of the legs, $c^2 = a^2 + b^2$.

Given: In figure 5.26, $\triangle ABC$ with right $\angle C$

Prove: $c^2 = a^2 + b^2$

Proof: Draw $\overline{CD} \perp \overline{AB}$, so that the projection of $\overline{AC}$ on the hypotenuse is $\overline{AD}$. Also, the projection of $\overline{CB}$ on $\overline{AB}$ is $\overline{DB}$. (See figure 5.26b.)

Denote $AD = x$ and $DB = y$. By Theorem 5.5.3,

$$\frac{c}{b} = \frac{b}{x} \quad \text{and} \quad \frac{c}{a} = \frac{a}{y}$$

Therefore $b^2 = cx$ and $a^2 = cy$

Using the Addition Property of Equality, we have

$$a^2 + b^2 = cy + cx$$
$$= c(y + x)$$

But $x + y = AD + DB = AB = c$. Thus

$$a^2 + b^2 = c(c) = c^2 \qquad ▲$$

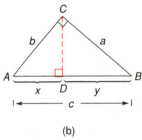

(b)

Figure 5.26

EXAMPLE 1   Given $\triangle RST$ with right $\angle S$, find:
(a) RT if $RS = 3$ and $ST = 4$
(b) RT if $RS = 4$ and $ST = 6$

(c) RS if $RT = 13$ and $ST = 12$

(d) ST if $RS = 6$ and $RT = 9$

Solution With right $\angle S$, the hypotenuse is $\overline{RT}$. Then $RT = c$, $RS = a$, and $ST = b$.

(a) $3^2 + 4^2 = c^2 \rightarrow 9 + 16 = c^2$
$$c^2 = 25 \quad \rightarrow c = 5$$
$$RT = 5$$

(b) $4^2 + 6^2 = c^2 \rightarrow 16 + 36 = c^2$
$$c^2 = 52 \quad \rightarrow c = \sqrt{52} = 2\sqrt{13}$$
$$RT = 2\sqrt{13}$$

(c) $a^2 + 12^2 = 13^2 \rightarrow a^2 + 144 = 169$
$$a^2 = 25 \quad \rightarrow a = 5$$
$$RS = 5$$

(d) $6^2 + b^2 = 9^2 \rightarrow 36 + b^2 = 81$
$$b^2 = 45 \quad \rightarrow b = \sqrt{45} = 3\sqrt{5}$$
$$ST = 3\sqrt{5}$$ ▲

The converse of the Pythagorean Theorem is also true.

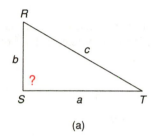

(a)

THEOREM 5.5.5 (Converse of Pythagorean Theorem) If a, b, and c are the lengths of the three sides of a triangle, with c the length of the longest side, and if $c^2 = a^2 + b^2$, then the triangle is a right triangle, with the right angle opposite the side of length c.

Given: $\triangle RST$ (figure 5.27) with sides a, b, and c so that $c^2 = a^2 + b^2$

Prove: $\triangle RST$ is a right triangle

Proof: We are given $\triangle RST$ for which $c^2 = a^2 + b^2$. Construct the right $\triangle ABC$, which has legs of lengths a and b and a hypotenuse of length x. (See figure 5.27b.) By the Pythagorean Theorem, $x^2 = a^2 + b^2$. By substitution, $x^2 = c^2$ and $x = c$. Thus $\triangle RTS \cong \triangle ABC$, by SSS. Then $\angle S$, opposite the side of length c, must be $\cong$ to $\angle C$, the right $\angle$ of $\triangle ABC$. Then $\angle S$ is a right $\angle$, and $\triangle RST$ is a right triangle. ▲

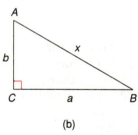

(b)

Figure 5.27

EXAMPLE 2 Which of the following can be the lengths of sides of a right triangle?

(a) $a = 5$, $b = 12$, $c = 13$

(b) $a = 15$, $b = 8$, $c = 17$

(c) $a = 7$, $b = 9$, $c = 10$

(d) $a = \sqrt{2}$, $b = \sqrt{3}$, $c = \sqrt{5}$

Solution **(a)** Because $5^2 + 12^2 = 13^2$ (that is, $25 + 144 = 169$), this is a right triangle.

(b) Because $15^2 + 8^2 = 17^2$ (that is, $225 + 64 = 289$), this triangle is a right triangle.

(c) $7^2 + 9^2 = 49 + 81 = 130$, which is not 10^2 (that is, 100), so this triangle is not a right triangle.

(d) Because $(\sqrt{2})^2 + (\sqrt{3})^2 = (\sqrt{5})^2$ (that is, $2 + 3 = 5$), this triangle is a right triangle. ▲

EXAMPLE 3   A ladder 12 ft long is leaning against a wall so that its base is 4 ft from the wall at ground level. How far up the wall does the ladder reach?

Solution   The desired height is represented by the hypotenuse h, so we have

$$4^2 + h^2 = 12^2$$
$$16 + h^2 = 144$$
$$h^2 = 128$$
$$h = \sqrt{128} = 8\sqrt{2}$$

The height is represented exactly by $h = 8\sqrt{2}$, which is approximately 11.3 ft.   ▲

EXAMPLE 4   One diagonal of a rhombus has the same length, 10 cm, as each side. How long is the other diagonal?

Solution   Because the diagonals are perpendicular bisectors of each other, four right Δs are formed. For each right Δ, a side of the rhombus is the hypotenuse. Half of each diagonal is a leg of each right triangle. Therefore

$$5^2 + b^2 = 10^2$$
$$25 + b^2 = 100$$
$$b^2 = 75$$
$$b = \sqrt{75}$$
$$= 5\sqrt{3}$$

Thus the length of the whole diagonal is $10\sqrt{3}$ cm.   ▲

In the next section, we will develop a shortcut for dealing with problems such as Example 4. In fact, there are two special cases to consider: the 45-45-90 relationship and the 30-60-90 relationship. In the previous example, it can be shown that the angles involved have measures of 30°, 60°, and 90°. Please refer back to the accompanying drawing to establish these angle measures.

While Example 5 uses the Pythagorean Theorem, it is considerably more complicated than Example 4. Indeed, it is one of those situations that may require some insight to solve. Note that the triangle described in Example 5 is *not* a right triangle because $4^2 + 5^2 \neq 6^2$.

EXAMPLE 5   A triangle has sides of lengths 4, 5, and 6 (as shown in figure 5.28). Find the length of the altitude to the side of length 6.

Solution   The altitude to the side of length 6 separates it into two parts; the lengths of these are given by x and $6 - x$. Using the two right triangles formed, we have by the Pythagorean Theorem:

$$x^2 + h^2 = 4^2 \qquad \text{and} \qquad (6 - x)^2 + h^2 = 5^2$$

Figure 5.28

Subtracting the first equation from the second, we can calculate x.

$$36 - 12x + x^2 + h^2 = 25$$
$$x^2 + h^2 = 16$$
$$36 - 12x = 9$$
$$-12x = -27$$
$$x = \frac{27}{12} = \frac{9}{4}$$

Now we use $x = \frac{9}{4}$ to find h:

$$x^2 + h^2 = 4^2$$

$$\left(\frac{9}{4}\right)^2 + h^2 = 4^2$$

$$\frac{81}{16} + h^2 = 16$$

$$\frac{81}{16} + h^2 = \frac{256}{16}$$

$$h^2 = \frac{175}{16}$$

$$h = \frac{\sqrt{175}}{4} = \frac{5\sqrt{7}}{4}$$ ▲

It is now possible to prove the HL method for the congruence of triangles.

THEOREM 5.5.6   If the hypotenuse and a leg of one right triangle are congruent to the hypotenuse and a leg of a second right triangle, then the triangles are congruent (HL).

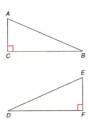

Given: Right $\triangle ABC$ with right $\angle C$ and right $\triangle DEF$ with right $\angle F$
$\overline{AB} \cong \overline{DE}$ and $\overline{AC} \cong \overline{EF}$

Prove: $\triangle ABC \cong \triangle EDF$

Proof: With right $\angle C$, the hypotenuse of $\triangle ABC$ is $\overline{AB}$; similarly, $\overline{DE}$ is the hypotenuse of right $\triangle EDF$. Because $\overline{AB} \cong \overline{DE}$, we denote the common length by c; that is, $AB = DE = c$. Because $\overline{AC} \cong \overline{EF}$, we also have $AC = EF = a$. Then

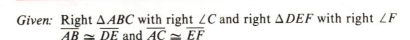

$$a^2 + (BC)^2 = c^2 \quad \text{and} \quad a^2 + (DF)^2 = c^2$$

which leads to

$$BC = \sqrt{c^2 - a^2} \quad \text{and} \quad DF = \sqrt{c^2 - a^2}$$

Then $BC = DF$ so that $\overline{BC} \cong \overline{DF}$. Hence, $\triangle ABC \cong \triangle EDF$ by SSS. ▲

Our work with the Pythagorean Theorem would be incomplete if we did not address two concerns. The first, Pythagorean Triples, involves natural (or counting) numbers. The second leads to a classification of triangles according to the lengths of their sides.

A **Pythagorean triple** is a set of three natural numbers (a, b, c) for which $a^2 + b^2 = c^2$. Three sets of triples encountered in this section so far are (3, 4, 5), (5, 12, 13), and (8, 15, 17). These numbers will always fit the sides of a right triangle.

Natural-number multiples of any of these triples will also constitute Pythagorean triples. For example, doubling (3, 4, 5) yields (6, 8, 10), which is also a Pythagorean triple. On the other hand, multiplying (3, 4, 5) by $\sqrt{3}$ yields $(3\sqrt{3}, 4\sqrt{3}, 5\sqrt{3})$, which is not a Pythagorean triple because $3\sqrt{3}$, $4\sqrt{3}$, and $5\sqrt{3}$ are not natural numbers. If a, b, c, and n are natural numbers and $a^2 + b^2 = c^2$, then

$$\begin{aligned}(na)^2 + (nb)^2 &= n^2a^2 + n^2b^2 \\ &= n^2(a^2 + b^2) \\ &= n^2c^2 \\ &= (nc)^2\end{aligned}$$

The Pythagorean triple (3, 4, 5) also leads to (9, 12, 15), (12, 16, 20), and (15, 20, 25). The Pythagorean triple (5, 12, 13) leads to triples such as (10, 24, 26) and (15, 36, 39). Basic Pythagorean triples that are used less frequently include (7, 24, 25), (9, 40, 41), and (20, 21, 29).

Pythagorean triples can be generated by using any of several formulas. One formula uses $2pq$ for one leg, $p^2 - q^2$ for the other leg, and $p^2 + q^2$ for the hypotenuse, where p and q are natural numbers and $p > q$.

The following table lists some Pythagorean triples corresponding to choices for p and q. The boldface triples are basic triples. While no exercises for this section include the phrase "Pythagorean triple," using these triples and their multiples will save you considerable time and effort.

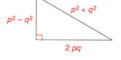

p	q	$p^2 - q^2$	$2pq$	$p^2 + q^2$
2	1	3	**4**	5
3	1	8	6	10
3	2	5	**12**	13
4	1	15	8	17
4	3	7	24	25
5	1	24	10	26
5	2	21	20	29
5	3	16	30	34
5	4	9	40	41

The Converse of the Pythagorean Theorem allows us to recognize a right triangle by knowing the lengths of sides. A variation on the Converse allows us to recognize whether a triangle is acute or obtuse. This theorem is stated without proof.

> **THEOREM 5.5.7** If *a*, *b*, and *c* are the lengths of the three sides of a triangle and *c* is the length of the longest side, and
>
> 1. $$c^2 > a^2 + b^2$$
>
> then the triangle is obtuse and the obtuse angle is opposite the side of length *c*.
> **or:**
>
> 2. $$c^2 < a^2 + b^2$$
>
> then the triangle is acute.

EXAMPLE 6   Determine the type of triangle represented if the lengths of its sides are as follows:

(a) 4, 5, 7
(b) 6, 7, 8
(c) 8, 12, 15
(d) 3, 4, 9

Solution   (a) Because *c* is the longest side, $c = 7$, and we have $7^2 > 4^2 + 5^2$ or $49 > 16 + 25$; the triangle is obtuse.

(b) Choosing $c = 8$, we have $8^2 < 6^2 + 7^2$ or $64 < 36 + 49$; the triangle is acute.

(c) Choosing $c = 15$, we have $15^2 = 9^2 + 12^2$ or $225 = 81 + 144$; the triangle is a right triangle.

(d) Because $9 > 3 + 4$, no triangle is possible. (*Recall:* The sum of the lengths of two sides of a triangle must be greater than the length of the third side.) ▲

5.5 EXERCISES

In Exercises 1 to 4, use the accompanying figure.

1. By naming the vertices in order, state three different triangles that are similar to each other.

2. Write a proportion in which
   (a) *SV* is used as a geometric mean
   (b) *RS* is a geometric mean
   (c) *TS* is a geometric mean

3. Find *RV* if
   (a) $ST = 10$ and $VT = 5$
   (b) $VT = 1$ and $RS = 2\sqrt{3}$

4. Find *VT* if
   (a) $VS = 15$ and $RV = 10$
   (b) $RS = \sqrt{6}$ and $RT = 4$

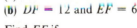

In Exercises 5 to 8, use the accompanying figure.

5. Find the length of $\overline{DF}$ if
   (a) $DE = 8$ and $EF = 6$
   (b) $DE = 5$ and $EF = 3$

6. Find the length of $\overline{DE}$ if
   (a) $DF = 13$ and $EF = 5$
   (b) $DF = 12$ and $EF = 6\sqrt{3}$

7. Find *EF* if
   (a) $DF = 17$ and $DE = 15$
   (b) $DF = 12$ and $DE = 8\sqrt{2}$

8. Find *DF* if
   (a) $DE = 12$ and $EF = 5$
   (b) $DE = 12$ and $EF = 6$

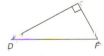

9. Determine the type of triangle represented if the lengths of its sides are

(a) $a = 4$, $b = 3$, and $c = 5$
(b) $a = 4$, $b = 5$, and $c = 6$
(c) $a = 2$, $b = \sqrt{3}$, and $c = \sqrt{7}$
(d) $a = 3$, $b = 8$, and $c = 15$

10. Determine the type of triangle represented if the lengths of its sides are

(a) $a = 1.5$, $b = 2$, and $c = 2.5$
(b) $a = 20$, $b = 21$, and $c = 29$
(c) $a = 10$, $b = 12$, and $c = 16$
(d) $a = 5$, $b = 7$, and $c = 9$

11. As shown in the drawing below, a guy wire 25 ft long supports an antenna at a point that is 20 ft from the base of the antenna. How far from the base of the antenna is the guy wire secured?

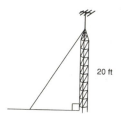

12. A strong wind keeps a kite that a girl is holding 30 ft above the earth in a position 40 ft across the ground, as shown in the accompanying drawing. How much string is out?

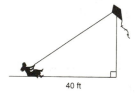

13. A boat is 6 m below the level of a pier and 12 m from the pier as measured across the water. How much rope is needed to reach the boat?

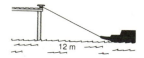

14. A hot air balloon is held in place by the ground crew at a point that is 21 ft from a point directly beneath the balloon, as shown in the accompanying illustration. If the rope is of length 29 ft, how far above ground level is the balloon?

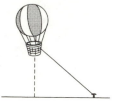

15. A rectangle has a width of 16 cm and a diagonal of length 20 cm. How long is the rectangle?

16. A right triangle has legs of lengths x and $2x + 2$ and a hypotenuse of length $2x + 3$. What are the lengths of its sides?

17. A rectangle has base $x + 3$, altitude $x + 1$, and diagonals $2x$ each. What are the lengths of its base, altitude, and diagonals?

18. The diagonals of a rhombus measure 6 m and 8 m. How long are each of the congruent sides?

19. Each side of a rhombus measures 12 in. If one diagonal is 18 in. long, how long is the other diagonal?

20. An isosceles right triangle has a hypotenuse of length 10 cm. How long is each leg?

21. An isosceles right triangle has each leg of length $6\sqrt{2}$ in. What is the length of the hypotenuse?

22. In right $\triangle ABC$ with right $\angle C$, $AB = 10$ and $BC = 8$. Find the length of $\overline{MB}$ if M is the midpoint of $\overline{AC}$.

23. In right $\triangle ABC$ with right $\angle C$, $AB = 17$ and $BC = 15$. Find the length of $\overline{MN}$ if M and N are the midpoints of $\overline{AB}$ and $\overline{BC}$, respectively.

24. Find the length of the altitude to the 10-in. side of a triangle whose sides are 6, 8, and 10 in. in length, respectively.

25. Find the length of the altitude to the 26-in. side of a triangle whose sides are 10, 24, and 26 in. in length, respectively.

26. In quadrilateral $ABCD$, $\overline{BC} \perp \overline{AB}$ and $\overline{DC} \perp$ diagonal $\overline{AC}$. If $AB = 4$, $BC = 3$, and $DC = 12$, determine DA.

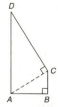

27. In quadrilateral $RSTU$, $\overline{RS} \perp \overline{ST}$ and $\overline{UT} \perp$ diagonal $\overline{RT}$. If $RS = 6$, $ST = 8$, and $RU = 15$, find UT.

28. *Given:* $\triangle ABC$ is not a right $\triangle$
Prove: $a^2 + b^2 \neq c^2$

29. If $a = p^2 - q^2$, $b = 2pq$, and $c = p^2 + q^2$, show that $c^2 = a^2 + b^2$.

30. Given that the segment shown in the accompanying drawing has length 1, construct a segment whose length is $\sqrt{2}$.

31. Using the same segment as in Exercise 30 (whose length is 1 unit), construct a segment of length 2 and then a second segment of length $\sqrt{5}$.

32. When the rectangle in the accompanying drawing (whose dimensions are 16 by 9) is cut into pieces and rearranged,

a square can be formed. What is the perimeter (sum of lengths of the four sides) of this square?

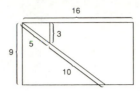

33. A, C, and F are three of the vertices of the cube shown in the accompanying figure. Given that each face of the cube is a square, what is the measure of angle ACF?

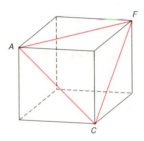

*34. Find the length of the altitude to the 8-in. side of a triangle whose sides are 4, 6, and 8 in. long. (*Hint:* Use the Pythagorean Theorem twice.)

5.6 SPECIAL RIGHT TRIANGLES

Certain right triangles occur so often that they deserve more attention than others. The special right triangles we will consider have angle measures of 45°, 45°, and 90° or of 30°, 60°, and 90°.

In the 45-45-90 triangle, the legs are opposite the congruent angles and are also congruent. Rather than using a and b to represent the lengths of the legs, we use a for both lengths, as shown in figure 5.29. It then follows by the Pythagorean Theorem that

$$c^2 = a^2 + a^2$$
$$= 2a^2$$
$$c = a\sqrt{2}$$

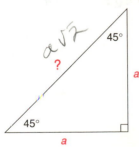

Figure 5.29

> **THEOREM 5.6.1 (45-45-90 Theorem)** In a triangle whose angles measure 45°, 45°, and 90°, the hypotenuse has a length equal to the product of $\sqrt{2}$ and the length of either leg.

It would be better to memorize the sketch in figure 5.30 than to repeat the steps preceding the 45-45-90 Theorem.

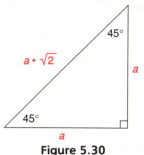

Figure 5.30

EXAMPLE 1   Find the lengths of the missing sides in each triangle in figure 5.31.

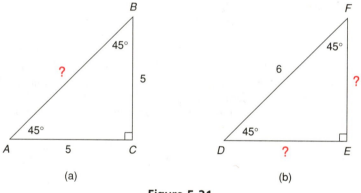

Figure 5.31

Solution   (a) The length of hypotenuse $\overline{AB}$ is $5\sqrt{2}$, the product of $\sqrt{2}$ and the length of either of the equal legs.

(b) Let a denote the length of $\overline{DE}$ and of $\overline{EF}$. Then $a\sqrt{2} = 6$, so $a = \dfrac{6}{\sqrt{2}}$.

Simplifying,

$$
\begin{aligned}
a &= \frac{6}{\sqrt{2}} \cdot \frac{\sqrt{2}}{\sqrt{2}} \\
&= \frac{6\sqrt{2}}{2} \\
&= 3\sqrt{2}
\end{aligned}
$$

Therefore $DE = EF = 3\sqrt{2}$.

Note: The solution in (a) can be found by solving the equation $5^2 + 5^2 = c^2$; the solution in (b) can be found by solving $a^2 + a^2 = 6^2$.   ▲

The second special triangle is the 30-60-90 triangle, shown in figure 5.32a. Its sides are related to each other in a way that can be seen by reflecting (as a

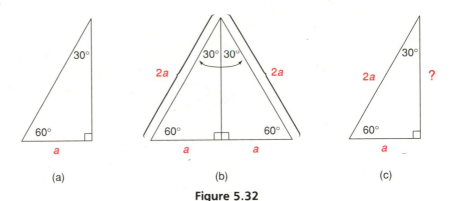

Figure 5.32

mirror might) the triangle across the side included by the 30° and 90° angles. As shown in figure 5.32b, the reflection produces an equiangular and therefore equilateral triangle in which the length of the side that was the hypotenuse of the 30-60-90 triangle is twice the length of the leg opposite the 30° angle. Extending the shorter leg while allowing for the 30° angle ensures that the two smaller triangles are congruent.

Take another look at the 30-60-90 triangle in figure 5.32c. Since we know that the length of the hypotenuse is twice the length of the shorter leg, we can again use the Pythagorean Theorem to represent the length of the longer leg. Where b is the length of the longer leg, we have

$$c^2 = a^2 + b^2$$
$$(2a)^2 = a^2 + b^2$$
$$4a^2 = a^2 + b^2$$
$$b^2 = 3a^2$$
$$b = a\sqrt{3}$$

> **THEOREM 5.6.2 (30-60-90 Theorem)** In a triangle whose angles measure 30°, 60°, and 90°, the hypotenuse has a length equal to twice the length of the shorter leg, while the length of the longer leg is the product of $\sqrt{3}$ and the length of the shorter leg.

It would be best to memorize the sketch in figure 5.33. So that you more easily recall which expression is used for each side, remember that the lengths of the sides follow the same order as the angles opposite them. Thus:

(smallest) opposite the 30° ∠ is a

(middle) opposite the 60° ∠ is $a\sqrt{3}$

(largest) opposite the 90° ∠ is $2a$

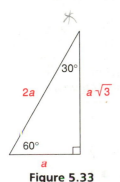

Figure 5.33

EXAMPLE 2   Find the lengths of the missing sides of each triangle in figure 5.34 on page 188.

Solution   (a) $RT = 2 \cdot RS = 2 \cdot 5 = 10$
$ST = RS\sqrt{3} = 5\sqrt{3}$

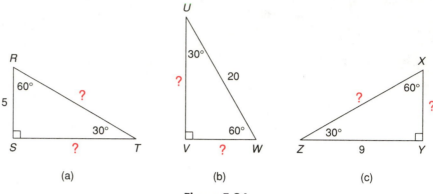

Figure 5.34

(b) $UW = 2 \cdot VW \rightarrow 20 = 2 \cdot VW \rightarrow VW = 10$
$UV = VW\sqrt{3} = 10\sqrt{3}$

(c) $ZY = XY\sqrt{3} \rightarrow 9 = XY \cdot \sqrt{3} \rightarrow XY = \dfrac{9}{\sqrt{3}} = \dfrac{9}{\sqrt{3}} \cdot \dfrac{\sqrt{3}}{\sqrt{3}}$

$$= \dfrac{9\sqrt{3}}{3} = 3\sqrt{3}$$

$XZ = 2 \cdot XY = 2 \cdot 3\sqrt{3} = 6\sqrt{3}$ ▲

Before we continue our work with geometry, we need to look back at some algebra. First recall that the FOIL method of multiplying binomials gives us this rule:

$$(a + b)(a - b) = a^2 - b^2$$

Then $(4 + \sqrt{2})(4 - \sqrt{2}) = 4^2 - 4\sqrt{2} + 4\sqrt{2} - (\sqrt{2})^2$, or $16 - 2$, which is 14. The product of such complicated expressions has a simple form. We call the expression $(4 - \sqrt{2})$ the **conjugate** of $(4 + \sqrt{2})$, and conversely. In general, the conjugate of $(a + \sqrt{b})$ is $(a - \sqrt{b})$, while the conjugate of $(a - \sqrt{b})$ is $(a + \sqrt{b})$. The table presented as Example 3 shows a few of these related values and their products.

EXAMPLE 3

Square Root Expression	Conjugate of Expression	Product of Expression and Conjugate
$5 + \sqrt{3}$	$5 - \sqrt{3}$	$25 - 3 = 22$
$\sqrt{7} - 2$	$\sqrt{7} + 2$	$7 - 4 = 3$
$2\sqrt{3} + \sqrt{5}$	$2\sqrt{3} - \sqrt{5}$	$12 - 5 = 7$
$-3 + \sqrt{2}$	$-3 - \sqrt{2}$	$9 - 2 = 7$
$\sqrt{a} + \sqrt{b}$	$\sqrt{a} - \sqrt{b}$	$a - b$

▲

Multiplication by the conjugate of a square root expression is used to eliminate square roots from the denominator of a fraction; this procedure is known as **rationalizing the denominator.**

EXAMPLE 4   Simplify the quotient $\dfrac{12}{4 + \sqrt{2}}$.

Solution   By multiplying *both* the numerator and the denominator of the fraction by the conjugate of the *denominator*, we can eliminate the radical from the denominator:

$$\frac{12}{4 + \sqrt{2}} = \frac{12}{4 + \sqrt{2}} \cdot \frac{4 - \sqrt{2}}{4 - \sqrt{2}}$$

$$= \frac{12(4 - \sqrt{2})}{14}$$

$$= \frac{6(4 - \sqrt{2})}{7}$$

▲

Recall Theorem 5.4.3:

If a ray bisects one angle of a triangle, then it divides the opposite side into segments that are proportional to the two sides which form that angle.

Combining facts from this theorem about angle bisectors and the 30-60-90 relationship, we can also find the lengths of sides of a right triangle whose angles measure 15°, 75°, and 90°. The method of solving the 15-75-90 triangle is suggested by the drawings in figure 5.35. In this method, we construct (or sketch) a 15° angle so that it appears as one of the two bisected parts of the 30° angle of a 30-60-90 triangle.

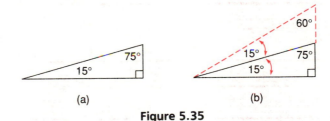

(a)   (b)

Figure 5.35

Although we illustrate the method for the 15-75-90 triangle in Example 5, neither this relationship nor the one involved in the 22.5-67.5-90 triangle (to be developed later, and illustrated in Example 6) are as important as the 30-60-90 and 45-45-90 relationships.

EXAMPLE 5   In right $\triangle RST$ in figure 5.36a, $m\angle S = 15°$. If $TR = 6$, find SR.

Solution   Construct an angle at vertex S measuring 15°, in such a way that its second side intersects $\overline{TR}$ extended upward. If we call the point of intersection V, $\triangle VSR$ must be a 30-60-90 triangle. (See figure 5.36b.) If $VT = x$, then

$$VR = x + 6$$
$$SR = (x + 6)\sqrt{3}$$
$$VS = 2(x + 6)$$

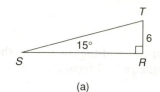

(a)

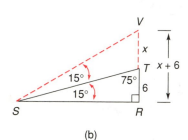

(b)

Figure 5.36

Using Theorem 5.4.3, we find

$$\frac{VT}{TR} = \frac{SV}{SR} \qquad \text{so} \qquad \frac{x}{6} = \frac{2\cancel{(x+6)}}{\cancel{(x+6)}\sqrt{3}}$$

Simplifying and using the Means-Extremes Property, we get

$$x\sqrt{3} = 12$$

$$x = \frac{12}{\sqrt{3}}$$

$$= \frac{12}{\sqrt{3}} \cdot \frac{\sqrt{3}}{\sqrt{3}}$$

$$= \frac{12\sqrt{3}}{3} = 4\sqrt{3}$$

That is, $VT = 4\sqrt{3}$; then $SR = (x + 6)\sqrt{3}$ becomes

$$SR = (4\sqrt{3} + 6)\sqrt{3}$$
$$= 12 + 6\sqrt{3} \qquad \blacktriangle$$

One variation combines the 45-45-90 triangle with the angle bisector theorem (Theorem 5.4.3). In this instance, the angle measures for the triangle are 22.5°, 67.5°, and 90°. (See figure 5.37.)

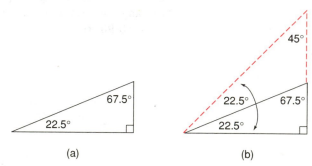

(a)   (b)

Figure 5.37

EXAMPLE 6   In $\triangle HJK$ shown in figure 5.38a, $m\angle H = 22.5°$ and $m\angle J = 67.5°$. If $HK = 10$, find JK.

Solution   Side $\overline{JK}$ is extended upward to meet $\overline{HP}$, which has been drawn so that $m\angle JHP = 22.5°$ (and $m\angle KHP = 45°$; see figure 5.38b). Now $\triangle PHK$ is a 45-45-90 triangle, so $PK = HK = 10$ and $PH = 10\sqrt{2}$.

Because $\overrightarrow{HJ}$ bisects $\angle PHK$, it follows that $\dfrac{PH}{HK} = \dfrac{PJ}{JK}$. Let $JK = x$ and $PJ = 10 - x$; then the proportion becomes

$$\frac{\cancel{10}\sqrt{2}}{\cancel{10}} = \frac{10 - x}{x}$$

$$\sqrt{2} = \frac{10 - x}{x}$$

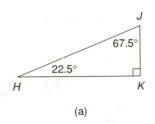

(a)

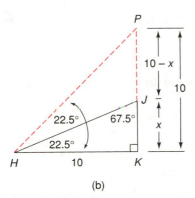

(b)

Figure 5.38

Then

$$\sqrt{2} \cdot x = 10 - x$$
$$x + (\sqrt{2} \cdot x) = 10$$
$$(1 + \sqrt{2})x = 10$$
$$x = \frac{10}{1 + \sqrt{2}}$$

Again, there is a radical in the denominator, so we must rationalize:

$$x = \frac{10}{1 + \sqrt{2}} \cdot \frac{1 - \sqrt{2}}{1 - \sqrt{2}}$$
$$= \frac{10(1 - \sqrt{2})}{1 - \sqrt{2} + \sqrt{2} - 2}$$
$$= \frac{10(1 - \sqrt{2})}{-1}$$
$$= -10(1 - \sqrt{2})$$
$$= 10(\sqrt{2} - 1)$$

Then $JK = 10(\sqrt{2} - 1) = 10\sqrt{2} - 10$.
Note: The solution is a positive number since $\sqrt{2} \approx 1.4$. ▲

5.6 EXERCISES

In Exercises 1 to 14, find the missing lengths. Leave your answers in simplest radical form.

Use the figure shown for Exercises 1 and 2.

1. *Given:* Right $\triangle XYZ$ with $m\angle X = 45°$ and $XZ = 8$
   Find: YZ and XY

2. *Given:* Right $\triangle XYZ$ with $\overline{XZ} \cong \overline{YZ}$ and $XY = 10$
   Find: XZ and YZ

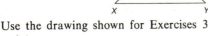

Use the drawing shown for Exercises 3 and 4.

3. *Given:* Right $\triangle DEF$ with $m\angle E = 60°$ and $DF = 5$
   Find: DF and FE

4. *Given:* Right $\triangle DEF$ with $m\angle E = 2 \cdot m\angle F$ and $EF = 12\sqrt{3}$
   Find: DE and DF

5. *Given:* Rectangle $HJKL$ with diagonals $\overline{HK}$ and $\overline{JL}$
   $m\angle HKL = 30°$
   Find: HL, HK, and MK

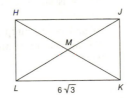

6. *Given:* Right $\triangle RST$ with $RT = 6\sqrt{2}$ and $m\angle STV = 150°$
   Find: RS and ST

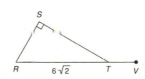

7. *Given:* $\triangle ABC$ with $m\angle A = m\angle B = 45°$ and $BC = 6$
   Find: AC and AB

8. *Given:* Right $\triangle MNP$ with $MP = PN$ and $MN = 10\sqrt{2}$
   Find: PM and PN

9. *Given:* $\triangle RST$ with $m\angle T = 30°$, $m\angle S = 60°$, and $ST = 12$
   Find: RS and RT

10. *Given:* $\triangle XYZ$ with $\overline{XY} \cong \overline{XZ} \cong \overline{YZ}$
   $\overline{ZW} \perp \overline{XY}$
   $YZ = 6$
   Find: ZW

11. *Given:* Square $ABCD$ with diagonals $\overline{DB}$ and $\overline{AC}$ intersecting at E
   $DC = 5\sqrt{3}$
   Find: DB

12. *Given:* $\triangle NQM$ with $\angle$s as shown in the accompanying drawing
   $\overline{MP} \perp \overline{NQ}$
   Find: NM, MP, MQ, PQ, and NQ

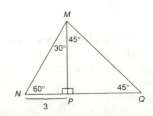

13. *Given:* $\triangle XYZ$ with ∠s as shown in the accompanying drawing
    Find: XY (*Hint:* Compare this drawing to the drawing in Exercise 12.)

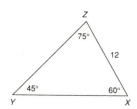

14. *Given:* Rhombus $ABCD$ in which diagonals $\overline{AC}$ and $\overline{DB}$ intersect at E
    $DB = AB$
    Find: AC

15. Give the conjugate of each square root expression:
    (a) $5 + \sqrt{3}$
    (b) $7 - \sqrt{2}$
    (c) $6 + 2\sqrt{5}$
    (d) $\sqrt{5} - \sqrt{3}$

16. Give the conjugate of each square root expression:
    (a) $a + \sqrt{b}$
    (b) $6\sqrt{2} + 3$
    (c) $5\sqrt{2} - 4\sqrt{3}$
    (d) $\sqrt{3} + 1$

17. Rationalize each denominator to simplify:
    (a) $\dfrac{6}{1 + \sqrt{3}}$ (b) $\dfrac{4}{\sqrt{5} - 1}$

18. Rationalize each denominator to simplify:
    (a) $\dfrac{10}{2\sqrt{3} + 1}$ (b) $\dfrac{15}{6 - \sqrt{3}}$

In Exercises 19 to 29, use the given information to find any indicated measures.

Use the figure shown for Exercises 19 and 20.

19. *Given:* In $\triangle ABC$, $\overrightarrow{AD}$ bisects ∠BAC
    m∠$B = 30°$ and $AB = 12$
    Find: DC and DB

20. *Given:* In $\triangle ABC$, $\overrightarrow{AD}$ bisects ∠BAC
    $AB = 20$ and $AC = 16$
    Find: DC and DB

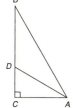

21. *Given:* $\triangle MNQ$ is equiangular
    $\overrightarrow{NR}$ bisects ∠$LMNQ$
    $\overrightarrow{QR}$ bisects ∠$LMQN$
    Find: NQ

22. *Given:* $\triangle STV$ is an isosceles right triangle
    M and N are midpoints of $\overline{ST}$ and $\overline{SV}$, respectively
    Find: MN

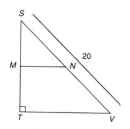

23. *Given:* ∠Y is a right ∠ in $\triangle XYZ$
    m∠$X = 15°$, m∠$Z = 75°$, and $XY = 6\sqrt{3}$
    Find: ZY

24. *Given:* ∠Y is a right ∠ in $\triangle XYZ$
    m∠$X = 22.5°$, m∠$Z = 67.5°$, and $XY = 12$
    Find: ZY

25. *Given:* The same angle measures as in Exercise 24, but with $XY = 20$
    Find: ZY

Use the drawing shown for Exercises 26 and 27.

*26. *Given:* $\triangle ABC$ with m∠$A = 45°$, m∠$B = 22.5°$, and m∠$C = 112.5°$
    $DB = 10\sqrt{2}$
    Find: AC

*27. *Given:* The same figure and angle measures as in Exercise 26, but with $DB = 12$
    Find: AC

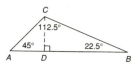

*28. *Given:* $\overline{HJ}$ is the altitude from vertex H to side $\overline{KL}$ of $\triangle HKL$
    m∠$K = 60°$, m∠$L = 67.5°$, and $HJ = 2$
    Find: KL

*29. *Given:* $\triangle ABC$ is isosceles with m∠$ABC = $ m∠$C = 72°$
    $\overrightarrow{BD}$ bisects ∠ABC and $AB = 1$
    Find: BC

▲ A LOOK BEYOND: AN UNUSUAL APPLICATION OF SIMILAR TRIANGLES

The following problem is one that can be solved in many ways. If methods of calculus are applied, the solution is found through many complicated and tedious calculations. The simplest solution, which follows, utilizes geometry and similar triangles.

PROBLEM

A hiker is at a location 450 ft downstream from his campsite. He is 200 ft away from the straight stream, and his tent is 100 ft away, as shown in figure 5.39a. Across the flat field,

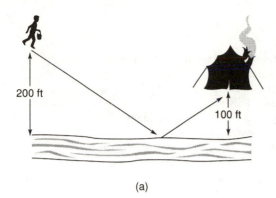

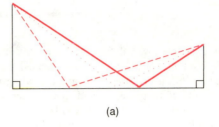

(a)

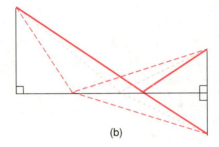

(b)

Figure 5.40

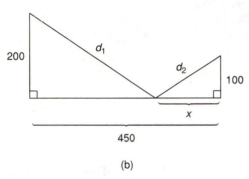

(b)

Figure 5.39

he sees that a spark from his campfire has ignited the tent. Taking the empty bucket he is carrying, he runs to the river to get water and then on to the tent. To what point on the river should he run to minimize the distance he travels?

We wish to determine x in figure 5.39b so that the total distance $D = d_1 + d_2$ is as small as possible. Consider three possible choices of this point on the river. These are suggested by dashed, dotted, and solid lines in figure 5.40a. Also consider the reflections of the triangles across the river. (See figure 5.40b.)

It is obvious that the minimum distance D occurs where the segments of lengths d_1 and d_2 form a straight line. That is, the configuration with the solid line segments minimizes the distance. In that case, the triangle at left and the reflected triangle at right are similar. (See figure 5.41.)

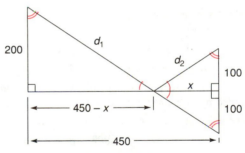

Figure 5.41

Thus
$$\frac{200}{100} = \frac{450 - x}{x}$$

$$200x = 100(450 - x)$$
$$200x = 45,000 - 100x$$
$$300x = 45,000$$
$$x = 150$$

Thus the desired point on the river is 300 ft (determined by $450 - x$) upstream from the hiker's location. ▲

▲ SUMMARY

A LOOK BACK AT CHAPTER 5

The goal of this chapter has been to define similarity for all polygons. We postulated a method for proving triangles similar and showed that proportions are a consequence of similar triangles, a line parallel to one side of a triangle, and a ray bisecting one angle of a triangle. The Pythagorean Theorem and its converse were proved. The 30-60-90 triangle, the 45-45-90 triangle, and other special triangles were discussed.

A Look Ahead to Chapter 6

In the next chapter, we will begin our work with the circle. Segments and lines of the circle will be defined, as will special angles in a circle. Several theorems dealing with the measurements of these angles and line segments will be proved. Our work with constructions will continue, to enable us to deal with locus and concurrency of lines.

Important Terms and Concepts of Chapter 5

5.1 Quadratic Equations
    Quadratic Formula
    Incomplete Quadratic Equations
    Radicals, Properties of Radicals
5.2 Ratio, Proportion
    Commensurable, Incommensurable

    Extremes and Means of a Proportion
    Geometric Mean
    Extended Ratio, Extended Proportion
5.3 Similar Polygons
5.4 Segments Divided Proportionally
    Projection of a Point on a Line
    Projection of a Line Segment on a Line
5.5 Pythagorean Theorem and Its Converse
    Pythagorean Triples
5.6 30-60-90 Triangle
    45-45-90 Triangle
    15-75-90 Triangle
    22.5-67.5-90 Triangle
A Look Beyond An Unusual Application of Similar Triangles

▲ REVIEW EXERCISES

Answer true or false for Review Exercises 1 to 7.

1. The ratio of 12 hr to 1 day is 2 to 1.

2. If the numerator and the denominator of a ratio are multiplied by 4, the new ratio equals the given ratio.

3. The value of a ratio must be less than one.

4. The three numbers 6, 14, and 22 are in a ratio of 3:7:11.

5. To correctly express a ratio, the terms must have the same unit of measure.

6. The ratio 3:4 is the same as the ratio 4:3.

7. If the second and third terms of a proportion are equal, then either is the geometric mean of the first and fourth terms.

8. Find the value(s) of x in each proportion:

(a) $\dfrac{x}{6} = \dfrac{3}{x}$

(b) $\dfrac{x-5}{3} = \dfrac{2x-3}{7}$

(c) $\dfrac{6}{x+4} = \dfrac{2}{x+2}$

(d) $\dfrac{x+3}{5} = \dfrac{x+5}{7}$

(e) $\dfrac{x-2}{x-5} = \dfrac{2x+1}{x-1}$

(f) $\dfrac{x(x+5)}{4x+4} = \dfrac{9}{5}$

(g) $\dfrac{x-1}{x+2} = \dfrac{10}{3x-2}$

(h) $\dfrac{x+7}{2} = \dfrac{x+2}{x-2}$

Use proportions to solve Review Exercises 9 to 11.

9. Four containers of fruit juice cost $2.52. How much do six containers cost?

10. Two packages of M&Ms cost 69¢. How many packages can you buy for $2.25?

11. A rug measuring 20 m² costs $132. How much would a 12-m² rug of the same material cost?

12. The ratio of the measures of the sides of a quadrilateral is 2:3:5:7. If the perimeter is 68, find the length of each side.

13. The length and width of a rectangle are 18 and 12, respectively. A similar rectangle has length 27. What is its width?

14. The sides of a triangle are 6, 8, and 9. The shortest side of a similar triangle is 15. How long are its other sides?

15. The ratio of the measure of the supplement of an angle to that of the complement of the angle is 5:2. Find the measure of the supplement.

16. *Given:* $ABCD$ is a parallelogram
      $\overline{DB}$ intersects $\overline{AE}$ at point F
Prove: $\dfrac{AF}{EF} = \dfrac{AB}{DE}$

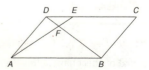

17. *Given:* $\angle 1 \cong \angle 2$

Prove: $\dfrac{AB}{AC} = \dfrac{BE}{CD}$

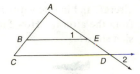

18. *Given:* $\triangle ABC \sim \triangle DEF$
  $m\angle A = 50°$, $m\angle E = 33°$
  $m\angle D = 2x + 40$
  Find: x, $m\angle F$

19. *Given:* In $\triangle ABC$ and $\triangle DEF$,
  $\angle B \cong \angle F$ and $\angle C \cong \angle E$
  $AC = 9$, $DE = 3$, $DF = 2$,
  $FE = 4$
  Find: AB, BC

In the accompanying figure for Review Exercises 20 to 22, $\overline{DE} \parallel \overline{AC}$.

20. $BD = 6$, $BE = 8$, $EC = 4$, $AD = ?$

21. $AD = 4$, $BD = 8$, $DE = 3$, $AC = ?$

22. $AD = 2$, $AB = 10$, $BE = 5$, $BC = ?$

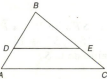

In the accompanying figure for Review Exercises 23 to 25, $\overrightarrow{GJ}$ bisects $\angle FGH$.

23. *Given:* $FG = 10$, $GH = 8$,
  $FJ = 7$
  Find: JH

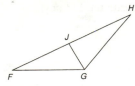

24. *Given:* $GF{:}GH = 1{:}2$, $FJ = 5$
  Find: JH

25. *Given:* $FG = 8$, $HG = 12$,
  $FH = 15$
  Find: FJ

26. *Given:* $\overleftrightarrow{EF} \parallel \overleftrightarrow{GO} \parallel \overleftrightarrow{HM} \parallel \overleftrightarrow{JK}$,
  with transversals $\overline{FJ}$ and $\overline{EK}$
  $FG = 2$, $GH = 8$, $HJ = 5$,
  $EM = 6$

Find: EO, EK

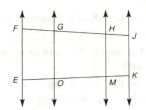

27. Prove that if a line bisects one side of a triangle and is parallel to a second side, then it bisects the third side.

28. Prove that the diagonals of a trapezoid divide themselves proportionally.

29. *Given:* $\triangle ABC$ with right $\angle BAC$
  $\overline{AD} \perp \overline{BC}$

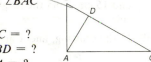

  (a) $BD = 3$, $AD = 5$, $DC = ?$
  (b) $AC = 10$, $DC = 4$, $BD = ?$
  (c) $BD = 2$, $BC = 6$, $BA = ?$
  (d) $BD = 3$, $AC = 3\sqrt{2}$, $DC = ?$

30. *Given:* $\triangle ABC$ with right $\angle ABC$
  $\overline{BD} \perp \overline{AC}$

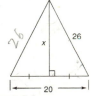

  (a) $BD = 12$, $AD = 9$, $DC = ?$
  (b) $DC = 5$, $BC = 15$, $AD = ?$
  (c) $AD = 2$, $DC = 8$, $AB = ?$
  (d) $AB = 2\sqrt{6}$, $DC = 2$, $AD = ?$

31. In the drawing shown, find x.

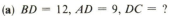

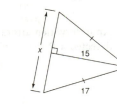

32. *Given:* $ABCD$ is a rectangle
  E is the midpoint of $\overline{BC}$
  $AB = 16$, $CF = 9$, $AD = 24$
  Find: AE, EF, AF

33. Find the length of a diagonal of a square whose side is 4 in. long.

34. Find the length of a side of a square whose diagonal is 6 cm long.

35. Find the length of a side of a rhombus whose diagonals are 48 cm and 14 cm long.

36. Find the length of an altitude of an equilateral triangle whose side is 10 in. long.

37. Find the length of a side of an equilateral triangle if an altitude is 6 in. long.

38. The lengths of three sides of a triangle are 13 cm, 14 cm, and 15 cm. Find the length of the altitude to the 14-cm side.

39. In the drawing, find x and y.

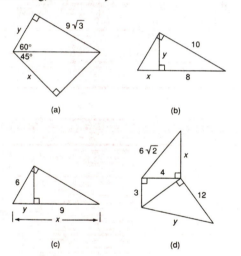

(a)

(b)

(c)

(d)

40. An observation aircraft flying at a height of 12 km has detected a Russian ship at a distance of 20 km from the aircraft and in line with an American ship that is 13 km from the aircraft. How far apart are the U.S. and Russian ships?

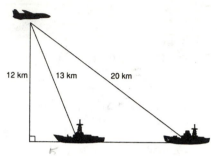

41. Tell whether each set of numbers represents the lengths of the sides of an acute triangle, of an obtuse triangle, of a right triangle, or of no triangle:

(a) 12, 13, 14   (e) 8, 7, 16

(b) 11, 5, 18   (f) 8, 7, 6

(c) 9, 15, 18   (g) 9, 13, 8

(d) 6, 8, 10   (h) 4, 2, 3

42. *Given:* $\triangle ABC$ with $m\angle A = 22.5°$, $m\angle B = 67.5°$, $AC = 5\sqrt{2}$

Find: BC

43. *Given:* Isosceles $\triangle RST$ with $\overline{RS} \cong \overline{RT}$
$m\angle SRT = 150°$, $ST = 16\sqrt{3}$
$\overline{RV}$ is an altitude to $\overline{ST}$

Find: RV

*44. Prove that in a 22.5°-67.5°-90° triangle, the length of the leg opposite the 22.5° angle is $(\sqrt{2} - 1)$ times the length of the leg opposite the 67.5° angle.

*45. Prove that in a 15°-75°-90° triangle, the length of the leg opposite the 75° angle is $(2 + \sqrt{3})$ times the length of the leg opposite the 15° angle.

6

CIRCLES

6.1
6.2

When we consider something as simple as a pancake, as functional as a gear or pulley, or as attractive as a wire-spoke wheel, we generally think of a circle. In this chapter, we deal with circles and develop their properties, which are logical consequences of the properties that have been developed in previous chapters. The gears, pulleys, and wheels of an automobile all illustrate the use of the circle in engineering.

CHAPTER OUTLINE

6.1 Circles and Related Segments and Angles

6.2 More Angle Measures in the Circle

6.3 Line and Segment Relationships in the Circle

6.4 Some Constructions and Inequalities for the Circle

6.5 Locus and Concurrency

A Look Beyond   The Value of π

6.1 CIRCLES AND RELATED SEGMENTS AND ANGLES

In this chapter, the focus of our attention is the circle. We will develop terminology related to the circle, some methods of measurement, and many properties of the circle.

DEFINITION: A **circle** is the set of all points in a plane that are at the same distance from a fixed point known as the center of the circle.

Pictured in figure 6.1 is circle P; a circle is named after its center point. The symbol for circle is ⊙, and the symbol for circles (plural) is ⑤. While P is the center of the circle, it is *not* a point of (or on) the circle. Points A, B, C, and D are points of (or on) the circle. Points P and R are in the *interior* of circle P; points G and H are in the *exterior* of the circle.

In ⊙Q in figure 6.2, $\overline{SQ}$ is a radius of the circle. A **radius** is a segment that joins the center of the circle to a point on the circle. $\overline{SQ}$, $\overline{TQ}$, $\overline{VQ}$, and $\overline{WQ}$ are **radii** (plural of "radius") of ⊙Q. Theorem 6.1.1 follows from the definition of "circle."

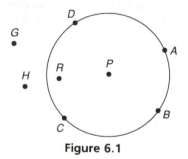

Figure 6.1

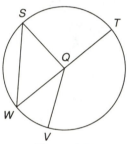

Figure 6.2

THEOREM 6.1.1   All radii of a circle are congruent.

A segment that joins two points of a circle (like $\overline{SW}$ in figure 6.2) is a **chord** of the circle. A **diameter** of a circle is a chord that contains the center of the circle; in figure 6.2, $\overline{TW}$ is a diameter of ⊙Q.

DEFINITION: **Congruent circles** are circles that have congruent radii.

Circles P and Q in figure 6.3 are congruent because their radii have equal lengths.

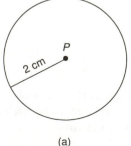

(a)

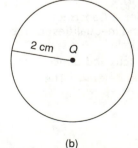

(b)

Figure 6.3

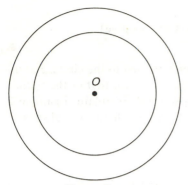

Figure 6.4

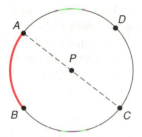

Figure 6.5

> **DEFINITION:** **Concentric circles** are coplanar circles that have a common center.

The circles shown in figure 6.4 are concentric and share the common center O. In $\odot P$, the part of the circle shown from point A to point B is **arc** AB; arc AB is symbolized by $\overset{\frown}{AB}$. If $\overline{AC}$ is a diameter, then $\overset{\frown}{ABC}$ (three letters are used for clarity) is a **semicircle.** Two phrases are now described informally. Refer to figure 6.5 for depictions of these terms. A **minor arc,** like $\overset{\frown}{AB}$, is one that is less than a semicircle; a **major arc,** like $\overset{\frown}{ABCD}$ (also denoted by $\overset{\frown}{ABD}$ or $\overset{\frown}{ACD}$) is one that is more than a semicircle.

> **DEFINITION:** A **central angle** of a circle is an angle whose vertex is the center of the circle and whose sides are rays containing radii of the circle.

In circle O in figure 6.6, $\angle NOP$ is a central angle. $\overset{\frown}{NP}$, which is cut off in the interior of the angle, is the **intercepted arc** of $\angle NOP$.

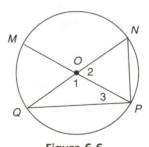

Figure 6.6

EXAMPLE 1 In figure 6.6, $\overline{MP}$ and $\overline{NQ}$ intersect at O, the center of the circle. Name
(a) All radii
(b) All diameters
(c) All chords
(d) One central angle
(e) One minor arc
(f) One semicircle
(g) One major arc
(h) One point not on $\odot O$

Solution (a) $\overline{OM}$, $\overline{OQ}$, $\overline{OP}$, and $\overline{ON}$
(b) $\overline{MP}$ and $\overline{QN}$
(c) $\overline{MP}$, $\overline{QN}$, $\overline{QP}$, and $\overline{NP}$
(d) $\angle QOP$ (other answers are possible)
(e) $\overset{\frown}{NP}$ (other answers are possible)

(f) $\widehat{MQP}$ (other answers are possible)

(g) $\widehat{MQN}$ (can be named $\widehat{MQPN}$; other answers are possible)

(h) point O　　　　　　　　　　　　　　　　　　　▲

We now consider some measures of segments related to the circle. For instance, the diameter of a circle has a length equal to twice that of the radius. This is a direct consequence of the Segment-Addition Postulate. Example 2 shows us that the related measures of segments in a circle are not always so apparent.

EXAMPLE 2　　$\overline{QN}$ is a diameter of $\odot O$ in figure 6.6 (page 199), and $PN = ON = 12$. Find the length of chord $\overline{QP}$.

Solution　　$PN = ON$ and $ON = OP$, so $\triangle NOP$ is equilateral and the measures of all $\angle$s of this $\triangle$ are 60°. Also, $OP = OQ$, so $\triangle POQ$ is isosceles with m$\angle 1 = 120°$, since this angle is supplementary to an interior angle of the equilateral triangle. Now m$\angle Q = $ m$\angle 3 = 30°$, since the sum of the measures of the angles of $\triangle POQ$ is 180°. If m$\angle N = 60°$ and m$\angle Q = 30°$, then $\triangle NPQ$ is a right $\triangle$ whose measures are 30-60-90. It follows that $QP = PN \cdot \sqrt{3} = 12\sqrt{3}$.　　▲

THEOREM 6.1.2　A radius drawn perpendicular to a chord bisects the chord.

Given: $\overline{OC} \perp \overline{AB}$ in $\odot O$

Prove: $\overline{OD}$ bisects $\overline{AB}$

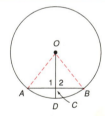

Proof: Draw radii $\overline{OA}$ and $\overline{OB}$. Now $\overline{OA} \cong \overline{OB}$ since all radii of a circle are $\cong$. Because $\angle 1$ and $\angle 2$ are right $\angle$s and $\overline{OC} \cong \overline{OC}$, we see that $\triangle OCA \cong \triangle OCB$ by HL. Then $\overline{AC} \cong \overline{CB}$ by CPCTC, and it follows that $\overline{OD}$ bisects $\overline{AB}$.　　▲

We also wish to consider measures of arcs and angles briefly here. To begin, notice that the sum of the measures of the angles about a point (angles determined by perpendicular diameters $\overline{AC}$ and $\overline{BD}$) is 360°. (See figure 6.7.) Similarly, the circle can be separated into 360 equal arcs, each of which measures 1 degree of arc measure; that is, each arc would be intercepted by a central angle measuring 1°. Thus the sum of the measures of all the arcs forming the circle is 360°. Some of the arc measures of circle O in figure 6.7 are m$\widehat{AB} = 90°$, m$\widehat{ABC} = 180°$, and m$\widehat{ABD} = 270°$.

Figure 6.7

POSTULATE 15　**(Central Angle Postulate)**　In a circle, the degree measure of a central angle is equal to the degree measure of its intercepted arc.

In circle Y in figure 6.8a, if m$\angle XYZ = 72°$, then m$\overset{\frown}{XZ} = 72°$ by the Central Angle Postulate. If two arcs have equal degree measures (as in figure 6.8b and c) but are parts of two circles with unequal radii, then these arcs will not coincide. This observation leads to the following definition.

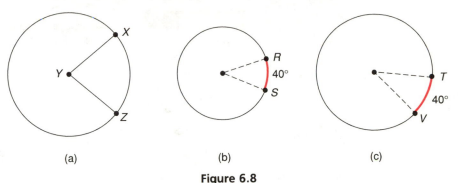

(a) (b) (c)

Figure 6.8

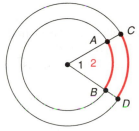

Figure 6.9

> **DEFINITION:** In a circle or congruent circles, **congruent arcs** are arcs with equal measures.

To fully appreciate this definition, consider the circles in figure 6.9, in which $\angle 1 \cong \angle 2$ and m$\overset{\frown}{AB}$ = m$\overset{\frown}{CD}$. However, it should be clear that $\overset{\frown}{AB} \not\cong \overset{\frown}{CD}$.

EXAMPLE 3 In $\odot O$ in figure 6.10, $\overrightarrow{OE}$ bisects $\angle AOD$. Using the measures indicated, find:
(a) m$\overset{\frown}{AB}$ (b) m$\overset{\frown}{BC}$ (c) m$\overset{\frown}{BD}$ (d) m$\angle AOD$
(e) m$\overset{\frown}{AE}$ (f) m$\overset{\frown}{ACE}$ (g) whether $\overset{\frown}{AE} \cong \overset{\frown}{ED}$

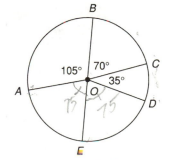

Figure 6.10

Solution (a) 105° (b) 70° (c) 105° (d) 150°, from $360 - (105 + 70 + 35)$ (e) 75° since the corresponding central angle ($\angle AOE$) is the result of bisecting $\angle AOD$, which was found to be 150° (f) 285° (from $360 - 75$, the measure of $\overset{\frown}{AE}$) (g) They are equal, since both measure 75° and are arcs in the same circle. ▲

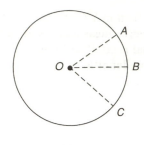

(a)

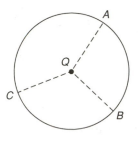

(b)

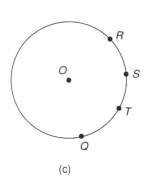

(c)

Figure 6.11

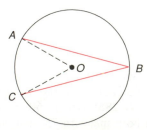

Figure 6.12

In Example 3(c), notice that the measure of $\widehat{BD}$ was equal to the sum of measures of $\widehat{BC}$ and $\widehat{CD}$ When B, C, and D lie in order on the circle, $m\widehat{BCD} = m\widehat{BC} + m\widehat{CD}$. Because this relationship between arc measures is true in general, it leads to the following postulate:

POSTULATE 16   **(Arc-Addition Postulate)**   If A, B, and C lie in that order on a circle, then $m\widehat{AB} + m\widehat{BC} = m\widehat{ABC}$.

The drawing in figure 6.11a further supports the claim in Postulate 16.

Given points A, B, and C on $\odot O$ as shown in figure 6.11a, suppose that radii $\overline{OA}$, $\overline{OB}$, and $\overline{OC}$ are drawn. Because

$$m\angle AOB + m\angle BOC = m\angle AOC$$

by the Angle-Addition Postulate, it seems reasonable that

$$m\widehat{AB} + m\widehat{BC} = m\widehat{ABC}$$

If you consider circle Q in figure 6.11b, you are likely to accept that the sum of $m\widehat{AB}$ and $m\widehat{BC}$ is $m\widehat{ABC}$. The reason for using $\widehat{ABC}$, rather than $\widehat{AB}$, in stating the Arc-Addition Postulate is to avoid confusing minor arc $\widehat{AC}$ with major arc $\widehat{ABC}$.

The Arc-Addition Postulate can easily be extended to include more than two arcs. For example, in figure 6.11c, it is easy to prove that $m\widehat{RS} + m\widehat{ST} + m\widehat{TQ} = m\widehat{RSTQ}$. It may happen that $\widehat{RS}$ and $\widehat{ST}$ have the same degree measures in figure 6.11c; in that case, point S is the **midpoint** of $\widehat{RT}$, and $\widehat{RT}$ is **bisected** at point S.

As we have seen, the measure of an arc can be used to measure a corresponding central angle. The measure of an arc can also be used to measure other types of angles related to the circle, including the inscribed angle.

DEFINITION: An **inscribed angle** of a circle is an angle whose vertex is a point on the circle and whose sides are chords of the circle.

The word "inscribed" can be associated with the word "inside" without serious difficulty, if we remember that the sides of the angle are chords. $\angle B$ in $\odot O$ of figure 6.12 is an inscribed angle whose sides are chords $\overline{BA}$ and $\overline{BC}$. If $m\widehat{AC} = 60°$, then $m\angle AOC = 60°$, but $m\angle ABC < m\angle AOC$. Now we seek a relationship between the measure of the intercepted arc and the measure of the inscribed angle. According to the following theorem, the measure of the inscribed angle is one-half the measure of its intercepted arc. Thus, in figure 6.12,

$$m\angle ABC = \frac{1}{2}m\widehat{AC} = \frac{1}{2}(60) = 30°$$

THEOREM 6.1.3   The measure of an inscribed angle of a circle is one-half the measure of its intercepted arc.

Proof of Theorem 6.1.3 must be divided into three cases:

1. One side of the inscribed angle is a diameter.
2. The diameter to the vertex of the inscribed angle lies in the interior of the angle.
3. The diameter to the vertex of the inscribed angle lies in the exterior of the angle.

The proof of case 1 is provided next, while those for the other cases are left as exercises for the student.

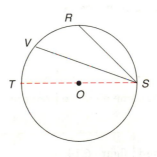

Given: $\odot O$ with inscribed $\angle RST$ diameter $\overline{ST}$

Prove: $m\angle S = \frac{1}{2} m \widehat{RT}$

Proof of case (1):   We begin by constructing radius $\overline{RO}$. It follows that $m\angle ROT$ equals $m\widehat{RT}$ since the central angle has a measure equal to its intercepted arc. With $\overline{OR} \cong \overline{OS}$, $\triangle ROS$ is isosceles and $m\angle R = m\angle S$. Now the exterior angle of the triangle is $\angle ROT$, so

$$m\angle ROT = m\angle R + m\angle S$$

Now $m\angle ROT = 2m\angle S$, and it follows that $m\angle S = \frac{1}{2}m\angle ROT$.

Because $m\angle ROT = m\widehat{RT}$, we have $m\angle S = \frac{1}{2}m\widehat{RT}$.   ▲

Drawings are provided for cases 2 and 3 in figure 6.13. Notice the auxiliary lines provided. While proofs in this chapter generally take the less formal paragraph form, it remains necessary for the student to provide a drawing, a Given, and a Prove. Further, the student must still justify each statement in the proof.

A number of theorems involving a circle or congruent circles are easily proved and often applied. These include the following four.

(a) Case 2

(b) Case 3

Figure 6.13

THEOREM 6.1.4   In a circle or in congruent circles, congruent minor arcs have congruent central angles.

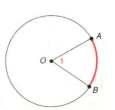
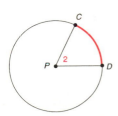

By Theorem 6.14, if $\widehat{AB} \cong \widehat{CD}$ in congruent circles O and P, then $\angle 1 \cong \angle 2$.

THEOREM 6.1.5   In a circle or in congruent circles, congruent central angles have congruent arcs.

THEOREM 6.1.6   In a circle or in congruent circles, congruent chords have congruent minor (major) arcs.

THEOREM 6.1.7   In a circle or in congruent circles, congruent arcs have congruent chords.

The distance from the center of a circle to a particular chord of the circle is the length of the perpendicular segment joining the center to that chord.

Theorems about circles that are more difficult to prove include the following two.

THEOREM 6.1.8   Chords at the same distance from the center of a circle are congruent.

THEOREM 6.1.9   Congruent chords are at the same distance from the center of a circle.

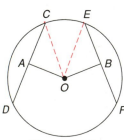

Figure 6.14

To prove Theorem 6.1.8, consider the drawing in figure 6.14.

Given: $\overline{OA} \perp \overline{CD}$ and $\overline{OB} \perp \overline{EF}$ in $\odot O$
$\overline{OA} \cong \overline{OB}$

Prove: $\overline{CD} \cong \overline{EF}$

Proof: With $\overline{OA} \perp \overline{CD}$ and $\overline{OB} \perp \overline{EF}$, $\angle OAC$ and $\angle OBE$ are right $\angle$s. $\overline{OA} \cong \overline{OB}$ is given; and $\overline{OC} \cong \overline{OE}$ since all radii of a circle are congruent. $\triangle OAC$ and $\triangle OBE$ are right triangles. Therefore by HL, $\triangle OAC$ and $\triangle OBE$ are congruent.

By CPCTC, we see that $\overline{CA} \cong \overline{BE}$ and $CA = BE$. Then, using the Multiplication Property of Equality, we get

$$2(CA) = 2(BE)$$

But $2(CA) = CD$ since A is the midpoint of chord $\overline{CD}$. (The perpendicular segment $\overline{OA}$ bisects the chord because it is part of a radius.) Likewise, $2(BE) = EF$, and it follows that

$$CD = EF \qquad \text{and} \qquad \overline{CD} \cong \overline{EF} \qquad \blacktriangle$$

Two final theorems are provided with drawings, but their proofs are left as exercises for the student.

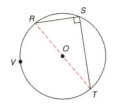

> **THEOREM 6.1.10**   An angle inscribed in a semicircle is a right angle.

In the figure, $\angle S$ is a right $\angle$. It is inscribed in the semicircle $\overset{\frown}{RST}$ because it is inside $\overset{\frown}{RST}$. $\angle S$ also intercepts semicircle $\overset{\frown}{RVT}$.

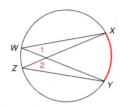

> **THEOREM 6.1.11**   If two inscribed angles intercept the same arc, then these angles are congruent.

$\angle 1$ and $\angle 2$ both intercept $\overset{\frown}{XY}$. Therefore, $\angle 1 \cong \angle 2$.

6.1 EXERCISES

1. *Given:* $\overline{AO} \perp \overline{OB}$ and $\overline{OC}$ bisects $\overset{\frown}{ACB}$ in $\odot O$
Find: (a) $m\overset{\frown}{AB}$
(b) $m\overset{\frown}{ACB}$
(c) $m\overset{\frown}{BC}$
(d) $m\angle AOC$

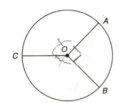

2. *Given:* $ST = \frac{1}{2}(SR)$ in $\odot Q$
$\overline{SR}$ is a diameter
Find: (a) $m\overset{\frown}{ST}$
(b) $m\overset{\frown}{TR}$
(c) $m\overset{\frown}{STR}$
(d) $m\angle S$

3. *Given:* $\odot Q$ in which $m\overset{\frown}{AB}{:}m\overset{\frown}{BC}{:}m\overset{\frown}{CA} = 2{:}3{:}4$
Find: (a) $m\overset{\frown}{AB}$ (h) $m\angle 5$
(b) $m\overset{\frown}{BC}$ (i) $m\angle 6$
(c) $m\overset{\frown}{CA}$
(d) $m\angle 1$
(e) $m\angle 2$
(f) $m\angle 3$
(g) $m\angle 4$

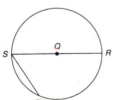

4. *Given:* $m\angle DOE = 76°$ and $m\angle EOG = 82°$ in $\odot O$
EF is a diameter
Find: (a) $m\overset{\frown}{DE}$
(b) $m\overset{\frown}{DF}$
(c) $m\angle F$
(d) $m\angle DGE$
(e) $m\angle EHG$
(f) whether $m\angle EHG = \frac{1}{2}(m\overset{\frown}{EG} + m\overset{\frown}{DF})$

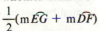

5. *Given:* $\odot O$ with $\overline{AB} \cong \overline{AC}$ and $m\angle BOC = 72°$
Find: (a) $m\overset{\frown}{BC}$
(b) $m\overset{\frown}{AB}$
(c) $m\angle A$
(d) $m\angle ABC$
(e) $m\angle ABO$

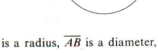

6. In $\odot O$ (not shown), $\overline{OA}$ is a radius, $\overline{AB}$ is a diameter, and $\overline{AC}$ is a chord.

(a) How does OA compare to AB?
(b) How does AC compare to AB?
(c) How does AC compare to OA?

7. *Given:* $\overline{OC} \perp \overline{AB}$ and $OC = 6$ in $\odot O$
Find: (a) AB
(b) BC

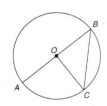

Use the figure shown in Exercises 8 and 9.

8. *Given:* Concentric circles with center Q
$SR = 3$ and $RQ = 4$
$\overline{QS} \perp \overline{TV}$ at R
Find: (a) RV
(b) TV

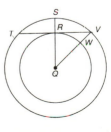

9. *Given:* Concentric circles with center Q
$TV = 8$ and $VW = 2$
$\overline{RQ} \perp \overline{TV}$
Find: RQ (*Hint:* Let $RQ = x$.)

In Exercises 10 and 11, use the figure shown.

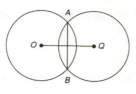

10. $\overline{AB}$ is the **common chord** of $\odot O$ and $\odot Q$. If $AB = 12$ and each circle has a radius of length 10, how long is $\overline{OQ}$?

11. Circles O and Q have the common chord $\overline{AB}$. If $AB = 6$, $\odot O$ has a radius of length 4, and $\odot Q$ has a radius of length 6, how long is $\overline{OQ}$?

12. Suppose that a circle is divided into three congruent arcs by points A, B, and C. What type of figure results when A, B, and C are joined by segments?

13. Suppose that a circle is divided by points A, B, C, and D into four congruent arcs. If these are joined in order, what type of quadrilateral results?

14. Following the pattern of Exercises 12 and 13, what type of figure results from dividing the circle by five points and joining those points in order?

15. Consider a circle or congruent circles, and explain why each statement is true:

(a) Congruent arcs have congruent central angles.
(b) Congruent central angles have congruent arcs.
(c) Congruent chords have congruent arcs.
(d) Congruent arcs have congruent chords.
(e) Congruent central angles have congruent chords.
(f) Congruent chords have congruent central angles.

In Exercises 16 to 19, provide a paragraph proof for each statement.

16. Congruent chords are at the same distance from the center of a circle.

17. A radius perpendicular to a chord bisects the arc of that chord.

18. An angle inscribed in a semicircle is a right angle.

19. If two inscribed angles intercept the same arc, then these angles are congruent.

In Exercises 20 to 23, provide a paragraph proof for each statement. You may use the theorems of this section, including Exercises 16 to 19.

20. *Given:* Diameters $\overline{AB}$ and $\overline{CD}$ intersecting at E in $\odot E$
Prove: $\widehat{AC} \cong \widehat{DB}$

21. *Given:* $\overline{MN} \parallel \overline{OP}$ in $\odot O$
Prove: $m\widehat{MQ} = 2(m\widehat{NP})$

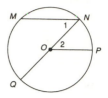

22. *Given:* $\overline{RS}$ and $\overline{TV}$ are diameters of $\odot W$
Prove: $\triangle RST \cong \triangle VTS$

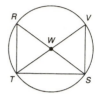

23. *Given:* Chords $\overline{AB}$, $\overline{BC}$, $\overline{CD}$, and $\overline{AD}$, as shown, in $\odot O$
Prove: $\triangle ABE \sim \triangle CED$

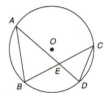

24. If $\overleftrightarrow{MN} \parallel \overleftrightarrow{PQ}$ as shown in $\odot O$, what can you conclude regarding quadrilateral $MNPQ$? (*Hint:* Draw a diagonal.)

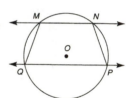

25. If $\widehat{ST} \cong \widehat{TV}$, what can you conclude regarding $\triangle STV$ in $\odot Q$?

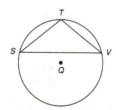

*26. Use a paragraph proof to complete this exercise.
Given: $\odot O$ with chords $\overline{AB}$ and $\overline{BC}$ radii $\overline{AO}$ and $\overline{OC}$ as shown
Prove: $m\angle ABC < m\angle AOC$

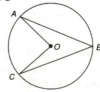

27. State the measure of the angle formed by the minute hand and the hour hand of a clock when the time is

(a) 1:30 P.M.   (b) 2:20 A.M.

28. Five points are equally spaced on a circle. A five-pointed star (pentagram) is formed by joining two nonconsecu-

tive points. What is the degree measure of an arc determined by two consecutive points?

29. Prove case 2 of Theorem 6.1.3, the theorem for measurement of an inscribed angle.

30. Prove case 3 of Theorem 6.1.3, the theorem for measurement of an inscribed angle.

6.2
MORE ANGLE
MEASURES
IN THE CIRCLE

We begin this section by considering lines, rays, and segments that have special
relationships to the circle.

> **DEFINITION:**   A **tangent** is a line that touches a circle at exactly one
> point; the point of intersection is the **point of contact** or **point of tan-
> gency.**

In the preceding definition, the term "tangent" also applies to a segment or ray
that touches a circle at one point and is part of a line tangent to a circle.

> **DEFINITION:**   A **secant** is a line (or segment or ray) that intersects a
> circle at exactly two points.

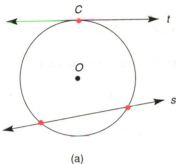

(a)

The notions of "tangent" and "secant" are illustrated in figure 6.15a and
b. In (a), line t is a tangent to $\odot O$, and point C is its point of contact; line s is
a secant to $\odot O$. In (b), $\overline{AB}$ is a tangent to $\odot Q$, and point T is its point of tangency. $\overrightarrow{CD}$ is a secant.

> **DEFINITION:**   A polygon is **inscribed in a circle** if its vertices are points
> on the circle and its sides are chords of the circle. In such a case, the
> circle is **circumscribed about the polygon.**

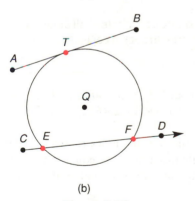

(b)

Figure 6.15

In figure 6.16, $\triangle ABC$ is inscribed in $\odot O$, and quadrilateral $RSTV$ is inscribed in $\odot Q$. Conversely, $\odot O$ is circumscribed about $\triangle ABC$, and $\odot Q$ is circumscribed about quadrilateral $RSTV$.

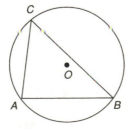

(a)

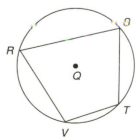

(b)

Figure 6.16

> **THEOREM 6.2.1**   If a quadrilateral is inscribed in a circle, then the opposite angles are supplementary.

A proof of Theorem 6.2.1 is now provided.

Given: *RSTV* is inscribed in $\odot O$, in figure 6.16b

Prove: $\angle R$ and $\angle T$ are supplementary

Proof: In Section 6.1, we saw that an inscribed $\angle$ (like $\angle R$) is equal in measure to one-half the measure of its intercepted arc. Then

$$m\angle R + m\angle T = \frac{1}{2}m\widehat{STV} + \frac{1}{2}m\widehat{SRV}$$

$$= \frac{1}{2}(m\widehat{STV} + m\widehat{SRV})$$

Because $\widehat{STV}$ and $\widehat{SRV}$ determine the entire circle, we see that

$$m\angle R + m\angle T = \frac{1}{2}(360) = 180°$$

Therefore, by definition, $\angle R$ and $\angle T$ are supplementary.   ▲

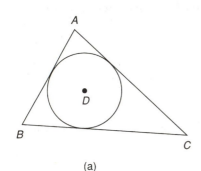

(a)

(b)

Figure 6.17

> **DEFINITION:**   A polygon is **circumscribed about a circle** if all sides of the polygon are segments tangent to the circle. In such a case, the circle is **inscribed in the polygon.**

In figure 6.17, $\triangle ABC$ is circumscribed about $\odot D$, while square *MNPQ* is circumscribed about $\odot T$. Conversely, $\odot D$ is inscribed in $\triangle ABC$, while $\odot T$ is inscribed in square *MNPQ*.

We wish to develop further methods for measuring angles related to the circle. At present, we know that a central angle has a measure equal to that of its intercepted arc, and that an inscribed angle has a measure one-half that of its intercepted arc.

> **THEOREM 6.2.2**   The measure of an angle formed by two intersecting chords is one-half the sum of the measures of the arcs intercepted by the angle and its vertical angle.

The claim made by this theorem is that

$$m\angle 1 = \frac{1}{2}(m\widehat{AC} + m\widehat{DB})$$

(See figure 6.18a.)

EXAMPLE 1

If $m\widehat{AC} = 84°$ and $m\widehat{DB} = 62°$, then find $m\angle 1$. (See figure 6.18a.)

Solution

By Theorem 6.2.2,

$$m\angle 1 = \frac{1}{2}(84 + 62)$$

$$= \frac{1}{2}(146) = 73°$$

▲

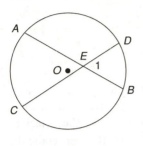

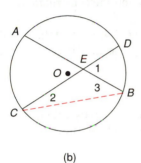

(a)

(b)

Figure 6.18

To prove Theorem 6.2.2, we repeat figure 6.18a but with an auxiliary line.

Given: Chords $\overline{AB}$ and $\overline{CD}$ intersecting at point E in $\odot O$

Prove: $m\angle 1 = \frac{1}{2}(m\widehat{AC} + m\widehat{DB})$

Proof: Draw $\overline{CB}$. (See figure 6.18b.) Now $m\angle 1 = m\angle 2 + m\angle 3$ because $\angle 1$ is an exterior $\angle$ of $\triangle CBE$. Since $\angle 2$ and $\angle 3$ are inscribed angles of $\odot O$, we see that

$$m\angle 2 = \frac{1}{2}m\widehat{DB} \text{ and } m\angle 3 = \frac{1}{2}m\widehat{AC}$$

It follows that

$$m\angle 1 = \frac{1}{2}m\widehat{DB} + \frac{1}{2}m\widehat{AC}$$

By the Distributive Property and the Commutative Property for Addition,

$$m\angle 1 = \frac{1}{2}(m\widehat{AC} + m\widehat{DB})$$

▲

Before we add to our list of theorems, recall that a circle separates points in the plane into points in the interior of the circle, points on the circle, and points in the exterior of the circle. In figure 6.19, point A and center O are in the interior of $\odot O$ because their distances from center O are less than the length of the radius. Point B is on the circle; but points C and D are in the exterior of $\odot O$ because their distances from O are greater than the length of the radius. In the proof of Theorem 6.2.3, we use the fact that a tangent to a circle cannot contain an interior point of the circle.

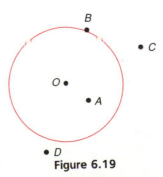

Figure 6.19

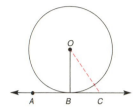

> **THEOREM 6.2.3**   The radius (or any other line through the center of a circle) drawn to a tangent at the point of tangency is perpendicular to the tangent at that point.

Given: $\odot O$ with tangent $\overleftrightarrow{AB}$ and radius $\overline{OB}$

Prove: $\overline{OB} \perp \overleftrightarrow{AB}$

Proof: Because a line contains an infinite set of points, we choose a point C (any point on $\overleftrightarrow{AB}$ except B is appropriate) on $\overleftrightarrow{AB}$, as shown in the accompanying figure. Now $OC > OB$ since C lies in the exterior of the circle. It also follows that $\overline{OB} \perp \overleftrightarrow{AB}$, because the perpendicular segment provides the shortest distance from a point to a line.   ▲

We will consider the converse of this theorem and the constructions of tangents in Section 6.4.

A consequence of Theorem 6.2.3 is Corollary 6.2.4, which follows. Of the three possible cases indicated in figure 6.20, only the first is proved; the remaining two are left as exercises for the student.

> **COROLLARY 6.2.4**   The measure of an angle formed by a tangent and a chord drawn to the point of tangency is one-half the measure of the intercepted arc. (See figure 6.20.)

For this proof, refer to the drawing of case 1 in figure 6.20a.

Given: Chord $\overline{CA}$ (which is a diameter) and tangent $\overrightarrow{CD}$

Prove: $m\angle 1 = \dfrac{1}{2}m\widehat{ABC}$

Proof: By the previous theorem, $\overline{AC} \perp \overrightarrow{CD}$. Then $\angle 1$ is a right angle and $m\angle 1 = 90°$. Because the intercepted arc $\widehat{ABC}$ is a semicircle, $m\widehat{ABC} = 180°$. Thus it follows that $m\angle 1 = \dfrac{1}{2}m\widehat{ABC}$.   ▲

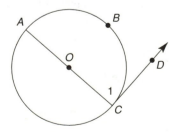

(a) Case 1
The chord is a diameter

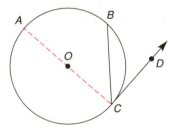

(b) Case 2
The diameter is in the exterior of the angle

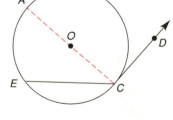

(c) Case 3
The diameter lies in the interior of the angle

Figure 6.20

EXAMPLE 2

Given: In figure 6.21, $\odot O$ with diameter $\overline{DB}$ and $m\widehat{DE} = 84°$

Find: (a) $m\angle 1$ (c) $m\angle ABD$
        (b) $m\angle 2$ (d) $m\angle ABE$

Solution

(a) $\angle 1$ is an inscribed angle; $m\angle 1 = \dfrac{1}{2}m\widehat{DE} = 42°$

(b) With $m\widehat{DE} = 84°$ and $\widehat{DEB}$ a semicircle, $m\widehat{BE} = 96°$. Then $m\angle 2 = 48°$ by Corollary 6.2.4.

(c) Since DB is perpendicular to $\overleftrightarrow{AB}$, $m\angle ABD = 90°$ ($\overline{DB} \perp \overleftrightarrow{AB}$)

(d) $m\angle ABE = m\angle ABD + m\angle 1 = 90 + 42 = 132°$ ▲

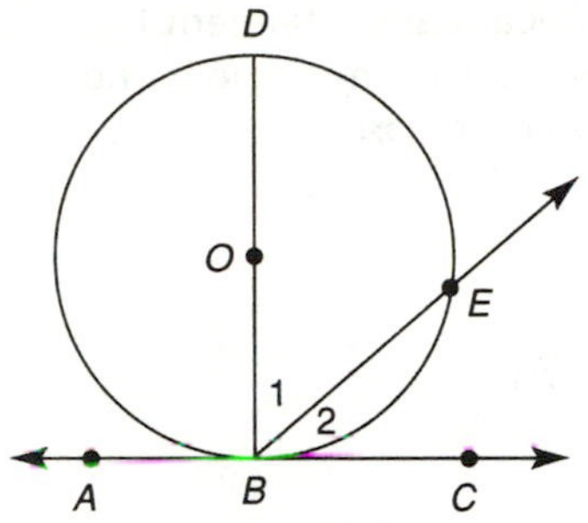

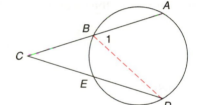

Figure 6.21

> **THEOREM 6.2.5** The measure of an angle formed when two secants intersect at a point outside the circle is one-half the difference of the measures of the two intercepted arcs.

Given: Secants $\overline{AC}$ and $\overline{DC}$ as shown in figure 6.22

Prove: $m\angle C = \dfrac{1}{2}(m\widehat{AD} - m\widehat{BE})$

Proof: Draw $\overline{BD}$ to form $\triangle BCD$. Then the measure of the exterior angle of $\triangle BCD$ is given by

$$m\angle 1 = m\angle C + m\angle D$$

Also,

$$m\angle C = m\angle 1 - m\angle D$$

$\angle 1$ and $\angle D$ are inscribed so that

$$m\angle C = \frac{1}{2}m\widehat{AD} - \frac{1}{2}m\widehat{BE}$$

$$= \frac{1}{2}(m\widehat{AD} - m\widehat{BE})$$
    ▲

Figure 6.22

EXAMPLE 3

Given: In figure 6.23, $m\angle AOB = 136°$ and $m\angle DOC = 46°$ in $\odot O$

Find: $m\angle E$

Solution $m\widehat{AB} = 136°$ and $m\widehat{DC} = 46°$, so

$$m\angle E = \frac{1}{2}(136 - 46)$$

$$= \frac{1}{2}(90) = 45°$$

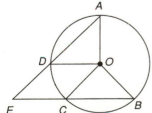

Figure 6.23   ▲

Any angle formed by two secants, a secant and tangent, or two tangents intersecting outside a circle has a measure equal to one-half the difference of the measures of the two intercepted arcs. While the next two theorems are not proved, the auxiliary lines shown will allow you to complete the proof.

> **THEOREM 6.2.6**   If an angle is formed by a secant and a tangent that intersect in the exterior of a circle, then the measure of the angle is one-half the difference of the measures of its intercepted arcs.

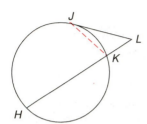

Figure 6.24

According to Theorem 6.2.6,

$$m\angle L = \frac{1}{2}(m\widehat{HJ} - m\widehat{JK})$$

in figure 6.24.

> **THEOREM 6.2.7**   If an angle is formed by two intersecting tangents, then the measure of the angle is one-half the difference of the measures of the intercepted arcs.

In figure 6.25a,

$$m\angle ABC = \frac{1}{2}(m\widehat{ADC} - m\widehat{AC})$$

(a)   **Figure 6.25**   (b)

EXAMPLE 4   *Given:* In figure 6.25b, $m\widehat{MN} = 70$, $m\widehat{NP} = 88$, $m\widehat{MR} = 46$, $m\widehat{RS} = 26$
Find: (a) $m\angle MTN$
(b) $m\angle NTP$
(c) $m\angle MTP$

Solution   (a) $m\angle MTN = \dfrac{1}{2}(m\widehat{MN} - m\widehat{MR})$

$$= \frac{1}{2}(70 - 46)$$

$$= \frac{1}{2}(24) = 12°$$

(b) $m\angle NTP = \dfrac{1}{2}(m\widehat{NP} - m\widehat{RS})$

$\qquad\qquad = \dfrac{1}{2}(88 - 26)$

$\qquad\qquad = \dfrac{1}{2}(62) = 31°$

(c) $m\angle MTP = m\angle MTN + m\angle NTP$
$\qquad\qquad\quad = 12 + 31 = 43°$ ▲

Before we go to our final example, we will review the methods used to measure different types of angles related to a circle. These methods can be summarized in Table 6.1.

Table 6.1   **Methods for Measuring Angles Related to a Circle**

Location of the Vertex of the Angle	*Rule for Measuring the Angle*
In the *interior* of the circle	One-half the *sum* of the measures of the intercepted arcs
On the circle	One-half the measure of the intercepted arc
In the *exterior* of the circle	One-half the *difference* of the measures of the two intercepted arcs

EXAMPLE 5   Given that $m\angle 1 = 46°$ in figure 6.26, find the measures of $\widehat{AB}$ and $\widehat{ACB}$.

Solution   Let $m\widehat{AB} = x$ and $m\widehat{ACB} = y$. Now

$$m\angle 1 = \frac{1}{2}(m\widehat{ACB} - m\widehat{AB})$$

so

$$46 = \frac{1}{2}(y - x)$$

which leads to $92 = y - x$.

Also, $y + x = 360$ since these two arcs form the entire circle. This gives us the system

$$
\begin{aligned}
y + x &= 360 \\
y - x &= 92 \\
\hline
2y &= 452 \\
y &= 226 \rightarrow x = 134
\end{aligned}
$$

Then $m\widehat{AB} = 134°$ and $m\widehat{ACB} = 226°$.   ▲

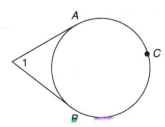

Figure 6.26

Because it is used frequently, one additional theorem is stated; the proof is left as an exercise.

THEOREM 6.2.8 If two parallel lines intersect a circle, then the intercepted arcs between these lines are congruent.

6.2 EXERCISES

In Exercises 1 and 2, use the drawing shown.

1. *Given:* $m\widehat{AB} = 92°$, $m\widehat{DA} = 114°$, $m\widehat{BC} = 138°$
   Find: (a) $m\angle 1$ (d) $m\angle 4$
   (b) $m\angle 2$ (e) $m\angle 5$
   (c) $m\angle 3$

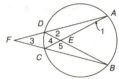

2. *Given:* $m\widehat{DC} = 30°$ and $\widehat{DABC}$ is trisected at points A and B
   Find: (a) $m\angle 1$ (d) $m\angle 4$
   (b) $m\angle 2$ (e) $m\angle 5$
   (c) $m\angle 3$

3. In the accompanying drawing, is it possible for all four conditions to hold simultaneously? Explain.

   i $\overline{RS}$ is a diameter.
   ii $\overline{TS}$ is a chord.
   iii $\overrightarrow{SW}$ is a tangent.
   iv $\overline{SW} \perp \overline{TS}$

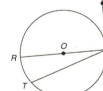

4. Is it possible for

   (a) an inscribed rectangle in a circle to have a diameter for a side? Explain.
   (b) a circumscribed rectangle about a circle to be a square? Explain.

5. *Given:* In $\odot Q$, $\overline{PR}$ contains Q
   $m\widehat{MP} = 112°$, $m\widehat{MN} = 60°$, $m\widehat{MT} = 46°$
   Find: (a) $m\angle MRP$
   (b) $m\angle 1$
   (c) $m\angle 2$

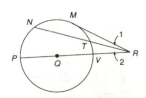

In Exercises 6 and 7, use the figure shown.

6. *Given:* $\overrightarrow{AB}$ and $\overrightarrow{AC}$ are tangent to $\odot O$; $m\widehat{BC} = 126°$
   Find: (a) $m\angle A$
   (b) $m\angle ABC$
   (c) $m\angle ACB$

7. *Given:* Tangents $\overrightarrow{AB}$
   and $\overrightarrow{AC}$ to $\odot O$
   $m\angle ACB = 68°$
   Find: (a) $m\widehat{BC}$
   (b) $m\widehat{BDC}$
   (c) $m\angle ABC$
   (d) $m\angle A$

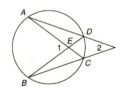

Use the accompanying figure in Exercises 8 and 9.

8. *Given:* $m\angle 1 = 72°$, $m\widehat{DC} = 34°$
   Find: (a) $m\widehat{AB}$
   (b) $m\angle 2$

9. *Given:* $m\angle 2 = 23°$
   $\widehat{DA} \cong \widehat{AB} \cong \widehat{BC}$
   Find: (a) $m\widehat{AB}$
   (b) $m\angle 1$

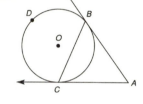

In Exercises 10 and 11, use the drawing shown.

10. *Given:* $m\angle 3 = 42°$
    Find: (a) $m\widehat{RT}$
    (b) $m\widehat{RST}$

11. *Given:* $\widehat{RS} \cong \widehat{ST} \cong \widehat{RT}$
    Find: (a) $m\widehat{RT}$
    (b) $m\widehat{RST}$
    (c) $m\angle 3$

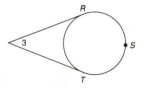

Use the drawing shown for Exercises 12 and 13.

12. *Given:* $m\angle 1 = 63°$,
    $m\widehat{RS} = 3x + 6$,
    $m\widehat{VT} = x$
    Find: $m\widehat{RS}$

13. *Given:* m∠2 = 124°,
       m$\widehat{TV}$ = x + 1,
       m$\widehat{SR}$ = 3(x + 1)
    Find: m$\widehat{TV}$

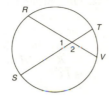

In Exercises 14 and 15, use the figure shown.

14. *Given:* $\overline{AC} \perp \overline{CE}$, m$\widehat{DE}$ = 116°,
       m$\widehat{BC}$ = 94°
    Find: m∠1

15. *Given:* m∠1 = 61°,
       m∠2 = 26°
    Find: m$\widehat{CE}$

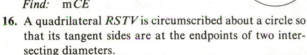

16. A quadrilateral *RSTV* is circumscribed about a circle so that its tangent sides are at the endpoints of two intersecting diameters.

   (a) What type of quadrilateral is *RSTV*?
   (b) If the diameters are also perpendicular, what type of quadrilateral is *RSTV*?

In Exercises 17 to 20, provide a paragraph proof for each statement.

17. *Given:* $\overline{AB}$ and $\overline{AC}$ are
       tangents to ⊙*O*
       from point *A*
    Prove: △*ABC* is isosceles

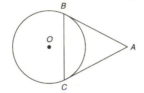

18. *Given:* $\overline{RS} \parallel \overline{TQ}$
    Prove: $\widehat{SRT} \cong \widehat{RSQ}$

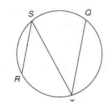

19. *Given:* Tangent $\overline{AB}$ to ⊙*O*
       at point *B*
       m∠A = m∠B
    Prove: m$\widehat{BD}$ = 2 · m$\widehat{BC}$

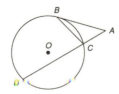

20. *Given:* Diameter $\overline{AB} \perp \overline{CE}$ at *D*
    Prove: *CD* is the geometric
       mean of *AD* and *DB*

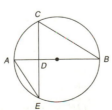

In Exercises 21 to 27, provide a paragraph proof for each statement. Be sure to provide a drawing, Given, and Prove.

21. If two parallel lines intersect a circle, then the intercepted arcs between these lines are congruent.

22. The line joining the centers of two circles that intersect at two points is the perpendicular bisector of the common chord.

23. If a trapezoid is inscribed in a circle, then it is an isosceles trapezoid.

24. If a parallelogram is inscribed in a circle, then it is a rectangle.

25. If one side of an inscribed triangle is a diameter, then the triangle is a right triangle.

26. Prove case 2 of Corollary 6.2.4: The measure of an angle formed by a tangent and a chord drawn to the point of tangency is one-half the measure of the intercepted arc.

27. Prove case 3 of Corollary 6.2.4.

In Exercises 28 to 30, use the accompanying drawing and find the missing measure(s).

*28. *Given:* m$\widehat{MQ}$ + m$\widehat{NP}$ = 172°
       m$\widehat{QP}$ − m$\widehat{MN}$ = 52°
    Find: m∠1 and m∠2

29. *Given:* m∠1 = 62° and m∠2 = 34°
    Find: m$\widehat{QP}$ and m$\widehat{MN}$

30. *Given:* m$\widehat{QP}$ = x and m$\widehat{MN}$ = y
    Find: An expression for m∠1 + m∠2 in terms of x and y

*31. *Given:* $\overline{AB}$ is a diameter in ⊙*O*
       M is the midpoint of chord $\overline{AC}$
       N is the midpoint of chord $\overline{CB}$
       MB = $\sqrt{73}$, *AN* = $2\sqrt{13}$
    Find: The length of diameter $\overline{AB}$

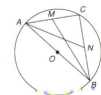

32. A surveyor can see a circular planetarium through a 60° angle. If the surveyor is 45 ft from the door, what is the diameter of the planetarium?

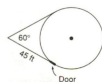

33. For the five-pointed star (pentagram) shown in the accompanying drawing, find the measures of ∠1 and ∠2.

*34. On a fitting for a hex wrench, the distance from the center O to a vertex is 5 mm. The length of radius $\overline{OB}$ of the circle is 10 mm. If $\overline{OC} \perp \overline{DE}$ at F, how long is $\overline{FC}$?

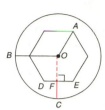

6.3
LINE AND SEGMENT RELATIONSHIPS IN THE CIRCLE

Several theorems are difficult to distinguish unless a drawing and a clearcut Given and Prove are provided.

THEOREM 6.3.1   If a line is drawn through the center of a circle perpendicular to a chord, then it bisects the chord and its arc. (Notice that "arc" generally refers to the minor arc, even though the major arc is also bisected.)

Given: $\overleftrightarrow{AB} \perp$ chord $\overline{CD}$ in circle A
Prove: $\overline{CB} \cong \overline{BD}$ and $\overset{\frown}{CE} \cong \overset{\frown}{ED}$

The proof is left as an exercise.
(*Hint:* Draw radii $\overline{AC}$ and $\overline{AD}$.)

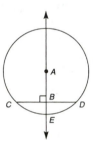

THEOREM 6.3.2   If a line through the center of a circle bisects a chord other than a diameter, then it is perpendicular to the chord.

Given: Circle O; $\overleftrightarrow{OM}$ is the bisector of chord $\overline{RS}$
Prove: $\overleftrightarrow{OM} \perp \overline{RS}$

Proof: Left as an exercise. (*Hint:* Draw radii $\overline{OR}$ and $\overline{OS}$.)

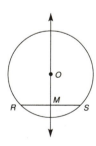

(a)

(b)

Figure 6.27

THEOREM 6.3.3   The perpendicular bisector of a chord contains the center of the circle.

Given:

Prove:

In figure 6.27, $\overleftrightarrow{QR}$, the perpendicular bisector of chord $\overline{TV}$ in $\odot O$

$\overleftrightarrow{QR}$ contains point O

Proof (by indirect method): Suppose O is not on $\overleftrightarrow{QR}$. Draw $\overline{OR}$ and radii $\overline{OT}$ and $\overline{OV}$. (See figure 6.27b, page 216.) Because QR is the perpendicular bisector of $\overline{TV}$, R must be the midpoint of $\overline{TV}$; then $\overline{TR} \cong \overline{RV}$. Also, $\overline{OT} \cong \overline{OV}$ (all radii of a $\odot$ are $\cong$). With $\overline{OR} \cong \overline{OR}$ by identity, we have $\triangle ORT \cong \triangle ORV$ by SSS.

Now $\angle ORT \cong \angle ORV$ by CPCTC. It follows that $\overline{OR} \perp \overline{TV}$ because these lines (segments) meet to form congruent adjacent angles.

Then $\overline{OR}$ is the perpendicular bisector of $\overline{TV}$. But $\overleftrightarrow{QR}$ is also the perpendicular bisector of $\overline{TV}$, which contradicts the uniqueness for the perpendicular bisector of a segment.

Then the supposition must be false, and it follows that center O is on $\overleftrightarrow{QR}$, the perpendicular bisector of chord $\overline{TV}$. ▲

EXAMPLE 1

Given: In figure 6.28, $\odot O$ has a radius of length 5
$\overline{OE} \perp \overline{CD}$ at B and $OB = 3$

Find: CD

Solution Draw radius $\overline{OC}$. By the Pythagorean Theorem,

$$(OC)^2 = (OB)^2 + (BC)^2$$
$$5^2 = 3^2 + (BC)^2$$
$$25 = 9 + (BC)^2$$
$$(BC)^2 = 16$$
$$BC = 4$$

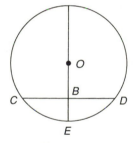

Figure 6.28

By the first theorem of this section, we know that $CD = 2 \cdot BC$; then it follows that $CD = 2 \cdot 4 = 8$. ▲

Recall that concentric circles are coplanar circles that share a common center. For the concentric circles shown in figure 6.29, the chord of the larger circle is a tangent of the smaller circle.

If two circles touch at one point, they are **tangent circles.** In figure 6.30, circles P and Q are internally tangent while circles O and R are externally tangent.

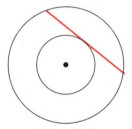

Figure 6.29

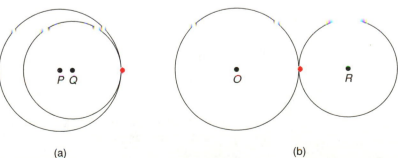

(a)   (b)

Figure 6.30

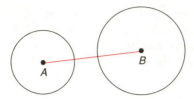

Figure 6.31

The line segment that joins the centers of two nonconcentric circles is often called the "line of centers" of the two circles. $\overline{AB}$ is the line of centers for circles A and B in figure 6.31.

A line that is tangent to each of two circles is a **common tangent** for these circles. If the common tangent does not intersect the line of centers, it is a **common external tangent.** In figure 6.32, circles P and Q have one common external tangent, $\overleftrightarrow{ST}$; circles A and B have two common external tangents, $\overleftrightarrow{WX}$ and $\overleftrightarrow{YZ}$.

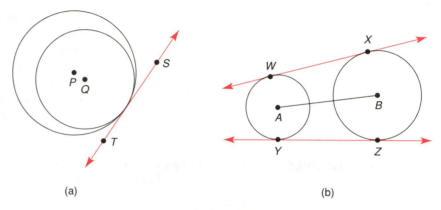

(a)   (b)

Figure 6.32

If the common tangent *does* intersect the line of centers for two circles, it is a **common internal tangent** for the two circles. In figure 6.33, $\overleftrightarrow{DE}$ is a common internal tangent for externally tangent circles O and R; $\overleftrightarrow{AB}$ and $\overleftrightarrow{CD}$ are common internal tangents for $\odot M$ and $\odot N$.

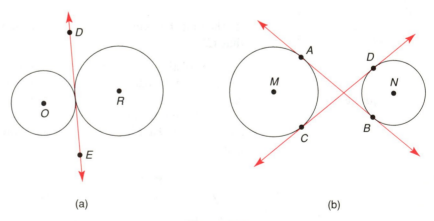

(a)   (b)

Figure 6.33

THEOREM 6.3.4   The tangent segments to a circle from an external point are congruent.

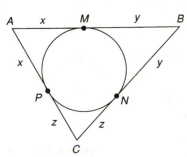

Figure 6.34

Given: In figure 6.34, $\overline{AB}$ and $\overline{AC}$ are tangents to $\odot O$ from point A

Prove: $\overline{AB} \cong \overline{AC}$

Proof: Draw $\overline{BC}$. Now $m\angle B = \frac{1}{2}m\widehat{BC}$ and $m\angle C = \frac{1}{2}m\widehat{BC}$. Then $\angle B \cong \angle C$ since these angles have equal measures. In turn, the opposite sides of the triangle are congruent and $\overline{AB} \cong \overline{AC}$. ▲

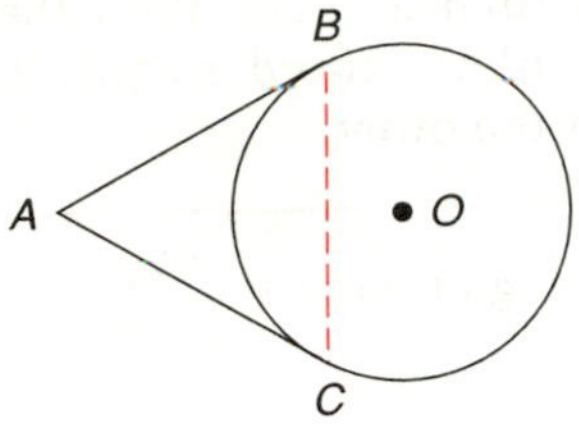

EXAMPLE 2

The circle shown in figure 6.35 is inscribed in $\triangle ABC$; $AB = 9$, $BC = 8$, and $AC = 7$. Find the lengths AM, MB, and NC.

Solution

Because the tangent segments from an external point are $\cong$, we can let

$$AM = AP = x$$
$$BM = BN = y$$
$$NC = CP = z$$

Now

$$
\begin{array}{ll}
x + y = 9 & \text{from } AB = 9 \\
y + z = 8 & \text{from } BC = 8 \\
x + z = 7 & \text{from } AC = 7
\end{array}
$$

Figure 6.35

Subtracting the second equation from the first, as shown, we get

$$
\begin{array}{r}
x + y \quad\;\; = 9 \\
y + z = 8 \\
\hline
x \quad\;\; - z = 1
\end{array}
$$

Now we use this new equation along with the third equation from the previous set and add:

$$
\begin{array}{r}
x - z = 1 \\
x + z = 7 \\
\hline
2x = 8 \to x = 4 \to AM = 4
\end{array}
$$

Because $x + y = 9$, we know that $4 + y = 9$; therefore $y = 5$, so $BM = 5$. From $x + z = 7$, we know that $4 + z = 7$; therefore $z = 3$, so $NC = 3$. Summarizing, we have $AM = 4$, $BM = 5$, and $NC = 3$. ▲

A simple check of the results found in Example 2 can be performed by inserting those lengths in their proper positions. For example, $AM + MB = 4 + 5 = 9$, which is the length of $\overline{AB}$.

We now consider three theorems that investigate some interesting and important relationships involving the lengths of segments that are chords, secants, or tangents. While the first of the theorems is proved, the second and third are left as exercises. In the drawings for Theorems 6.3.6 and 6.3.7, auxiliary lines have been drawn as hints for completing the proofs.

> **THEOREM 6.3.5** If two chords intersect within a circle, then the product of the lengths of the segments (parts) of one chord is equal to the product of the lengths of the segments of the other.

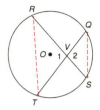

Given: Circle O with chords $\overline{RS}$ and $\overline{TQ}$ intersecting at point V

Prove: $RV \cdot VS = TV \cdot VQ$

Proof: Draw $\overline{RT}$ and $\overline{QS}$. In $\triangle RTV$ and $\triangle QSV$, we have $\angle 1 \cong \angle 2$ (vertical $\angle$s). Also, $\angle R$ and $\angle Q$ are inscribed angles which intercept the same arc (namely $\overset{\frown}{TS}$), so $\angle R \cong \angle Q$. By AA, we know that $\triangle RTV \sim \triangle QSV$. Then it follows that

$$\frac{RV}{VQ} = \frac{TV}{VS}$$

since corresponding sides of $\sim$ $\triangle$s are proportional. Finally, using the Means-Extremes Property, we obtain $RV \cdot VS = TV \cdot VQ$. ▲

EXAMPLE 3

In figure 6.36, $HP = 4$, $PJ = 5$, and $LP = 8$. Find PM.

Solution From Theorem 6.3.5,

$$HP \cdot PJ = LP \cdot PM$$

Then

$$4 \cdot 5 = 8 \cdot PM$$
$$8 \cdot PM = 20$$
$$PM = 2.5$$

Figure 6.36 ▲

EXAMPLE 4

In figure 6.36, $HP = 6$, $PJ = 4$, and $LM = 11$. Find LP.

Solution Since $LP + PM = LM$ or $LP + PM = 11$, we represent the lengths as $LP = x$ and $PM = 11 - x$. Now $HP \cdot PJ = LM \cdot PM$ becomes

$$6 \cdot 4 = x(11 - x)$$
$$24 = 11x - x^2$$
$$x^2 - 11x + 24 = 0$$
$$(x - 3)(x - 8) = 0$$
$$x = 3 \quad \text{or} \quad x = 8$$

Therefore

$$LP = 3 \quad \text{or} \quad LP = 8 \qquad ▲$$

Note: In Example 4, if $LP = 3$, then $PM = 8$; conversely, if $LP = 8$, then $PM = 3$. That is, the segments of chord $\overline{LM}$ have lengths of 3 and 8.

> **THEOREM 6.3.6** If two secant segments are drawn to a circle from an external point, then the products of the lengths of each secant and its external segment are equal.

Given: Secants $\overline{AB}$ and $\overline{AC}$ for the circle in figure 6.37

Prove: $AB \cdot RA = AC \cdot TA$

Proof: Left as an exercise for the student.

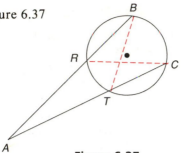

Figure 6.37

EXAMPLE 5 Use the circle in figure 6.37.

Given: $AB = 14$, $BR = 5$, and $TC = 5$

Find: AC

Solution Let $AC = x$, so $TA = x - 5$. If $AB = 14$ and $BR = 5$, then $AR = 9$. The statement $AB \cdot RA = AC \cdot TA$ becomes

$$14 \cdot 9 = x(x - 5)$$
$$126 = x^2 - 5x$$
$$x^2 - 5x - 126 = 0$$
$$(x - 14)(x + 9) = 0$$
$$x = 14 \quad \text{or} \quad x = -9 \quad (x = -9 \text{ is discarded.})$$

Thus $AC = 14$, so $TA = 9$. ▲

THEOREM 6.3.7 If a tangent segment and secant segment are drawn to a circle from an external point, then the square of the length of the tangent equals the product of the lengths of the secant and its external segment.

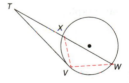

Figure 6.38

Given: Tangent $\overline{TV}$ and secant $\overline{TW}$ in figure 6.38

Prove: $(TV)^2 = TW \cdot TX$

Proof: Left as an exercise for the student.

Notice the auxiliary lines needed for the proof of this theorem. These lines are used to prove that triangles are similar. A proportion that follows from the similar triangles places the tangent in the position of a mean proportional.

EXAMPLE 6 *Given:* In figure 6.39, $SV = 3$ and $VR = 9$

Find: ST

Solution

$$(ST)^2 = SR \cdot SV$$
$$(ST)^2 = 12 \cdot 3$$
$$(ST)^2 = 36$$
$$ST = 6 \quad \text{or} \quad -6$$

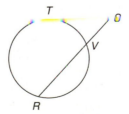

Figure 6.39

Again the negative choice, $ST = -6$, is discarded, so $ST = 6$. ▲

6.3 EXERCISES

In Exercises 1 and 2, use the accompanying figure.

1. *Given:* ⊙O with $\overline{OE} \perp \overline{CD}$
   $CD = OC$
   Find: $m\widehat{CF}$

2. *Given:* $OC = 8$ and $OE = 6$
   $\overline{OE} \perp \overline{CD}$ in ⊙O
   Find: CD

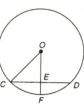

In Exercises 3 and 4, use the drawing shown.

3. *Given:* $\overline{OV} \perp \overline{RS}$ in ⊙O
   $OV = 9$ and $OT = 6$
   Find: RS

4. *Given:* V is the midpoint of $\widehat{RS}$ in ⊙O
   $m\angle S = 15°$ and $OT = 6$
   Find: OR

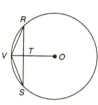

5. Sketch two circles that have:
   (a) No common tangents
   (b) Exactly one common tangent
   (c) Exactly two common tangents
   (d) Exactly three common tangents
   (e) Exactly four common tangents

6. Two intersecting circles B and D (not shown) have a line (segment) of centers $\overline{BD}$ and a common chord $\overline{AC}$ that are congruent. What may you conclude regarding quadrilateral $ABCD$?

Use the accompanying figure for Exercises 7 to 12. O is the center of the circle.

7. *Given:* $AE = 6, EB = 4, DE = 8$
   Find: EC

8. *Given:* $DE = 12, EC = 5, AE = 8$
   Find: EB

9. *Given:* $AE = 8, EB = 6, DC = 16$
   Find: DE and EC

10. *Given:* $AE = 7, EB = 5, DC = 12$
    Find: DE and EC

11. *Given:* $AE = 6, EC = 3, AD = 8$
    Find: CB

12. *Given:* $AD = 10, BC = 4, AE = 7$
    Find: EC

Use the accompanying figure for Exercises 13 to 16.

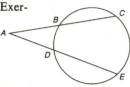

13. *Given:* $AB = 6, BC = 8, AE = 15$
    Find: DE

14. *Given:* $AC = 12, AB = 6, AE = 14$
    Find: AD

15. *Given:* $AB = 4, BC = 5, AD = 3$
    Find: DE

16. *Given:* $AB = 5, BC = 6, AD = 6$
    Find: AE

Use the accompanying figure for Exercises 17 to 20. $\overline{RS}$ is tangent to the circle at S.

17. *Given:* $RS = 8$ and $RV = 12$
    Find: RT

18. *Given:* $RT = 4$ and $TV = 6$
    Find: RS

19. *Given:* $\overline{RS} \cong \overline{TV}$ and $RT = 6$
    Find: RS

20. *Given:* $RT = \dfrac{1}{2} \cdot RS$ and $TV = 9$
    Find: RT

In Exercises 21 to 24, provide a paragraph proof.

21. *Given:* ⊙O and ⊙Q at point F
    secant $\overline{AC}$ to ⊙O
    secant $\overline{AE}$ to ⊙Q
    common internal tangent $\overline{AF}$
    Prove: $AC \cdot AB = AE \cdot AD$

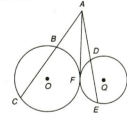

22. *Given:* ⊙O with $\overline{OM} \perp \overline{AB}$ and $\overline{ON} \perp \overline{BC}$
    $\overline{OM} \cong \overline{ON}$
    Prove: $\triangle ABC$ is isosceles

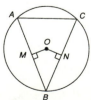

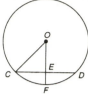

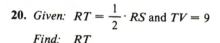

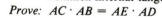

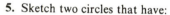

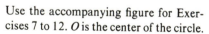

23. *Given:* ⊙Q with tangents
$\overline{MN}$ and $\overline{MP}$ so
that $\overline{MN} \perp \overline{MP}$
Prove: MNQP is a square

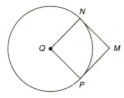

In Exercises 24 and 25, use the figure shown.

24. *Given:* $\overline{AB} \cong \overline{DC}$ in ⊙P
Prove: $\triangle ABD \cong \triangle CDB$

25. Does it follow from Exercise
24 that $\triangle ADE$ is also congruent to $\triangle CBE$? What may
you conclude about $\overline{AE}$ and
$\overline{CE}$ in the drawing? What
may you conclude about $\overline{DE}$
and $\overline{EB}$?

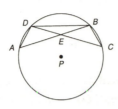

26. In ⊙O, $\overline{RS}$ is a diameter and T is the midpoint of
semicircle $\overarc{RTS}$. What is the value of the ratio $\dfrac{RT}{RS}$?

The ratio $\dfrac{RT}{RO}$?

27. *Given:* Tangents $\overline{AB}$, $\overline{BC}$,
and $\overline{AC}$ to ⊙O
at points M, N,
and P, respectively
$AB = 14$, $BC = 16$,
$AC = 12$
Find: AM, PC, and BN

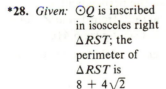

*28. *Given:* ⊙Q is inscribed
in isosceles right
$\triangle RST$; the
perimeter of
$\triangle RST$ is
$8 + 4\sqrt{2}$
Find: TM

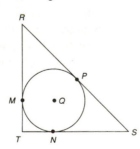

*29. *Given:* $\overline{AB}$ is an external tangent to ⊙O and ⊙Q at points
A and B; radii for ⊙O
and ⊙Q are 4 and 9,
respectively
Find: AB (*Hint:* The line of
centers $\overline{OQ}$ contains
point C, the point at
which ⊙O and ⊙Q
are tangent.)

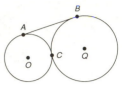

In Exercises 30 and 31, use the accompanying drawing.

*30. If the larger gear has 30 teeth and the smaller gear has
18, then the gear ratio (larger to smaller) is 5:3. When
the larger gear rotates through an angle of 60°, through
what angle measure does the smaller gear rotate?

*31. In Exercise 30, suppose that the larger gear has 20 teeth
and the smaller gear has 10 (the gear ratio is 2:1). If the
smaller gear rotates through an angle of 90°, through
what angle measure does the larger gear rotate?

In Exercises 32 to 35, prove the stated theorems.

32. If a line is drawn through the center of a circle perpendicular to a chord, then it bisects the chord and its minor
arc. (*Note:* The major arc is also bisected by the line.)

33. If a line is drawn through the center of a circle to the
midpoint of a chord other than a diameter, then it is perpendicular to the chord.

34. If two secant segments are drawn to a circle from an external point, then the products of the lengths of each
secant with its external segment are equal.

35. If a tangent segment and a secant segment are drawn to
a circle from an external point, then the square of the
length of the tangent equals the product of the lengths
of the secant and its external segment.

6.4

**SOME
CONSTRUCTIONS
AND INEQUALITIES
FOR THE CIRCLE**

In Section 6.3, we proved that the radius drawn to a tangent at the point of
contact is perpendicular to the tangent at that point. We now prove the converse
of that theorem by the indirect method; the contradiction in the proof involves
the fact that there is only one perpendicular to a line at a point on the line.

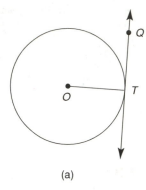

(a)

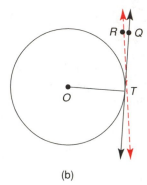

(b)

Figure 6.40

THEOREM 6.4.1   The line that is perpendicular to the radius of a circle at its endpoint on the circle is a tangent to the circle.

Given: In figure 6.40, $\odot O$ with radius $\overline{OT}$
$\overleftrightarrow{QT} \perp \overline{OT}$

Prove: $\overleftrightarrow{QT}$ is a tangent to $\odot O$ at point T

Proof: Suppose that $\overleftrightarrow{QT}$ is not a tangent to $\odot O$ at T. Then the tangent—call it $\overleftrightarrow{RT}$—can be drawn at T, the point of tangency. (See figure 6.40b.)

Now $\overline{OT}$ is the radius to tangent $\overleftrightarrow{RT}$ at T, and since a radius drawn to a tangent at the point of contact of the tangent is perpendicular to the tangent, $\overline{OT} \perp \overleftrightarrow{RT}$. But $\overline{OT} \perp \overleftrightarrow{QT}$ by hypothesis. Thus two lines are perpendicular to $\overline{OT}$ at point T, and this is a contradiction of the known fact mentioned earlier. Therefore $\overleftrightarrow{QT}$ must be a tangent to $\odot O$ at point T.   ▲

CONSTRUCTION 8   To construct a tangent to a circle at a point on the circle.

Given: In figure 6.41, $\odot P$ and point X on the circle

Construct: A tangent $\overleftrightarrow{XW}$ to $\odot P$ at point X

Construction: First draw radius $\overline{PX}$ and extend it to form $\overrightarrow{PX}$. On $\overrightarrow{PX}$, using X as the center and any radius length less than XP, draw arcs to intersect $\overrightarrow{PX}$, as shown in figure 6.41b.

Now complete the construction of the perpendicular to $\overrightarrow{PX}$ at point P; that is, mark arcs with equal radii from Y and Z. Join this point of intersection W to point X. (See figure 6.41c.)

Now draw $\overleftrightarrow{XW}$, the desired tangent to $\odot P$ at point X.   ▲

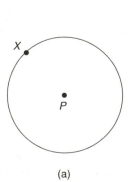

(a)

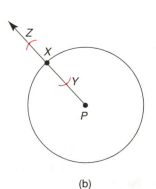

(b)

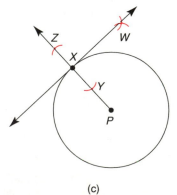

(c)

Figure 6.41

EXAMPLE 1   Points A, B, C, and D are on $\odot O$, in that order. If tangents are constructed at points A, B, C, and D, what type of quadrilateral would be formed by the tangent segments if:

(a) $m\widehat{AB} = m\widehat{CD}$ and $m\widehat{BC} = m\widehat{AD}$?

(b) All arcs $\widehat{AB}$, $\widehat{BC}$, $\widehat{CD}$, and $\widehat{DA}$ are equal?

Solution   (a) A rhombus (opposite $\angle$s are $\cong$; all sides $\cong$)

(b) A square (all four $\angle$s are right $\angle$s; all sides $\cong$)   ▲

We now consider a more difficult construction.

(a)

CONSTRUCTION 9   To construct a tangent to a circle from an external point.

Given: $\odot Q$ and external point E

Construct: A tangent $\overline{ET}$, with T as the point of tangency

Construction: Draw $\overline{EQ}$. Construct the perpendicular bisector of $\overline{EQ}$, to intersect $\overline{EQ}$ at its midpoint M. (See figure 6.42b.)

With M as center and MQ (or ME) as the length of radius, construct a circle. The points of intersection of circle M with circle Q are designated by T and V.

Draw $\overline{ET}$, the desired tangent. (See figure 6.42c.) (Note that $\overline{EV}$, if drawn, would also be a tangent to $\odot Q$).   ▲

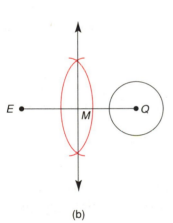

(b)

In the preceding construction, $\overline{QT}$ is a radius of the smaller circle Q. In the larger circle M, $\angle ETQ$ is an inscribed angle that intercepts a semicircle. Thus $\angle ETQ$ is a right angle and $\overline{ET} \perp \overline{TQ}$. Since the line drawn perpendicular to the radius of a circle at its endpoint on the circle is a tangent to the circle, $\overline{ET}$ is a tangent to circle Q.

While our earlier work has focused on congruent angles, congruent arcs, and congruent chords, the circle has several significant inequality relationships, too. These relationships deal with angles, arcs, and chords that are noncongruent. By "noncongruent," we mean that the measures are not equal, and we refer to these angles, arcs, and chords as "unequal."

THEOREM 6.4.2   In a circle (or in congruent circles) containing two unequal central angles, the larger angle corresponds to the larger intercepted arc.

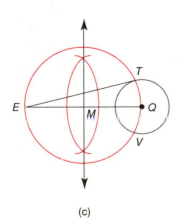

(c)

Figure 6.42

Given: $\odot O$ with central angles $\angle 1$ and $\angle 2$ in figure 6.43 (next page)
$m\angle 1 > m\angle 2$

Prove: $m\widehat{AB} > m\widehat{CD}$

Proof: In $\odot O$, $m\angle 1 > m\angle 2$. By the Central Angle Postulate, $m\angle 1 = m\widehat{AB}$ and $m\angle 2 = m\widehat{CD}$. By substitution, $m\widehat{AB} > m\widehat{CD}$.   ▲

The converse of the previous theorem is also easy to prove.

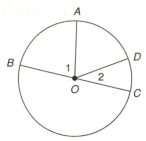

Figure 6.43

> **THEOREM 6.4.3** In a circle (or in congruent circles) containing two unequal arcs, the larger arc corresponds to a larger central angle.

Given: In figure 6.43, $\odot O$ with $\overset{\frown}{AB}$ and $\overset{\frown}{CD}$
   $m\overset{\frown}{AB} > m\overset{\frown}{CD}$

Prove: $m\angle 1 > m\angle 2$

Proof: Left as an exercise for the student.

EXAMPLE 2   In figure 6.44, $\odot Q$ with $m\overset{\frown}{RS} > m\overset{\frown}{TV}$.
(a) Using Theorem 6.4.3, what may you conclude regarding the measures of $\angle RQS$ and $\angle TQV$?
(b) What does intuition suggest regarding RS and TV?

Solution   (a) $m\angle RQS > m\angle TQV$
(b) $RS > TV$ ▲

   The following statement can be proved in much the same way as Theorem 6.4.5.

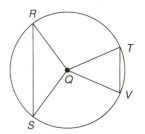

Figure 6.44

> **THEOREM 6.4.4** In a circle (or in congruent circles) containing two unequal chords, the shorter chord is at a greater distance from the center of the circle.

EXAMPLE 3   In circle P, the radius has a length of 6 cm, and chords $AB = 4$ cm, $DC = 6$ cm, and $EF = 10$ cm. Let $\overline{PR}$, $\overline{PS}$, and $\overline{PT}$ name perpendicular segments to these chords. (See figure 6.45.)
(a) Of PR, PS, and PT, which is greatest?
(b) Which is least?

Solution   (a) PR is greatest, according to Theorem 6.4.4.
(b) PT is least.   ▲

> **THEOREM 6.4.5** In a circle (or in congruent circles) containing two unequal chords, the chord nearer the center of the circle has a greater length.

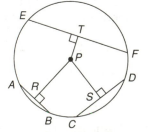

Figure 6.45

Given: In figure 6.46, $\odot Q$ with chords $\overline{AB}$ and $\overline{CD}$
   $\overline{QM} \perp \overline{AB}$ and $\overline{QN} \perp \overline{CD}$
   $QM < QN$

Prove: $AB > CD$

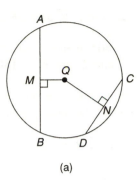

(a)

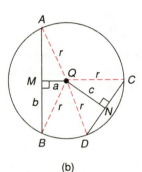

(b)

Figure 6.46

Proof: In figure 6.46b, denote lengths of $\overline{QM}$ and $\overline{QN}$ by a and c, respectively. Draw radii $\overline{QA}$, $\overline{QB}$, $\overline{QC}$, and $\overline{QD}$, and denote all lengths by r. $\overline{QM}$ is the perpendicular bisector of $\overline{AB}$, and $\overline{QN}$ is the perpendicular bisector of $\overline{CD}$, since a radius perpendicular to a chord bisects the chord and its arc. Let $MB = b$ and $NC = d$.

With right angles at M and N, we may say that $\triangle QMB$ and $\triangle QNC$ are right triangles.

According to the Pythagorean Theorem, $r^2 = a^2 + b^2$ and $r^2 = c^2 + d^2$, so $b^2 = r^2 - a^2$ and $d^2 = r^2 - c^2$. If $QM < QN$, then $a < c$ and $a^2 < c^2$. Therefore $-a^2 > -c^2$. Adding r^2, we have $r^2 - a^2 > r^2 - c^2$ or $b^2 > d^2$, which implies that $b > d$. If $b > d$, then $2b > 2d$. But $AB = 2b$ while $CD = 2d$. Therefore $AB > CD$. ▲

It is important that the phrase "minor arc" be used in our final theorems. For the second theorem, the proof is provided because it is more involved.

THEOREM 6.4.6 In a circle (or in congruent circles) containing two unequal chords, the longer chord corresponds to the greater minor arc.

THEOREM 6.4.7 In a circle (or in congruent circles) containing two unequal minor arcs, the greater minor arc corresponds to the longer of the chords related to these arcs.

Given: In figure 6.47a, $\odot O$ with $\text{m}\widehat{AB} > \text{m}\widehat{CD}$
chords $\overline{AB}$ and $\overline{CD}$

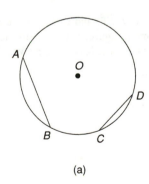

(a)

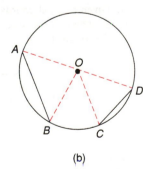

(b)

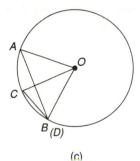

(c)

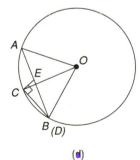

(d)

Figure 6.47

Prove: $AB > CD$

Proof: In circle O, draw radii $\overline{OA}$, $\overline{OB}$, $\overline{OC}$, and $\overline{OD}$. $\text{m}\widehat{AB} > \text{m}\widehat{CD}$ leads to $\text{m}\angle AOB > \text{m}\angle COD$ since the larger arc in a circle corresponds to a larger central angle. (See figure 6.47b.)

We now rotate $\triangle COD$ to the position on the circle for which D coincides with *B*, as shown in figure 6.46c. Because radii $\overline{OC}$ and $\overline{OB}$ are congruent, ΔCOD is isosceles; also, m∠*C* = m∠*ODC*.

In ΔCOD, m∠*COD* + m∠*C* + m∠*CDO* = 180°. Because m∠*COD* is positive, we have m∠*C* + m∠*CDO* < 180° and 2 · m∠*C* < 180° by substitution. Therefore m∠*C* < 90°.

Now construct the perpendicular segment to $\overline{CD}$ at point *C*, as shown in figure 6.46d. Denote the intersection of the perpendicular segment and $\overline{AB}$ by point *E*.

Because ΔDCE is a right Δ with hypotenuse $\overline{EB}$, we know that *EB* > *CD* (*).

Because *AB* = *AE* + *EB*, in which *AE* > 0, we have *AB* > *EB* (*).

Using the starred (*) statements and the Transitive Property of Inequality, we have the desired claim, *AB* > *CD*.   ▲

Note: In the preceding proof, $\overline{CE}$ must intersect $\overline{AB}$ at some point between *A* and *B*. If it were to intersect at *A*, the measure of inscribed ∠*BCA* would have to be more than 90°; this follows from the facts that $\widehat{AB}$ is a minor arc and that the intercepted arc for ∠*BCA* would have to be a major arc.

6.4 EXERCISES

1. Construct a circle *O* and choose some point *D* on the circle. Now construct the tangent to circle *O* at point *D*.

2. Construct a circle *P* and choose three points *R*, *S*, and *T* on the circle. Construct the triangle that has its sides tangent at *R*, *S*, and *T*.

3. *X*, *Y*, and *Z* are on circle *O* so that m$\widehat{XY}$ = 120°, m$\widehat{YZ}$ = 130°, and m$\widehat{XZ}$ = 110°. Suppose that triangle *XYZ* is drawn and that the triangle *ABC* is constructed with its sides tangent to circle *O* at *X*, *Y*, and *Z*. Are ΔXYZ and ΔABC similar triangles?

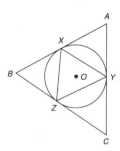

4. Construct the two tangent segments to circle *P* from external point *E*.

5. Point *V* is in the exterior of circle *Q* so that $\overline{VQ}$ is equal in length to the diameter of circle *Q*. Construct the two tangents to circle *Q* from point *V*. Then determine the measure of the angle that has vertex *V* and has the tangents as sides.

6. Given circle *P* and points *R*–*P*–*T* so that *R* and *T* are in the exterior of circle *P*, suppose that tangents are constructed from *R* and *T* to form a quadrilateral (as shown). Identify the type of quadrilateral formed

   (a) when *RP* > *PT*
   (b) when *RP* = *PT*

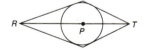

7. Given parallel chords $\overline{AB}$, $\overline{CD}$, $\overline{EF}$, and $\overline{GH}$ in circle *O* as shown in the accompanying figure. Which chord has the greatest length? Which has the least length? Why?

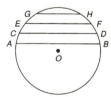

8. Given chords $\overline{MN}$, $\overline{RS}$, and $\overline{TV}$ in $\odot Q$ so that *QZ* > *QY* > *QX*. Which chord has the greatest length? Which has the least length? Why?

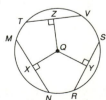

In Exercises 9 and 10, use the accompanying figure.

9. In circle O, points A, B, and C are on the circle so that $m\widehat{AB} = 60°$ and $m\widehat{BC} = 40°$.

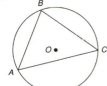

   (a) How are $\angle AOB$ and $\angle BOC$ related?
   (b) How are AB and BC (lengths of chords) related?

10. In $\odot O$, $AB = 6$ cm and $BC = 4$ cm.

   (a) How are $m\angle AOB$ and $m\angle BOC$ related?
   (b) How are $m\widehat{AB}$ and $m\widehat{BC}$ related?

11. In $\odot O$, $m\angle AOB = 70°$ and $m\angle BOC = 30°$.

   (a) How are $m\widehat{AB}$ and $m\widehat{BC}$ related?
   (b) How are AB and BC related?

In Exercises 12 and 13, use the accompanying drawing.

12. Triangle ABC is inscribed in circle O; $AB = 5$, $BC = 6$, and $AC = 7$.

   (a) Which is the largest minor arc of $\odot O$—$\widehat{AB}$, $\widehat{BC}$, or $\widehat{AC}$?
   (b) Which side of the triangle is nearest point O?

13. Given circle O with $m\widehat{BC} = 120°$ and $m\widehat{AC} = 130°$.

   (a) Which angle of triangle ABC is smallest?
   (b) Which side of triangle ABC is nearest point O?

14. Circle O has a diameter of length 20 cm. Chord $\overline{AB}$ has length 12 cm, and chord $\overline{CD}$ has length 10 cm. How much closer is $\overline{AB}$ than $\overline{CD}$ to point O?

15. Circle P has a radius of length 8 in. Points A, B, C, and D are on circle P so that $m\angle APB = 90°$ while $m\angle CPD = 60°$. How much closer is chord $\overline{AB}$ than $\overline{CD}$ to point P?

16. A tangent $\overline{ET}$ is constructed to circle Q from external point E. Which angle and side of triangle QTE are largest? Which angle and side are smallest?

17. Two congruent circles $\odot O$ and $\odot P$ do not intersect. Construct a common external tangent for $\odot O$ and $\odot P$.

18. Explain why the following statement is incorrect:

   In a circle (or in congruent circles) containing two unequal chords, the longer chord corresponds to the greater major arc.

19. Prove that in a circle containing two unequal arcs, the larger arc corresponds to a larger central angle.

20. Prove that in a circle containing two unequal chords, the longer chord corresponds to a larger central angle. (*Hint:* You may use any theorems stated in this section.)

6.5 LOCUS AND CONCURRENCY

At times we need to describe a set of points that satisfy a given condition or set of conditions. The term used to describe the resulting geometric figure is "locus," the plural of which is "loci" (pronounced lō-sī). The English word "location" is derived from the Latin word "locus."

> **DEFINITION:** A **locus** is the set of all points and only those points that satisfy a given condition or set of conditions.

In this definition, the phrase "all points and only those points" has a double meaning:

1. All points of the locus satisfy the given condition.

2. All points satisfying the given condition are included in the locus.

Generally, the set of points satisfying a given locus description results in some geometric figure with which you are acquainted. In Examples 1 through 5, you will find the locus to be a circle, a line, a ray, a plane, and a circle, respectively.

EXAMPLE 1   Describe the locus of points in a plane that are at a fixed distance (*r*) from a given point (*P*).

Solution   The locus is the circle with center *P* and radius *r*. (See figure 6.48.)   ▲

EXAMPLE 2   Describe the locus of points in a plane that are equidistant from two fixed points (*P* and *Q*).

Solution   The locus is a line that is the perpendicular bisector of $\overline{PQ}$. (See figure 6.49.) (Notice that $PX = QX$ for any point *X* on line *t*.)   ▲

EXAMPLE 3   Describe the locus of points in a plane that are equidistant from the sides of an angle in that plane.

Solution   The locus is the ray that bisects ∠*ABC*. (See figure 6.50.)   ▲

Two observations are appropriate here. First, be aware that some definitions are given as locus descriptions; for example, the following is an alternative definition of "circle."

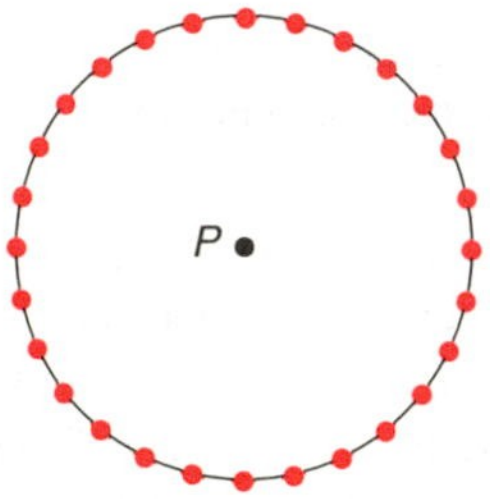

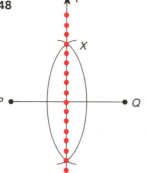

Figure 6.48

> **DEFINITION:**   A **circle** is the locus of points in a plane that are at a fixed distance from a given point.

Figure 6.49

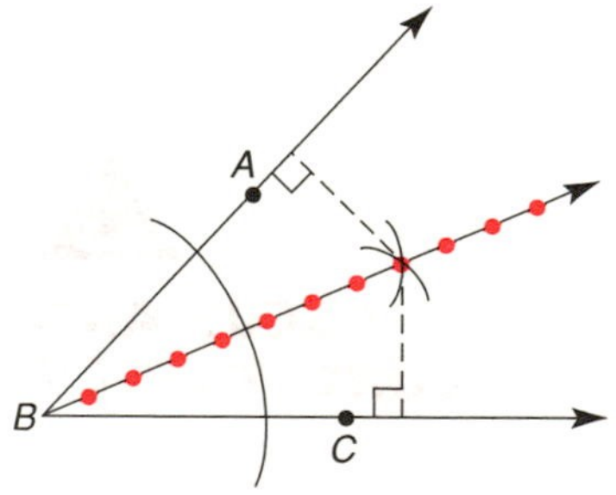

Figure 6.50

Second, each solution in Examples 1, 2, and 3 includes the phrase "in a plane." If that phrase is omitted, the locus will change. For instance, the locus of points that are at a fixed distance from a given point is actually a sphere (the three-dimensional object in figure 6.51); the sphere has the fixed point as center, and the fixed distance determines the length of the radius. Unless otherwise stated, we will consider the locus to be restricted to a plane.

Figure 6.51

EXAMPLE 4   Describe the locus of points *in space* that are equidistant from two parallel planes (*P* and *Q*).

Solution   The locus is the plane parallel to each of the given planes and midway between them. (See figure 6.52.)   ▲

Our next example mentions neither plane nor space; thus we consider the locus in a plane. Suppose that a given line segment in a fixed location is to be used as the hypotenuse of a right triangle. How might you locate possible positions for the vertex of the right angle? One method might be to construct 30° and 60° angles at the endpoints so that the remaining angle formed must measure 90°. This is only one possibility, but it actually provides four permissible points, due to symmetry; these points are indicated in figure 6.53b.

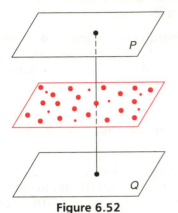

Figure 6.52

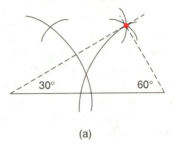

(a)

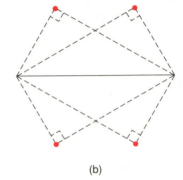

(b)

Figure 6.53

EXAMPLE 5   Find the locus of the vertex of the right angle of a right triangle if the hypotenuse is the given segment $\overline{AB}$ in figure 6.54a.

Solution   Rather than use the "hit or miss" approach for locating the possible vertices (as suggested just before this example), recall that an angle inscribed in a semicircle is a right angle.

Thus, we construct the circle whose center is the midpoint *M* of the hypotenuse and whose radius equals one-half the length of the hypotenuse. First, the midpoint *M* of the hypotenuse $\overline{AB}$ is located. (See figure 6.54b.)

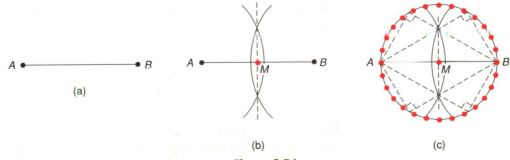

Figure 6.54

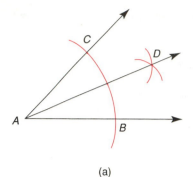

(a)

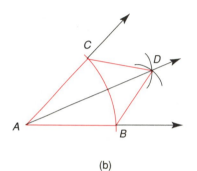

(b)

Figure 6.55

With the radius of the circle equal to one-half the hypotenuse, the circle with center M is drawn in figure 6.54c.

The locus of the vertex of a right angle whose hypotenuse is given is the circle whose center is at the midpoint of the given segment and whose radius is equal in length to half the length of the given segment.   ▲

In Example 5, the construction involves locating the midpoint M of $\overline{AB}$, and this is found by the method for the perpendicular bisector. The compass is then opened to a radius whose length is MA or MB, and the circle is drawn. When a construction is performed, it falls into one of two categories:

1. A basic construction method

2. A construction problem that may require several steps and may involve several basic construction methods

In Example 7, we will consider one of the category 2 problems.

When earlier constructions were presented, we made no attempt to justify them. Consider the basic method for constructing an angle bisector. After arcs of equal radii are drawn from the vertex of the angle to intersect the two sides, those points are used in turn as centers of arcs. Arcs of equal radii are marked with the earlier points of intersection as centers. Each such point of intersection determines a point on the angle bisector, which is then drawn. (See figure 6.55a.) Why is this method valid?

EXAMPLE 6   Verify the angle bisector method.

Proof (verification)   Figure 6.55a has been redrawn with $\overline{CD}$ and $\overline{DB}$. (See figure 6.55b.) By the method of construction and the fact that all radii of a circle are equal in length, we know that

$$\overline{AB} \cong \overline{AC} \qquad \text{and} \qquad \overline{BD} \cong \overline{CD}$$

Also, $\overline{AD} \cong \overline{AD}$, so $\triangle ADC \cong \triangle ADB$ by SSS. Then $\angle 1 \cong \angle 2$ by CPCTC, and $\overrightarrow{AD}$ bisects $\angle CAB$ by the definition of angle bisector.   ▲

When we verify the locus theorems, we *must* establish two results:

1. If a point is in the locus, then it satisfies the condition.

2. If a point satisfies the condition, then it is a point of the locus.

> **THEOREM 6.5.1**   The locus of points in a plane and equidistant from the sides of an angle is the angle bisector.

Proof   (Notice that *both* parts i and ii are necessary):
i. If a point is on the angle bisector, then it is equidistant from the sides of the angle.

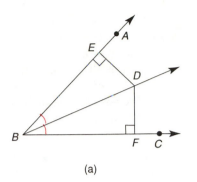

(a)

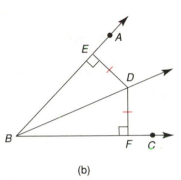

(b)

Figure 6.56

Given: $\overrightarrow{BD}$ bisects $\angle ABC$
$\overline{DE} \perp \overline{BA}$ and $\overline{DF} \perp \overline{BC}$

Prove: $\overline{DE} \cong \overline{DF}$

Proof: In figure 6.56a, $\overrightarrow{BD}$ bisects $\angle ABC$; thus $\angle ABD \cong \angle CBD$. $\overline{DE} \perp \overline{BA}$ and $\overline{DF} \perp \overline{BC}$, so $\angle DEB$ and $\angle DFB$ are $\cong$ right $\angle$s. $\overline{BD} \cong \overline{BD}$. By AAS, $\triangle DEB \cong \triangle DFB$. Then $\overline{DE} \cong \overline{DF}$ by CPCTC.

ii. If a point is equidistant from the sides of an angle, then it is on the angle bisector.

Given: $\angle ABC$ so that $\overline{DE} \perp \overrightarrow{BA}$ and $\overline{DF} \perp \overrightarrow{BC}$
$\overline{DE} \cong \overline{DF}$

Prove: $\overrightarrow{BD}$ bisects $\angle ABC$ (That is, D is on the bisector of $\angle ABC$)

Proof: In figure 6.56b, $\overline{DE} \perp \overrightarrow{BA}$ and $\overline{DF} \perp \overrightarrow{BC}$, so $\angle DEB$ and $\angle DFB$ are right triangles. $\overline{DE} \cong \overline{DF}$ by hypothesis. Also, $\overline{BD} \cong \overline{BD}$. $\triangle DEB \cong \triangle DFB$ by HL. Then $\angle ABD \cong \angle CBD$ by CPCTC, and $\overrightarrow{BD}$ bisects $\angle ABC$ by definition. ▲

In both construction problems and locus problems, we can verify results. In locus problems, however, we must remember to demonstrate two relationships! The following example combines the notion of locus with that of construction.

EXAMPLE 7 (Construction Problem): Construct a rectangle for which $\overline{RT}$ is a diagonal so that the angle formed by the diagonals measures 30°.

Solution Because the diagonals of a rectangle bisect each other, we begin by constructing the perpendicular bisector of $\overline{RT}$; our purpose is to find the midpoint M of $\overline{RT}$. (See figure 6.57b.)

Now with $\overline{RT}$ as a diagonal of the rectangle, the locus of the vertex of one of the right angles is along a circle that has $\overline{RT}$ as a diameter. In this case, $\overline{RT}$ is also a hypotenuse for the possible right triangles. Thus we construct a circle with M as center and a radius equal in length to MR or MT. (See figure 6.57b.)

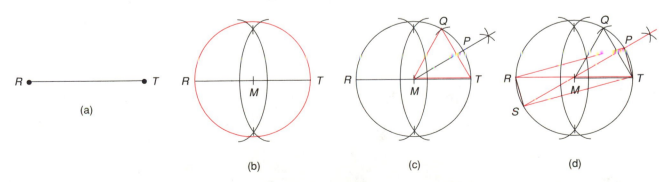

Figure 6.57

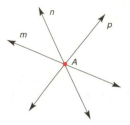

(a) *m*, *n*, and *p* are concurrent

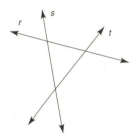

(b) *r*, *s*, and *t* are *not* concurrent

Figure 6.58

So that the angle of intersection will measure 30°, we construct an equilateral $\triangle QMT$ by using the length of $\overline{MT}$ for each side of the triangle. $\angle QMT$ is then bisected to form an angle whose measure is 30°. (See figure 6.57c.) Now the point of intersection P on the circle and the side of the 30° angle is the vertex of the right angle, which allows us to complete the construction. The angle bisector $\overline{MP}$ is extended to become diameter $\overline{PS}$. We now draw segments from S to R, R to P, P to T, and T to S to form the desired rectangle $RPTS$. (See figure 6.57d.) ▲

While this problem requires a strategy, the plan that results is correct if it can be justified.

Next we consider the term "concurrent."

DEFINITION: A number of lines are **concurrent** if they have exactly one point in common.

The lines in figure 6.58a are concurrent, while those in figure 6.58b are not.

THEOREM 6.5.2 The three angle bisectors of the angles of a triangle are concurrent.

Proof In figure 6.59a, the bisectors of $\angle BAC$ and $\angle ABC$ are shown intersecting at point E.

Because the angle bisector of $\angle BAC$ is the locus of points equidistant from the sides of $\angle BAC$, we know that $\overline{EM} \cong \overline{EN}$ in figure 6.59b. Similarly, we know that $\overline{EM} \cong \overline{EP}$.

By the Transitive Property of Congruence, it follows that $\overline{EP} \cong \overline{EN}$.

Because the angle bisector is the locus of points equidistant from the sides of the angle, we know that E is also on the bisector of the third angle, $\angle ACB$. Thus, the angle bisectors are concurrent. ▲

The point E at which the angle bisectors meet in Example 5 is the **incenter** of the triangle. As the following example shows, the term "incenter" is appropriate because this point is the *center* of the *in*scribed circle of the triangle.

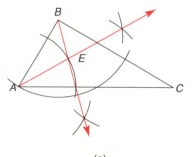

(a)

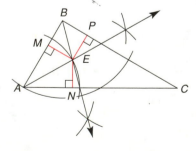

(b)

Figure 6.59

EXAMPLE 8

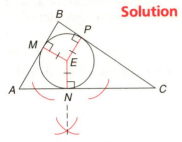

Figure 6.60

Solution

Complete the construction of the inscribed circle for △*ABC* in figure 6.59b.

Having found the incenter *E*, we need to find the length of the radius; this is done by constructing a segment from *E* perpendicular to any side ($\overline{EN} \perp \overline{AC}$ is shown in figure 6.60). The length of $\overline{EN}$ is the desired radius, and the circle is completed. ▲

Note: The sides of the triangle are tangents for the inscribed circle.

It is also possible to circumscribe a circle about a given triangle. The construction depends on the following theorem, whose proof is sketched afterward.

THEOREM 6.5.3   The three perpendicular bisectors of the sides of a triangle are concurrent.

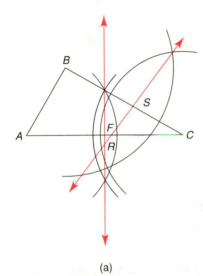

(a)

Suppose that we wish to construct the circumscribed circle for △*ABC*. We need a center that is equidistant from vertices *A*, *B*, and *C*. This point of concurrency is found by constructing the perpendicular bisectors of at least two sides of the triangle. (See figure 6.61a.) We are reminded that $\overline{AF} \cong \overline{FC}$ (where $\overline{FR}$ is the ⊥ bisector of $\overline{AC}$) and that $\overline{BF} \cong \overline{FC}$ (where $\overline{FS}$ is the ⊥ bisector of $\overline{BC}$). It follows that $\overline{AF} \cong \overline{BF}$ by the Transitive Property of Congruence. Then *F* is on the ⊥ bisector of $\overline{AB}$ and is equidistant from vertices *A* and *B*.

To complete the construction of the circumscribed circle, we use *F* as center and a length of radius equal to that from *F* to any vertex. (See figure 6.61b.)

The point at which the perpendicular bisectors of the sides of a triangle meet is the **circumcenter** of the triangle. This is easily remembered as the *center* of the *circum*scribed circle.

The incenter and the circumcenter of a triangle are generally distinct points. However, it is possible for the two centers to coincide in a special type of triangle.

For completeness (but without proof), we state these final theorems and related terms.

THEOREM 6.5.4   The three altitudes of a triangle are concurrent.

The point of concurrency for the three altitudes of a triangle is the **orthocenter** of the triangle. In the accompanying drawing, point *N* is the orthocenter of △*DEF*.

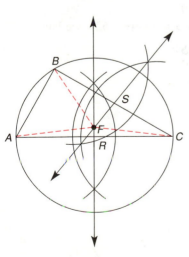

(b)

Figure 6.61

Recall that a median of a triangle joins a vertex to the midpoint of the opposite side.

> **THEOREM 6.5.5** The three medians of a triangle are concurrent at a point that is two-thirds the distance from any vertex to the midpoint of the opposite side.

The point of concurrency for the three medians is the **centroid** of the triangle. Point Q in figure 6.62a of Example 9 is the centroid of $\triangle RST$.

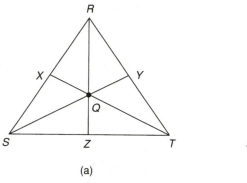

(a)   (b)

Figure 6.62

EXAMPLE 9

Given: In figure 6.62a, isosceles $\triangle RST$ with $RS = RT = 15$, while $ST = 18$ medians $\overline{RZ}$, $\overline{TX}$, and $\overline{SY}$ meet at centroid Q

Find: RQ

Solution

Median $\overline{RZ}$ leads to congruent right triangles $\triangle RZS$ and $\triangle RZT$; this follows from SSS. With Z the midpoint of $\overline{ST}$, $SZ = 9$.

Using the Pythagorean Theorem and $\triangle RZS$ in figure 6.62b, we have

$$(RS)^2 = (RZ)^2 + (SZ)^2$$
$$15^2 = (RZ)^2 + 9^2$$
$$225 = (RZ)^2 + 81$$
$$(RZ)^2 = 144$$
$$RZ = 12$$

By Theorem 6.5.5,

$$RQ = \frac{2}{3}RZ$$

$$= \frac{2}{3}(12)$$

$$= 8$$ ▲

In Chapter 8, we will prove the theorem regarding the concurrency of the medians of a triangle by the analytic method.

6.5 EXERCISES

In Exercises 1 to 8, perform each basic construction.

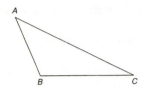

In Exercises 1 to 6, use the drawing above.

1. *Given:* Obtuse $\triangle ABC$
   Construct: The bisector of $\angle ABC$

2. *Given:* Obtuse $\triangle ABC$
   Construct: The bisector of $\angle BAC$

3. *Given:* Obtuse $\triangle ABC$
   Construct: The perpendicular bisector of $\overline{AB}$

4. *Given:* Obtuse $\triangle ABC$
   Construct: The perpendicular bisector of $\overline{AC}$

5. *Given:* Obtuse $\triangle ABC$
   Construct: The altitude from A to $\overline{BC}$ (*Hint:* Extend $\overline{BC}$.)

6. *Given:* Obtuse $\triangle ABC$
   Construct: The altitude from B to $\overline{AC}$

Use the figure shown below for Exercises 7 and 8.

7. *Given:* Right $\triangle RST$
   Construct: The median from S to $\overline{RT}$

8. *Given:* Right $\triangle RST$
   Construct: The median from R to $\overline{ST}$

In Exercises 9 to 20, sketch and describe each locus in the plane.

9. Find the locus of points that are at a given distance from a fixed line.

10. Find the locus of points that are equidistant from two given parallel lines.

11. Find the locus of points that are at a distance of 3 in. from a fixed point O.

12. Find the locus of points that are equidistant from two fixed points A and B.

13. Find the locus of points that are equidistant from three noncollinear points D, E, and F.

14. Find the locus of the midpoints of the radii of a circle O that has a radius of length 8 cm.

15. Find the locus of the midpoints of all chords of circle Q, if the chords are all parallel to diameter $\overline{PR}$.

16. Find the locus of points in the interior of a right triangle with sides of 6 in., 8 in., and 10 in. and at a distance of 1 in. from the triangle.

17. Find the locus of points that are equidistant from two given intersecting lines.

*18. Find the locus of points that are equidistant from a fixed line and a point not on that line. (*Note:* This figure is known as a parabola.)

19. Given that lines p and q intersect, find the locus of points that are at a distance of 1 cm from line p and also at a distance of 2 cm from line q.

20. Given that congruent circles O and P have radii of length 4 in. and that the line of centers has length 6 in., find the locus of points that are 1 in. from each circle.

In Exercises 21 to 28, sketch and describe the locus of points in space.

21. Find the locus of points that are at a given distance from a fixed line.

22. Find the locus of points that are equidistant from two fixed points.

23. Find the locus of points that are at a distance of 2 cm from a sphere whose radius is 5 cm.

24. Find the locus of points that are at a given distance from a given plane.

25. Find the locus of points that are the midpoints of the radii of a sphere whose center is point O and whose radius has a length of 5 m.

26. Find the locus of points that are equidistant from three noncollinear points D, E, and F.

27. In a room, find the locus of points that are equidistant from the ceiling and floor, which are 8 ft apart.

28. Find the locus of points that are equidistant from all points on the surface of a sphere whose center is point Q.

In Exercises 29 and 30, use the method of Example 6 to justify each construction method.

29. The perpendicular bisector method.

30. The construction of a perpendicular to a line from a point outside the line.

In Exercises 31 to 34, refer to the following line segments shown in the drawing.

31. Construct an isosceles right triangle that has hypotenuse $\overline{AB}$.

32. Construct a rhombus whose sides are equal in length to *AB*, while one diagonal has length *CD*.

33. Construct an isosceles triangle in which each leg has length *CD* and the altitude to the base has length *AB*.

34. Construct an equilateral triangle in which the altitude to any side has length *AB*.

Use the following figure for Exercises 35 and 36.

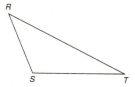

35. Construct the inscribed circle for obtuse △*RST*.

36. Construct the circumscribed circle for obtuse △*RST*.

37. Use the following theorem to locate the center of the circle of which $\widehat{RT}$ is a part.

Theorem: The perpendicular bisector of a chord passes through the center of a circle.

*38. Use the following theorem to construct the geometric mean of the numerical lengths of the segments $\overline{WX}$ and $\overline{YZ}$.

Theorem: The length of the altitude to the hypotenuse of a right triangle is the geometric mean between the lengths of the segments of the hypotenuse.

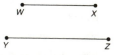

39. Use the following theorem to construct a triangle similar to the given triangle but with sides that are twice the length of those of the given triangle.

Theorem: If the three pairs of sides for two triangles are in proportion, then those triangles are similar.

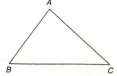

40. Classify the following statements as true or false:

   (a) The perpendicular bisectors of the sides of a rectangle are concurrent.

   (b) The locus of the midpoints of the chords of a circle is also a circle.

   (c) The angle bisectors of the angles of a rectangle are concurrent.

   (d) In space, the locus of the midpoints of the radii of a sphere is also a sphere.

*41. Verify this locus theorem:

   The locus of points equidistant from two fixed points is the perpendicular bisector of the segment joining those points.

42. To locate the orthocenter, is it necessary to construct the three altitudes of a right triangle? Explain.

43. For what type of triangle are the angle bisectors, the medians, the perpendicular bisectors of sides, and the altitudes all the same?

44. Is the incenter always located in the interior of the triangle?

45. Is the circumcenter always located in the interior of the triangle?

46. Find the radius of the inscribed circle for a right triangle whose legs measure 6 and 8.

47. Find the distance from the circumcenter to each vertex of an equilateral triangle whose sides measure 10.

48. A triangle has angles measuring 30°, 30°, and 120°. If the congruent sides measure 6 units each, find the radius of the circumscribed circle.

For Exercises 49 and 50, use the accompanying drawing.

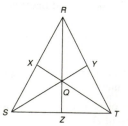

49. *Given:* Isosceles △*RST*
   RS = *RT* = 17 and *ST* = 16
   medians $\overline{RZ}$, $\overline{TX}$, and $\overline{SY}$ meet at centroid *Q*
   Find: *RQ* and *SQ*

50. *Given:* Isosceles △*RST*
   RS = *RT* = 10 and *ST* = 16
   medians $\overline{RZ}$, $\overline{TX}$, and $\overline{SY}$ meet at *Q*
   Find: *RQ* and *QT*

▲ A Look Beyond: The Value of π

In geometry, any two figures that have the same shape are described as similar. Because all circles have the same shape, we say that all circles are similar to each other. Just as a proportionality exists among the corresponding sides of similar triangles, we also assume a proportionality among the circumferences (distances around) and diameters (distances across) of circles. By representing the circumferences of the circles in figure 6.63 by C_1, C_2, and C_3 and their corresponding lengths of diameters by d_1, d_2, and d_3, we are claiming that

$$\frac{C_1}{d_1} = \frac{C_2}{d_2} = \frac{C_3}{d_3} = k$$

for some constant of proportionality k.

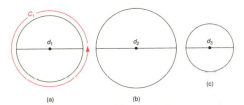

Figure 6.63

POSTULATE 17   The ratio of the circumference of a circle to its length of diameter is a constant.

We denote the constant k—which is described in the earlier work and in Postulate 17—by the Greek letter π.

DEFINITION: π is the ratio between the circumference and the length of diameter in any circle; that is, $\pi = \dfrac{C}{d}$, where C is the circumference and d the length of diameter in a given circle.

From the definition of π, it follows that $C = \pi d$ or $C = 2\pi r$ (since $d = 2r$ in any circle). In applying these formulas for the circumference of a circle, we often wish to leave π in the answer so that the result is exact. When an approximation for the circumference (and later for the area) of a circle is needed, several common substitutions are used for π.

Commonly used approximations of π are $\dfrac{22}{7}$, 3.14, and 3.1416.

Because π is needed in many applications involving the circumference or area of a circle, its approximation is often necessary; but finding an accurate approximation of π was not quickly or easily done. While the formula for circumference can be expressed as $C = 2\pi r$, the formula for the area of the circle is $A = \pi r^2$. This and other area formulas will be given greater attention in Chapter 7.

Several references to the value of π are made in literature. One of the earliest comes from the Bible; the passage from I Kings, verse 23 describes the distance around a vat as three times the distance across the vat (which suggests that π equals 3—a very rough approximation). Perhaps no greater accuracy was needed in applications at that time.

In the content of the Rhind papyrus (a document over 3000 years old), the Egyptian scribe Ahmes gives the formula for the area of a circle as $\left(d - \dfrac{1}{9}d\right)^2$. To determine the Egyptian approximation of π, we need to expand this expression as follows:

$$\left(d - \frac{1}{9}d\right)^2 = \left(\frac{8}{9}d\right)^2 = \left(\frac{8}{9} \cdot 2r\right)^2$$
$$= \left(\frac{16}{9}r\right)^2 = \frac{256}{81}r^2$$

In the formula for the area of the circle, the value of π is expressed by the coefficient of r^2. Because this coefficient is $\dfrac{256}{81}$ (which has the decimal equivalent of 3.1604), we can see that the Egyptians had a better approximation of π than was given in the book of I Kings.

Archimedes, the brilliant Greek geometer, knew that the formula for the area of a circle was $A = \dfrac{1}{2}Cr$ (with C the circumference and r the length of radius). His formula was equivalent to the one which we use today and is developed as follows:

$$A = \frac{1}{2}Cr$$
$$= \frac{1}{2}(2\pi r)r$$
$$= \pi r^2$$

The second proposition of Archimedes' work *Measure of the Circle* develops a relationship between the area of a circle and the area of the square in which it is inscribed. (See figure 6.64.) Specifically, Archimedes claimed that the ratio of the area of the circle to that of the square was 11:14. This leads to the following set of equations and an approximation of the value of π:

$$\frac{\pi r^2}{(2r)^2} \approx \frac{11}{14}$$

$$\frac{\pi r^2}{4r^2} \approx \frac{11}{14}$$

$$\frac{\pi}{4} \approx \frac{11}{14}$$

$$\pi \approx 4 \cdot \frac{11}{14} \approx \frac{22}{7}$$

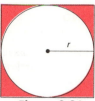

Figure 6.64

Archimedes later improved his approximation of π by showing that

$$3\frac{10}{71} < \pi < 3\frac{1}{7}$$

Today's calculators provide excellent approximations for the irrational number π. We should recall, however, that π is an irrational number which can be expressed exactly only by the unique symbol π.

▲ **SUMMARY**

A LOOK BACK AT CHAPTER 6

One goal in this chapter has been to classify angles inside, on, and outside the circle. Formulas for finding the measures of these angles were developed. Line and line segments related to a circle were defined, and some ways of finding the measures of these segments were described. Theorems involving inequalities in a circle were proved. Using the concept of locus, we justified several of the basic constructions found earlier in our work.

A LOOK AHEAD TO CHAPTER 7

Our goal in the next chapter is to deal with the area of triangles, certain quadrilaterals, and regular polygons. The area of a circle and the area of a sector of a circle will be discussed. Special right triangles will play an important role in determining the area of these plane figures.

IMPORTANT TERMS AND CONCEPTS OF CHAPTER 6

6.1 Circle, Congruent Circles, Concentric Circles
     Center of the Circle
     Radius, Diameter, Chord
     Semicircle
     Arc, Major Arc, Minor Arc, Congruent Arcs
     Central Angle, Inscribed Angle
6.2 Tangent, Point of Tangency, Secant
     Polygon Inscribed in a Circle, Circumscribed Circle
     Polygon Circumscribed About a Circle, Inscribed
       Circle
     Interior and Exterior of a Circle
6.3 Common External Tangents
     Common Internal Tangents
     Internally Tangent Circles
     Externally Tangent Circles
     Line of Centers
6.4 Constructions of Tangents to a Circle
     Inequalities for the Circle
6.5 Locus
     Concurrent Lines
     Incenter, Circumcenter, Orthocenter, Centroid
A Look Beyond The Value of π

▲ **REVIEW EXERCISES**

1. The radius of a circle is 15 mm. The length of a chord that is not the diameter is 24 mm. Find the distance from the center of the circle to the chord.

2. Find the length of a chord that is 8 cm from the center of a circle that has a radius of 17 cm.

3. Two circles intersect and have a common chord 10 in.

long. The radius of one circle is 13 in. and the centers of the circles are 16 in. apart. Find the radius of the other circle.

4. Two circles intersect and have a common chord 12 cm long. The measure of the angles formed by the common chord and a radius of each circle to the points of intersection of the circles is 45°. Find the radius of each circle.

In Review Exercises 5 to 10, $\overrightarrow{BA}$ is tangent to the circle in the figure shown.

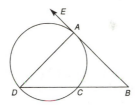

5. $m\angle B = 25°$, $m\widehat{AD} = 140°$, $m\widehat{DC} = ?$

6. $m\widehat{DC} = 190°$, $m\widehat{AD} = 120°$, $m\angle B = ?$

7. $m\angle EAD = 70°$, $m\angle B = 30°$, $m\widehat{AC} = ?$

8. $m\angle D = 40°$, $m\widehat{DC} = 130°$, $m\angle B = ?$

9. *Given:* C is the midpoint of $\widehat{ACD}$ and $m\angle B = 40°$
   Find: $m\widehat{AD}$, $m\widehat{AC}$, $m\widehat{DC}$

10. *Given:* $m\angle B = 35°$ and $m\widehat{DC} = 70°$
    Find: $m\widehat{AD}$, $m\widehat{AC}$

In Review Exercises 11 and 12, use the accompanying figure.

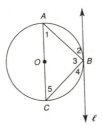

11. *Given* ⊙O with tangent ℓ and $m\angle 1 = 46°$
    Find: $m\angle 2$, $m\angle 3$, $m\angle 4$, $m\angle 5$

12. *Given:* ⊙O with tangent ℓ and $m\angle 5 = 40°$
    Find: $m\angle 1$, $m\angle 2$, $m\angle 3$, $m\angle 4$

13. Two circles are concentric. A chord of the larger circle is also tangent to the smaller circle. The radius of one circle is 20, and the radius of the other is 16. Find the length of the chord.

14. Two parallel chords of a circle each have length 16. The distance between them is 12. Find the radius of the circle.

In Review Exercises 15 to 22, state whether the statements are always true (A), sometimes true (S), or never true (N).

15. In a circle, congruent chords are equidistant from the center.

16. If a triangle is inscribed in a circle and one of its sides is a diameter, then the triangle is an isosceles triangle.

17. If a central angle and an inscribed angle of a circle intercept the same arc, then they are congruent.

18. A trapezoid can be inscribed in a circle.

19. If a parallelogram is inscribed in a circle, then each of its diagonals must be a diameter.

20. If two chords of a circle are not congruent, then the shorter chord is nearer the center of the circle.

21. Tangents to a circle at the endpoints of a diameter are parallel.

22. Two concentric circles have at least one point in common.

23. (a) $m\widehat{AB} = 80°$, $m\angle AEB = 75°$, $m\widehat{CD} = ?$
    (b) $m\widehat{AC} = 62°$, $m\angle DEB = 45°$, $m\widehat{BD} = ?$
    (c) $m\widehat{AB} = 88°$, $m\angle P = 24°$, $m\angle CED = ?$
    (d) $m\angle CED = 41°$, $m\widehat{CD} = 20°$, $m\angle P = ?$
    (e) $m\angle AEB = 65°$, $m\angle P = 25°$, $m\widehat{AB} = ?$, $m\widehat{CD} = ?$
    (f) $m\angle CED = 50°$, $m\widehat{AC} + m\widehat{BD} = ?$

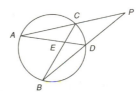

24. Given that $\overline{CF}$ is a tangent:
    (a) $CF = 6$, $AC = 12$, $BC = ?$
    (b) $AG = 3$, $BE = 10$, $BG = 4$, $DG = ?$
    (c) $AC = 12$, $BC = 4$, $DC = 3$, $CE = ?$
    (d) $AG = 8$, $GD = 5$, $BG = 10$, $GE = ?$
    (e) $CF = 6$, $AB = 5$, $BC = ?$
    (f) $EG = 2$, $GB = 4$, $AD = 9$, $GD = ?$
    (g) $AC = 30$, $BC = 3$, $CD = ED$, $ED = ?$
    (h) $AC = 9$, $BC = 5$, $ED = 12$, $CD = ?$
    (i) $ED = 8$, $DC = 4$, $FC = ?$
    (j) $FC = 6$, $ED = 9$, $CD = ?$

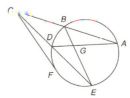

In Review Exercises 25 and 26, use the drawing shown at the top of the next page.

25. *Given:* $\overline{DF} \cong \overline{AC}$ in $\odot O$
$OE = 5x + 4$
$OB = 2x + 19$
Find: OE

26. *Given:* $\overline{OE} \cong \overline{OB}$ in $\odot O$
$DF = x(x - 2)$
$AC = x + 28$
Find: DE and AC

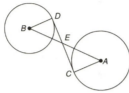

In Review Exercises 27 to 29, give a proof for each statement.

27. *Given:* $\overline{DC}$ is tangent to circles B and A at points D and C, respectively
Prove: $AC \cdot ED = CE \cdot BD$

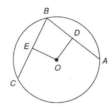

28. *Given:* $\odot O$ with $\overline{EO} \perp \overline{BC}, \overline{DO} \perp \overline{BA}, \overline{EO} \cong \overline{OD}$
Prove: $\widehat{BC} \cong \widehat{BA}$

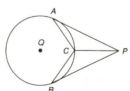

29. *Given:* $\overline{AP}$ and $\overline{BP}$ are tangent to $\odot Q$ at A and B
C is the midpoint of $\widehat{AB}$
Prove: $\overrightarrow{PC}$ bisects $\angle APB$

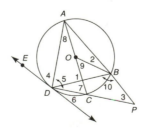

30. *Given:* $\odot O$ with diameter $\overline{AC}$ and tangent $\overleftrightarrow{DE}$
$m\widehat{AD} = 136°$ and $m\widehat{BC} = 50°$
Find: The measures of the angles, $\angle 1$ through $\angle 10$

31. A square is inscribed in a circle with a radius of 6 cm. Find the perimeter of the square.

32. A 30-60-90 triangle is inscribed in a circle with a radius of 5 cm. Find the perimeter of the triangle.

33. A circle is inscribed in a right triangle. The radius of the circle is 6 cm, and the hypotenuse is 29 cm. Find the length of the two segments of the hypotenuse.

34. *Given:* $\odot O$ is inscribed in $\triangle ABC$
$AB = 9, BC = 13, AC = 10$
Find: AD, BE, FC

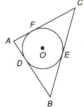

35. In $\odot Q$ with $\triangle ABQ$ and $\triangle CDQ$, $m\widehat{AB} > m\widehat{CD}$. Also, $\overline{QP} \perp \overline{AB}$ and $\overline{QR} \perp \overline{CD}$.

(a) How are AB and CD related?
(b) How are QP and QR related?
(c) How are $\angle A$ and $\angle C$ related?

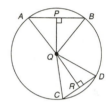

36. Given $\odot O$ with secant $\overleftrightarrow{AB}$ intersecting the circle at A and B; C is a point on $\overleftrightarrow{AB}$ in the exterior of the circle.

(a) Construct the tangent to $\odot O$ at point B.
(b) Construct the tangents to $\odot O$ from point C.

In Review Exercises 37 and 38, use the figure shown.

37. Construct a right triangle so that one leg has length AB and the other has length twice AB.

38. Construct a rhombus with side $\overline{AB}$ and $\angle ABC$.

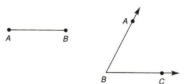

In Review Exercises 39 to 41, sketch and describe the locus in a plane.

39. Find the locus of the midpoints of the radii of a circle.

40. Find the locus of the centers of all circles passing through two given points.

41. What is the locus of the center of a penny that rolls around a half-dollar?

In Exercises 42 and 43, sketch and describe the locus in space.

42. Find the locus of points less than 3 units from a given point.

43. Find the locus of points equidistant from two parallel planes.

In Review Exercises 44 to 49, use construction methods with the accompanying figure.

44. *Given:* $\triangle ABC$
   Find: The incenter

45. *Given:* $\triangle ABC$
   Find: The circumcenter

46. *Given:* $\triangle ABC$
   Find: The orthocenter

47. *Given:* $\triangle ABC$
   Find: The centroid

48. Use Exercise 44 to inscribe a circle in $\triangle ABC$.

49. Use Exercise 45 to circumscribe the circle about the triangle.

In Review Exercises 50 and 51, use the figure shown.

50. *Given:* $\triangle ABC$ with medians $\overline{AE}$, $\overline{DC}$, $\overline{BF}$
   Find: (a) BG if $BF = 18$
       (b) GE if $AG = 4$
       (c) DG if $CG = 4\sqrt{3}$

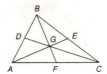

51. *Given:* $\triangle ABC$ with medians $\overline{AE}$, $\overline{DC}$, $\overline{BF}$
   $AG = 2x + 2y$, $GE = 2x - y$
   $BG = 3y + 1$, $GF = x$
   Find: BF and AE

AREAS OF POLYGONS AND CIRCLES

The part of a line known as a segment can be measured through the use of linear units such as the foot, inch, or meter; but we also wish to measure the part of a plane enclosed by some plane figure such as a polygon. This measure, known as the area of the region, has many practical appiications. The units of measure for area are known as the square foot, the square inch, the square meter, and so on. The acre is a unit of area measure used in agriculture.

The accompanying drawing shows the floor plan of the office of a business administrator who plans to have new carpet installed in the office. How many square yards of carpet will be needed? This and many other questions involving area are covered in Chapter 7.

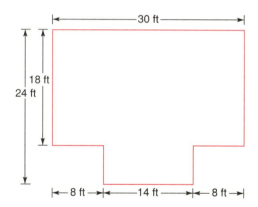

CHAPTER OUTLINE

7.1 Area and Initial Postulates

7.2 More with Perimeter and Area

7.3 Regular Polygons and Area

7.4 The Area of a Circle

A Look Beyond Another Look at the Pythagorean Theorem

 AREA AND INITIAL POSTULATES

When a line segment is measured, we express the result in linear units such as inches, centimeters, or yards. If a line segment measures 5 centimeters, we will write $AB = 5$ cm or simply $AB = 5$ (if the units are apparent or are not stated). The instrument of measure is the ruler.

Lines are one-dimensional; that is, we may speak only of the dimension length in measuring the part of a line known as a segment.

While a plane is an infinite two-dimensional surface, a closed or bounded portion of the plane is a **region;** the plane figure that creates the

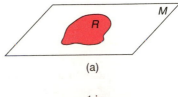

R

M

(a)

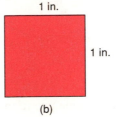

1 in.

1 in.

(b)

Figure 7.1

boundary is considered to be included in the "closed" region. When a region like R in plane M of figure 7.1a is measured, we refer to the measure as the "area of the region." Square units, rather than linear units, are used in area measure. Areas are measured in units such as square inches and square centimeters, symbolized by in.^{2} and cm^2. Because the region in figure 7.1b is a square, with each side having a length of 1 in., the region has an area of 1 in.^{2}

One application of area involves measuring the floor area to be covered by carpeting, often measured in square yards (yd^2). Another involves calculating the number of squares of shingles needed to cover a roof. (A "square" refers to the number of shingles needed to cover a 100-ft^2 section of the roof.)

As shown in figure 7.2, some of the regions that have measurable areas are bounded by figures that we have encountered in earlier chapters. Each of the regions in figure 7.2 is bounded in the sense that we can distinguish between interior and exterior. In the case of the triangular region in figure 7.2b, it is important to recognize that we are actually finding the area of the region within its boundaries (the sides of the triangle). While the triangle itself does not have area (the region bounded has the area), we often speak of "the area of the triangle."

The previous discussion does not formally define a region or its area. These are accepted as the undefined terms on which the following postulate depends.

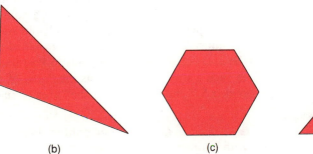

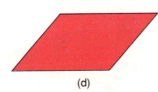

(a)   (b)   (c)   (d)

Figure 7.2

POSTULATE 18 **(Area Postulate)** Corresponding to every bounded region is a unique positive number A, known as the area of that region.

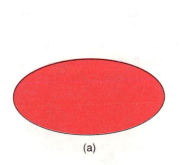

Figure 7.3

One way to estimate the area of a region is to place it in a grid, as shown in figure 7.3. Counting only the number of whole squares inside the region gives an approximation that is less than the actual area. On the other hand, counting squares that are inside or partially inside provides an approximation that is greater than the actual area. A fair estimate of the area of a region is often given by the average of the smaller and larger approximations just described. If the area of the circle shown in figure 7.3 is between 9 and 21, we might estimate its area to be 15.

For a stronger sense of the notion of area, consider $\triangle ABC$ and $\triangle DEF$ (which are congruent) in the following figure. Because the one can be placed over the other so that they coincide, the two triangles have equal areas, even though they

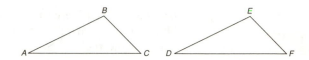

do *not* enclose the same regions (since the sets of points enclosed are not the same).

If two closed plane figures are congruent, then their areas are equal.

EXAMPLE 1   In figure 7.4, points B and C trisect $\overline{AD}$; $\overline{EC} \perp \overline{AD}$. Name two triangles with equal areas.

Solution $\triangle ECB \cong \triangle ECD$ by SAS. Then $\triangle ECB$ and $\triangle ECD$ have equal areas by Postulate 19.   ▲

Note: $\triangle EBA$ is also equal in area to $\triangle ECB$ and $\triangle ECD$, but this cannot be established until the material in Section 7.2 is considered.

Consider figure 7.5. The entire region is bounded by a curve and then subdivided by a segment into smaller regions R and S. These regions have a common boundary and do not overlap. Since a numerical area can be associated with each region R and S, it is intuitively obvious that the area of $R \cup S$ (read as "R union S" and meaning region R joined to region S) is equal to the sum of the areas of R and S. This leads to Postulate 20 in which subscripting is used in such a way that A_R means the "area of region R," A_S means the "area of region S," and $A_{R \cup S}$ means the "area of region $R \cup S$."

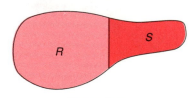

Figure 7.4

Figure 7.5

(Area-Addition Postulate)   Let R and S be two regions with a common boundary and for which the intersection is empty (no overlap). Then

$$A_{R \cup S} = A_R + A_S$$

EXAMPLE 2   In figure 7.6, the pentagon $ABCDE$ is formed by square $ABCD$ and $\triangle ADE$. If the area of the square is 36 in.² while that of $\triangle ADE$ is 12 in.², find the area of pentagon $ABCDE$.

Solution   Because square $ABCD$ and $\triangle ADE$ do not overlap and have common boundary $\overline{AD}$, we have Area (pentagon $ABCDE$) = area (square $ABCD$) + area ($\triangle ADE$). By the Area-Addition Postulate,

Area (pentagon $ABCDE$) = 36 in.² + 12 in.² = 48 in.²   ▲

At this time, it is convenient to subscript A (area) with letters that name the figure whose area is indicated. The principle used in Example 2 is more conveniently and compactly stated in the form

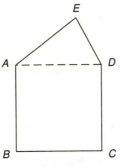

Figure 7.6

$$A_{ABCDE} = A_{ABCD} + A_{ADE}$$

Study the rectangle *MNPQ* in the accompanying figure, and note that it has dimensions of 3 cm and 4 cm. The number of squares, 1 cm on a side, in the rectangle is obviously 12. But rather than count the number of squares in the figure, we could have completed a multiplication. In area calculations, we multiply units in the same manner that we multiply like bases. Thus, just as $3x \cdot 4x = 12x^2$, so also does $3 \text{ cm} \cdot 4 \text{ cm} = 12 \text{ cm}^2$.

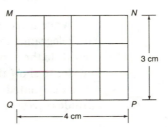

If units are indicated but are *not* the same, they must be converted into like units. For instance, if measuring the area requires multiplying 2 ft by 6 in., we first note that 2 ft = 24 in., so

$$2 \text{ ft} \cdot 6 \text{ in.} = 24 \text{ in.} \cdot 6 \text{ in.} = 144 \text{ in.}^2$$

Alternatively, 6 in. $= \dfrac{1}{2}$ ft, so

$$2 \text{ ft} \cdot \frac{1}{2} \text{ ft} = 1 \text{ ft}^2$$

Note: $1 \text{ ft}^2 = 144 \text{ in.}^2$

POSTULATE 21   The area *A* of a rectangle whose base has length *b* and whose altitude has length *h* is given by $A = b \cdot h$.

EXAMPLE 3   Find the area of rectangle *ABCD* if $AB = 12$ cm and $AD = 7$ cm.

Solution   Because it makes little difference which dimension is chosen as base *b* and which as altitude *h*, we arbitrarily choose $AB = b = 12$ cm and $AD = h = 7$ cm. Then

$$\begin{aligned}
A &= b \cdot h \\
&= 12 \text{ cm} \cdot 7 \text{ cm} \\
&= 84 \text{ cm}^2
\end{aligned}$$

▲

If no units are named in a problem, we assume that the units are alike. In such a case, we simply give the area as a number.

THEOREM 7.1.1   The area *A* of a square whose sides are each of length *s* is given by $A = s^2$.

No proof is given for Theorem 7.1.1, which follows immediately from Postulate 21.

While a rectangle's altitude is one of its dimensions, that is not true of a parallelogram. An **altitude** of a parallelogram is a perpendicular segment from one side to the opposite side, known as the **base**. A side may have to be extended in order to show this altitude-base relationship in a drawing. In figure 7.7a, if $\overline{RS}$ is designated as the base, then any of the segments $\overline{ZR}$, $\overline{VX}$, or $\overline{YS}$ is an altitude corresponding to that base (or to base $\overline{VT}$).

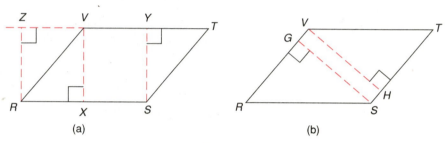

Figure 7.7

Another look at $\square RSTV$ (in figure 7.7b) shows that $\overline{ST}$ (or $\overline{VR}$) could just as well have been chosen as the base. Possible choices for the altitude then become $\overline{VH}$ or $\overline{GS}$. In the theorem that follows, it is necessary to select a base and an altitude drawn to that base!

THEOREM 7.1.2   The area *A* of a parallelogram with a base of length *b* and whose corresponding altitude has length *h* is given by

$$A = b \cdot h$$

Given: In figure 7.8, $\square RSTV$ with $\overline{VX} \perp \overline{RS}$
$RS = b$ and $VX = h$

Prove: $A_{RSTV} = b \cdot h$

Proof: Construct $\overline{YS} \perp \overline{VT}$ and $\overline{RZ} \perp \overline{VT}$, in which Z lies on an extension of $\overline{VT}$, as shown in figure 7.8b. Right $\angle Z$ and right $\angle SYT$ are $\cong$. Also, $\overline{ZR} \cong \overline{SY}$ since parallel lines are everywhere equidistant.

Because $\angle 1$ and $\angle 2$ are $\cong$ corresponding angles for parallel segments $\overline{VR}$ and $\overline{TS}$, $\triangle RZV \cong \triangle SYT$ by AAS. Then $A_{RZV} = A_{SYT}$ because congruent $\triangle$s have equal areas.

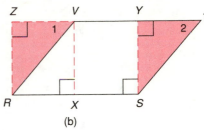

Figure 7.8

Because $A_{RSTV} = A_{RSYV} + A_{SYT}$, it follows that

$$A_{RSTV} = A_{RSYV} + A_{RZV}$$

But $RSYV \cup RZV$ is rectangle $RSYZ$, which has the area $b \cdot h$. Therefore $A_{RSTV} = A_{RSYZ} = b \cdot h$.   ▲

EXAMPLE 4   Given that all dimensions in figure 7.9 are in inches, find the area of $\square MNPQ$ by using base
(a) MN (b) PN

Solution   (a) $MN = b = 8$, while the corresponding altitude is of length $QT = h = 5$. Then

$$A = 8 \text{ in.} \cdot 5 \text{ in.}$$
$$= 40 \text{ in.}^2$$

(b) $PN = b = 6$, so the corresponding altitude length is $MR = h = 6\frac{2}{3}$.

Then

$$A = 6 \cdot 6\frac{2}{3}$$

PN as base

$$= 6 \cdot \frac{20}{3}$$

$$= 40 \text{ in.}^2 \quad ▲$$

Figure 7.9

Note: The area of the $\square$ is not changed when a different base and its corresponding altitude are used.

EXAMPLE 5   *Given:* $\square MNPQ$ with $PN = 8$ and $QP = 10$
                 Altitude $\overline{QR}$ to base $\overline{MN}$ has length $QR = 6$

Find: SN, the length of the altitude between $\overline{QM}$ and $\overline{PN}$

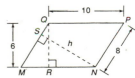

Solution   Choosing $MN = b = 10$ and $QR = h = 6$, we see that

$$A = b \cdot h = 10 \cdot 6 = 60$$

Now we choose $PN = b = 8$ and $SN = h$, so $A = 8h$. Because the area of the parallelogram is unique, it follows that

$$8 \cdot h = 60$$

$$h = \frac{60}{8} = 7.5 \qquad \blacktriangle$$

The last formula we consider in this section is the area of a triangle. It follows easily from the formula for the area of a parallelogram.

THEOREM 7.1.3 The area A of a triangle whose base has length b and whose corresponding altitude has length h is given by

$$A = \frac{1}{2} \cdot b \cdot h$$

Following is a proof of the formula for the area of a triangle.

Given: In figure 7.10, $\triangle ABC$ with $\overline{CD} \perp \overline{AB}$
$AB = b$ and $CD = h$

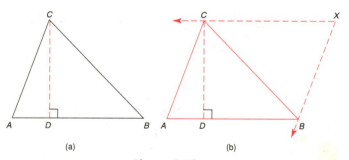

Figure 7.10

Prove: $A = \frac{1}{2} \cdot b \cdot h$

Proof: Construct a line through point C parallel to $\overline{AB}$ and a line through B parallel to $\overline{AC}$. These lines meet at a point X. (See figure 7.10b.) Now $ABXC$ is a parallelogram, and $\overline{CB}$ is its diagonal. Then $\triangle ABC \cong \triangle XCB$ since a diagonal separates the $\square$ into two $\cong$ $\triangle$s, and so $A_{ABC} = A_{XCB}$. Further, $A_{ABXC} = A_{ABC} + A_{XCB}$ by the Area-Addition Postulate. So

$$A_{ABXC} = A_{ABC} + A_{ABC} = 2 \cdot A_{ABC}$$

Then

$$A_{ABC} = \frac{1}{2} \cdot A_{ABXC} = \frac{1}{2}(b \cdot h)$$

Dropping subscripts, we have $A = \frac{1}{2} \cdot b \cdot h$. $\qquad \blacktriangle$

EXAMPLE 6   *Given:* Right $\triangle MPN$ with $PN = 8$ and $MN = 17$

Find: A_{MNP}

Solution   With $\overline{PN}$ as base, we need altitude $\overline{PM}$. By the Pythagorean Theorem,

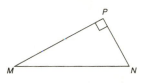

$$17^2 = (PM)^2 + 8^2$$
$$289 = (PM)^2 + 64$$

Then $(PM)^2 = 225$, so $PM = 15$.
   With $PN = b = 8$ and $PM = h = 15$, we have

$$A = \frac{1}{2} \cdot 8 \cdot 15 = 60$$ ▲

> ▲ **WARNING:** The phrase "area of a polygon" really means the area of the region enclosed by the polygon.

7.1 EXERCISES

1. Suppose that two triangles have equal areas. Are the triangles congruent? Why or why not? Are two squares with equal areas necessarily congruent? Why or why not?

2. In the accompanying drawing, the area of the square is 12, while the area of the circle is 30. Does the shaded area equal 42? Why or why not?

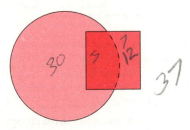

3. Consider the same drawing and information as in Exercise 2, but suppose you know that the area of the region defined by the intersection of the square and the circle measures 5. What is the area of the shaded region?

4. If $MNPQ$ is a rhombus as shown, which formula should be used to calculate its area?

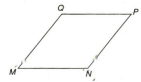

5. In rhombus $MNPQ$, how does the length of the altitude to $\overline{PN}$ compare to the length of the altitude to $\overline{MN}$? Explain.

6. When the diagonals of a rhombus are drawn, how do the areas of the four resulting smaller triangles compare to each other and to the given rhombus?

In Exercises 7 to 16, find areas of the figures shown or described.

7. A rectangle's length is 6 cm, and its width is 9 cm.

8. A right triangle has one leg measuring 20 in. and a hypotenuse measuring 29 in.

9. A 45-45-90 triangle has a hypotenuse measuring 6 m.

10. A triangle's altitude to the 15-in. side measures 8 in.

11.

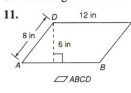

$\square ABCD$

12.

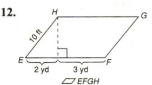

$\square EFGH$

13.

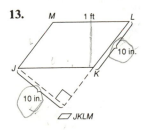

$\square JKLM$

14.

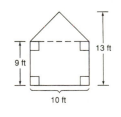

15.

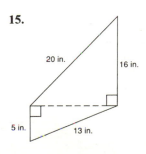

16.

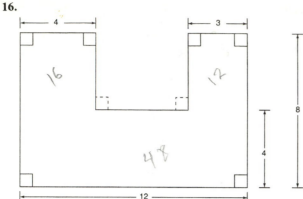

In Exercises 17 to 22, find the area of the shaded region.

17.

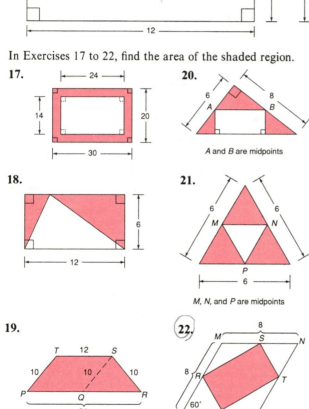

18.

19.

20.

A and B are midpoints

21.

M, N, and P are midpoints

22.

▱ $LPNM$

R, S, T, and Q are midpoints

▱ $PQST$

23. A triangular corner of a store has been roped off to be used as an area for displaying Christmas ornaments. Find the area of the display section.

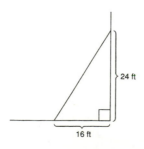

24. Carpeting is to be purchased for the family room and hallway shown in the accompanying drawing. What is the area to be covered?

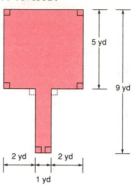

25. The exterior wall (the gabled end of the house shown in the accompanying drawing) remains to be painted.

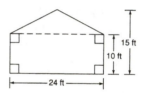

(a) What is the area of the outside wall?
(b) If each gallon of paint covers approximately 105 ft², how many gallons of paint must be purchased?
(c) If each gallon of paint is on sale for $15.50, what is the total cost of the paint?

26. The roof of the house shown in the accompanying drawing needs to be shingled.

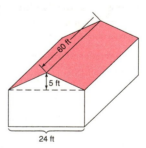

(a) Considering that both the front and back sections of the roof have equal areas, find the total area to be shingled.
(b) If roofing is sold in squares (each covering 100 ft²), how many squares are needed to complete the work?
(c) If each square costs $22.50 and an extra square is allowed for trimming around vents, what is the total cost of the shingles?

27. A beach tent is designed as shown in the accompanying drawing so that one side is open. Find the number of square feet of canvas needed to make the tent.

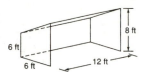

28. Gary and Carolyn plan to build the deck shown.

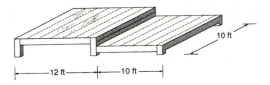

(a) Find the total floor space (area) for the deck.
(b) Find the approximate cost of building the deck if the estimated cost is $3.20 per ft².

In Exercises 29 to 34, provide a paragraph proof for each exercise.

29. *Given:* $\triangle ABC$ with midpoints M, N, and P of the sides
Prove: $A_{ABC} = 4 \cdot A_{MNP}$

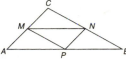

30. *Given:* $\triangle RST$ with median $\overline{RV}$
Prove: $A_{RSV} = A_{RVT}$

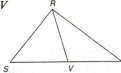

31. *Given:* Rectangle $MNPQ$
Prove: $A_{MNS} = \dfrac{1}{2} \cdot A_{MNPQ}$

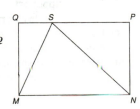

32. *Given:* Right $\triangle ABC$
Prove: $h = \dfrac{ab}{c}$

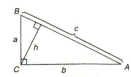

33. *Given:* Square $HJKL$ with $LJ = d$
Prove: $A_{HJKL} = \dfrac{d^2}{2}$

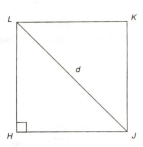

34. *Given:* $\square RSTV$ with $\overline{VW} \cong \overline{VT}$ as shown
Prove: $A_{RSTV} = (RS)^2$

35. *Given:* The area of right $\triangle ABC$ is 30 in.²
$m\angle C = 90°$
$AC = x$
$BC = x + 2$
Find: x

36. The lengths of the legs of a right triangle are consecutive even integers. The numerical value of the area is three times that of the longer leg. Find the lengths of the legs of the triangle.

*37. *Given:* $\triangle ABC$, whose sides are 13 in., 14 in., and 15 in.
Find: (a) BD, the length of the altitude to the 14-in. side (*Hint:* Use the Pythagorean Theorem twice.)
(b) The area of $\triangle ABC$, using the result from part (a)

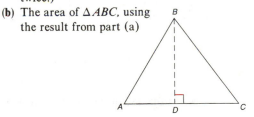

*38. *Given:* $\triangle ABC$, whose sides are 10 cm, 17 cm, and 21 cm
Find: (a) BD, the length of the altitude to the 21-cm side
(b) The area of $\triangle ABC$, using the result from part (a)

39. If the base of a rectangle is increased by 20 percent and the altitude is increased by 30 percent, by what percentage is the area increased?

40. If the base of a rectangle is increased by 20 percent but the altitude is decreased by 30 percent, by what percentage is the area changed? Is this an increase or decrease in area?

41. *Given:* Region $R \cup S$
Prove: $A_{R \cup S} > A_R$

42. *Given:* Region $R \cup S \cup T$
Prove: $A_{R \cup S \cup T} = A_R + A_S + A_T$

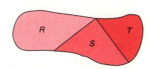

43. The algebra method of FOIL multiplication is illustrated geometrically in the accompanying drawing. Use the drawing with rectangular regions to complete the following rule:
$(a + b)(c + d) = $ _____

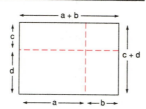

44. Use the square configuration in the accompanying drawing to complete the following algebra rule:
$(a + b)^2 = $ _____
Note: Simplify where possible.

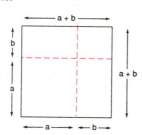

In Exercises 45 to 48, use the fact that the area of the polygon is unique.

45. In the right triangle shown in the accompanying drawing, find the length of the altitude drawn to the hypotenuse.

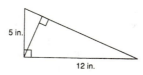

46. In the accompanying triangle, whose sides are 13, 20, and 21 cm long, the length of the altitude drawn to the 21-cm side is 12 cm. Find the lengths of the remaining altitudes of the triangle.

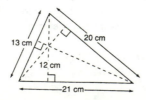

47. In $\square MNPQ$, $QP = 10$ and $QM = 5$. The length of altitude $\overline{QR}$ (to side $\overline{MN}$) is 4. Find the length of altitude $\overline{QS}$ from Q to $\overline{PN}$.

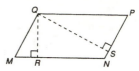

48. In $\square ABCD$, $AB = 7$ and $BC = 12$. The length of altitude $\overline{AF}$ (to side $\overline{BC}$) is 5. Find the length of altitude $\overline{AE}$ from A to $\overline{DC}$.

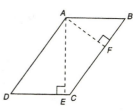

49. Consider the accompanying figure.
   (a) Find a lower estimate of the area of the figure by counting whole squares within the figure.
   (b) Find an upper estimate of the area of the figure by counting whole and partial squares within the figure.
   (c) Use the average of the results in parts (a) and (b) to provide a better estimate of the area of the figure.
   (d) Does intuition suggest that the area estimate of part (c) is the exact answer?

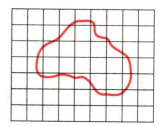

50. Consider the accompanying figure.
   (a) Find a lower estimate of the area of the figure by counting whole squares within the figure.
   (b) Find an upper estimate of the area of the figure by counting whole and partial squares within the figure.
   (c) Use the average of the results in parts (a) and (b) to provide a better estimate of the area of the figure.
   (d) Does intuition suggest that the area estimate of part (c) is the exact answer?

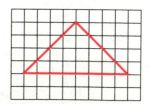

From time to time, we have used the notion of a perimeter. We now generalize this concept for all polygons.

> **DEFINITION:** The **perimeter** of a polygon is the sum of lengths of all sides of the polygon.

EXAMPLE 1 Find the perimeter of $\triangle ABC$ if:
(a) $AB = 5$ in., $AC = 6$ in., and $BC = 7$ in.
(b) Altitude $AD = 8$ cm, $BC = 6$ cm, and $\overline{AB} \cong \overline{AC}$

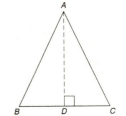

Solution (a) $P_{ABC} = AB + AC + BC$
$$= 5 + 6 + 7$$
$$= 18 \text{ in.}$$

(b) With $\overline{AB} \cong \overline{AC}$, $\triangle ABC$ is isosceles. Then $\overline{AD}$ is the $\perp$ bisector of $\overline{BC}$. If $BC = 6$, it follows that $DC = 3$. Using the Pythagorean Theorem, we have

$$(AD)^2 + (DC)^2 = (AC)^2$$
$$8^2 + 3^2 = (AC)^2$$
$$64 + 9 = (AC)^2$$
$$AC = \sqrt{73}$$

Now $P_{ABC} = 6 + \sqrt{73} + \sqrt{73} = 6 + 2\sqrt{73}$.

Note: Because $x + x = 2x$, $\sqrt{73} + \sqrt{73} = 2\sqrt{73}$.

Table 7.1 summarizes the formulas for perimeters of triangles.

Table 7.1 **Perimeter of a Triangle**

Scalene Triangle	Isosceles Triangle	Equilateral Triangle
$P = a + b + c$	$P = b + 2 \cdot s$	$P = 3 \cdot s$

EXAMPLE 2 Find the perimeter of square $ABCD$, which is inscribed in $\odot O$ of radius length 5.

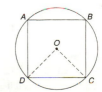

Solution   Each central angle is of measure 90°. Because $\overline{OC} \cong \overline{OD}$, $\triangle DOC$ is 45-45-90. Then $CD = 5\sqrt{2}$. Now

$$\begin{aligned} P_{ABCD} &= AB + BC + CD + DA \\ &= 5\sqrt{2} + 5\sqrt{2} + 5\sqrt{2} + 5\sqrt{2} \\ &= 20\sqrt{2} \end{aligned}$$

▲

Note: In combining terms, the preceding form is the same as $5x + 5x + 5x + 5x$, where x has the value $\sqrt{2}$.

Table 7.2 summarizes perimeters of quadrilaterals.

Table 7.2   **Perimeter of a Quadrilateral**

Rectangle	Square (or Rhombus)	Parallelogram	Trapezoid
$P = 2b + 2h$ $P = 2(b + h)$	$P = 4 \cdot s$	$P = 2b + 2s$ $P = 2(b + s)$	$P = s + b + s' + b'$

EXAMPLE 3   While remodeling, the Gibsons have decided to replace the old woodwork with Colonial-style oak woodwork.
(a) Using the floorplan in figure 7.11, find the amount of baseboard (in linear ft) needed for the room. Do *not* make any allowances for doors!
(b) Find the cost of the baseboard if the price is $1.32 per linear foot.

Solution   (a) The perimeter or "distance around" the room is

$$12 + 6 + 8 + 12 + 20 + 18 = 76 \text{ linear feet}$$

(b) The cost is $76 \cdot \$1.32 = \$100.32$.

▲

If the lengths of the three sides of a triangle are known, the formula for area uses the notion of a **semiperimeter,** which means one-half the perimeter of the figure. The formula that follows is known as Heron's Formula in honor of Heron of Alexandria.

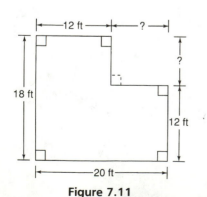

Figure 7.11

THEOREM 7.2.1 (Heron's Formula)   If the three sides of a triangle have lengths *a, b,* and *c,* then the area *A* of the triangle is given by

$$A = \sqrt{s(s - a)(s - b)(s - c)}$$

where

$$s = \frac{1}{2}(a + b + c)$$

We accept and apply Heron's Formula without proof.

EXAMPLE 4   Find the area of the triangle whose sides have lengths 13, 14, and 15. (This was Exercise 37 in Section 7.1.)

Solution   For $a = 13$, $b = 14$, and $c = 15$, the semiperimeter is given by

$$s = \frac{1}{2}(13 + 14 + 15) = \frac{1}{2}(42) = 21$$

Therefore

$$A = \sqrt{s(s - a)(s - b)(s - c)}$$
$$= \sqrt{21(21 - 13)(21 - 14)(21 - 15)}$$
$$= \sqrt{21(8)(7)(6)} \quad \text{or} \quad \sqrt{7 \cdot 3 \cdot 2 \cdot 2 \cdot 2 \cdot 7 \cdot 2 \cdot 3}$$
$$= \sqrt{7^2 \cdot 3^2 \cdot 2^2 \cdot 2^2} = 7 \cdot 3 \cdot 2 \cdot 2 = 84 \qquad \blacktriangle$$

If we draw a diagonal, we can calculate the areas of some quadrilaterals through the use of Heron's Formula. Other quadrilaterals, such as the trapezoid, have area formulas that can be derived by using the formula for the area of a triangle:

$$A = \frac{1}{2} \cdot b \cdot h$$

Recall that the two parallel sides of a trapezoid are known as its bases. The altitude is any segment that is a perpendicular from one base to the other. In the accompanying drawing, $\overline{AB}$ and $\overline{DC}$ are bases, while $\overline{AE}$ is an altitude for the trapezoid.

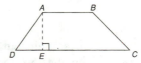

In the formula given in Theorem 7.2.2, numerical subscripts are used. The notation b_1, read "b sub 1," is used to indicate the length of the first base; similarly, b_2 is the length of the second base.

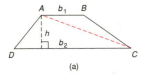

(a)

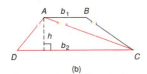

(b)

THEOREM 7.2.2   The area A of a trapezoid whose bases have lengths b_1 and b_2 and whose altitude has length h is given by

$$A = \frac{1}{2}h(b_1 + b_2)$$

Given: Trapezoid $ABCD$ with $\overline{AB} \parallel \overline{DC}$

Prove: $A_{ABCD} = \frac{1}{2}h(b_1 + b_2)$

Proof: Draw $\overline{AC}$. Now $\triangle ADC$ has an altitude of length h and a base of length b_2. Therefore,

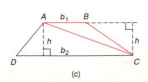

(c)

$$A_{ADC} = \frac{1}{2} h \cdot b_2$$

Also, $\triangle ABC$ has an altitude of length h and a base of length b_1. Then

$$A_{ABC} = \frac{1}{2} h \cdot b_1$$

Thus

$$A_{ABCD} = A_{ABC} + A_{ADC}$$

$$= \frac{1}{2} h \cdot b_1 + \frac{1}{2} h \cdot b_2$$

$$= \frac{1}{2} h (b_1 + b_2)$$ ▲

EXAMPLE 5   Find the area of the trapezoid in figure 7.12 if $RS = 5$, $TV = 13$, and $RW = 6$.

Solution   Let $RS = 5 = b_1$ and $TV = 13 = b_2$. Also, $RW = h = 6$. Now,

$$A = \frac{1}{2} h (b_1 + b_2)$$

becomes

$$A = \frac{1}{2} \cdot 6 \cdot (5 + 13)$$

$$= \frac{1}{2} \cdot 6 \cdot (18)$$

$$= 3 \cdot (18) = 54$$ ▲

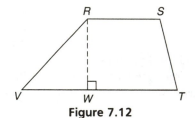

Figure 7.12

Recall that a rhombus is a parallelogram with two congruent adjacent sides; in fact, all four sides are congruent. To develop a formula for the area of a rhombus, we focus on the lengths of the two diagonals; these lengths are represented by d_1 and d_2.

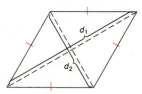

THEOREM 7.2.3   The area A of a rhombus whose diagonals have lengths d_1 and d_2 is given by

$$A = \frac{1}{2} \cdot d_1 \cdot d_2$$

Given: Rhombus $HJKL$ with $HK = d_1$ and $LJ = d_2$

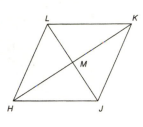

Prove: $A_{HJKL} = \dfrac{1}{2} \cdot d_1 \cdot d_2$

Proof: The diagonals of a rhombus are perpendicular bisectors of each other. Then $\triangle HML$, $\triangle KML$, $\triangle HMJ$, and $\triangle KMJ$ are all congruent to each other by SAS. Each

of these $\triangle$s has one leg of length $\dfrac{1}{2}d_1$ and the other leg of length $\dfrac{1}{2}d_2$.

Therefore each triangle has an area equal to $\dfrac{1}{2}\left(\dfrac{1}{2}d_1\right)\left(\dfrac{1}{2}d_2\right)$ by the

formula $A = \dfrac{1}{2} \cdot b \cdot h$. Then each of the four congruent triangles has

an area of $\dfrac{1}{8} \cdot d_1 \cdot d_2$, and the four together have an area of

$4\left(\dfrac{1}{8} \cdot d_1 \cdot d_2\right)$. Simplifying, we get

$$A_{HJKL} = \dfrac{1}{2} \cdot d_1 \cdot d_2 \qquad \blacktriangle$$

EXAMPLE 6   Find the area of the rhombus $MNPQ$ if $MP = 12$ and $NQ = 16$.

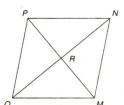

Solution   By Theorem 7.2.3, we have

$$A_{MNPQ} = \dfrac{1}{2} \cdot d_1 \cdot d_2 = \dfrac{1}{2} \cdot 12 \cdot 16 = 96 \qquad \blacktriangle$$

In problems involving the rhombus, we can often use the fact that diagonals are perpendicular to find needed measures. If the length of a side and the length of either diagonal are known, the length of the other diagonal can be found. If the size of an angle is known, that may also allow us to find a needed length.

Recall that a kite is a quadrilateral in which one and only one diagonal is the perpendicular bisector of the other. (See figure 7.13.)

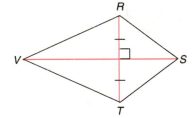

Figure 7.13

THEOREM 7.2.4   The area A of a kite whose diagonals have lengths of d_1 and d_2 is given by

$$A = \dfrac{1}{2} \cdot d_1 \cdot d_2$$

Given: Kite $ABCD$ so that $\overline{AC} \perp \overline{BD}$ and $\overline{AE} \cong \overline{EC}$
          $BD = d_1$ and $AC = d_2$

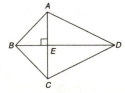

Prove: $A_{ABCD} = \dfrac{1}{2} \cdot d_1 \cdot d_2$

Proof: Consider $\triangle ABD$ and $\triangle CBD$. Because $\overline{AC} \perp \overline{BD}$, the base for $\triangle ABD$ can be chosen to be $\overline{BD}$ if its altitude is $\overline{AE}$. For $\triangle CBD$, we use base $\overline{BD}$ and its altitude $\overline{EC}$. Then

$$A_{ABCD} = A_{ABD} + A_{CBD}$$

$$= \left(\frac{1}{2} \cdot BD \cdot AE \right) + \left(\frac{1}{2} \cdot BD \cdot EC \right)$$

$$= \left(\frac{1}{2} \cdot d_1 \cdot AE \right) + \left(\frac{1}{2} \cdot d_1 \cdot EC \right)$$

$$= \frac{1}{2} \cdot d_1 \cdot (AE + EC)$$

$$= \frac{1}{2} \cdot d_1 \cdot (AC)$$

$$= \frac{1}{2} \cdot d_1 \cdot d_2 \qquad \blacktriangle$$

EXAMPLE 7   Find the length of $\overline{RT}$ if the area of the kite $RSTV$ is 360 in.² and $SV = 30$ in.

Solution $A = \dfrac{1}{2} \cdot d_1 \cdot d_2$ becomes $360 = \dfrac{1}{2}(30)d$, in which d is the length of the remaining diagonal $\overline{RT}$. Then $360 = 15d$, which means that $d = 24$ in. Therefore, $RT = 24$ in.

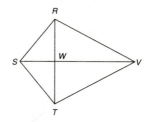

7.2 EXERCISES

In Exercises 1 to 8, find the perimeter of each figure.

1.

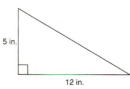

5 in.

12 in.

2.

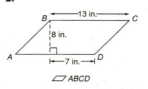

—13 in.—

B   C

8 in.

A

—7 in.—D

▱ ABCD

3.

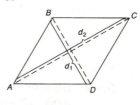

B   C

d_2

d_1

A   D

▱ ABCD with $\overline{AB} \cong \overline{BC}$

$d_1 = 4$ m

$d_2 = 10$ m

4.

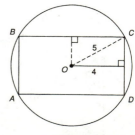

B   C

5

O   4

A   D

▱ ABCD in ⊙O

5.

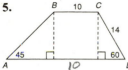

Trapezoid $ABCD$ with $\overline{BC} \parallel \overline{AD}$
$m\angle A = 45$, $m\angle D = 60$,
$BC = 10$, $CD = 14$

6.

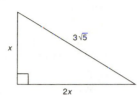

7.

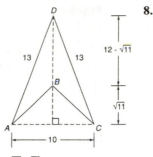

$\overline{AB} \cong \overline{BC}$ in concave
quadrilateral $ABCD$

8.

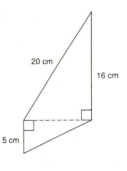

In Exercises 9 and 10, use Heron's Formula.

9. Find the area of a triangle whose sides measure 13, 14, and 15 in.

10. Find the area of a triangle whose sides measure 10, 17, and 21 cm.

In Exercises 11 to 16, find the area of the given polygon.

11.

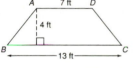

Trapezoid $ABCD$ with $\overline{AB} \cong \overline{DC}$

14.

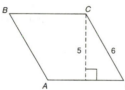

$\square\, ABCD$ with $\overline{BC} \cong \overline{CD}$

12.

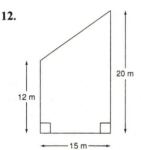

15.

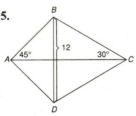

Kite $ABCD$ with $BD = 12$
$m\angle BAC = 45°$, $m\angle BCA = 30°$

13.

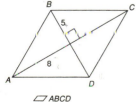

$\square\, ABCD$

16.

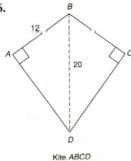

Kite $ABCD$

17. In a triangle of perimeter 76 in., the length of the first side is twice the length of the second side, and the length of the third side is 12 in. more than the length of the second side. Find the lengths of the three sides.

18. In a triangle whose area is 72 in.², the base has a length of 8 in. Find the length of the altitude.

19. A trapezoid has an area of 96 cm². If the altitude has a length of 8 cm and one base has a length of 9 cm, find the length of the other base.

20. The numerical difference between the area of a square and the perimeter of that square is 32. Find the length of a side of the square.

In Exercises 21 to 24, give a paragraph form of proof. Provide drawings as needed.

21. *Given:* Equilateral $\triangle ABC$ with each side of length s

   Prove: $A_{ABC} = \dfrac{s^2}{4}\sqrt{3}$ (*Hint:* Use Heron's Formula.)

22. *Given:* Isosceles $\triangle MNQ$ with $QM = QN = s$ and $MN = 2a$

   Prove: $A_{MNQ} = 4\sqrt{s^2 - a^2}$ (*Note:* $s > a$.)

23. *Given:* $\triangle ABC$ with sides of lengths a, b, and c
   $\overline{CD}$ is the altitude from C to $\overline{AB}$

   Prove: $CD = \dfrac{2\sqrt{s(s-a)(s-b)(s-c)}}{c}$

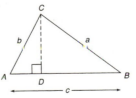

24. *Given:* Kite $RSTV$ with
   $\overline{SK} \perp \overline{ST}$ and
   $\overline{RV} \cong \overline{TV}$

   Prove: $A_{SRV} = \dfrac{1}{4}(RT)(SV)$

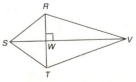

In Exercises 25 to 28, find the area of the figure shown.

25. *Given:* In $\odot O$, $OA = 5$
   $BC = 6$ and $CD = 4$
   Find: A_{ABCD}

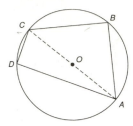

26. *Given:* Hexagon $RSTVWX$ with
   $\overline{WV} \parallel \overline{XT} \parallel \overline{RS}$
   $RS = 10$
   $ST = 8$
   $TV = 5$
   $WV = 16$
   and $\overline{WX} \cong \overline{VT}$
   Find: A_{RSTVWX}

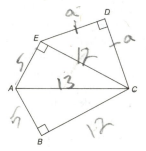

27. *Given:* Pentagon $ABCDE$,
   with $\overline{DC} \cong \overline{DE}$,
   $AE = AB = 5$,
   $BC = 12$
   Find: A_{ABCDE}

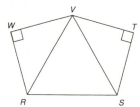

28. *Given:* Pentagon $RSTVW$ with
   $m\angle VRS = m\angle VSR = 60°$,
   $RS = 8\sqrt{2}$, $\overline{RW} \cong$
   $\overline{WV} \cong \overline{VT} \cong \overline{TS}$
   Find: A_{RSTVW}

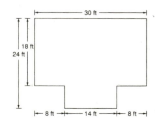

29. Mary Frances has a rectangular garden plot that encloses an area of 48 yd². If 28 yd of fencing are purchased to enclose the garden, what are the dimensions of the rectangular plot?

30. The perimeter of a right triangle is 12 m. If the hypotenuse has a length of 5 m, find the lengths of the two legs.

31. Farmer Watson wishes to fence a rectangular plot of ground measuring 245 ft by 140 ft.

   (a) What amount of fencing is needed?
   (b) What is the total cost of the fencing if it costs $0.59 per ft?

32. The farmer in Exercise 31 has decided to take the fencing purchased and use it to enclose the subdivided plots shown in the accompanying illustration.

   (a) What are the overall dimensions of the rectangular enclosure shown?
   (b) What is the total area of the enclosures shown?

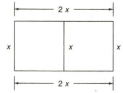

33. Find the area of the room whose floor plan is shown in the accompanying drawing.

34. Find the perimeter of the room in Exercise 33.

35. Examine several rectangles, each with a perimeter of 40 in., and find the dimensions of the rectangle that has the largest area. What type of figure has the largest area?

36. Examine several rectangles, each with an area of 36 in.², and find the dimensions of the rectangle that has the smallest perimeter. What type of figure has the smallest perimeter?

37. Prove that the area of a trapezoid whose altitude has length h and whose median has length m is given by $A = h \cdot m$.

7.3

REGULAR POLYGONS AND AREA

Suppose that we seek to inscribe a polygon in or to circumscribe a polygon about a given circle. This goal is usually rather easy to achieve. For instance, a pentagon has been inscribed in the circle in figure 7.14 merely by choosing five points on the circle and joining these by segments that become chords of the circle. A somewhat more interesting situation arises in the following example.

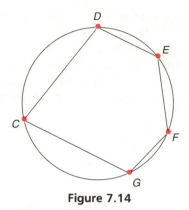

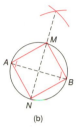

Figure 7.14

EXAMPLE 1   Inscribe a square in the given circle.

Solution   Begin by drawing any diameter in the circle; call the diameter $\overline{AB}$. Now construct the perpendicular bisector of diameter $\overline{AB}$; it intersects the circle at M and N.

    If we now join A to M, M to B, B to N, and N to A, the resulting quadrilateral is square $AMBN$.   ▲

    It is also easy to circumscribe a polygon about a circle. In figure 7.15, a pentagon has been circumscribed about the circle through the following procedure:

1. Choose any five points on the circle; these become points of contact for the tangent sides of the pentagon.
2. Draw radii of the circle to each of the five points.
3. Construct perpendicular lines to each radius at its point on the circle; these form the pentagon shown in figure 7.15.

Figure 7.15

EXAMPLE 2   Inscribe a regular hexagon in the circle in figure 7.16.

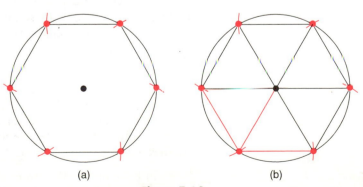

        (a)   (b)

Figure 7.16

Solution   With the compass opened to a length equaling that of the circle's radius and with any point on the circle as a center, mark off an arc to intersect the circle at a second point. Continue this procedure, using the second point as a center to locate a third point. When six points that determine six congruent arcs have been located, draw chords through each consecutive pair of points. The six chords are congruent because they correspond to congruent arcs. Thus the hexagon is equilateral. (*)

But the length of each congruent chord is equal to the length of the radius of the circle. Thus each triangle shown in figure 7.16b is equilateral (and equiangular); each angle of each triangle measures 60°. It follows that each interior angle of the hexagon measures 120° so the hexagon is also equiangular. (**)

From the starred statements, it follows that the hexagon is regular.   ▲

Note: By joining every other point on the circle, we could have inscribed an equilateral triangle in the circle.

Let us now reverse the order in which the polygon and the circle are given.

EXAMPLE 3   For the rectangle *ABCD*:
(**a**) Circumscribe a circle.
(**b**) Inscribe a circle.

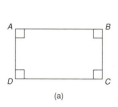
(a)

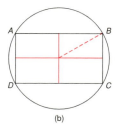
(b)

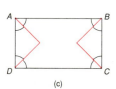
(c)

Solution   (**a**) Because the ⊥ bisectors of the four sides can be shown to be concurrent, some point is equidistant from the four vertices of the rectangle. Using this point of concurrency as center and a radius equal in length to the distance from the center to a vertex, we can complete the circle.
(**b**) Intuitively, we see that inscribing a circle is impossible unless the rectangle is a square—that is, unless its angle bisectors are concurrent. As the drawing suggests, the angle bisectors of rectangle *ABCD* are *not* concurrent.   ▲

In Example 3, it was possible to circumscribe a circle about the rectangle but not to inscribe the circle in the rectangle. For the rhombus, it is not possible to circumscribe the circle, but it is possible to inscribe the circle. While the angle bisectors are concurrent for rhombus *HJKL* in figure 7.17, the perpendicular bisectors of the sides are not concurrent. Thus the rhombus allows an inscribed circle but no circumscribed circle.

Note: In Example 3a, the center of the circumscribed circle could have been located by using the diagonals of the rectangle. Because the diagonals are con-

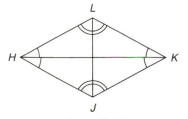
Figure 7.17

gruent and bisect each other, they determine a point that is equidistant from the vertices of the rectangle.

Let us recall now that a regular polygon, like the regular hexagon in Example 2, must be both equilateral and equiangular. A few of these are now illustrated.

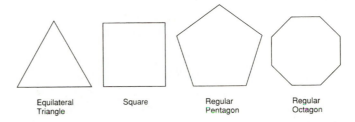

| Equilateral Triangle | Square | Regular Pentagon | Regular Octagon |

We saw earlier that the sum of measures of the interior angles of a polygon is given by the expression

$$(n - 2)180$$

where n is the number of sides of the polygon. The sum of the measures of the exterior angles is always $360°$.

EXAMPLE 4 (a) Find the measure of each interior angle of a regular polygon with 15 sides.
(b) Find the number of sides of a regular polygon if each interior angle measures $144°$.

Solution (a) Because each of the n angles have equal measures,

$$\frac{(n - 2)180}{n} \quad \text{becomes} \quad \frac{(15 - 2)180}{15}$$

which simplifies to $156°$.
(b) We can determine the number of sides by solving the equation

$$\frac{(n - 2)180}{n} = 144$$

Then
$$(n - 2)180 = 144n$$
$$180n - 360 = 144n$$
$$36n = 360$$
$$n = 10 \qquad \blacktriangle$$

Note: In Example 4a, we could have found the measure of each exterior angle and then used the fact that the interior angle is its supplement. In Example 4b, the supplement of the interior angle is the exterior angle; then we could have used the expression $\dfrac{360}{n}$ to find n.

Regular polygons, without fail, allow us to inscribe and to circumscribe a circle. The proof of this claim establishes several things:

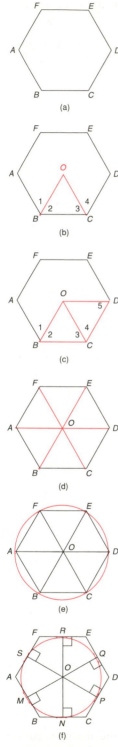

Figure 7.18

1. The centers of the inscribed and circumscribed circles are the same.
2. A standardized method can be used to locate this common center.
3. A standardized method can be used to determine radius length for each of the circles.

THEOREM 7.3.1   A circle can be circumscribed about (or inscribed in) any regular polygon.

Given: Regular polygon *ABCDEF*. (See figure 7.18a.)

Prove: A circle *O* can be circumscribed about *ABCDEF* and a circle can be inscribed in *ABCDEF*.

Proof: Let point *O* be the point at which the angle bisectors for ∠*ABC* and ∠*BCD* meet. (See figure 7.18b.) Then ∠1 ≅ ∠2 and ∠3 ≅ ∠4.

Because ∠*ABC* ≅ ∠*BCD* (by definition of regular polygons), it follows that

$$\frac{1}{2}m\angle ABC = \frac{1}{2}m\angle BCD$$

In turn, m∠2 = m∠3, so ∠2 ≅ ∠3. Then $\overline{OB} \cong \overline{OC}$ (sides opposite ≅ ∠s of a Δ).

From the facts that ∠3 ≅ ∠4, $\overline{OC} \cong \overline{OC}$, and $\overline{BC} \cong \overline{CD}$, it follows that Δ*OCB* ≅ Δ*OCD* by SAS. (See figure 7.18c.) In turn, $\overline{OC} \cong \overline{OD}$ by CPCTC, so ∠4 ≅ ∠5 since these lie opposite $\overline{OC}$ and $\overline{OD}$.

Because ∠5 ≅ ∠4 and $m\angle 4 = \frac{1}{2}m\angle BCD$, it follows that $m\angle 5 = \frac{1}{2}m\angle BCD$. But ∠*BCD* ≅ ∠*CDE* since these are angles of a regular polygon. For that reason, $m\angle 5 = \frac{1}{2}m\angle CDE$, and $\overline{OD}$ bisects ∠*CDE*.

By continuing this procedure, we can show that $\overline{OE}$ bisects ∠*DEF*, $\overline{OF}$ bisects ∠*EFA*, and $\overline{OA}$ bisects ∠*FAB*. Therefore the resulting Δ*AOB*, Δ*BOC*, Δ*COD*, Δ*DOE*, Δ*EOF*, and Δ*FOA* are congruent by ASA. (See figure 7.18d.) By CPCTC, $\overline{OA} \cong \overline{OB} \cong \overline{OC} \cong \overline{OD} \cong \overline{OE} \cong \overline{OF}$. With *O* as center and $\overline{OA}$ as radius, circle *O* can be circumscribed about *ABCDEF*, as shown in figure 7.18e.

Because corresponding altitudes of ≅ Δs are also congruent, we see that $\overline{OM} \cong \overline{ON} \cong \overline{OP} \cong \overline{OQ} \cong \overline{OR} \cong \overline{OS}$, where these are the altitudes to the bases of the triangles.

Again with *O* as center, but now with a radius equal in length to *OM*, we complete the inscribed circle in *ABCDEF*. (See figure 7.18f.)   ▲

In the proof of Theorem 7.3.1, a regular hexagon was drawn. The method of proof would not change, regardless of the number of sides chosen.

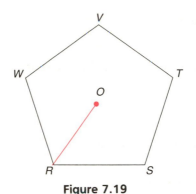

Figure 7.19

DEFINITION: The **center of a regular polygon** is the point that constitutes a common center for the inscribed and circumscribed circles of the polygon.

Note: The preceding definition may be misleading in that the center of a regular polygon can be located as the intersection of the angle bisectors of two consecutive angles; alternatively, the intersection of the perpendicular bisectors of two consecutive sides can also be used to locate the center of the regular polygon. Furthermore, a regular polygon has a center, whether or not any of the related circles are shown.

In figure 7.19, point O is the center of the regular pentagon $RSTVW$. In this figure, $\overline{OR}$ is called a "radius" of the regular pentagon.

DEFINITION: A **radius of a regular polygon** is any segment that joins the center of a regular polygon to one of its vertices.

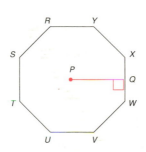

DEFINITION: An **apothem** of a regular polygon is any segment drawn from the center of that polygon perpendicular to one of the sides.

In regular octagon $RSTUVWXY$, whose center is point P, the segment $\overline{PQ}$ is an apothem. Any regular polygon of n sides has n apothems and n radii.

DEFINITION: A **central angle of a regular polygon** is an angle formed by two consecutive radii of the regular polygon.

The following theorem is stated without proof.

THEOREM 7.3.2   The measure of the central angle of a regular polygon of n sides is given by $c = \dfrac{360}{n}$.

EXAMPLE 5   (a) Find the measure of the central angle of a regular polygon of 9 sides.
(b) Find the number of sides of a regular polygon whose central angle measures 72°.

Solution   (a) $c = \dfrac{360}{9} = 40°$

(b) $72 = \dfrac{360}{n} \rightarrow 72n = 360 \rightarrow n = 5$ sides   ▲

Two other theorems are stated without proof:

THEOREM 7.3.3 Any radius of a regular polygon bisects the angle at the vertex to which it is drawn.

THEOREM 7.3.4 Any apothem to a side of a regular polygon bisects that side of the polygon.

Many of the preceding statements lay the groundwork for an important area formula. In the proof of Theorem 7.3.5, the drawing shown happens to be a regular pentagon. However, the proof applies to regular polygons of any number of sides.

It is also worth noting that the perimeter P of a regular polygon is the sum of its equal sides. Thus, if there are n sides of length s each, then

$$P = n \cdot s$$

THEOREM 7.3.5 The area A of a regular polygon whose apothem has length a and whose perimeter is P is given by

$$A = \frac{1}{2} \cdot a \cdot P$$

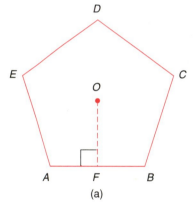

(a)

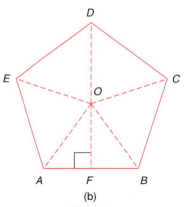

(b)

Figure 7.20

Given: Regular polygon $ABCDE$ in figure 7.20a so that $OF = a$ and the perimeter of $ABCDE$ is P

Prove: $A_{ABCDE} = \dfrac{1}{2} \cdot a \cdot P$

Proof: From center O, draw radii $\overline{OA}, \overline{OB}, \overline{OC}, \overline{OD},$ and $\overline{OE}$. (See figure 7.20b.) Now $\triangle AOB, \triangle BOC, \triangle COD, \triangle DOE,$ and $\triangle EOA$ are all $\cong$ by SSS. Where s represents the length of each of the congruent sides of the regular polygon, we see that the area of each $\triangle$ is $\dfrac{1}{2} \cdot s \cdot a$ $\left(\text{from } A = \dfrac{1}{2} \cdot b \cdot h\right)$. Therefore the area of the pentagon is given by

$$A_{ABCDE} = \left(\frac{1}{2} \cdot s \cdot a\right) + \left(\frac{1}{2} \cdot s \cdot a\right) + \left(\frac{1}{2} \cdot s \cdot a\right)$$
$$+ \left(\frac{1}{2} \cdot s \cdot a\right) + \left(\frac{1}{2} \cdot s \cdot a\right)$$
$$= \frac{1}{2}a(s + s + s + s + s)$$

But since the sum $s + s + s + s + s$ represents the perimeter of the polygon, we have

$$A_{ABCDE} = \frac{1}{2} \cdot a \cdot P \qquad \blacktriangle$$

EXAMPLE 6 In figure 7.20a, find the area of the regular pentagon $ABCDE$, whose center is O, if $OF = 4$ and $AB = 5.9$.

Solution $OF = a = 4$ and $AB = 5.9$. Therefore $P = 5(5.9)$ or $P = 29.5$. Consequently, we have

$$A_{ABCDE} = \frac{1}{2} \cdot 4(29.5)$$

$$= 59 \text{ square units} \qquad \blacktriangle$$

7.3 EXERCISES

1. Describe, if possible, how you would inscribe a circle within kite $ABCD$.

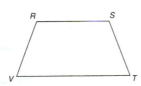

2. What condition must be satisfied for it to be possible to circumscribe a circle about kite $ABCD$?

3. Inscribe a circle in rhombus $JKLM$.

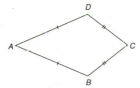

4. What condition must be satisfied for it to be possible to circumscribe a circle about trapezoid $RSTV$?

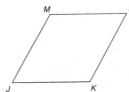

5. Inscribe a regular octagon within a circle.

6. Inscribe an equilateral triangle within a circle.

7. Circumscribe a square about a circle.

8. Circumscribe an equilateral triangle about a circle.

9. Find the lengths of the apothem and the radius of a square whose sides have length 10 in.

10. Find the lengths of the apothem and the radius of a regular hexagon whose sides have length 6 cm.

11. Find the lengths of the side and the radius of an equilateral triangle whose apothem's length is 8 ft.

12. Find the lengths of the side and the radius of a regular hexagon whose apothem's length is 10 m.

13. Find the measure of the central angle of a regular polygon of

    (a) Three sides   (c) Five sides
    (b) Four sides   (d) Six sides

14. Find the number of sides of a regular polygon that has a central angle measuring

    (a) 30   (c) 36
    (b) 72   (d) 20

15. Find the area of a regular hexagon whose sides have length 6 cm.

16. Find the area of a square whose apothem measures 5 cm.

17. Find the area of an equilateral triangle whose radius measures 10 in.

18. Find the approximate area of a regular pentagon whose apothem measures 6 in. and each of whose sides measures approximately 8.9 in.

19. In a regular polygon of 12 sides, the measure of each side is 2 in., while the measure of an apothem is $(2 + \sqrt{3})$ inches. Find the area.

20. In a regular octagon, the measure of each apothem is 4 cm, while each side measures $8(\sqrt{2} - 1)$ cm. Find the area.

21. Find the ratio of the area of a square circumscribed about a circle to the area of a square inscribed in the circle.

22. Regular octagon *ABCDEFGH* is inscribed in a circle whose radius is $\frac{7}{2}\sqrt{2}$ cm.

Considering that the area of the octagon is less than the area of the circle and greater than the area of the square *ACEG*, find the two integers between which the area of the octagon must lie.

$\left(Note:\ \text{Let}\ \pi \approx \frac{22}{7}.\right)$

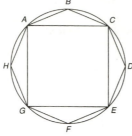

23. *Given:*   Regular pentagon *RSTVQ* with equilateral △*PQR*
   Find: m∠*VPS*

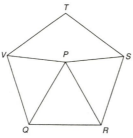

24. *Given:*   Regular pentagon *JKLMN* with diagonals $\overline{LN}$ and $\overline{KN}$
   Find: m∠*LNK*

25. *Given:*   Quadrilateral *ABCD* is circumscribed about ⊙*O*
   Prove: *AB* + *CD* = *DA* + *BC*

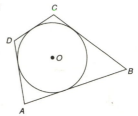

26. *Given:*   Quadrilateral *RSTV* inscribed in ⊙*Q*
   Prove: m∠*R* + m∠*T* = m∠*V* + m∠*S*

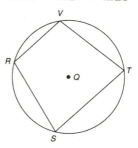

*27. *Prove:*   If a circle is divided into *n* congruent arcs (*n* ≥ 3), the chords determined by joining consecutive endpoints of these arcs form a regular polygon.

*28. *Prove:*   If a circle is divided into *n* congruent arcs (*n* ≥ 3), the tangents drawn at the endpoints of these arcs form a regular polygon.

THE AREA OF A CIRCLE

Recall that a circle's circumference is its "distance around"; circumference is to a circle what perimeter is to a polygon. In the Look Beyond section of Chapter 6, we saw that π was defined as the ratio of the circumference to the length of the diameter of a circle. From that definition, recall the following facts about circles.

$$\frac{C}{d} = \pi \rightarrow C = \pi \cdot d \rightarrow C = 2 \cdot \pi \cdot r$$

Note: *C* is the measure of the circumference, *d* is the length of the diameter, and *r* is the length of the radius.

Some commonly used approximations for the irrational number π are $\frac{22}{7}$, 3.14, and 3.1416.

EXAMPLE 1   In ⊙O, $OA = 7$ cm. Using $\pi \approx \dfrac{22}{7}$:

(a) Find the circumference C of ⊙O.
(b) Find the length of the minor arc $\widehat{AB}$.

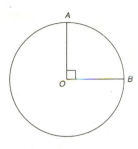

Solution   (a) $C = 2 \cdot \pi \cdot r$

$$= 2 \cdot \frac{22}{\cancel{7}} \cdot \cancel{7}$$

$$= 44 \text{ cm}$$

(b) Because the degree measure of this arc is 90°, we have $\dfrac{90}{360}$ or $\dfrac{1}{4}$ of the circumference for the arc length. Then

$$\text{length of } \widehat{AB} = \frac{90}{360} \cdot 44 = \frac{1}{4} \cdot 44 = 11 \text{ cm} \qquad \blacktriangle$$

EXAMPLE 2   If the exact circumference of a circle is 17π in.,
(a) Find the length of the radius.
(b) Find the length of the diameter.

Solution   (a) $C = 2 \cdot \pi \cdot r$

$$17\pi = 2 \cdot \pi \cdot r$$

$$\frac{17\pi}{2\pi} = \frac{2 \cdot \pi \cdot r}{2\pi}$$

$$r = \frac{17}{2} = 8.5 \text{ in.}$$

(b) Because $d = 2 \cdot r$, $d = 17$ in. $\qquad \blacktriangle$

Two observations should be made with regard to part b of Example 1:

1. The ratio of the degree measure of the arc to 360 is the same as the ratio of the length of the arc to the circumference.
2. As m$\widehat{AB}$ denotes the degree measure of an arc, $\ell\widehat{AB}$ denotes the length of the arc.

THEOREM 7.4.1   In a circle whose circumference is C, the length ℓ of an arc of a circle whose measure is m is given by

$$\ell = \frac{m}{360} \cdot C$$

In the work that follows, we use the undefined term "limit." Informally, a limit represents the largest possible number for a measure. The following example illustrates this notion.

EXAMPLE 3   Find the limit (largest possible number) for the length of a chord in a circle whose length of radius is 5 cm.

Solution   By considering several chords, we see that the greatest possible length of chord is that of a diameter. Thus the limit of the length of a chord is 10 cm.   ▲

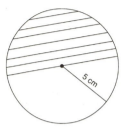

Now consider the problem of finding the area of a circle. To do so, let a regular polygon of n sides be inscribed in the circle. As we allow n to grow larger (often written as $n \to \infty$ and read as "n approaches infinity"), two observations can be made:

1. The length of an apothem of the regular polygon approaches the length of a radius of the circle as its limit ($a \to r$).

2. The perimeter of the regular polygon approaches the circumference of the circle as its limit ($P \to C$).

From figure 7.21, it is clear that the area of the inscribed regular polygon approaches the area of the circle as its limit. Using observations 1 and 2, we make the following claim: Because the expression for the area of a regular polygon is given by

$$A = \frac{1}{2} \cdot a \cdot P$$

the area of the circumscribed circle is given by

$$A = \frac{1}{2} \cdot r \cdot C$$

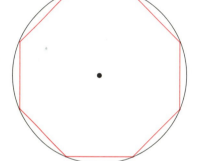

Figure 7.21

But this is the same as

$$A = \frac{1}{2}r(2 \cdot \pi \cdot r) \quad \text{or} \quad A = \pi r^2$$

While this argument is more appeal than proof, it does provide the correct formula for the area of a circle.

THEOREM 7.4.2   The area A of a circle whose radius is of length r is given by $A = \pi r^2$.

EXAMPLE 4   Find the area of a circle whose radius has a length of 10 in. (Use $\pi \approx 3.14$.)

Solution $A = \pi r^2$ becomes $A = 3.14(10)^2$. Then

$$A = 3.14(100) = 314 \text{ in.}^2 \qquad ▲$$

EXAMPLE 5   The area of a circle is 38.5 cm². Find the length of the radius of the circle. $\left(\text{Use } \pi \approx \frac{22}{7}. \right)$

Solution $A = \pi r^2$ becomes $38.5 = \frac{22}{7} \cdot r^2$ or $\frac{77}{2} = \frac{22}{7} \cdot r^2$. Multiplying each side of the

equation by the reciprocal of $\frac{22}{7}$, we have

$$\frac{7}{22} \cdot \frac{77}{2} = \frac{7}{22} \cdot \frac{22}{7} \cdot r^2$$

$$r^2 = \frac{49}{4}$$

Taking the positive square root for the radius, we have

$$r = \sqrt{\frac{49}{4}} = \frac{\sqrt{49}}{\sqrt{4}} = \frac{7}{2} = 3.5 \text{ cm}$$ ▲

DEFINITION: A **sector** of a circle is a region bounded by two radii of a circle and the arc intercepted by those radii.

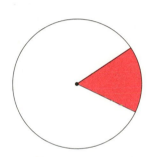

A sector such as the one shown in the accompanying drawing will generally be shaded to avoid confusion about whether the arc is intended to be a major or minor arc. In simple terms, the sector of a circle generally has the shape of a piece of pie.

THEOREM 7.4.3   The area A of a sector of a circle whose arc measure is m and whose radius has length r is given by

$$A = \frac{m}{360} \cdot \pi \cdot r^2$$

EXAMPLE 6   Find the area of the 100° sector shown in the accompanying circle, whose radius has length 10 in. (Use $\pi \approx 3.1416$.)

Solution $A = \frac{m}{360} \cdot \pi \cdot r^2$ becomes

$$A = \frac{100}{360}(3.1416) \cdot 10^2$$

$$= \frac{5}{18}(3.1416)(100) = 87.27 \text{ in.}^2$$ ▲

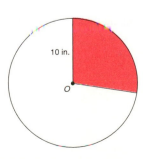

In work with circles, it is often convenient to leave exact answers for circumference and area. In such cases, we simply leave π in the result. For instance, in a circle of radius length 5 in., the exact circumference is 10π in., and the exact area is expressed as 25π in.^{2}.

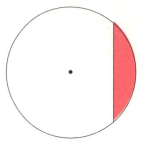

Figure 7.22

> **DEFINITION:** A **segment** of a circle is a region bounded by a chord and its arc.

In figure 7.22, the segment shown is bounded by chord $\overline{AB}$ and its minor arc $\widehat{AB}$. As in the case of a sector, we can avoid confusion by shading the segment whose area we are interested in. Otherwise, two segments of a circle are determined by a chord and its arcs.

EXAMPLE 7   Find the exact area of the segment bounded by a chord and an arc whose measure is 90°, if the diameter has length 24 in.

Solution   Because

$$A_\triangle + A_{\text{segment}} = A_{\text{sector}}$$
$$A_{\text{segment}} = A_{\text{sector}} - A_\triangle$$
$$= \frac{90}{360} \cdot \pi \cdot 12^2 - \frac{1}{2} \cdot 12 \cdot 12$$
$$= (36\pi - 72) \text{ in.}^2$$

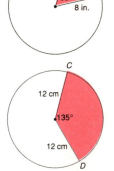

7.4 EXERCISES

1. Find the exact circumference and area of a circle whose radius has length 8 cm.

2. Find the exact circumference and area of a circle whose diameter has length 10 in.

3. Find the approximate circumference and area of a circle whose radius has length $10\frac{1}{2}$ in. Use $\pi \approx \frac{22}{7}$.

4. Find the approximate circumference and area of a circle whose diameter is 20 cm. Use $\pi \approx 3.14$.

5. Find the exact lengths of a radius and a diameter of a circle whose circumference is:

   (a) 44π in.   (b) 60π ft

6. Find the approximate lengths of a radius and a diameter of a circle whose circumference is:

   (a) 88 in. $\left(\text{Use } \pi \approx \frac{22}{7}.\right)$ (b) 157 m (Use $\pi \approx 3.14$.)

7. Find the exact lengths of a radius and a diameter of a circle whose area is:

   (a) 25π in.^{2} (b) 2.25π cm^2

8. Find the exact length of a radius and the exact circumference of a circle whose area is:

   (a) 36π m^2 (b) 6.25π ft^2

9. In the accompanying drawing, find the exact length $\ell \widehat{AB}$, where $\widehat{AB}$ refers to the minor arc of the circle.

10. In the drawing for Exercise 9, find the exact perimeter and area of the sector shown.

11. In the accompanying drawing, find the exact length $\ell \widehat{CD}$ of the minor arc shown.

12. In the drawing for Exercise 11, find the exact perimeter and area of the sector shown.

13. In the accompanying drawing, find the approximate perimeter of the sector shown. Use $\pi \approx 3.14$.

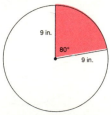

14. In the drawing for Exercise 13, find the approximate area of the sector shown. Use $\pi \approx \frac{22}{7}$.

15. In the accompanying drawing, find the exact perimeter and area of the segment shown, given that m∠O = 60° and OA = 12 in.

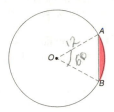

16. In the drawing for Exercise 15, find the exact perimeter and area of the segment shown, given that m∠O = 120° and AB = 10 in.

17. A rectangle has a perimeter of 16 in. What is the limit (largest possible value) of the area of the rectangle?

18. Two sides of a triangle measure 5 in. and 7 in., respectively. What is the limit of the length of the third side?

19. Let N be any point on side $\overline{BC}$ of the right triangle ABC. Find the limit of the length of $\overline{AN}$.

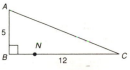

20. What is the limit of m∠RTS if T lies in the shaded region of the accompanying drawing?

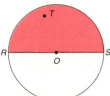

In Exercises 21 to 26, find exact areas of the shaded regions.

21.

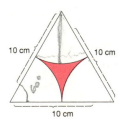

22.

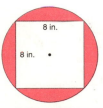

Square inscribed in a circle

23.

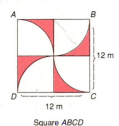

Square ABCD

24.

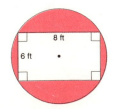

25.

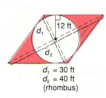

$d_1 = 30$ ft
$d_2 = 40$ ft
(rhombus)

26.

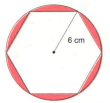

Regular hexagon inscribed in a circle

27. Using $\pi \approx \frac{22}{7}$, find the length of the radius of a circle whose area is 154 cm².

28. Using $\pi \approx 3.14$, find the length of the diameter of a circle whose circumference is 157 in.

29. Assuming that a 90° arc has an exact length of 4π in., find the length of the radius of the circle.

30. Assuming that the exact area of a sector determined by a 40° arc and its chord is $\frac{9}{4}\pi$ cm², find the length of the radius of the circle.

31. For concentric circles with radii of lengths 3 in. and 6 in., find the area of the segment determined by a chord of the larger circle that is also tangent to the smaller circle.

32. The ratio of the circumferences of two circles is 2:1. What is the ratio of their areas?

*33. A circle can be inscribed in the trapezoid shown in the accompanying drawing. Find the area of that circle.

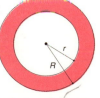

*34. A circle can be inscribed in an equilateral triangle, each of whose sides has length 10 cm. Find the area of that circle.

In Exercises 35 and 36, provide a paragraph proof for each exercise.

35. *Given:* Concentric circles as shown, with radii of lengths R and r, where $R > r$
Prove: $\text{Area}_{\text{ring}} = \pi(R + r)(R - r)$

36. *Given:* A circle with diameter of length d
Prove: $A_{\text{circle}} = \frac{1}{4} \cdot \pi \cdot d^2$

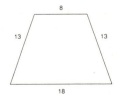

37. The radii of two concentric circles differ in length by exactly 1 in. If their areas differ by exactly 7π in.², find the lengths of the radii of the two circles.

38. In a circle whose radius has length 12 m, the length of an arc is 6π m. What is the degree measure of that arc?

39. The carpet in the circular entryway of a church needs to be replaced. The diameter of the circular region to be carpeted is 18 ft.

   (a) What length (in ft) of a metal protective strip is needed to bind the circumference of the carpet?
   (b) If the metal strips are sold in lengths of 6 ft, how many will be needed? (*Note:* Assume that these can be bent to follow the circle *and* that they can be placed end-to-end.)
   (c) If the cost of the metal strip is $1.59 per linear ft, find the cost of the metal strips needed.

40. At center court on the gymnasium floor, a large circular emblem is to be painted. The circular design has a radius of 8 ft.

   (a) What is the area to be painted? Use $\pi \approx 3.14$.
   (b) If a pint of paint covers 70 ft², how many pints of paint are needed to complete the job?
   (c) If each pint of paint costs $2.95, find the cost of the paint needed.

41. A track is to be constructed around the football field at a junior high school. If the straightaways are 100 yd in length, what length of radius is needed for each of the

semicircles shown in the accompanying drawing if the total length around the track is to be 440 yd? (Use $\pi \approx 3.14$.)

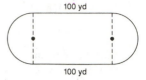

42. A circular grass courtyard at a shopping mall has a 40-ft diameter. This area needs to be reseeded. (Use $\pi \approx 3.14$.)

   (a) What is the total area to be reseeded?
   (b) If 1 lb of seed is to be used to cover a 60-ft² region, how many pounds of seed will be needed?
   (c) If the cost of 1 lb of seed is $1.65, what is the total cost of the grass seed needed?

43. Find the approximate area of a regular polygon that has 20 sides if the length of its radius is 7 cm. $\left(\textit{Hint:} \text{ Let } \pi \approx \dfrac{22}{7}.\right)$

44. Find the approximate perimeter of a regular polygon that has 20 sides if the length of its radius is 7 cm. $\left(\textit{Hint:} \text{ Let } \pi \approx \dfrac{22}{7}.\right)$

▲ A LOOK BEYOND: ANOTHER LOOK AT THE PYTHAGOREAN THEOREM

Some of the many proofs of the Pythagorean Theorem depend on area relationships. One such proof was devised by President James A. Garfield (1831–1881), twentieth president of the United States.

In his proof, the right triangle with legs a and b and hypotenuse c is introduced into a trapezoid, as shown in figure 7.23b.

If the drawing is **perceived as a trapezoid (as shown in** figure 7.24), the area is given by

$$A = \frac{1}{2}h(b_1 + b_2)$$

$$= \frac{1}{2}(a + b)(a + b)$$

$$= \frac{1}{2}(a + b)^2$$

$$= \frac{1}{2}(a^2 + 2ab + b^2)$$

$$= \frac{1}{2}a^2 + ab + \frac{1}{2}b^2$$

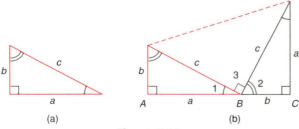

(a)   (b)

Figure 7.23

In figure 7.23b, the points A, B, and C are collinear. With $\angle 1$ and $\angle 2$ being complementary and the sum of angles about point B being 180°, it follows that $\angle 3$ is a right angle.

$b_1 = a$

$b_2 = b$

$h = a + b$

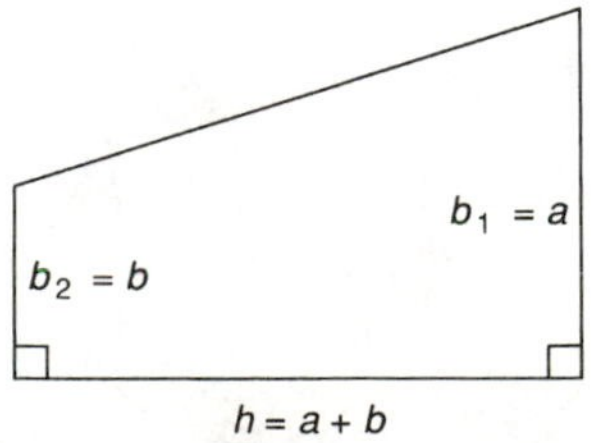

Figure 7.24

Now we treat the
trapezoid as a composite
of three triangles
in figure 7.25.

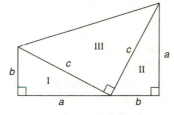

Figure 7.25

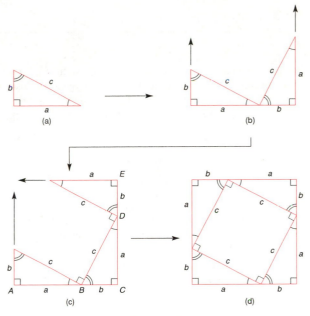

(a)

(b)

(c)

(d)

Figure 7.26

The area of regions (triangles) I, II, and III is given by

$$A = A_I + A_{II} + A_{III}$$

$$= \frac{1}{2}ab + \frac{1}{2}ab + \left(\frac{1}{2}c \cdot c\right)$$

$$= ab + \frac{1}{2}c^2$$

Equating the areas of the trapezoid in figure 7.24 and the composite in figure 7.25, we find that

$$\frac{1}{2}a^2 + ab + \frac{1}{2}b^2 = ab + \frac{1}{2}c^2$$

$$\frac{1}{2}a^2 + \frac{1}{2}b^2 = \frac{1}{2}c^2$$

Multiplying by 2, we get

$$a^2 + b^2 = c^2$$

The earlier proof (over 2000 years earlier!) of this theorem by the Greek mathematician Pythagoras is found in many historical works on geometry. It is not difficult to see the relationship between the two proofs.

In the proof credited to Pythagoras, a right triangle with legs of lengths a and b and a hypotenuse of length c is reproduced several times to form a square. Again, points A, B, C (and C, D, E; and so on) must be collinear. (See figure 7.26c.)

The area of the large square in figure 7.27a is given by

$$A = (a + b)^2$$
$$= a^2 + 2ab + b^2$$

Considering the composite in figure 7.27b, we find that

$$A = A_I + A_{II} + A_{III} + A_{IV} + A_V$$
$$= 4 \cdot A_I + A_V$$

since the four right triangles are congruent.

Then $$A = 4\left(\frac{1}{2}ab\right) + c^2$$

$$= 2ab + c^2$$

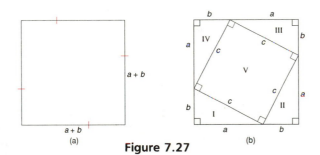

(a)

(b)

Figure 7.27

Again, due to the uniqueness of area, the results (area of square and area of composite) must be equal. Then

$$a^2 + 2ab + b^2 = 2ab + c^2$$
$$a^2 + b^2 = c^2$$

Another look at the proofs by President Garfield and by Pythagoras makes it clear that the results must be consistent. In figure 7.28, observe that Garfield's trapezoid must have one-half the area of Pythagoras' square, while maintaining the relationship that

$$c^2 = a^2 + b^2$$

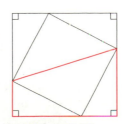

Figure 7.28

▲ SUMMARY

A LOOK BACK AT CHAPTER 7

Our goal in this chapter was to determine the area of triangles, certain quadrilaterals, and regular polygons. We also worked with the circumference and area of a circle and the area of a sector of a circle. The area of a circle is sometimes approximated by using $\pi \approx 3.14$ or $\pi \approx \frac{22}{7}$. At other times, the exact area is given by leaving π in the answer.

A LOOK AHEAD TO CHAPTER 8

Our goal in the next chapter is to deal with analytic or coordinate geometry. This type of geometry relates algebra and geometry. Formulas for the midpoint of a segment, the length of a segment, and the slope of a line will be developed. Lines, circles, and parabolas will be graphed. Concrete examples will lead to more general proofs using analytic geometry.

IMPORTANT TERMS AND CONCEPTS OF CHAPTER 7

7.1 Area of Parallelograms and Triangles
7.2 Perimeter
    Heron's Formula
    Area of a Trapezoid
    Area of a Rhombus
    Area of a Kite
7.3 Regular Polygon
    Center and Central Angle of a Regular Polygon
    Radius and Apothem of a Regular Polygon
7.4 Circumference and Area of a Circle
    Limit
    Sector
    Segment of a Circle
    Exact Area
A Look Beyond Another Look at the Pythagorean Theorem

▲ REVIEW EXERCISES

In Review Exercises 1 to 3, use the figure shown.

1. *Given:* $\square ABCD$ with $BD = 34$ and $BC = 30$ $\quad m\angle C = 90°$
    Find: A_{ABCD}

2. *Given:* $\square ABCD$ with $AB = 8$ and $AD = 10$
    Find: A_{ABCD} if:
    (a) $m\angle A = 30°$
    (b) $m\angle A = 60°$
    (c) $m\angle A = 45°$

3. *Given:* $\square ABCD$ with $\overline{AB} \cong \overline{BD}$ and $AD = 10$ $\quad \overline{BD} \perp \overline{DC}$
    Find: A_{ABCD}

In Review Exercises 4 and 5, use the drawing shown.

4. *Given:* $AB = 26$, $BC = 25$, $AC = 17$
    Find: A_{ABC}

5. *Given:* $AB = 30$, $BC = 26$, and $AC = 28$
    Find: A_{ABC}

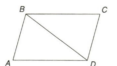

In Review Exercises 6 and 7, use the figure shown.

6. *Given:* Trapezoid $ABCD$, with $\overline{AB} \cong \overline{CD}$, $BC = 6$, $AD = 12$, and $AB = 5$
    Find: A_{ABCD}

7. *Given:* Trapezoid $ABCD$, with $AB = 6$ and $BC = 8$ $\overline{AB} \cong \overline{CD}$
    Find: A_{ABCD} if:
    (a) $m\angle A = 45°$
    (b) $m\angle A = 30°$
    (c) $m\angle A = 60°$

8. Find the area and the perimeter of a rhombus whose diagonals have lengths 18 in. and 24 in.

9. Tom Morrow wants to buy some fertilizer for his yard. His rectangular lot size is 140 ft by 160 ft. The outside measurements of his house are 80 ft by 35 ft. The driveway measures 30 ft by 20 ft.

    (a) What is the square footage of his yard that needs to be fertilized?
    (b) If each bag of fertilizer covers 5000 ft², how many bags should he buy?

(c) If the fertilizer costs $18 per bag, what is his total cost?

10. Alice's mother wants to wallpaper two adjacent walls in Alice's bedroom. She also wants to put a border around all four walls. The bedroom is 9 ft by 12 ft by 8 ft high.

(a) If each double roll covers approximately 60 ft² and is sold in double rolls only, how many double rolls are needed?

(b) If the border is sold in rolls of 5 yd each, how many rolls of the border are needed?

11. *Given:* Isosceles trapezoid *ABCD*
     equilateral △*FBC*
     right △*AED*
     BC = 12, *AB* = 5,
     ED = 16
   Find: (a) A_{EAFD}
           (b) Perimeter of *EAFD*

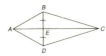

12. *Given:* Kite *ABCD* with
     AB = 10, *BC* = 17,
     and *BD* = 16
   Find: A_{ABCD}

13. One side of a rectangle is 2 cm more than the other side. If the area is 35 cm², find the dimensions of the rectangle.

14. One side of a triangle is 10 more than a second side and the third side is 5 more than the second side. The perimeter of the triangle is 60 cm.

(a) Find the lengths of the three sides.
(b) Find the area of the triangle.

15. In the accompanying drawing, find the area of △*ABD*.

16. Find the area of an equilateral triangle, each of whose sides has length 12 cm.

17. In the accompanying drawing, find the area of the shaded triangle, with $\overline{AC}$ a diameter of circle *O*.

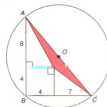

18. For a regular pentagon, find the measure of each:

(a) Central angle
(b) Interior angle
(c) Exterior angle

19. Find the area of a regular hexagon each of whose sides has length 8 ft.

20. The area of an equilateral triangle is $108\sqrt{3}$ in.². If the length of each side of the triangle is $12\sqrt{3}$ in., find the length of the apothem.

21. Find the area of a regular hexagon whose apothem has length 9 in.

22. A regular polygon has each central angle equal to 45°.

(a) How many sides does the regular polygon have?
(b) If each side is 5 cm and each apothem is approximately 6 cm, what is the approximate area of the polygon?

23. Can a circle be circumscribed about every example of the following figures? Why or why not?

(a) Parallelogram   (c) Rectangle
(b) Rhombus   (d) Square

24. Can a circle be inscribed in every example of the following figures? Why or why not?

(a) Parallelogram   (c) Rectangle
(b) Rhombus   (d) Square

25. Inscribe a regular 12-sided polygon in a circle.

26. The Keiths want to carpet the cement around their pool. The dimensions for the entire area are 20 ft by 30 ft. The pool is 12 ft by 24 ft.

(a) How many square feet need to be covered?
(b) Since carpet is sold only by square yards, approximately how many square yards does the area in part (a) represent?
(c) If the carpet is $9.97 per square yard, what will the carpet cost?

Find the exact area of the shaded regions in Exercises 27 to 31.

27.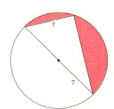

Square

29.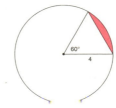

28.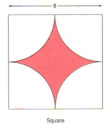

30.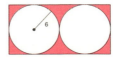

Two ≅ tangent circles, inscribed in a rectangle

31.

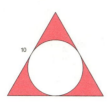

Equilateral triangle

32. An arc of a sector measures 40°. Find the exact length of the arc and the exact area of the sector if the radius is $3\sqrt{5}$ cm.

33. The circumference of a circle is 66 ft.

(a) Find the diameter of the circle, using $\pi \approx \dfrac{22}{7}$.

(b) Find the area of the circle, using $\pi \approx \dfrac{22}{7}$.

34. A circle has an exact area of 27π ft².

(a) What is the area of a sector of this circle if the arc of the sector measures 80°?

(b) What is the exact perimeter of the sector in part (a)?

35. An isosceles right triangle is inscribed in a circle that has a diameter of 12 in. Find the exact area between one of the legs of the triangle and its corresponding arc.

36. *Given:* Concentric circles with radii of lengths R and r, with $R > r$

Prove: Area$_{\text{ring}} = \pi(BC)^2$

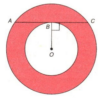

37. Prove that the area of a circle circumscribed about a square is twice the area of the circle inscribed within the square.

38. Prove that if semicircles are constructed on each of the sides of a right triangle, then the area of the semicircle on the hypotenuse is equal to the sum of the areas of the semicircles on the two legs.

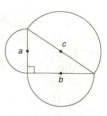

39. Jeff and Helen want to carpet their family room, except for the entranceway and the semicircle in front of the fireplace, which they want to tile.

(a) How many square yards of carpeting are needed? (Use $\pi \approx 3.14$.)

(b) How many square feet are to be tiled?

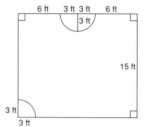

40. Sue and Dave's semicircular driveway is to be resealed, and then flowers are to be planted on either side.

(a) What is the number of square feet to be resealed? (Use $\pi \approx 3.14$.)

(b) If the cost of resealing is $0.18 per square foot, what is the total cost?

(c) If individual flowers are to be planted 1 ft from the edge of the driveway at intervals of approximately every 1 ft on both sides of the driveway, how many flowers are needed?

8 ANALYTIC GEOMETRY

The French mathematician, René Descartes, is considered the father of analytic geometry. His ingenious device for relating algebra and geometry, the Cartesian coordinate system, was a major breakthrough in the development of much of mathematics. Some have written that Descartes was inspired while watching a spider weave its web in the corner of his room. Others say that the brilliant idea came to Descartes as he was dreaming. In any event, the branch of mathematics known as calculus would not have grown as rapidly without analytic geometry.

The periscope reminds us of the coordinate system for which Descartes is well known.

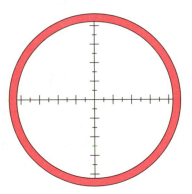

CHAPTER OUTLINE

8.1 The Rectangular Coordinate System

8.2 Graphs of Linear Equations and Slope

8.3 Equations of Lines

8.4 Circles and Parabolas

8.5 Preparing to Do Analytic Proofs

8.6 Analytic Proofs

A Look Beyond The Banach-Tarski Paradox

 8.1 THE RECTANGULAR COORDINATE SYSTEM

In Chapter 1, we considered equations and inequalities in one variable. Graphing the solution sets for $3x - 2 = 7$ and $3x - 2 > 7$ required a single number line to indicate the values of x. In this chapter we deal with equations containing two variables. To relate the algebra involved to geometry, we need two number lines.

The study of the relationships between number pairs and points is usually referred to as **analytic geometry.** The **Cartesian coordinate system** or **rectangular coordinate system** is the plane that results when two number

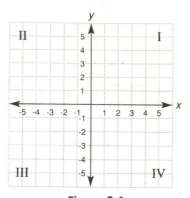

Figure 8.1

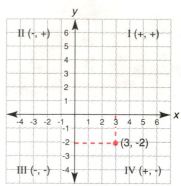

Figure 8.2

lines intersect perpendicularly at the origin (the point corresponding to the number 0) of each line. The horizontal number line is known as the *x* **axis,** and its numerical coordinates increase from left to right. On the vertical number line, the *y* **axis,** values increase from bottom to top, as shown in figure 8.1. The common scale separates the plane into four sections known as **quadrants;** the quadrants are numbered I, II, III, and IV, as shown. The point that marks the common origin of the two number lines is the **origin** of the rectangular coordinate system. It is convenient to identify the origin as (0, 0); this notation indicates that the *x* **value** (listed first) is 0 and also that the *y* **value** (listed second) is 0.

In the coordinate system in figure 8.2, the point $(3, -2)$ is shown. For each point, we have the order (x, y); these pairs are referred to as **ordered pairs** since we require *x* before *y*. To plot (or locate) this point, we see that $x = 3$ and that $y = -2$. Thus the point is located by moving 3 units to the right of the origin and then 2 units down from the *x* axis. The dashed lines shown are used to emphasize the reason for the name "rectangular" coordinate system. Notice that this point $(3, -2)$ could also have been located by first moving down 2 units and then moving 3 units to the right of the *y* axis. This point is located in Quadrant IV. Ordered pairs of plus and minus signs characterize each quadrant.

EXAMPLE 1   Plot points $A(-3, 4)$ and $B(2, 4)$, and find the distance between them.

Solution   Point *A* is located by moving 3 units to the left of the origin and then 4 units up from the *x* axis. *B* is located by moving 2 units to the right of the origin and then 4 units up from the *x* axis. In figure 8.3, $\overline{AB}$ is a horizontal segment. Projections of *A* and *B* on the *x* axis complete a rectangle *ABCD* for which $DC = 5$. Because opposite sides of a rectangle are congruent, $AB = 5$.   ▲

> **DEFINITION:**   For any real number *a*, the **absolute value** of *a* is represented by $|a|$, where $|a| = \begin{cases} a \text{ if } a \geq 0 \\ -a \text{ if } a < 0. \end{cases}$

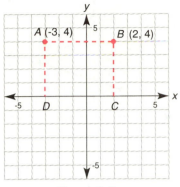

Figure 8.3

Using the preceding definition, we can see that $|5| = 5$, $|-3| = 3$, and $|0| = 0$.

In Example 1, the points $(-3, 4)$ and $(2, 4)$ have the same y values; the distance between them is merely $|2 - (-3)|$ or $|(-3) - 2|$ or 5. It is also easy to use absolute values to find the distance between the two points when they lie on the same vertical line.

DEFINITION: Given points A (x_1, y_1) and B (x_2, y_1) on horizontal segment $\overline{AB}$, the **distance** between these points is $AB = |x_2 - x_1|$.

Note: If $x_2 > x_1$, then $|x_2 - x_1| = x_2 - x_1$.

DEFINITION: Given points C (x_1, y_1) and D (x_1, y_2) on vertical segment $\overline{CD}$, the **distance** between these points is $CD = |y_2 - y_1|$.

Note: If $y_2 > y_1$, then $|y_2 - y_1| = y_2 - y_1$.

EXAMPLE 2

Name the coordinates of points C and D, and find the distance between them.

Solution

C is the point $(0, 1)$ since C is 1 unit above the origin; similarly, D is the point $(0, 5)$.

Since $\overline{CD}$ is a vertical segment, we designate the coordinates of point C by $x_1 = 0$ and $y_1 = 1$. Similarly, the coordinates of point D are $x_1 = 0$ and $y_2 = 5$. Using the preceding definition, we find that

$$CD = |y_2 - y_1| = |5 - 1| = 4$$

We now turn our attention to the more general problem of finding the distance between any two points.

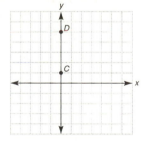

THEOREM 8.1.1 (Distance Formula) The distance between two points (x_1, y_1) and (x_2, y_2) is given by the formula

$$d = \sqrt{(x_2 - x_1)^2 + (y_2 - y_1)^2}$$

Proof

In the coordinate system in figure 8.4 are points P_1 (x_1, y_1) and P_2 (x_2, y_2). In addition to drawing the segment joining these points, we draw an auxiliary horizontal segment through P_1 and an auxiliary vertical segment through P_2; these

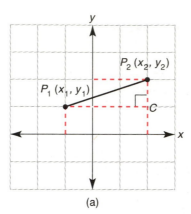
(a)

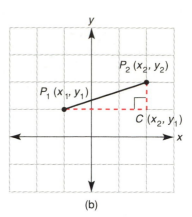
(b)

Figure 8.4

meet at point C, as shown in figure 8.4a. Using figure 8.4b and the definitions for lengths of horizontal and vertical segments,

$$P_1C = x_2 - x_1 \quad \text{and} \quad P_2C = y_2 - y_1$$

In right triangle P_1P_2C in figure 8.4b, let $d = P_1P_2$. By the Pythagorean Theorem,

$$d^2 = (x_2 - x_1)^2 + (y_2 - y_1)^2$$

Taking the positive square root for length,

$$d = \sqrt{(x_2 - x_1)^2 + (y_2 - y_1)^2} \qquad \blacktriangle$$

EXAMPLE 3 Find the distance between points $A\,(5, -1)$ and $B\,(-1, 7)$.

Solution Using the Distance Formula and choosing $x_1 = 5$ and $y_1 = -1$ (from A) and $x_2 = -1$ and $y_2 = 7$ (from B), we obtain

$$d = \sqrt{(-1 - 5)^2 + [7 - (-1)]^2}$$
$$= \sqrt{(-6)^2 + (8)^2}$$
$$= \sqrt{100} = 10$$

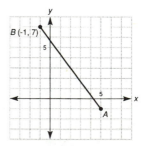

$\blacktriangle$

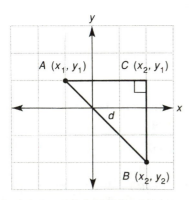

Figure 8.5

Note: If the coordinates of point A had been designated as $x_2 = 5$ and $y_2 = -1$ while those for point B were designated as $x_1 = -1$ and $y_1 = 7$, the distance computed would have remained the same.

In the proof of the Distance Formula, figure 8.4 shows only one of several possible placements of points. If the placement had been like the one in figure 8.5, then $AC = x_2 - x_1$, since $x_2 > x_1$, while $BC = y_1 - y_2$, since $y_1 > y_2$. The Pythagorean Theorem leads to a different result:

$$d^2 = (x_2 - x_1)^2 + (y_1 - y_2)^2$$

but this can be converted to the earlier formula by using the fact that

$$(y_1 - y_2)^2 = (y_2 - y_1)^2$$

This follows from the fact that $(-a)^2 = a^2$ for any real number a.

The following example reminds us of the form of a **linear equation**—an equation whose graph is a straight line. In general, this form is $Ax + By = C$ for constants A, B, and C (where A and B do not both equal 0). We will consider the graphing of linear equations in Section 8.2, and we will turn to the determination of the equation of a line when its graph is provided (or described) in Section 8.3.

EXAMPLE 4

Find the equation that describes all points (x, y) that are equidistant from A $(5, -1)$ and B $(-1, 7)$.

Solution

In Chapter 6, we saw that the locus of points equidistant from two fixed points was a line. This line ($\overleftrightarrow{MX}$, as shown in figure 8.6) is the perpendicular bisector of $\overline{AB}$.

If X is on the locus, then $BX = AX$. By the Distance Formula, we have

$$\sqrt{(x - 5)^2 + [y - (-1)]^2} = \sqrt{[x - (-1)]^2 + (y - 7)^2}$$

or

$$(x - 5)^2 + (y + 1)^2 = (x + 1)^2 + (y - 7)^2$$

after simplifying and squaring. Then

$$x^2 - 10x + 25 + y^2 + 2y + 1 = x^2 + 2x + 1 + y^2 - 14y + 49$$

Eliminating equal square terms by subtraction leads to the equation

$$-12x + 16y = 24$$

and dividing by 4, we get

$$-3x + 4y = 6$$

Equivalently (we could have divided by -4), a correct solution is

$$3x - 4y = -6 \qquad \blacktriangle$$

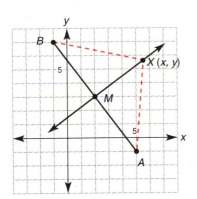

Figure 8.6

In figure 8.6, point M was the midpoint of segment $\overline{AB}$. It will be shown in Example 5a that M is the point $(2, 3)$.

A generalized midpoint formula is given in Theorem 8.1.2. The result shows that the coordinates of the midpoint of the line segment are averages of the coordinates of the endpoints.

To prove the Midpoint Formula, we need to establish two things:

1. $BM + MA = BA$, which establishes that the three points A, M, and B are collinear.

2. $BM = MA$, which establishes that point M is midway between A and B.

> **THEOREM 8.1.2 (Midpoint Formula)**   The midpoint M of the line seg-ment joining (x_1, y_1) and (x_2, y_2) has coordinates x_M and y_M, where
>
> $$(x_M, y_M) = \left(\frac{x_1 + x_2}{2}, \quad \frac{y_1 + y_2}{2} \right)$$
>
> that is,
>
> $$M = \left(\frac{x_1 + x_2}{2}, \quad \frac{y_1 + y_2}{2} \right)$$

EXAMPLE 5   Use the Midpoint Formula to find the midpoint of the segment joining
(a) $(5, -1)$ and $(-1, 7)$ (b) (a, b) and (c, d)

Solution   (a) By using the Midpoint Formula and setting $x_1 = 5$, $y_1 = -1$, $x_2 = -1$, and $y_2 = 7$, we have

$$M = \left(\frac{5 + (-1)}{2}, \quad \frac{-1 + 7}{2} \right) \qquad \text{or} \qquad M = (2, 3)$$

(b) Using the Midpoint Formula and setting $x_1 = a$, $y_1 = b$, $x_2 = c$, and $y_2 = d$, we have

$$M = \left(\frac{a + c}{2}, \quad \frac{b + d}{2} \right) \qquad\qquad ▲$$

In part (a) of Example 5, it may be helpful to make a sketch of the seg-ment; this will allow you to test whether your solution is reasonable! In part (b), we are generalizing the coordinates in preparation for the analytic geometry proofs that appear later in the chapter. In those sections, we will choose the x and y values of each point in such a way as to be as general as possible. When the Midpoint Formula is to be used, it may be a good idea to select such co-ordinates as $(2a, 2b)$ for a point so that division by 2 will not introduce fractions.

Proof of the Midpoint Formula   For the segment joining P_1 and P_2, we designate the midpoint by M, as shown in figure 8.7. Let the coordinates of M be designated by (x_M, y_M). Now con-struct horizontal segments through P_1 and M and vertical segments through M and P_2 to intersect at points A and B, as shown in figure 8.7b.
   Since $\overline{P_1A}$ and $\overline{MB}$ are both horizontal, these segments are parallel. Then $\angle 1 \cong \angle 2$. With $\overline{P_1M} \cong \overline{MP_2}$ by the definition of a midpoint, we have $\triangle P_1AM \cong \triangle MBP_2$ by AAS. Because A is the point (x_M, y_1), we have $P_1A = x_M - x_1$. Likewise, the coordinates of B are (x_2, y_M), so $MB = x_2 - x_M$. Because $\overline{P_1A} \cong \overline{MB}$ by CPCTC, we represent the common length of the segments $\overline{P_1A}$ and $\overline{MB}$ by a. From the first equation, $x_M - x_1 = a$, so $x_M = x_1 + a$. From the second equation, $x_2 - x_M = a$, so $x_2 = x_M + a$. Substituting $x_1 + a$ for x_M, we have

$$(x_1 + a) + a = x_2$$
$$x_1 + 2a = x_2$$

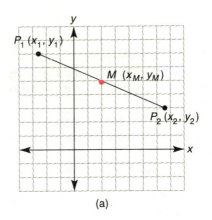

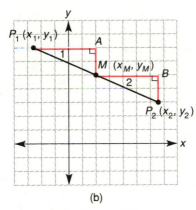

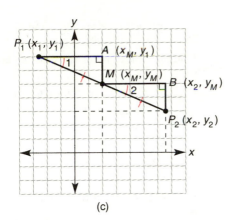

(a)   (b)   (c)

Figure 8.7

Then

$$2a = x_2 - x_1$$

so

$$a = \frac{x_2 - x_1}{2}$$

It follows that

$$x_M = x_1 + a$$

$$= x_1 + \frac{x_2 - x_1}{2}$$

$$= \frac{2 \cdot x_1}{2} + \frac{x_2 - x_1}{2}$$

$$= \frac{x_1 + x_2}{2}$$

The y coordinate of the midpoint can be determined in a similar manner, to show that

$$M = \left(\frac{x_1 + x_2}{2}, \quad \frac{y_1 + y_2}{2} \right) \qquad \blacktriangle$$

8.1 EXERCISES

1. Plot and label the points $A\ (0, -3)$, $B\ (3, -4)$, $C\ (5, 6)$, $D\ (-2, -5)$, and $E\ (-3, 5)$.

2. Give the coordinates of each point $A, B, C, D,$ and E shown in the accompanying figure. Also name the quadrant in which each point lies.

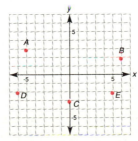

3. Find the distance between each pair of points:

   (a) $(5, -3)$ and $(5, 1)$
   (b) $(-3, 4)$ and $(5, 4)$
   (c) $(0, 2)$ and $(0, -3)$
   (d) $(-2, 0)$ and $(7, 0)$

4. If the distance between $(-2, 3)$ and $(-2, a)$ is 5 units, find all possible values of a.

5. If the distance between $(b, 3)$ and $(7, 3)$ is 3.5 units, find all possible values of b.

6. Find an expression for the distance between (a, b) and (a, c) if $b > c$.

7. Find the distance between each pair of points:

   (a) $(0, -3)$ and $(4, 0)$ (c) $(3, 2)$ and $(5, -2)$
   (b) $(-2, 5)$ and $(4, -3)$ (d) $(a, 0)$ and $(0, b)$

8. Find the distance between each pair of points:

   (a) $(-3, -7)$ and $(2, 5)$ (c) $(-a, -b)$ and (a, b)
   (b) $(0, 0)$ and $(-2, 6)$ (d) $(2a, 2b)$ and $(2c, 2d)$

9. Find the midpoint of the segment that joins each pair of points:

   (a) $(0, -3)$ and $(4, 0)$ (c) $(3, 2)$ and $(5, -2)$
   (b) $(-2, 5)$ and $(4, -3)$ (d) $(a, 0)$ and $(0, b)$

10. Find the midpoint of the segment that joins each pair of points:

    (a) $(-3, -7)$ and $(2, 5)$ (c) $(-a, -b)$ and (a, b)
    (b) $(0, 0)$ and $(-2, 6)$ (d) $(2a, 2b)$ and $(2c, 2d)$

11. The origin $(0, 0)$ is the midpoint of $\overline{AB}$. Find the coordinates of B if A is the point:

    (a) $(3, -4)$ (c) $(a, 0)$
    (b) $(0, 2)$ (d) (b, c)

12. The x axis is the perpendicular bisector of $\overline{AB}$. Find the coordinates of B if A is the point:

    (a) $(3, -4)$ (c) $(0, a)$
    (b) $(0, 2)$ (d) (b, c)

13. The y axis is the perpendicular bisector of $\overline{AB}$. Find the coordinates of B if A is the point:

    (a) $(3, -4)$ (c) $(a, 0)$
    (b) $(2, 0)$ (d) (b, c)

14. $M (3, -4)$ is the midpoint of $\overline{AB}$, in which A is the point $(-5, 7)$. Find the coordinates of B. (*Note:* Algebra can be used!)

15. $M (2.1, -5.7)$ is the midpoint of $\overline{AB}$, in which A is the point $(1.7, 2.3)$. Find the coordinates of B.

16. A circle has its center at the point $(-2, 3)$. If one endpoint of the diameter is at $(3, -5)$, find the other endpoint.

17. A rectangle $ABCD$ has three of its vertices at $A (2, -1)$, $B (6, -1)$, and $C (6, 3)$. Find the fourth vertex D and the area of rectangle $ABCD$.

18. A rectangle $MNPQ$ has three of its vertices at $M (0, 0)$, $N (a, 0)$, and $Q (0, b)$. Find the fourth vertex P and the area of the rectangle $MNPQ$.

19. Use the Distance Formula to determine the type of triangle that has these vertices:

(a) $A (0, 0)$, $B (4, 0)$, and $C (2, 5)$
(b) $D (0, 0)$, $E (4, 0)$, and $F (2, 2\sqrt{3})$
(c) $G (-5, 2)$, $H (-2, 6)$, and $K (2, 3)$

20. Use the method of Example 4 to find the equation of the line that describes all points equidistant from the points $(-3, 4)$ and $(3, 2)$.

21. Use the method of Example 4 to find the equation of the line that describes all points equidistant from the points $(1, 2)$ and $(4, 5)$.

22. For coplanar points A, B, and C, suppose that you have used the Distance Formula to show that $AB = 5$, $BC = 10$, and $AC = 15$. What may you conclude regarding points A, B, and C?

23. If two vertices of an equilateral triangle are at $(0, 0)$ and $(2a, 0)$, what is the third vertex? (*Hint:* See the accompanying illustration.)

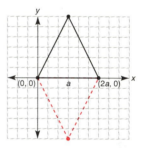

24. The rectangle whose vertices are $A (0, 0)$, $B (a, 0)$, $C (a, b)$, and $D (0, b)$ is shown. Use the Distance Formula to draw a conclusion concerning the lengths of the diagonals $\overline{AC}$ and $\overline{BD}$.

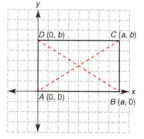

*25. There are two points on the y axis at a distance of 6 units from the point $(3, 1)$. Determine the coordinates of each point.

*26. There are two points on the x axis at a distance of 6 units from the point $(3, 1)$. Determine the coordinates of each point.

27. The triangle that has vertices at $M (-4, 0)$, $N (3, -1)$, and $Q (2, 4)$ has been boxed in as shown in the accompanying figure. Find the area of $\triangle MNQ$.

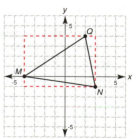

28. Use the method suggested in Exercise 27 to find the area of $\triangle RST$, with $R (-2, 4)$, $S (-1, -2)$, and $T (6, 5)$.

8.2

GRAPHS OF LINEAR EQUATIONS AND SLOPE

In Section 8.1, we were reminded that the general form of a linear equation is $Ax + By = C$ (where A and B do not both equal 0). Some examples of linear equations are $2x + 3y = 12$, $3x - 4y = 12$, and $3x = -6$. As we shall see, the graph of each linear equation is a line.

> **DEFINITION:** The **graph of an equation** is the set of all points (x, y) in the rectangular coordinate system whose ordered pairs satisfy the equation.

EXAMPLE 1   Draw the graph of the equation $2x + 3y = 12$.

Solution   We begin by completing a table. It is convenient to use one point for which $x = 0$, one point for which $y = 0$, and a third point as a check:

$$x = 0 \rightarrow 2(0) + 3y = 12 \rightarrow y = 4$$
$$y = 0 \rightarrow 2x + 3(0) = 12 \rightarrow x = 6$$
$$x = 3 \rightarrow 2(3) + 3y = 12 \rightarrow y = 2$$

x	y	(x, y)
0	4	(0, 4)
6	0	(6, 0)
3	2	(3, 2)

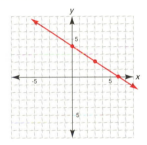

Upon plotting the third point, we see that the three points are collinear; the graph of a linear equation must be a straight line.   ▲

For the equation in Example 1, the number 6 is known as the **x intercept** because $(6, 0)$ is the point at which the graph crosses the x axis; similarly, the number 4 is known as the **y intercept**. Most linear equations have two intercepts; these are generally represented by a (the x intercept) and b (the y intercept).

To determine the x intercept, let $y = 0$ in the given equation and solve for x. The y intercept can similarly be found by choosing $x = 0$ and solving for y.

EXAMPLE 2   Find the x and y intercepts of the equation $3x - 4y = -12$, and use them to graph the equation.

Solution   The x intercept is found when $y = 0$: $3x - 4(0) = -12$, so $x = -4$. The x intercept is -4, so $(-4, 0)$ is on the graph. The y intercept results when $x = 0$: $3(0) - 4y = -12$ so $y = 3$. The y intercept is 3, so $(0, 3)$ is on the

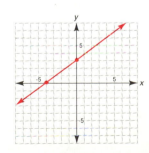

graph. Once $(-4, 0)$ and $(0, 3)$ are plotted, the graph can be completed by drawing the line through the two points.   ▲

As we see in Example 3, a linear equation may have only one intercept. Is it possible for a linear equation to have no intercepts at all?

EXAMPLE 3   Draw the graphs of the following equations:
(a) $x = -2$ (b) $y = 3$

Solution   First note that each equation is a linear equation:

$x = -2$ is equivalent to $(1 \cdot x) + (0 \cdot y) = -2$
$y = 3$ is equivalent to $(0 \cdot x) + (1 \cdot y) = 3$

(a) The equation $x = -2$ claims that the value of x is -2 regardless of the value of y; this leads to the following table:

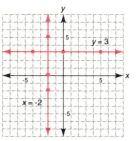

x	y	→	(x, y)
-2	-2	→	$(-2, -2)$
-2	0	→	$(-2, 0)$
-2	5	→	$(-2, 5)$

(b) The equation $y = 3$ claims that the value of y is 3 regardless of the value of x; this leads to the following table:

x	y	→	(x, y)
-4	3	→	$(-4, 3)$
0	3	→	$(0, 3)$
5	3	→	$(5, 3)$

▲

Note: When an equation can be written in the form $x = a$ (for constant a), its graph is the vertical line containing the point $(a, 0)$. When an equation can be written in the form $y = b$ (for constant b), its graph is the horizontal line containing the point $(0, b)$.

DEFINITION (Slope Formula)   The **slope** of the line that contains the points (x_1, y_1) and (x_2, y_2) is given by

$$m = \frac{y_2 - y_1}{x_2 - x_1} \quad \text{for } x_1 \neq x_2$$

Note: When $x_1 = x_2$, meaning that the line in question is vertical, we say that the slope is undefined.

While the uppercase italic M means midpoint, we use the lowercase italic m for slope. Other terms used to describe the slope of a line include "pitch"

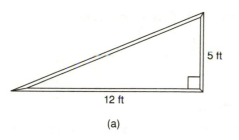

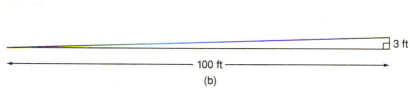

(a)

(b)

Figure 8.8

and "grade." A carpenter may say that a roofline has a $\frac{5}{12}$ pitch. In constructing

a stretch of roadway, an engineer may say that there is a grade of $\frac{3}{100}$ or 3

percent. (See figure 8.8.)

Whether in geometry, carpentry, or engineering, the term **slope** refers to the ratio of the change along the vertical to the change along the horizontal, for any two points on the line in question.

Any horizontal line has slope 0; any vertical line has an undefined slope. Figure 8.9 shows an example of each of these types of lines.

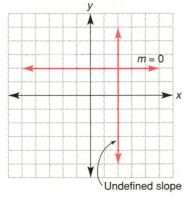

Figure 8.9

A line that "rises" from left to right has a *positive* slope.

A line that "falls" from left to right has a *negative* slope.

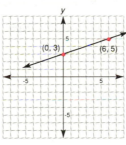

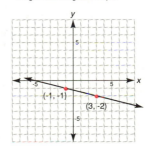

$$m = \frac{y_2 - y_1}{x_2 - x_1}$$

$$m = \frac{5 - 3}{6 - 0} = \frac{2}{6} = \frac{1}{3}$$

(a)

$$m = \frac{y_2 - y_1}{x_2 - x_1}$$

$$m = \frac{-2 - (-1)}{3 - (-1)} = -\frac{1}{4}$$

(b)

EXAMPLE 1 Without graphing, find the slope of the line that contains:

(a) (2, 2) and (5, 3) (b) (1, −1) and (1, 3)

Solution (a) Using the Slope Formula and choosing $x_1 = 2$, $y_1 = 2$, $x_2 = 5$, and $y_2 = 3$, we have

$$m = \frac{3 - 2}{5 - 2} = \frac{1}{3}$$

Note: If drawn, this line will slant upward from left to right.

(b) Let $x_1 = 1$, $y_1 = -1$, $x_2 = 1$, and $y_2 = 3$. Then we calculate

$$m = \frac{3 - (-1)}{1 - 1} = \frac{4}{0}$$

which is undefined.   ▲

Note: If drawn, this line will be vertical.

The slope of a line is unique; that is, the slope does not change when the following changes occur:

1. The order of the two points is reversed.

2. Different points of the line are selected.

The first situation is clear from the fact that $\dfrac{-a}{-b} = \dfrac{a}{b}$. The second situation is more difficult but depends on similar triangles.

For an explanation of point 2, consider figure 8.10, in which points P_1, P_2, and P_3 are collinear. What we wish to show is that the slope of line ℓ is the same whether P_1 and P_2, or P_1 and P_3, are used in the Slope Formula. If horizontal and vertical segments are drawn as shown in figure 8.10, we can show that triangles P_1P_2A and P_2P_3B are similar.

The similarity follows from the facts that $\angle 1 \cong \angle 2$ (since $\overline{P_1A} \parallel \overline{P_2B}$) and that $\angle A$ and $\angle B$ are right angles. Then $\dfrac{P_2A}{P_3B} = \dfrac{P_1A}{P_2B}$ since these are corresponding sides of similar triangles. By interchanging the means, we have $\dfrac{P_2A}{P_1A} = \dfrac{P_3B}{P_2B}$. But

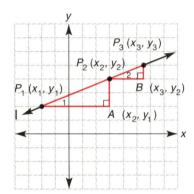

Figure 8.10

$$\frac{P_2A}{P_1A} = \frac{y_2 - y_1}{x_2 - x_1} \quad \text{and} \quad \frac{P_3B}{P_2B} = \frac{y_3 - y_2}{x_3 - x_2}$$

so

$$\frac{y_2 - y_1}{x_2 - x_1} = \frac{y_3 - y_2}{x_3 - x_2}$$

Thus the slope is not changed by having used either pair of points. For that matter, a fourth point could have been shown, in which case the slope of the line would not be changed if P_1 and P_2, or P_3 and P_4, were used; in every case, the slopes agree because of similar triangles.

In summary, if points P_1, P_2, and P_3 are collinear, then the slopes of $\overline{P_1P_2}$, $\overline{P_1P_3}$, and $\overline{P_2P_3}$ are the same. The converse of this statement is also true.

EXAMPLE 5   Are the points $A\,(2, -3)$, $B\,(5, 1)$, and $C\,(-4, -11)$ collinear?

Solution   Let $m_{\overline{AB}}$ and $m_{\overline{BC}}$ represent the slopes of $\overline{AB}$ and $\overline{BC}$, respectively. By the Slope Formula, we have

$$m_{\overline{AB}} = \frac{1 - (-3)}{5 - 2} = \frac{4}{3} \quad \text{and} \quad m_{\overline{BC}} = \frac{-11 - 1}{-4 - 5} = \frac{-12}{-9} = \frac{4}{3}$$

Because $m_{\overline{AB}} = m_{\overline{BC}}$, it follows that A, B, and C are collinear. ▲

The Slope Formula tells us that

$$m = \frac{\text{change in } y}{\text{change in } x} \quad \text{or} \quad m = \frac{\text{vertical change}}{\text{horizontal change}}$$

from one point to the other. This notion is used in Example 6.

EXAMPLE 6   Draw the line through $(-1, 5)$, with slope $m = -\dfrac{2}{3}$.

Solution   First we plot the point $(-1, 5)$. The slope can be written as $m = \dfrac{-2}{3}$. Thus we let the change in y from the first to the second point be -2 while the change in x is 3. From the first point $(-1, 5)$, we locate the second point by moving 2 units down and 3 units to the right. The line is then drawn as shown in figure 8.11. ▲

Two theorems are now stated without proof. However, drawings are provided, and these are to be used in the following exercises. Each proof depends on the fact that similar triangles are created through the use of the auxiliary segments included in the drawings.

THEOREM 8.2.1   If two nonvertical lines are parallel, then their slopes are equal.

Note: If $\ell_1 \parallel \ell_2$, then $m_1 = m_2$.

In figure 8.12, notice that $\overline{AC} \parallel \overline{DF}$. Also, $\overline{AB}$ and $\overline{DE}$ are horizontal, while $\overline{BC}$ and $\overline{EF}$ are auxiliary vertical segments. In the proof of Theorem 8.2.1, the goal is to show that $m_{\overline{AC}} = m_{\overline{DF}}$.

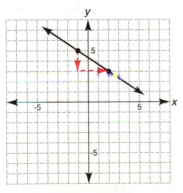

Figure 8.11

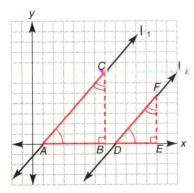

Figure 8.12

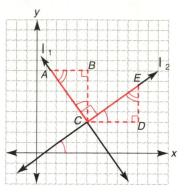

Figure 8.13

> **THEOREM 8.2.2** If two lines (neither horizontal nor vertical) are perpendicular, then the product of their slopes is -1.
>
> *Note:* If $\ell_1 \perp \ell_2$, then $m_1 \cdot m_2 = -1$.

In figure 8.13, auxiliary segments have again been added. The goal in proving Theorem 8.2.2 is to show that $m_{\overline{AC}} \cdot m_{\overline{CE}} = -1$. This relationship indicates that the slopes are **negative reciprocals.** The proof of this theorem is left as an exercise.

EXAMPLE 7

Given the points $A\,(-2, 3)$, $B\,(2, 1)$, $C\,(-1, 8)$, and $D\,(7, 3)$, are $\overline{AB}$ and $\overline{CD}$ parallel, perpendicular, or neither?

Solution

$$m_{\overline{AB}} = \frac{1 - 3}{2 - (-2)} = \frac{-2}{4} = -\frac{1}{2}$$

$$m_{\overline{CD}} = \frac{3 - 8}{7 - (-1)} = \frac{-5}{8} \text{ or } -\frac{5}{8}$$

Since $m_{\overline{AB}} \neq m_{\overline{CD}}$, $\overline{AB} \not\parallel \overline{CD}$. The slopes are not negative reciprocals, so $\overline{AB}$ is not perpendicular to $\overline{CD}$. Neither relationship holds for $\overline{AB}$ and $\overline{CD}$. ▲

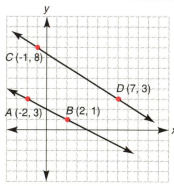

Figure 8.14

In Example 7, it would have been worthwhile to sketch the lines described. It is apparent from figure 8.14 that no special relationship exists between the lines. While a sketch may help show that a relationship does not exist in Example 7, it is not a dependable method for showing that lines *are* parallel or perpendicular.

EXAMPLE 8

Are the lines that are the graphs of $2x + 3y = 6$ and $3x - 2y = 12$ parallel, perpendicular, or neither?

Solution

Because $2x + 3y = 6$ contains the points $(3, 0)$ and $(0, 2)$, its slope is $\frac{2 - 0}{0 - 3} = -\frac{2}{3}$. The line $3x - 2y = 12$ contains $(0, -6)$ and $(4, 0)$; thus its slope is equal to $\frac{0 - (-6)}{4 - 0}$ or $\frac{6}{4}$ or $\frac{3}{2}$. Because the product of the slopes is -1, the lines described are perpendicular. ▲

EXAMPLE 9

Determine the value of a for which the line through $(2, -3)$ and $(5, a)$ is perpendicular to the line $3x + 4y = 12$.

Solution

The line $3x + 4y = 12$ contains the points $(4, 0)$ and $(0, 3)$; this line has the slope

$$m = \frac{3 - 0}{0 - 4} = -\frac{3}{4}$$

For the two lines to be perpendicular, the second line must have slope $\frac{4}{3}$. Using the Slope Formula, the second line has the slope

$$\frac{a - (-3)}{5 - 2}$$

so

$$\frac{a + 3}{3} = \frac{4}{3}$$

Multiplying by 3, we obtain $a + 3 = 4$, so $a = 1$. ▲

EXAMPLE 10   Show that the quadrilateral $ABCD$, with vertices $A\,(0, 0)$, $B\,(a, 0)$, $C\,(a, b)$, and $D\,(0, b)$, is a rectangle.

Solution   By applying the Slope Formula, we see that

$$m_{\overline{AB}} = \frac{0 - 0}{a - 0} = 0 \quad \text{and} \quad m_{\overline{DC}} = \frac{b - b}{a - 0} = 0$$

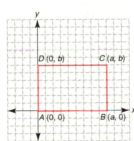

Then $\overline{AB}$ and $\overline{DC}$ are horizontal and parallel to each other.

For $\overline{DA}$ and $\overline{CB}$, the slopes are undefined since the denominators in the Slope Formula each equal 0. Then $\overline{DA}$ and $\overline{CB}$ are vertical and parallel to each other.

Thus $ABCD$ is a parallelogram. With $\overline{AB}$ being horizontal and $\overline{DA}$ vertical, it follows that $\overline{DA} \perp \overline{AB}$. Therefore $ABCD$ is a rectangle by definition. ▲

8.2 EXERCISES

In Exercises 1 to 8, draw the graph of each equation. Name any intercepts.

1. $3x + 4y = 12$

2. $3x + 5y = 15$

3. $x - 2y = 5$

4. $x - 3y = 4$

5. $2x + 6 = 0$

6. $3y - 9 = 0$

7. $\frac{1}{2}x + y = 3$ (*Hint:* Multiply each term by 2, then graph the resulting equation.)

8. $\frac{2}{3}x - y = 1$

9. Find the slopes of lines containing:
   (a) $(2, -3)$ and $(4, 5)$
   (b) $(3, -2)$ and $(3, 7)$
   (c) $(1, -1)$ and $(2, -2)$
   (d) $(-2.7, 5)$ and $(-1.3, 5)$
   (e) (a, b) and (c, d)
   (f) $(a, 0)$ and $(0, b)$

10. Find the slopes of the lines containing:
   (a) $(3, -5)$ and $(-1, 2)$
   (b) $(-2, -3)$ and $(-5, -7)$
   (c) $(2\sqrt{2}, -3\sqrt{6})$ and $(3\sqrt{2}, 5\sqrt{6})$
   (d) $(\sqrt{2}, \sqrt{7})$ and $(\sqrt{2}, \sqrt{3})$
   (e) $(a, 0)$ and $(a + b, c)$
   (f) (a, b) and $(-b, -a)$

11. Find x so that $\overline{AB}$ has slope m, where:
   (a) A is $(2, -3)$, B is $(x, 5)$, and $m = 1$
   (b) A is $(x, -1)$, B is $(3, 5)$, and $m = -0.5$

12. Find y so that $\overline{CD}$ has slope m, where:
   (a) C is $(2, -3)$, D is $(4, y)$, and $m = \frac{3}{2}$
   (b) C is $(-1, -4)$, D is $(3, y)$, and $m = -\frac{2}{3}$

13. Are these points collinear?

  (a) $A(-2, 5)$, $B(0, 2)$, and $C(4, -4)$
  (b) $D(-1, -1)$, $E(2, -2)$, and $F(5, -5)$

14. Are these points collinear?

  (a) $A(-1, -2)$ $B(3, 2)$, and $C(5, 5)$
  (b) $D(a, c - d)$, $E(b, c)$, and $F(2b - a, c + d)$

15. Parallel lines ℓ_1 and ℓ_2 have slopes m_1 and m_2, respectively. Find m_2 if m_1 equals:

  (a) $\dfrac{3}{4}$

  (c) -2

  (b) $-\dfrac{5}{3}$

  (d) $\dfrac{a - b}{c}$

16. Perpendicular lines ℓ_1 and ℓ_2 have slopes m_1 and m_2, respectively. Find m_2 if m_1 equals:

  (a) 5

  (c) $-\dfrac{1}{2}$

  (b) $-\dfrac{5}{3}$

  (d) $\dfrac{a - b}{c}$

In Exercises 17 to 20, state whether the lines are parallel, perpendicular, the same, or none of these.

17. $2x + 3y = 6$ and $2x - 3y = 12$
18. $2x + 3y = 6$ and $4x + 6y = -12$
19. $2x + 3y = 6$ and $3x - 2y = 12$
20. $2x + 3y = 6$ and $4x + 6y = 12$

21. Find x so that the points $A(x, 5)$, $B(2, 3)$, and $C(4, -5)$ are collinear.

22. Find a so that the points $A(1, 3)$, $B(4, 5)$, and $C(a, a)$ are collinear.

23. Find x so that the line through $(2, -3)$ and $(3, 2)$ is perpendicular to the line through $(-2, 4)$ and $(x, -1)$.

24. Find x so that the line through $(2, -3)$ and $(3, 2)$ is parallel to the line through $(-2, 4)$ and $(x, -1)$.

In Exercises 25 to 30, draw the line described:

25. Through $(3, -2)$ and with $m = 2$

26. Through $(-2, -5)$ and with $m = \dfrac{5}{7}$

27. With y intercept 5 and with $m = -\dfrac{3}{4}$

28. With x intercept -3 and with $m = 0.25$

29. Through $(-2, 1)$ and parallel to the line $2x - y = 6$

30. Through $(-2, 1)$ and perpendicular to the line that has intercepts $a = -2$ and $b = 3$

31. Use slopes to decide whether the triangle with vertices at $(6, 5)$, $(-3, 0)$, and $(4, -2)$ is a right triangle.

32. If $A(2, 2)$, $B(7, 3)$, and $C(4, x)$ are the vertices of a right triangle with right angle C, find the value of x.

*33. If $(2, 3)$, $(5, -2)$, and $(7, 2)$ are three vertices (not necessarily consecutive) of a parallelogram, find the possible locations of the fourth vertex.

34. Three consecutive vertices of rectangle $ABCD$ are $A(-5, 1)$, $B(-2, -3)$, and $C(6, y)$. Find the value of y and also the fourth vertex.

35. Show that quadrilateral $RSTV$ is an isosceles trapezoid.

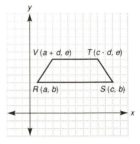

36. Show that quadrilateral $ABCD$ is a parallelogram.

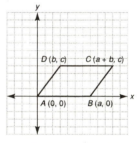

37. Quadrilateral $EFGH$ has vertices $E(0, 0)$, $F(a, 0)$, $G(b, c)$, and $H(d, e)$, as shown in the accompanying drawing. Verify that the quadrilateral formed by joining the consecutive midpoints of $\overline{EF}$, $\overline{FG}$, $\overline{GH}$, and $\overline{HE}$ is a parallelogram.

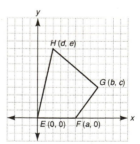

38. Find an equation involving a, b, c, d, and e if $\overleftrightarrow{AC} \perp \overleftrightarrow{BC}$. (*Hint:* Use slopes.)

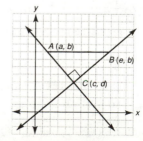

39. Prove that if two nonvertical lines are parallel, then their slopes are equal. (*Hint:* See figure 8.12.)

*****40.** Prove that if two lines (neither horizontal nor vertical) are perpendicular, then the product of their slopes is −1. (*Hint:* See figure 8.13, and use the fact that $\triangle ABC \sim \triangle EDC$.)

EQUATIONS OF LINES

The graph of $2x + 3y = 6$ and the graph of $4x + 6y = 12$ are both lines. Because the intercepts are $a = 3$ and $b = 2$ for both lines, the lines are the same and are described as **coincident lines.** In general, the graphs of $Ax + By = C$ and $kAx + kBy = kC$ (in which k is a nonzero multiplier of the first equation) are the same. For that reason, we may replace the equation $\frac{1}{2}x + y = 3$ with the equation $x + 2y = 6$ when graphing. (See Exercise 7, Section 8.2.) Just as there is a graph for every linear equation, there is also an equation for every line drawn in the rectangular coordinate system. Because $4x + 6y = 12$ and $2x + 3y = 6$ represent the same line, we will designate the equation with reduced coefficients—namely, $2x + 3y = 6$—as the one we seek.

EXAMPLE 1   Write an equation whose graph is the same as that of $4x - 12y = 60$.

Solution   Dividing by 4, we have $x - 3y = 15$. If we divide by −4, the result is $-x + 3y = -15$. Either is correct; these equations are described as "equivalent equations" because their graphs and solutions are identical.   ▲

Many exercises in this section ask for an equation of a line in the form $Ax + By = C$. Your answer is correct if either:

1. It matches perfectly the solution equation provided.

2. It is a nonzero multiple of the solution equation.

Note: In general, an answer expressed in integers is preferable to one that uses fractions for A, B, and C. It may be that an integer equation is impossible, as in the case of the equation $\sqrt{2}x + \sqrt{5}y = 7$. When integer choices are possible for A, B, and C, the smallest choices are generally used; thus, $12x + 16y = 36$ should be replaced by $3x + 4y = 9$.

We now turn our attention to methods for finding the equation of a line. In the first technique, the equation can be found if the slope and the y intercept of the line are known.

THEOREM 8.3.1 (Slope-Intercept Form of a Line) The line whose slope is m and whose y intercept is b has the equation $y = mx + b$.

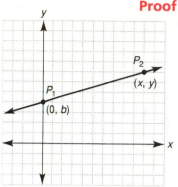

Figure 8.15

Proof Consider the line whose slope is m, as shown in figure 8.15. Using the Slope Formula,

$$m = \frac{y_2 - y_1}{x_2 - x_1}$$

we designate (x, y) as P_2 and $(0, b)$ as P_1. Then

$$m = \frac{y - b}{x - 0} \qquad \text{or} \qquad m = \frac{y - b}{x}$$

Multiplying by x, we have $mx = y - b$. Then $mx + b = y$ and, by the Symmetric Property of Equality, $y = mx + b$. ▲

EXAMPLE 2 Find the general equation $Ax + By = C$ for the line with slope $m = -\dfrac{2}{3}$ and y intercept -2.

Solution With $y = mx + b$, we have

$$y = -\frac{2}{3}x - 2$$

Multiplying by 3, we obtain

$$3y = -2x - 6 \qquad \text{or} \qquad 2x + 3y = -6 \qquad ▲$$

Note: An equivalent and correct solution is $-2x - 3y = 6$.

It is often easier to graph an equation if it is in the form $y = mx + b$. Sometimes the equation can quickly be changed to this form, at which time we know that its graph is a line with slope m and containing $(0, b)$.

EXAMPLE 3 Draw the graph of $\dfrac{1}{2}x + y = 3$.

Solution Solving for y, we have $y = -\dfrac{1}{2}x + 3$. Then $m = -\dfrac{1}{2}$, and the y intercept is 3. We first plot the point $(0, 3)$. Because $m = -\dfrac{1}{2}$ or $\dfrac{-1}{2}$, the vertical change -1 corresponds to a horizontal change $+2$. Thus the second point is located 1 unit down from and then 2 units to the right of the first point. The line is drawn in figure 8.16. ▲

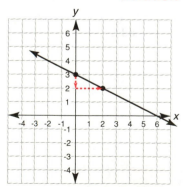

Figure 8.16

If a point other than the y intercept of the line is known, we generally do not use Slope-Intercept Form to find the equation of the line. Instead, the Point-Slope Form of the equation of a line is used. This form is also used when the coordinates of two points of the line are known; in that case, the value of m is found by the Slope Formula.

> **THEOREM 8.3.2 (Point-Slope Form of a Line)**   The line with slope m
> and containing the point (x_1, y_1) has the equation
> $$y - y_1 = m(x - x_1)$$

Proof   Let P_1 be the given point (x_1, y_1) on the line, and let P_2 be (x, y), which represents any other point on the line. (See figure 8.17.) Using the Slope Formula, we have

$$m = \frac{y - y_1}{x - x_1}.$$

Multiplying the equation by $(x - x_1)$, we get

$$m(x - x_1) = y - y_1$$

By the Symmetric Property of Equality, it follows that

$$y - y_1 = m(x - x_1) \qquad \blacktriangle$$

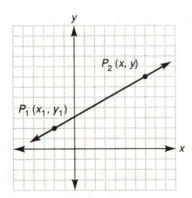

Figure 8.17

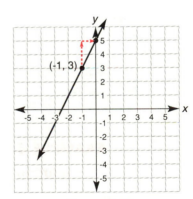

Figure 8.18

EXAMPLE 4   Find the general equation, $Ax + By = C$, for the line with $m = 2$ and containing $(-1, 3)$.

Solution   By the Point-Slope Form, the line shown in figure 8.18 has the equation

$$
\begin{aligned}
y - 3 &= 2[x - (-1)] \\
&= 2(x + 1) \\
&= 2x + 2 \\
-2x + y &= 5
\end{aligned}
$$

Note: An equivalent and correct answer is the equation $2x - y = -5$.   ▲

EXAMPLE 5 Find an equation for the line containing the points $(-1, 2)$ and $(4, 1)$.

Solution To use the Point-Slope Form, we need to know the slope of the line shown in figure 8.19. Choosing P_1 $(-1, 2)$ and P_2 $(4, 1)$, we have

$$m = \frac{1 - 2}{4 - (-1)}$$

$$= \frac{-1}{5} = -\frac{1}{5}$$

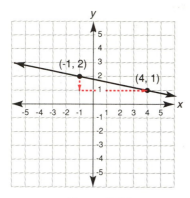

Therefore

$$y - 2 = -\frac{1}{5}[x - (-1)]$$

$$= -\frac{1}{5}x - \frac{1}{5}$$

Multiplying the equation by 5, we get

$$5y - 10 = -1x - 1 \qquad \text{or} \qquad x + 5y = 9 \qquad \blacktriangle$$

Figure 8.19

Note: Other correct forms of the answer are $-x - 5y = -9$ and $y = -\frac{1}{5}x + \frac{9}{5}$. In any correct form, the given points must satisfy the equation.

We now use the Point-Slope Form of the equation of a line as a means of drawing the graph of a linear equation. Note that the form of the equation in Example 6 makes it easy for us to recognize the slope of the line and a point on it.

EXAMPLE 6 Draw the graph of the equation $y + 4 = 2(x - 1)$.

Solution Comparing this equation to the Point-Slope Form, we can write it as

$$y - (-4) = 2(x - 1)$$

Then it follows that $y_1 = -4$ and $x_1 = 1$, while $m = 2$. Therefore the line contains $(1, -4)$, which we plot first.

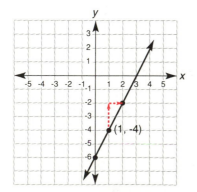

Because $m = 2 = \frac{2}{1}$, the vertical change of 2 corresponds to a horizontal change of 1. Starting at $(1, 4)$, a second point is plotted by moving 2 units up and 1 unit to the right. The line is then drawn through the two points, as shown in figure 8.20. $\blacktriangle$

In earlier chapters, we solved systems of equations such as

$$\begin{cases} x + 2y = 6 \\ 2x - y = 7 \end{cases}$$

Figure 8.20

by using the Addition Property or the Subtraction Property of Equality. We review the method in Example 7.

EXAMPLE 7   Solve the following system by algebra:

$$\begin{cases} x + 2y = 6 \\ 2x - y = 7 \end{cases}$$

Solution   Multiplying the second equation by 2, the system becomes

$$\begin{cases} x + 2y = 6 \\ 4x - 2y = 14 \end{cases}$$

Adding these equations yields $5x = 20$ or $x = 4$. Substituting $x = 4$ into the first equation, we get $4 + 2y = 6$, so $2y = 2$ or $y = 1$. The solution is the ordered pair $(4, 1)$.   ▲

Another method for solving a system of equations is geometric and requires graphing. Solving by graphing amounts to searching for the point of intersection of the linear graphs. That point is the ordered pair that constitutes the common solution (when one exists) for the two equations.

EXAMPLE 8   Solve the following system by graphing:

$$\begin{cases} x + 2y = 6 \\ 2x - y = 7 \end{cases}$$

Solution   Each equation is changed to the form $y = mx + b$ so that the slope and the y intercept are used in graphing:

$$x + 2y = 6 \rightarrow 2y = -1x + 6 \rightarrow y = -\frac{1}{2}x + 3$$

$$2x - y = 7 \rightarrow -y = -2x + 7 \rightarrow y = 2x - 7$$

The graph of $y = -\frac{1}{2}x + 3$ is a line with y intercept 3 and slope $m = -\frac{1}{2}$. The graph of $y = 2x - 7$ is a line with y intercept -7 and slope $m = 2$.

The graphs are drawn in figure 8.21 in the same coordinate system. The point of intersection $(4, 1)$ is the common solution for each of the given equations and thus is the solution of the system.   ▲

Note: To check the result of this example, we show that $(4, 1)$ satisfies both of the given equations:

$$x + 2y = 6 \rightarrow 4 + 2(1) = 6 \text{ is } true$$
$$2x - y = 7 \rightarrow 2(4) - 1 = 7 \text{ is } true$$

The check of the solution in Examples 7 and 8 requires that both statements resulting from substitution into the given equations be true. If either or both statements are false, we do not have the solution of the system.

The graphing method of solving a system has both advantages and disadvantages. We begin with the disadvantages:

1. Graphing the two equations can be very time-consuming.

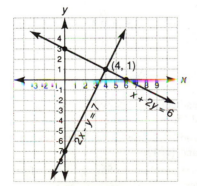

Figure 8.21

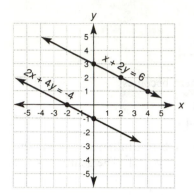

Figure 8.22

2. Sometimes it is difficult to read the solution from the graphs—particularly when solutions involve coordinates with fractions or decimals.

Advantages of the method of solving a system of equations by graphing include the following:

1. It is easy to understand why a system such as

$$\begin{cases} x + 2y = 6 \\ 2x - y = 7 \end{cases} \quad \text{can be replaced by} \quad \begin{cases} x + 2y = 6 \\ 4x - 2y = 14 \end{cases}$$

when solving by addition or subtraction. The graphs of $2x - y = 7$ and $4x - 2y = 14$ are the same.

2. It is easy to understand why a system such as

$$\begin{cases} x + 2y = 6 \\ 2x + 4y = -4 \end{cases}$$

has no solution. The graphs are parallel, as shown in figure 8.22.

3. If the graph of one (or both) of the equations is not a straight line, the method of graphing can still be used. In some instances, the purpose of graphing nonlinear systems is to determine the number of solutions rather than the actual values of particular solutions. We consider this in Section 8.4.

8.3 EXERCISES

In Exercises 1 to 4, write an equation of the form $Ax + By = C$ that is equivalent to the one provided. Then write the given equation in the form $y = mx + b$.

1. $8x + 16y = 48$

2. $15x - 35y = 105$

3. $-6x + 18y = -240$

4. $27x - 36y = 108$

In Exercises 5 to 10, draw the graphs of each equation, using the methods of Examples 3 and 6.

5. $y = 2x - 3$

6. $y = -2x + 5$

7. $\frac{2}{5}x + y = 6$

8. $3x - 2y = 12$

9. $y - 3 = \frac{3}{4}(x - 1)$

10. $y + 3 = -2(x + 4)$

In Exercises 11 to 24, find the equation of the line described.

11. The line has slope $m = -\frac{2}{3}$ and contains $(0, 5)$.

12. The line has slope $m = -3$ and contains $(0, -2)$.

13. The line contains $(2, 4)$ and $(0, 6)$.

14. The line contains $(-2, 5)$ and $(2, -1)$.

15. The line contains $(0, -1)$ and $(3, 1)$.

16. The line contains $(-2, 0)$ and $(4, 3)$.

17. The line has intercepts $a = 2$ and $b = -2$.

18. The line has intercepts $a = -3$ and $b = 5$.

19. The line goes through $(-1, 5)$ and is parallel to the line $5x + 2y = 10$.

20. The line goes through $(0, 3)$ and is parallel to the line $3x + y = 7$.

21. The line goes through $(0, -4)$ and is perpendicular to $y = \frac{3}{4}x - 5$.

22. The line goes through $(2, -3)$ and is perpendicular to $2x - 3y = 6$.

23. The line is the perpendicular bisector of the line segment that joins $(3, 5)$ and $(5, -1)$.

24. The line is the perpendicular bisector of the line segment that joins $(-4, 5)$ and $(1, 1)$.

In Exercises 25 to 30, use graphing to find the point of intersection of the two lines.

25. $y = \frac{1}{2}x - 3$ and $y = \frac{1}{3}x - 2$.

26. $y = 2x + 3$ and $y = 3x$.

27. $2x + y = 6$ and $3x - y = 19$.

28. $\frac{1}{2}x + y = -3$ and $\frac{3}{4}x - y = 8$.

29. $4x + 3y = 18$ and $x - 2y = 10$.

30. $2x + 3y = 3$ and $3x - 2y = 24$.

31. For constants a and b, the graphs of $ax + by = 7$ and $ax - by = 13$ intersect at $(5, -1)$. Using algebra, find the values of a and b.

32. For constants a and b, the graphs of $ax + by = 13$ and $ax - by = 7$ intersect at $(2, -1)$. Using algebra, find the values of a and b.

*33. The graphs of the lines $2x - 3y = 12$, $x + 2y = -1$, and $kx + y = 13$ all contain the same point. Find that point and also the value of k.

*34. The graphs of the lines $3x + 4y = 2$, $5x - y = 11$, and $x + ky = -2$ all contain the same point. Find that point and also the value of k.

35. Prove: For $C \neq D$, the lines $Ax + By = C$ and $Ax + By = D$ are parallel.

36. Prove: The lines $Ax + By = C$ and $Bx - Ay = D$ are perpendicular.

CIRCLES AND PARABOLAS

The general form of the equation of a line $Ax + By = C$ has variable terms of degree 1; that is, the exponent understood for each variable term is 1 since $x^1 = x$ and $y^1 = y$. We will say that the equation itself has degree 1.

The **degree** of each term of a variable expression is the sum of the exponents of its variable factors. For example, the term y^2z^3 has degree 5 (five variable factors) and $5n^4$ has degree 4. The **degree of an equation** equals the degree of its highest-degree term. Equations containing terms of degree 2 and larger are nonlinear and generally have graphs possessing the quality of curvature. For instance, the graphs of $x^2 + y^2 = 9$ (a circle) and $y = x^2$ (a parabola) are shown in figure 8.23.

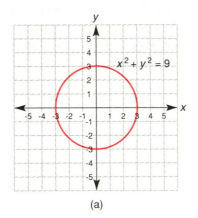

(a)

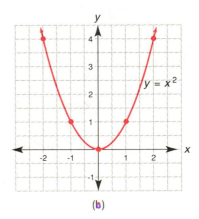

(b)

Figure 8.23

THEOREM 8.4.1 The circle whose center is (h, k) and whose radius has length r, where $r > 0$, has the equation

$$(x - h)^2 + (y - k)^2 = r^2$$

Proof

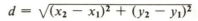

Let (x, y) represent the general point on the circle. Then the distance from (h, k) to (x, y) is r. The Distance Formula,

$$d = \sqrt{(x_2 - x_1)^2 + (y_2 - y_1)^2}$$

becomes

$$r = \sqrt{(x - h)^2 + (y - k)^2}$$

when we designate (x, y) as P_2 and (h, k) as P_1. Squaring each side of this equation leads to the form

$$r^2 = (x - h)^2 + (y - k)^2 \quad \text{or} \quad (x - h)^2 + (y - k)^2 = r^2 \quad \blacktriangle$$

EXAMPLE 1

Find the equation for the circle whose center is $(0, 0)$ and whose radius has length 5.

Solution

We compute

$$(x - h)^2 + (y - k)^2 = r^2$$
$$(x - 0)^2 + (y - 0)^2 = 5^2$$
$$x^2 + y^2 = 25 \quad \blacktriangle$$

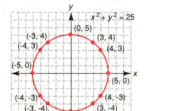

Notice that the graph for Example 1 indicates several solutions for the equation obtained.

The *general form* of the equation of a circle is

$$x^2 + y^2 + Dx + Ey + F = 0$$

where D, E, and F are constants. This form indicates a sum of terms involving x and y, both variables having exponents of 2. In the following example, the equation of a circle is left in its general form.

EXAMPLE 2

Find and simplify the equation of the circle with center $(2, -3)$ and radius of length 4.

Solution

We have

$$(x - h)^2 + (y - k)^2 = r^2$$
$$(x - 2)^2 + [y - (-3)]^2 = 4^2$$
$$(x - 2)^2 + (y + 3)^2 = 16$$
$$x^2 - 4x + 4 + y^2 + 6y + 9 = 16$$
$$x^2 + y^2 - 4x + 6y - 3 = 0 \quad \blacktriangle$$

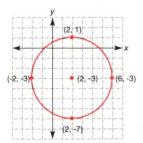

If we were asked to graph $x^2 + y^2 - 4x + 6y - 3 = 0$, we would know from Example 2 to expect the graph to be a circle. Usually, an equation of the form $x^2 + y^2 + Dx + Ey + F = 0$ will be a circle; however, two counterexamples follow Example 3.

Before graphing the circle whose equation is in general form, we must usually complete squares with the expressions $x^2 + Dx$ and $y^2 + Ey$. The goal is to find a number to add to $x^2 + Dx$ so that the result can be factored as the square of a binomial; that is, we want to find a number G so that $x^2 + Dx + G = (x + N)^2$. The procedure follows.

In solving an equation containing the form $x^2 + Dx$, we complete the square as follows:

1. Take one-half of the linear coefficient D, to get $\dfrac{1}{2}D$.

2. Square the result found in step 1 so that $\left(\dfrac{1}{2}D\right)^2 = \dfrac{1}{4}D^2$.

3. Add the result in step 2 to each side of the equation.

EXAMPLE 3   Draw the graph of $x^2 + y^2 - 4x + 6y - 3 = 0$.

Solution   We write the equation in the form appropriate for completing the squares:

$$(x^2 - 4x \qquad) + (y^2 + 6y \qquad) = 3$$

Now $\left(\dfrac{-4}{2}\right)^2 = (-2)^2$ or 4, and $\left(\dfrac{6}{2}\right)^2 = 3^2$ or 9; therefore, 4 and 9 are added to each side of the equation:

$$(x^2 - 4x + 4) + (y^2 + 6y + 9) = 3 + 4 + 9$$
$$(x - 2)^2 + (y + 3)^2 = 16$$
$$(x - 2)^2 + [y - (-3)]^2 = 4^2$$

The graph is a circle whose center is $(2, -3)$ and whose radius has length 4. See Example 2 for the graph.   ▲

> ▲ **WARNING:** *An equation of the form* $x^2 + y^2 + Dx + Ey + F = 0$ *is* not *always a circle.*

Suppose that the equation in Examples 2 and 3 had read

$$x^2 + y^2 - 4x + 6y + 13 = 0$$

When completing the square, we would have obtained

$$(x^2 - 4x + 4) + (y^2 + 6y + 9) = -13 + 4 + 9$$
$$(x - 2)^2 + (y + 3)^2 = 0$$

It appears that we have a circle centered at $(2, -3)$ and with a radius of length 0. The graph, of course, is only the point $(2, -3)$.

If the equation in Examples 2 and 3 had read

$$x^2 + y^2 - 4x + 6y + 14 = 0$$

we would have obtained

$$(x - 2)^2 + (y + 3)^2 = -1$$

Since the square of the radius cannot possibly equal -1, no graph is possible. We may also say that the graph is the empty set.

Summarizing the above results, we see that the graph of the equation $x^2 + y^2 + Dx + Ey + F = 0$ could be a circle, a point, or the empty set.

Table 8.1 compares the general forms of line and circle equations with the forms that are convenient for graphing.

Figure Given	General Form	Form for Graphing
Line	$Ax + By = C$	$\left\{ \begin{array}{c} y = mx + b \\ \text{or} \\ y - y_1 = m(x - x_1) \end{array} \right.$
Circle	$x^2 + y^2 + Dx + Ey + F = 0$	$(x - h)^2 + (y - k)^2 = r^2$

Table 8.1

Another type of figure is the **parabola.** This figure contains all points in the plane that are at equal distances from a fixed line and a point not on that line. The parabolas we consider in this textbook are the result of graphing an equation of the form $y = ax^2 + bx + c$, in which a, b, and c are constants and $a \neq 0$. Some parabolas are shown in figure 8.24. Each has a line of symmetry known as its **axis of symmetry.** The point of intersection of this axis and the parabola is the **vertex** of the parabola.

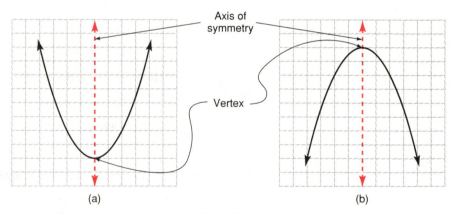

Axis of symmetry

Vertex

(a)   (b)

Figure 8.24

EXAMPLE 4   Use a table to draw the graph of $y = x^2$.

Solution   Plotting the points given in the following table, we see that the y axis is the axis of symmetry. In the graph shown in figure 8.25, the vertex of the parabola is the point $(0, 0)$; in this case, the vertex is the low point of the parabola.

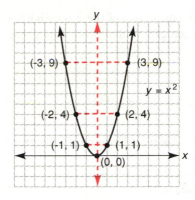

Figure 8.25

x	y
−3	9
−2	4
−1	1
0	0
1	1
2	4
3	9

▲

Note: In the table, we chose values of x that made it easy to evaluate the corresponding y values. Fraction and decimal values for x, although not used in the table, would improve the graph by providing additional points on the figure.

EXAMPLE 5 Draw the graph of $y = -x^2 + 4x$.

Solution By plotting the points given in the following table, we see that the axis of symmetry is along the line $x = 2$. The vertex of the parabola is at the point $(2, 4)$; in this case, the vertex is the high point of the parabola. (See figure 8.26.)

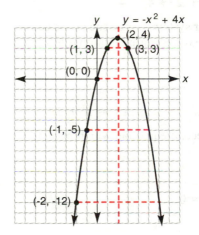

Figure 8.26

x	y
−3	−21
−2	−12
−1	−5
0	0
1	3
2	4
3	3
4	0

▲

For the graph of the equation $y = ax^2 + bx + c$, where $a \neq 0$:

1. The result is always a parabola with y intercept c.
2. The graph opens up (holds water) when $a > 0$ and opens down when $a < 0$.
3. The parabola has a vertical axis of symmetry.
4. The vertex is the point of intersection of the parabola with its axis of symmetry.
5. The graph becomes narrower when $|a| > 1$ and widens when $|a| < 1$.

When two distinct lines are drawn in the same coordinate system, they can intersect in no more than one point. However, a line and circle can intersect in zero, one, or two points. Similarly, a line and a parabola can intersect in zero, one, or two points. While Examples 6 and 8 illustrate two of these possibilities, we will see that the actual points of intersection can be determined by means of geometry or of algebra.

EXAMPLE 6   Using geometry (graphing), find the intersection of the parabola $y = x^2$ and the line $y = x + 2$.

Solution   The graph of $y = x^2$ is drawn by recalling the table from Example 4. The graph of $y = x + 2$ is a line with y intercept 2 and slope $m = 1$.

The parabola and line are graphed in the same coordinate system and intersect at $(-1, 1)$ and $(2, 4)$ as shown in figure 8.27.   ▲

Note: The solutions found in this problem are easily checked. Each ordered pair must make *both* equations true. Checking:

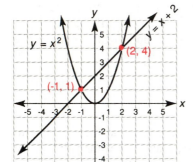

Figure 8.27

$$(-1, 1): \quad y = x^2 \quad \to 1 = (-1)^2 \quad \to 1 = 1$$
$$y = x + 2 \to 1 = (-1) + 2 \to 1 = 1$$
$$(2, 4): \quad y = x^2 \quad \to 4 = 2^2 \quad \to 4 = 4$$
$$y = x + 2 \to 4 = 2 + 2 \to 4 = 4$$

Examples 6 and 7 involve solving nonlinear systems, since one of the equations is of degree 2. We now solve the nonlinear system of Example 6 by algebraic means.

EXAMPLE 7   Using algebra, find the points of intersection of $y = x^2$ and $y = x + 2$.

Solution   We substitute x^2 for y (since $y = x^2$) in the second equation, which becomes

$$x^2 = x + 2$$

Therefore

$$x^2 - x - 2 = 0$$
$$(x - 2)(x + 1) = 0$$
$$x = 2 \quad \text{or} \quad x = -1$$

Now each value of x corresponds to a value of y at a point of intersection. Recalling that $y = x + 2$ (either equation can be used), we see that

$$x = 2 \to y = 4 \quad \text{and} \quad x = -1 \to y = 1$$

As we found in Example 6, the points of intersection are $(2, 4)$ and $(-1, 1)$.   ▲

Example 8   Using geometry, find the intersection of the circle $x^2 + y^2 - 4x + 6y - 3 = 0$ and the line $y = x + 2$.

Solution   In Example 3, we found the graph of the first equation in this system to be a circle with center at $(2, -3)$ and radius of length $r = 4$. The second graph is the line with y intercept 2 and slope $m = 1$. (See figure 8.28.) The graphs do not appear to intersect, so the solution is the empty set.   ▲

One difficulty of solving by the graphing method is that it is often difficult to read the graph. For this reason, we repeat Example 8—this time solving by the algebraic method. The result should underscore the need to understand both the algebraic method and the geometric method and how they are related.

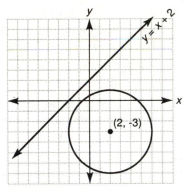

Figure 8.28

EXAMPLE 9   Using algebra, find the intersection of the circle $x^2 + y^2 - 4x + 6y - 3 = 0$ and the line $y = x + 2$.

Solution   Using $x + 2$ for y (since $y = x + 2$), we substitute into the first equation to obtain

$$x^2 + (x + 2)^2 - 4x + 6(x + 2) - 3 = 0$$
$$x^2 + (x^2 + 4x + 4) - 4x + 6x + 12 - 3 = 0$$
$$2x^2 + 6x + 13 = 0$$

But $2x^2 + 6x + 13$ cannot be factored. To use the Quadratic Formula, let $a = 2$, $b = 6$, and $c = 13$. Then we find

$$x = \frac{-b \pm \sqrt{b^2 - 4ac}}{2a}$$
$$= \frac{-6 \pm \sqrt{6^2 - 4(2)(13)}}{2(1)}$$
$$= \frac{-6 \pm \sqrt{-68}}{2}$$

Because $\sqrt{-68}$ is not a real number, we have no points of intersection for the graphs of the equations in the given system. This verifies our claim in Example 8.   ▲

8.4 EXERCISES

In Exercises 1 to 14, draw the graph of each equation.

1. $x^2 + y^2 = 25$
2. $x^2 + y^2 = 16$
3. $(x - 2)^2 + (y + 3)^2 = 9$
4. $(x + 1)^2 + (y - 2)^2 = 16$
5. $x^2 + y^2 - 2x + 4y + 1 = 0$

6. $x^2 + y^2 - 4x + 6y - 12 = 0$
7. $x^2 + y^2 - 6x - 4y - 3 = 0$
8. $x^2 + y^2 - 8x + 2y - 19 = 0$
9. $y = x^2$
10. $y = -x^2$
11. $y = x^2 - 4x - 5$
12. $y = -x^2 + 6x$

13. $y = -2x^2 + 4x$
14. $y = \frac{1}{2}x^2 + 2x$

In Exercises 15 to 20, use geometry to find the intersection of the two graphs; that is, draw graphs.

15. $y = x^2$ and $y = -1x + 2$

16. $y = -x^2$ and $y = 2x - 3$

17. $x^2 + y^2 = 25$ and $x + 3y = 5$

18. $x^2 + y^2 = 16$ and $x + y = 4$

19. $x^2 + y^2 = 4$ and $y = x - 4$

20. $y = x^2$ and $x - y = 3$

For Exercises 21 to 26, use algebra to solve the given systems of equations.

21. $y = x^2$ and $y = -1x + 2$

22. $y = -x^2$ and $y = 2x - 3$

23. $x^2 + y^2 = 25$ and $x + 3y = 5$

24. $x^2 + y^2 = 16$ and $x + y = 4$

25. $x^2 + y^2 = 4$ and $y = x - 4$

26. $y = x^2$ and $x - y = 3$

In Exercises 27 to 34, find the general equation for the circle described.

27. Center is $(2, -5)$; radius length is $r = 4$.

28. Center is $(3, -1)$; $r = 2$.

29. Center is $(-3, -1)$; $r = 5$.

30. Center is $(-2, 6)$; $r = 3$.

31. Endpoints of the diameter are $(-3, 2)$ and $(5, -4)$.

32. Endpoints of the diameter are $(8, 5)$ and $(-2, -1)$.

33. Radius length is $r = 5$; circle lies in Quadrant II and is tangent to both axes.

34. Radius length is $r = 4$; circle lies in Quadrant I and is tangent to both axes.

35. The parabola $y = ax^2$ contains the point $(1, 2)$. Find the value of the constant a.

36. The parabola $y = ax^2 + c$ has the y intercept -3 and also contains the point $(2, 5)$. Find the values of the constants a and c.

In Exercises 37 to 39, use the figure shown.

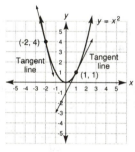

Write the equation of the tangent line at $(3, 4)$ in the form $Ax + By = C$. (*Hint:* The tangent line and the radius drawn to it are $\perp$.)

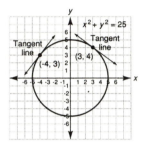

37. The tangent line to the graph of the parabola $y = x^2$ at the point $(1, 1)$ has slope $m = 2$. Write the equation of the tangent line in the form $Ax + By = C$.

38. The tangent line to the graph of the parabola $y = x^2$ at the point $(-2, 4)$ has slope $m = -4$. Write the equation of the tangent line in the form $Ax + By = C$.

*39. The tangent line to the graph of the circle $x^2 + y^2 = 25$ at the point $(3, 4)$ is as shown.

*40. In the form $Ax + By = C$, write the equation of the tangent line to the circle $x^2 + y^2 = 25$ at the point $(-4, 3)$.

*41. How are the two tangent lines to the circle in Exercises 39 and 40 related? Explain.

42. Use graphing to find the solution(s) of the nonlinear system

$$\begin{cases} x^2 + y^2 = 25 \\ y = x^2 - 5 \end{cases}$$

43. Use graphing to find the solution(s) of the nonlinear system

$$\begin{cases} y = x^2 + 1 \\ y = 9 - x^2 \end{cases}$$

8.5

PREPARING TO DO ANALYTIC PROOFS

In this section, our goal is to lay the groundwork for analytic proofs of geometry theorems. An analytic proof requires use of the coordinate system and application of the formulas found in earlier sections of this chapter. Because of the need for these formulas, a summary of them is now provided. Be sure that you have these memorized and know when to use them.

FORMULAS OF ANALYTIC GEOMETRY	
Distance	$d = \sqrt{(x_2 - x_1)^2 + (y_2 - y_1)^2}$
Midpoint	$M = \left(\dfrac{x_1 + x_2}{2}, \dfrac{y_1 + y_2}{2} \right)$
Slope	$m = \dfrac{y_2 - y_1}{x_2 - x_1}$
Equations of lines	$y = mx + b$ $y - y_1 = m(x - x_1)$ $Ax + By = C$
Equations of circles	$(x - h)^2 + (y - k)^2 = r^2$ $x^2 + y^2 + Dx + Ey + F = 0$
Equation of a parabola	$y = ax^2 + bx + c$
Special relationships for lines	$\ell_1 \parallel \ell_2 \leftrightarrow m_1 = m_2$ $\ell_1 \perp \ell_2 \leftrightarrow m_1 \cdot m_2 = -1$

Note: Neither ℓ_1 nor ℓ_2 is a vertical line in the preceding claims.

To see how the preceding list might be used in this and the next section, consider the following examples.

EXAMPLE 1 Identify the geometric figure that is the graph of each equation:
(a) $x^2 + y^2 = 4$ (b) $y = 2x - 3$ (c) $y = x^2 - 3$

Solution (a) $x^2 + y^2 = 4$ can be written as $(x - 0)^2 + (y - 0)^2 = 2^2$. The graph is a **circle** centered at $(0, 0)$ and with a radius of length 2.
(b) $y = 2x - 3$ has the form $y = mx + b$. It is the **line** whose slope is $m = 2$ and whose y intercept is -3.
(c) $y = x^2 - 3$ has the form $y = ax^2 + bx + c$, in which $a = 1$, $b = 0$, and $c = -3$. It is a **parabola.** ▲

EXAMPLE 2 Suppose you are trying to prove the following relationships:
(a) Two lines are parallel.   (b) Two lines are perpendicular.
(c) Two line segments are congruent.
Which formula(s) would you need to use? How would you complete your proof?

Solution (a) Use the Slope Formula first, to find the slope of each line. Then show that the slopes are equal.
(b) Use the Slope Formula first, to find the slope of each line. Then show that $m_1 \cdot m_2 = -1$.

(c) Use the Distance Formula first, to find the length of each segment. Then show that the resulting lengths are equal. ▲

The following example has been separated from Example 2 because its proof is subtle. A drawing is provided to help you understand the concept.

EXAMPLE 3   How can the Midpoint Formula be used to show that two line segments bisect each other?

Solution   If $\overline{AB}$ bisects $\overline{CD}$, and conversely, then M is the common midpoint of the two segments. The Midpoint Formula is used to find the midpoint of each segment, and the results are then shown to be the same point. This establishes that each segment has been **bisected** by a point that is on the other segment. ▲

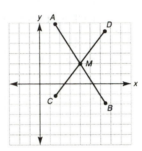

EXAMPLE 4   Suppose that line ℓ_1 has slope $\dfrac{c}{d}$. Use this fact to identify the slopes of the following lines:
(a) ℓ_2 if $\ell_1 \parallel \ell_2$ (b) ℓ_3 if $\ell_1 \perp \ell_3$

Solution   (a) $m_2 = \dfrac{c}{d}$ because $m_1 = m_2$ when $\ell_1 \parallel \ell_2$.

(b) $m_3 = -\dfrac{d}{c}$ because $m_1 \cdot m_3 = -1$ when $\ell_1 \perp \ell_3$. ▲

EXAMPLE 5   What may you conclude if you know that the point (a, b) lies on the circle $x^2 + y^2 = r^2$?

Solution   Since (a, b) is on the circle, it is also a solution for the equation $x^2 + y^2 = r^2$. Therefore $a^2 + b^2 = r^2$. ▲

To construct proofs of geometry theorems by analytic methods, we must use the hypothesis to determine the drawing. Unlike earlier proof problems, the figure must be placed in the coordinate system. Making the drawing requires careful placement of the figure and proper naming of the vertices, using coordinates of the rectangular system. The following chart is helpful in positioning the figure and in naming its vertices.

MAKING THE DRAWING FOR ANALYTIC PROOFS

Five considerations are relevant in making the drawing:

1. Coordinates of the vertices must be general; for instance, you may use (a, b) as a vertex, but do *not* use $(2, 3)$.

2. Make the drawing satisfy the hypothesis without providing any additional qualities; if the theorem describes a rectangle, draw and label a rectangle but *not* a square.

3. For simplicity in your calculations, drop the figure into the rectangular coordinate system in such a manner that

    (a) As many 0 coordinates are used as possible.
    (b) The remaining coordinates represent positive numbers, due to your positioning of the figure in Quadrant I.

    Note: In some cases, it is convenient to place a figure so that it has symmetry with respect to the y axis, in which case some negative coordinates are present.

4. When possible, utilize horizontal and vertical segments, since you know of their parallel (like two vertical lines) and perpendicular qualities.

5. Use as few variable names in the coordinates as possible.

Now we consider Example 6, which clarifies the preceding list of suggestions. As you observe the drawing in each part of the example, imagine that $\triangle ABC$ has been cut out of a piece of cardboard and dropped into the coordinate system in each position indicated. Since we have freedom of placement, we want to choose the positioning that allows us the simplest possible solution.

EXAMPLE 6   Suppose that you are asked to make a drawing for the following theorem, which is to be proved analytically: "The midpoint of the hypotenuse of a right triangle is equidistant from the three vertices of the triangle." Explain why the placement of right $\triangle ABC$ in each part of figure 8.29 (see page 314) is poor.

Solution   (a) The choice of vertices causes $AB = BC$, so the triangle is also an isosceles triangle. This contradicts point 2 of the list of suggestions.

(b) Coordinates are too specific! This contradicts point 1 of the chart. A proof for these coordinates would *not* establish the general case.

(c) The drawing does *not* make use of horizontal and vertical lines to obtain the right angle. This violates point 4 of the chart.

(d) This placement fails point 3 of the chart, because b is a negative number. The length of $\overline{AB}$ would be $-b$, which is more confusing than we want.

(e) This placement fails point 3 because we have not used as many 0 coordinates as we could have used. As we shall see, it also fails point 5.

(f) This fails point 2. It seems that we do not have a right triangle unless $a = b$.   ▲

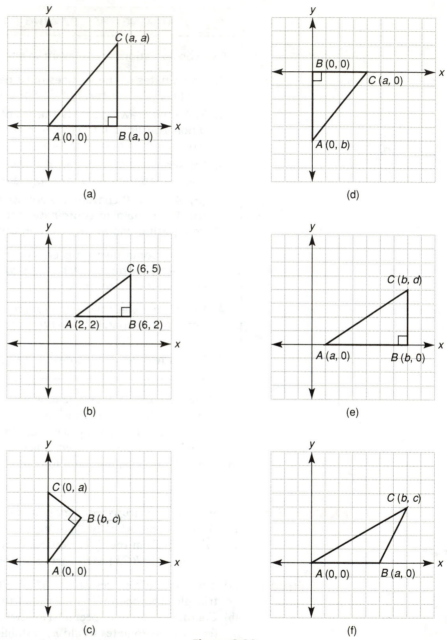

Figure 8.29

With reference to Example 6, what we need to do is place the right $\triangle ABC$ so that we satisfy as many of the conditions given in the previous chart as possible. Two convenient placements are shown in figure 8.30. The triangle in figure 8.30b is slightly better than the one in 8.30a, in that it uses four 0 coordinates rather than three. Another advantage of figure 8.30b is that the placement forces angle B to be a right angle since the x and y axes are perpendicular.

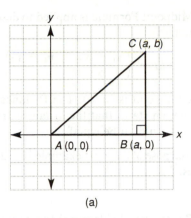

(a)

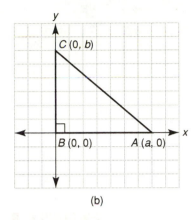

(b)

Figure 8.30

We now turn our attention to the conclusion of the theorem. A second chart examines some considerations in proving statements analytically.

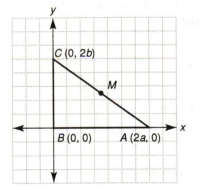

Figure 8.31

USING THE CONCLUSION TO DO ANALYTIC PROOFS

Three considerations are relevant for using the conclusion as a guide:

1. If the conclusion is a conjunction "P and Q," be sure to verify both parts of the conclusion.

2. The following pairings indicate how to prove statements of the type shown on the left:

To prove:	Use:
(a) Segments have equal lengths (like $AB = CD$)	Distance Formula
(b) Segments are parallel (like $\overline{AB} \parallel \overline{CD}$)	Slope Formula (need $m_{\overline{AB}} = m_{\overline{CD}}$)
(c) Segments are perpendicular (like $\overline{AB} \perp \overline{CD}$)	Slope Formula (need $m_{\overline{AB}} \cdot m_{\overline{CD}} = -1$)
(d) A segment is bisected	Midpoint Formula

3. Anticipate the proof by thinking of the steps of the proof in reverse order.

EXAMPLE 7   (a) Provide an ideal drawing for the following theorem: The midpoint of the hypotenuse of a right triangle is equidistant from the three vertices of the triangle.

(b) By studying the theorem, name at least two of the formulas that will be used to complete the proof.

Solution   (a) We improve figure 8.30b by giving the value $2a$ to the x coordinate of A and the value $2b$ to the y coordinate of C. (This makes it easy to describe the midpoint M of $\overline{AC}$.) (See figure 8.31.)

(b) The Midpoint Formula is applied to describe the midpoint of $\overline{AC}$. Using the formula,

$$M = \left(\frac{x_1 + x_2}{2}, \quad \frac{y_1 + y_2}{2} \right) = \left(\frac{2a + 0}{2}, \quad \frac{0 + 2b}{2} \right)$$

so the midpoint is (a, b). The Distance Formula will also be needed since the theorem states that the distances from M to A, from M to B, and from M to C should all be equal. ▲

The purpose of our next example is to demonstrate efficiency in the labeling of vertices.

EXAMPLE 8   If $MNPQ$ is the parallelogram shown in figure 8.32, provide the coordinates of point P.

Solution   Consider $\square MNPQ$ in figure 8.32, as shown. For the moment, we refer to point P as (x, y).

Because $\overline{MN} \parallel \overline{QP}$, we have $m_{\overline{MN}} = m_{\overline{QP}}$. But $m_{\overline{MN}} = \dfrac{0 - 0}{a - 0} = 0$, while $m_{\overline{QP}} = \dfrac{y - d}{x - c}$, so we are led to the equation

$$\frac{y - d}{x - c} = 0 \rightarrow y - d = 0 \rightarrow y = d$$

Now P is described by (x, d). Because $\overline{MQ} \parallel \overline{NP}$, we are also led to the equation equating slopes of these segments. But

$$m_{\overline{MQ}} = \frac{d - 0}{c - 0} = \frac{d}{c}$$

while

$$m_{\overline{NP}} = \frac{d - 0}{x - a} = \frac{d}{x - a}$$

so

$$\frac{d}{c} = \frac{d}{x - a}$$

Thus, by using the Means-Extremes Property, we have

$$d(x - a) = d \cdot c \qquad \text{with } d \neq 0$$
$$x - a = c$$
$$x = a + c$$

Therefore P is the point $(a + c, d)$. ▲

In Example 8, we name the vertices of a parallelogram in the fewest possible letters. We now extend our result in Example 8 to allow for a rhombus— a parallelogram with two congruent adjacent sides.

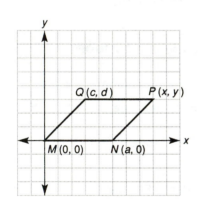

Figure 8.32

EXAMPLE 9   Find an equation that relates a, b, c, and d if quadrilateral $ABCD$ is a rhombus.

Solution As we saw in Example 8, the coordinates of the vertices of *ABCD* define a parallelogram. For emphasis, we note that

$$m_{\overline{AB}} = m_{\overline{DC}} = 0 \quad \text{and} \quad m_{\overline{AD}} = m_{\overline{BC}} = \frac{d}{c}$$

For figure 8.33 to represent a rhombus too, we need $AB = AD$. Now $AB = a - 0 = a$ because $\overline{AB}$ is a horizontal segment. To find an expression for the length of $\overline{AD}$, we need to use the Distance Formula:

$$d = \sqrt{(x_2 - x_1)^2 + (y_2 - y_1)^2}$$
$$AD = \sqrt{(c - 0)^2 + (d - 0)^2}$$
$$= \sqrt{c^2 + d^2}$$

Because $AB = AD$, we are led to $a = \sqrt{c^2 + d^2}$. Squaring, we have the desired relationship: $a^2 = c^2 + d^2$. ▲

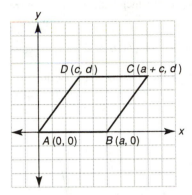

Figure 8.33

EXAMPLE 10 If $\ell_1 \perp \ell_2$, find a relationship among the variables a, b, c, and d.

Solution First we find the slopes of the lines shown in figure 8.34. For ℓ_1, we have

$$m_1 = \frac{0 - d}{a - 0} = -\frac{d}{a}$$

For ℓ_2, we have

$$m_2 = \frac{c - 0}{b - 0} = \frac{c}{b}$$

With $\ell_1 \perp \ell_2$, it follows that $m_1 \cdot m_2 = -1$. From the slopes found above, we have

$$-\frac{d}{a} \cdot \frac{c}{b} = -1$$

so

$$-\frac{dc}{ab} = -1$$

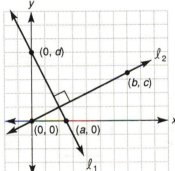

Figure 8.34

Equivalently, $\frac{dc}{ab} = 1$ and $dc = ab$. ▲

8.5 EXERCISES

1. Find an expression for the
   (a) Distance between $(a, 0)$ and $(0, a)$
   (b) Slope of the segment joining (a, b) and (c, d)

2. Find the coordinates of the midpoint of the segment that joins the points
   (a) $(a, 0)$ and $(0, b)$ (b) $(2a, 0)$ and $(0, 2b)$

3. Find the slope-intercept form of the equation of the line containing the points
   (a) $(a, 0)$ and $(0, a)$ (b) $(a, 0)$ and $(0, b)$

4. Find the slope of the line that is
   (a) Parallel to the line containing $(a, 0)$ and $(0, b)$
   (b) Perpendicular to the line through $(a, 0)$ and $(0, b)$

In Exercises 5 to 10, supply the missing coordinates for the vertices, using as few variables as possible.

5.

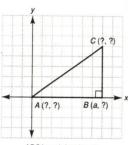

ABC is a right triangle

8.

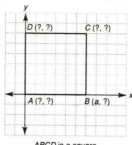

ABCD is a square

6.

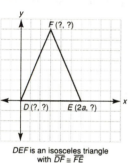

DEF is an isosceles triangle
with $\overline{DF} \cong \overline{FE}$

9.

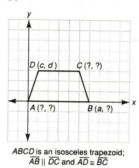

ABCD is an isosceles trapezoid;
$\overline{AB} \parallel \overline{DC}$ and $\overline{AD} \cong \overline{BC}$

7.

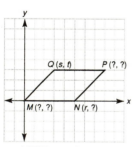

MNPQ is a parallelogram

10.

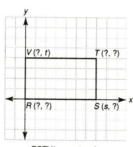

RSTV is a rectangle

In Exercises 11 to 16, draw an ideally placed figure in the coordinate system; then name the coordinates of each vertex for the figure named.

11. (a) A square
(b) A square for which midpoints of sides are needed
12. (a) A rectangle
(b) A rectangle for which midpoints of sides are needed
13. (a) A parallelogram
(b) A parallelogram for which midpoints of sides are needed
14. (a) A triangle
(b) A triangle for which midpoints of sides are needed

15. (a) An isosceles triangle
(b) An isosceles triangle for which midpoints of sides are needed
16. (a) A trapezoid
(b) A trapezoid for which midpoints of sides are needed

In Exercises 17 to 22, find the equation (relationship) requested, and eliminate fractions or radicals from the equation.

17. If $\square MNPQ$ is a rhombus, state an equation that relates r, s, and t.

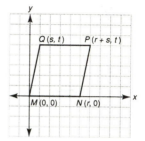

18. For $\square tRSTV$, suppose that $RT = VS$. State an equation that relates s, t, and v.

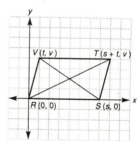

19. For $\square ABCD$, suppose that diagonals $\overline{AC}$ and $\overline{DB}$ are $\perp$. State an equation that relates a, b, and c.

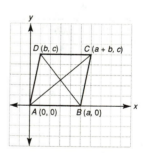

20. For quadrilateral $RSTV$, suppose that $\overline{RV} \parallel \overline{ST}$. State an equation that relates m, n, p, q, and r.

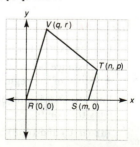

21. Suppose that $\triangle BC$ is an equilateral triangle. State an equation that relates variables a and b.

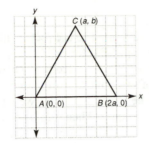

C (a, b)

A (0, 0)   *B (2a, 0)*

22. Suppose that $\triangle RST$ is isosceles, with $\overline{RS} \cong \overline{RT}$. State an equation that relates s, t, and v.

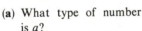

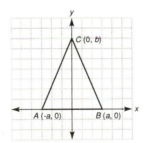

T (t, v)

R (0, 0)   *S (s, 0)*

23. The accompanying drawing could be used to represent an isosceles $\triangle ABC$ in which $\overline{AC} \cong \overline{BC}$.

(a) What type of number is a?

(b) What type of number is $-a$?

(c) Find an expression for the length of $\overline{AB}$.

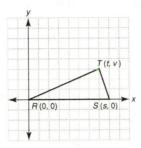

C (0, b)

A (-a, 0)   *B (a, 0)*

24. The accompanying drawing shows a parallelogram *RSTV*.

(a) What type of number is r?

(b) Find an expression for *RS*.

(c) Describe the coordinate t in terms of the other variables.

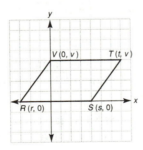

V (0, v)   *T (t, v)*

R (r, 0)   *S (s, 0)*

25. Which formula would you use to establish each of the following claims?

(a) $\overline{AC} \perp \overline{DB}$

(b) $AC = DB$

(c) $\overline{DB}$ bisects $\overline{AC}$

(d) $\overline{AD} \parallel \overline{BC}$

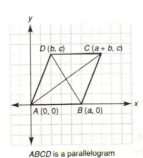

D (b, c)   *C (a + b, c)*

A (0, 0)   *B (a, 0)*

ABCD is a parallelogram

26. Which formula would you use to establish each of the following claims?

(a) The coordinates of X are (d, c)

(b) $m_{\overline{VT}} = 0$

(c) $\overline{VT} \parallel \overline{RS}$

(d) The length of $\overline{RV}$ is $2\sqrt{d^2 + c^2}$

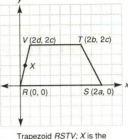

V (2d, 2c)   *T (2b, 2c)*

X

R (0, 0)   *S (2a, 0)*

Trapezoid *RSTV*; X is the midpoint of $\overline{RV}$

In Exercises 27 to 30, draw and label a well-placed figure in the coordinate system for each theorem. Do not attempt to prove the theorem!

27. The line segment joining the midpoints of the two non-parallel sides of a trapezoid is parallel to the bases of the trapezoid.

28. If the midpoints of the sides of a quadrilateral are joined in order, the resulting quadrilateral is a parallelogram.

29. An angle inscribed in a semicircle is a right angle. (*Hint:* Place the center of the circle at the origin.)

30. The diagonals of a rhombus are perpendicular to each other.

31. Find the point of intersection of the line $y = \dfrac{a}{b}x$ and the vertical line $x = c$.

32. At the point on the parabola $y = x^2$ for which $x = a$, the slope of the tangent line is $m = 2a$. Show that the equation of the tangent line is $y = 2ax - a^2$.

*33. Let $\triangle DEF$ have its vertices at $D\ (0, 0)$, $E\ (a, b)$, and $F\ (c, 0)$.

(a) Show that an equation for the altitude from D to $\overline{EF}$ is $y = \dfrac{c - a}{b}x$.

(b) Show that the general equation for the altitude from F to $\overline{DE}$ is $ax + by = ac$.

*34. Show that the equation for the circle whose center is (h, k) and whose radius has length r is

$$x^2 + y^2 + Dx + Ey + F = 0$$

in which $D = -2h,\ E = -2k$
and $F = h^2 + k^2 - r^2$

*35. Show that the point-slope form of the tangent to the circle with center (h, k) and radius length r at point of tangency (a, b) is

$$y - b = \frac{a - h}{k - b}(x - a)$$

*36. Show that the general equation of the perpendicular bisector of the segment joining $A\ (0, 0)$ to $B\ (2c, 2d)$ is $cx + dy = c^2 + d^2$.

8.6

ANALYTIC PROOFS

When we use the algebra related to the rectangular coordinate system to prove a geometric theorem, the proof is termed "analytic." Some theorems proved by analytic methods here are repeated from an earlier chapter.

In Section 8.5, we saw that a parallelogram can be labeled as shown in figure 8.35. By using the Slope Formula, we can show that the slopes of the opposite pairs of sides are equal. Then it follows that both pairs of opposite sides are parallel, so the figure is indeed a parallelogram. In the following example, choosing coordinates 2a, 2b, 2c, and so on makes computation of the midpoints of the diagonals easier.

EXAMPLE 1   Prove the following theorem by the analytic method:

THEOREM 8.6.1   The diagonals of a parallelogram bisect each other.

Proof   As shown and labeled in figure 8.36, the quadrilateral $ABCD$ is a parallelogram. The diagonals intersect at point P. By the Midpoint Formula, we have

$$M_{\overline{AC}} = \left(\frac{0 + (2a + 2b)}{2}, \quad \frac{0 + 2c}{2} \right)$$

$$= (a + b, c)$$

But this point is also on $\overline{DB}$ and is, in fact, the midpoint of $\overline{DB}$, since

$$M_{\overline{DB}} = \left(\frac{2a + 2b}{2}, \quad \frac{0 + 2c}{2} \right)$$

$$= (a + b, c)$$

Thus $(a + b, c)$ is the common midpoint of the two diagonals and must be the point of intersection of $\overline{AC}$ and $\overline{DB}$. Then $\overline{AC}$ and $\overline{DB}$ bisect each other at point P.   ▲

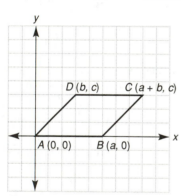

Figure 8.35

The proof of Theorem 8.6.1 suggests that other approaches to the proof are possible. In some cases, the alternative approach may be better! In Example 1, we could have used a three-step proof:

1. Find the equations of the two lines.

2. Determine the point of intersection.

3. Show that this point is the common midpoint.

But the statement of the theorem implied the use of the Midpoint Formula. Our approach to Example 1 was far easier and just as correct. The use of the Midpoint Formula is necessary and generally quick to apply when the word "bisect" appears in the statement of a theorem.

We now outline the method of analytic proof.

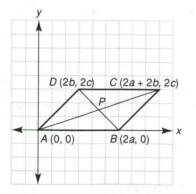

Figure 8.36

COMPLETING AN ANALYTIC PROOF

1. Read the theorem carefully to distinguish the hypothesis and the conclusion.

2. Use the hypothesis (and nothing more) to determine a convenient placement of the figure described in the rectangular coordinate system. Then label the figure.

3. If any special quality is provided by the hypothesis, be sure to state this early in the proof. (For example, a rhombus should be described as a parallelogram that has two congruent adjacent sides.)

4. Study the conclusion, and devise a plan to prove this claim; this may involve reasoning back from the conclusion step by step until the hypothesis is reached.

5. Write the proof; be careful to properly order and justify each statement discovered in step 4.

EXAMPLE 2 Prove Theorem 8.6.2 by the analytic method.

> **THEOREM 8.6.2** The diagonals of a rhombus are perpendicular.

Proof In figure 8.37, $ABCD$ has the coordinates of a parallelogram. Because $\square ABCD$ is a rhombus, $AB = AD$. Then $a = \sqrt{b^2 + c^2}$ by the Distance Formula, and squaring gives $a^2 = b^2 + c^2$. Now

$$m_{\overline{AC}} = \frac{c}{a + b} \quad \text{and} \quad m_{\overline{DB}} = \frac{-c}{a - b}$$

by the Slope Formula. Then the product of the slopes of the diagonals is

$$m_{\overline{AC}} \cdot m_{\overline{DB}} = \frac{c}{a + b} \cdot \frac{-c}{a - b}$$

$$= \frac{-c^2}{a^2 - b^2}$$

$$= \frac{-c^2}{(b^2 + c^2) - b^2}$$

$$= \frac{-c^2}{c^2} = -1$$

Then $\overline{AC} \perp \overline{DB}$ since the product of their slopes equals -1. ▲

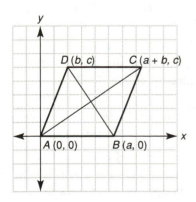

Figure 8.37

In Example 2, we had to use the condition that two adjacent sides of the rhombus were congruent in order to complete the proof. Had that condition been omitted, the product of slopes could not have been shown to equal -1. In general, the diagonals of a parallelogram are *not* perpendicular.

In our next example, we consider the proof of the converse of an earlier theorem. While it is easy to complete an analytic proof of the statement, "The diagonals of a rectangle are equal in length," the proof of the converse is not as straightforward.

EXAMPLE 3   Prove Theorem 8.6.3 by the analytic method.

> **THEOREM 8.6.3**   If the diagonals of a parallelogram are equal in length, then the parallelogram is a rectangle.

Proof   In parallelogram $ABCD$ shown in figure 8.38, we know that $AC = DB$. Applying the Distance Formula,

$$AC = \sqrt{[(a + b) - 0]^2 + (c - 0)^2} \quad \text{and} \quad DB = \sqrt{(a - b)^2 + (0 - c)^2}$$

Then it follows that

$$\sqrt{(a + b)^2 + c^2} = \sqrt{(a - b)^2 + (-c)^2}$$

Squaring, $(a + b)^2 + c^2 = (a - b)^2 + (-c)^2$

Simplifying, $a^2 + 2ab + b^2 + c^2 = a^2 - 2ab + b^2 + c^2$

$$4ab = 0$$

Dividing by 4, $a \cdot b = 0$

Thus $a = 0 \quad \text{or} \quad b = 0$

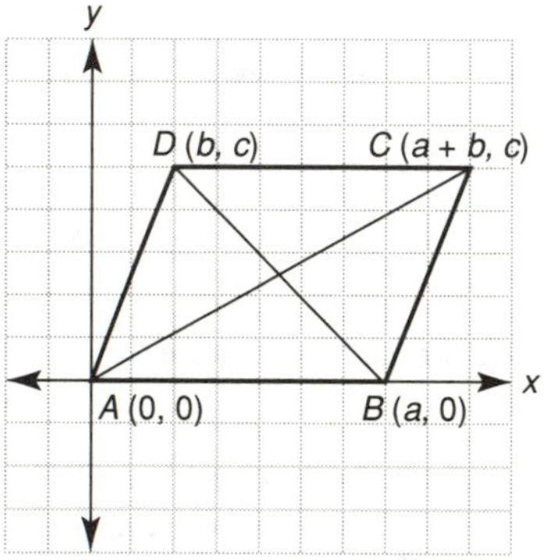

Because $a \neq 0$ (otherwise points A and B would coincide), it is necessary that $b = 0$, so point D is on the y axis. The resulting coordinates of the figure are $A\,(0, 0)$, $B\,(a, 0)$, $C\,(a, c)$, and $D\,(0, c)$. Then $ABCD$ must be a rectangle because $\overline{AB}$ is horizontal while $\overline{AD}$ is vertical.   ▲

Our final example illustrates how the analytic method can be used to derive a relationship rather than to prove one that is already known.

Figure 8.38

EXAMPLE 4   Use the analytic method to find the point of intersection of medians $\overline{AX}$ and $\overline{BY}$ in $\triangle ABC$ shown in figure 8.39a.

Solution   First we find the coordinates of X and Y since these were not given. By the definition of a median, these two points are the midpoints of the sides $\overline{BC}$ and $\overline{AC}$, respectively.

Now we repeat the drawing, with coordinates of X and Y calculated by the Midpoint Formula, as shown in figure 8.39b.

We have to find the coordinates of Z, the intersection of $\overline{AX}$ and $\overline{BY}$. The equations of the lines are needed; to find that of $\overline{AX}$, we use $y = mx + b$. By the Slope Formula,

$$m_{\overline{AX}} = \frac{c - 0}{(a + b) - 0} = \frac{c}{a + b}$$

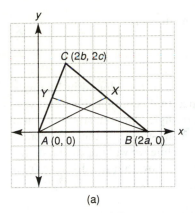

(a)

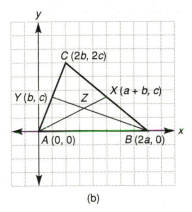

(b)

Figure 8.39

Because the y intercept of $\overline{AX}$ is 0, the equation of $\overline{AX}$ is

$$y = \frac{c}{a+b}x$$

To find the equation of $\overline{BY}$, we use the Point-Slope form $y - y_1 = m(x - x_1)$. The Slope Formula gives

$$m_{\overline{BY}} = \frac{0 - c}{2a - b} = \frac{-c}{2a - b}$$

A point on $\overline{BY}$ is $(2a, 0)$, so the equation becomes

$$y - 0 = \frac{-c}{2a - b}(x - 2a)$$

$$y = \frac{-c}{2a - b}x + \frac{2ac}{2a - b}$$

Now we substitute the expression $\frac{c}{a+b}x$ for y (the result in finding the equation of $\overline{AX}$) in the equation for $\overline{BY}$:

$$\frac{c}{a+b}x = \frac{-c}{2a - b}x + \frac{2ac}{2a - b}$$

$$\frac{c}{a+b}x + \frac{c}{2a - b}x = \frac{2ac}{2a - b}$$

$$\left(\frac{c}{a+b} + \frac{c}{2a - b}\right)x = \frac{2ac}{2a - b}$$

$$\frac{c(2a - b) + c(a + b)}{(a + b)(2a - b)}x = \frac{2ac}{2a - b}$$

In order to clear the equation of fractions, we now multiply each side by $(a + b)(2a - b)$. The cleared equation is

$$[c(2a - b) + c(a + b)]x = 2ac(a + b)$$
$$[2ac - bc + ac + bc]x = 2ac(a + b)$$
$$3ac \cdot x = 2ac(a + b)$$

Dividing by $3ac$, we get

$$x = \frac{2}{3}(a + b)$$

Because $y = \frac{c}{a+b}x$, we have

$$y = \frac{c}{a+b} \cdot \frac{2}{3} \cdot (a + b)$$

$$= \frac{2}{3}c$$

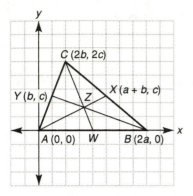

Figure 8.40

Therefore, the point of intersection of the two medians is

$$Z\left(\frac{2}{3}(a + b), \frac{2}{3}c\right)$$ ▲

If the equation of the median from C to $\overline{AB}$ had been found in Example 4, the coordinates of point Z would have satisfied that equation. That fact verifies the concurrency of the three medians of a triangle. An even stronger claim is that the point of concurrency is at a distance two-thirds of the way from a vertex toward the midpoint of the opposite side. While we will not establish this distance relationship for all three medians, we will show that the distance from A to Z is two-thirds that from A to X. (See figure 8.40.)

By the Distance Formula,

$$AZ = \sqrt{\left[\frac{2}{3}(a + b) - 0\right]^2 + \left[\frac{2}{3}c - 0\right]^2}$$

$$= \sqrt{\frac{4}{9}(a + b)^2 + \frac{4}{9}c^2}$$

$$= \sqrt{\frac{4}{9}[(a + b)^2 + c^2]}$$

$$= \sqrt{\frac{4}{9}} \cdot \sqrt{(a + b)^2 + c^2}$$

$$= \frac{2}{3}\sqrt{(a + b)^2 + c^2}$$

Since
$$AX = \sqrt{(a + b)^2 + c^2}$$

we have the desired relationship. For emphasis, we restate the result as a theorem.

THEOREM 8.6.4　The three medians of a triangle are concurrent at a point that is two-thirds the distance from any vertex to the midpoint of the opposite side.

8.6 EXERCISES

In Exercises 1 to 19, complete an analytic proof for each theorem.

1. The diagonals of a rectangle are equal in length.

2. The opposite sides of a parallelogram are equal in length.

3. The diagonals of a square are perpendicular bisectors of each other.

4. The diagonals of an isosceles trapezoid are equal in length.

5. The median from the vertex of an isosceles triangle to the base is perpendicular to the base.

6. The medians to the congruent sides of an isosceles triangle are equal in length.

7. The segments that join the midpoints of the consecutive sides of a quadrilateral form a parallelogram.

8. The segments that join the midpoints of the opposite sides of a quadrilateral bisect each other.

9. The segments that join the midpoints of the consecutive sides of a rectangle form a rhombus.

10. The segments that join the midpoints of the consecutive sides of a rhombus form a rectangle.

11. The midpoint of the hypotenuse of a right triangle is equidistant from the three vertices of the triangle.

12. The median of a trapezoid is parallel to the bases of the trapezoid and has a length equal to one-half the sum of the lengths of the two bases.

13. The segment that joins the midpoints of two sides of a triangle is parallel to the third side and has a length equal to one-half the length of the third side.

14. The perpendicular bisector of the base of an isosceles triangle contains the vertex of the triangle.

*15. If the diagonals of a parallelogram are perpendicular, then the parallelogram is a rhombus.

*16. If the median to one side of a triangle is also an altitude of the triangle, then the triangle is isosceles.

*17. If an angle is inscribed in a semicircle, then the angle is a right angle. (*Hint:* Place the center of the circle at $(0, 0)$, and find the equation of the circle of radius r.)

*18. The perpendicular bisectors of the sides of a triangle are concurrent.

*19. The altitudes of a triangle are concurrent.

20. Use the analytic method to decide what type of quadrilateral results when segments join the midpoints of the consecutive sides of a parallelogram.

21. Use the analytic method to decide what type of triangle results when the midpoints of the sides of an isosceles triangle are joined.

22. Describe the steps of the procedure that allows us to find the distance from a point $P(a, b)$ to the line $Ax + By = C$. Use the accompanying drawing for reference.

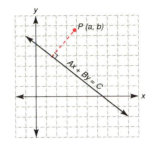

23. Would the theorem of Exercise 7 remain true if the quadrilateral involved happened to be concave?

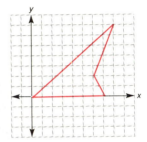

*24. Use the accompanying drawing to complete an analytic proof of the following theorem: In a triangle that has sides of lengths a, b, and c, if $c^2 = a^2 + b^2$, then the triangle is a right triangle.

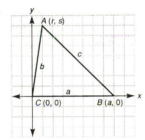

▲ A LOOK BEYOND: THE BANACH-TARSKI PARADOX

In the 1920s, two Polish mathematicians proposed a mathematical dilemma to their colleagues. Known as the Banach-Tarski paradox, their proposal has puzzled students of geometry for decades. What was most baffling was that the proposal indicated that matter could be created through rearrangement of the pieces of a figure. The following steps outline the Banach-Tarski paradox.

First consider the square whose sides are each of length 8. (See figure 8.41a.) By counting squares or by applying a formula, it is clearly the case that the 8-by-8 square must have an area of 64 square units. We now subdivide the square (as shown) to form two right triangles and two trapezoids. Notice the dimensions indicated on each piece of the square in figure 8.41b.

The parts of the square are now rearranged to form a rectangle (figure 8.42) whose dimensions are 13 and 5. This rectangle clearly has an area that measures 65 square units—

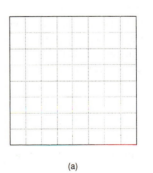

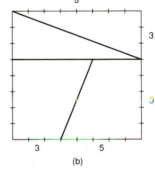

Figure 8.41

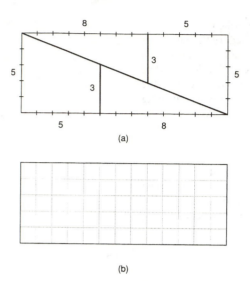

(a)

(b)

Figure 8.42

1 square unit more than the given square. How is it possible that the second figure has an area greater than the first?

While the puzzle is real, you may also realize that something is wrong. This paradox can be explained by considering the slopes of lines. The triangles, which have legs of lengths 3 and 8, determine a hypotenuse whose slope is $-\frac{3}{8}$. Although the side of the trapezoid appears to be collinear with the hypotenuse, it actually has a slope of $-\frac{2}{5}$. It was easy to accept that the segments were collinear because the slopes are nearly equal; in fact, $-\frac{3}{8} = -0.375$, while $-\frac{2}{5} = -0.400$. In figure 8.43 (which is somewhat exaggerated), a very thin parallelogram appears in the space between the original segments of the cut-up square. One may quickly conclude that the area of that parallelogram is 1 square unit, and the paradox has been resolved once more!

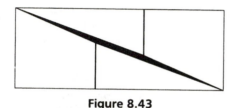

Figure 8.43

▲ SUMMARY

A LOOK BACK AT CHAPTER 8

Our goal in this chapter was to relate algebra and geometry. This relationship is called analytic geometry or coordinate geometry. Formulas for the length of a segment, the midpoint of a segment, and the slope of a line were developed. We found the general equation for a line, a circle, and a parabola, and we used these forms for graphing. Analytic proofs of general statements followed analytic proofs of concrete examples.

A LOOK AHEAD TO CHAPTER 9

Our goal of the next chapter is to deal with a type of geometry known as solid geometry. We will find the surface areas of solids with polygonal or circular bases. We will also find the volumes of these solid figures.

IMPORTANT TERMS AND CONCEPTS OF CHAPTER 8

8.1 Analytic Geometry

Cartesian Coordinate System
Rectangular Coordinate System
x Axis, y Axis, Quadrants, Origin
Ordered Pair
Distance Formula
Midpoint Formula
8.2 Graphs of Equations
Slope, Slope Formula
8.3 Equations of Lines
Slope-Intercept Form of a Line
Point-Slope Form of a Line
8.4 Equations of Circles and Parabolas
Graphs of Circles and Parabolas
8.5 Preparation for Analytic Proofs
8.6 Analytic Proofs
A Look Beyond The Banach-Tarski Paradox

▲ REVIEW EXERCISES

1. Find the distance between each pair of points:

   (a) $(6, 4)$ and $(6, -3)$ (c) $(-5, 2)$ and $(7, -3)$
   (b) $(1, 4)$ and $(-5, 4)$ (d) $(x - 3, y + 2)$ and $(x, y - 2)$

2. Find the distance between each pair of points:

   (a) $(2, -3)$ and $(2, 5)$ (c) $(-4, 1)$ and $(4, 5)$
   (b) $(3, -2)$ and $(-7, -2)$ (d) $(x - 2, y - 3)$ and $(x + 4, y + 5)$

3. Find the midpoint of the segment that joins each pair of points in Exercise 1.

4. Find the midpoint of the segment that joins each pair of points in Exercise 2.

5. Find the slope of the line joining each pair of points in Exercise 1.

6. Find the slope of the line joining each pair of points in Exercise 2.

7. $(2, 1)$ is the midpoint of $\overline{AB}$, in which A has coordinates $(8, 10)$. Find the coordinates of B.

8. The y axis is the perpendicular bisector of $\overline{RS}$. Find the coordinates of R if S is the point $(-3, 7)$.

9. If A has coordinates $(2, 1)$ and B has coordinates $(x, 3)$, find x so that the slope of $\overleftrightarrow{AB}$ is -3.

10. If R has coordinates $(-5, 2)$ and S has coordinates $(2, y)$, find y so that the slope of $\overleftrightarrow{RS}$ is $\dfrac{-6}{7}$.

11. Without graphing, determine whether the pairs of lines are parallel, perpendicular, the same, or none of these:

    (a) $x + 3y = 6$ and $3x - y = -7$
    (b) $2x - y = -3$ and $y = 2x - 14$
    (c) $y + 2 = -3(x - 5)$ and $2y = 6x + 11$
    (d) $0.5x + y = 0$ and $2x - y = 10$

12. Determine whether the points $(-6, 5)$, $(1, 7)$, and $(16, 10)$ are collinear.

13. Find x so that $(-2, 3)$, $(x, 6)$, and $(8, 8)$ are collinear.

14. Draw the graph of $3x + 7y = 21$, and name the intercepts.

15. Draw the graph of $4x - 3y = 9$ by changing the equation to Slope-Intercept Form.

16. Draw the graph of $y + 2 = \dfrac{-2}{3}(x - 1)$.

17. Write the equation of the figure described:

    (a) The line through $(2, 3)$ and $(-3, 6)$
    (b) The line through $(-2, -1)$ and parallel to the line through $(6, -3)$ and $(8, -9)$
    (c) The line through $(3, -2)$ and perpendicular to $x + 2y = 4$
    (d) The line through $(-3, 5)$ and parallel to the x axis
    (e) The circle with center $(3, -1)$ and radius 4
    (f) The circle with center $(2, -3)$ and tangent to the y axis

18. Show that the triangle whose vertices are $(-2, -3)$, $(4, 5)$, and $(-4, 1)$ is a right triangle.

19. Show that the triangle whose vertices are $(3, 6)$, $(-6, 4)$, and $(1, -2)$ is an isosceles triangle.

20. Show that the quadrilateral whose vertices are $(-5, -3)$, $(1, -11)$, $(7, -6)$, and $(1, 2)$ is a parallelogram.

Draw the graph of each equation in Exercises 21 and 22.

21. (a) $x^2 + y^2 = 49$
    (b) $(x - 1)^2 + (y + 3)^2 = 25$
    (c) $y = -2x^2 + 3$
    (d) $x^2 + y^2 - 4x + 10y + 20 = 0$

22. (a) $x^2 + y^2 = 18$
    (b) $(x + 2)^2 + (y - 1)^2 = 16$
    (c) $y = x^2 + 4x$
    (d) $x^2 + y^2 + 10x + 2y - 23 = 0$

In Exercises 23 to 26, find the intersection of the two equations by graphing.

23. $4x - 3y = -3$
    $x + 2y = 13$

24. $x^2 + y^2 = 40$
    $3x - y = 20$

25. $y = x^2 + 3$
    $y = 4x$

26. $x^2 + y^2 = 36$
    $x^2 + (y + 3)^2 = 9$

In Exercises 27 to 30, solve the systems of equations in Exercises 23 to 26 by using algebraic methods.

27. Refer to Exercise 23.

28. Refer to Exercise 24.

29. Refer to Exercise 25.

30. Refer to Exercise 26.

31. The center of circle O is $(5, 3)$. Point A, with coordinates $(9, 6)$, lies on this circle. Find the equation of circle O.

32. Find the equation of the tangent to circle O through point A, as described in Exercise 31.

33. Three of four vertices of a parallelogram are $(0, -2)$, $(6, 8)$, and $(10, 1)$. Find the possibilities for the coordinates of the remaining vertex.

34. Find the center and radius of a circle whose equation is $x^2 + y^2 = 8x - 14y + 35$.

35. Find the center and radius of a circle whose equation is $x^2 + y^2 = 12y - 10x - 10$.

36. A (3, 1), B (5, 9), and C (11, 3) are the vertices of a triangle.

   (a) Find the length of the median from B to $\overline{AC}$.
   (b) Find the slope of the altitude from B to $\overline{AC}$.
   (c) Find the slope of a line through B parallel to $\overline{AC}$.

In Exercises 37 to 40, supply the missing coordinates for the vertices, using as few variables as possible.

37.

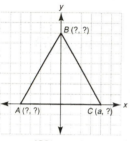

$\triangle ABC$ is isosceles,
with base $\overline{AC}$

38.

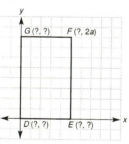

Rectangle $DEFG$ with
$\overline{DG} = 2 \cdot \overline{DE}$

39.

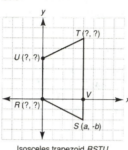

Isosceles trapezoid $RSTU$
with $\overline{RV} \cong \overline{RU}$

40.

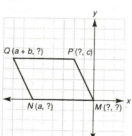

Parallelogram $MPQN$

41. A (2a, 2b), B (2c, 2d), and C (0, 2e) are the vertices of a triangle.

   (a) Find the length of the median from C to $\overline{AB}$.
   (b) Find the slope of the altitude from B to $\overline{AC}$.
   (c) Find the equation of the altitude from B to $\overline{AC}$.

Prove the statements in Exercises 42 to 46, using analytic geometry.

42. The segments that join the midpoints of consecutive sides of a parallelogram form another parallelogram.

43. If the diagonals of a rectangle are perpendicular, then the rectangle is a square.

44. If the diagonals of a trapezoid are equal in length, then the trapezoid is an isosceles trapezoid.

45. If two medians of a triangle are equal in length, then the triangle is isosceles.

46. The segments joining the midpoints of consecutive sides of an isosceles trapezoid form a rhombus.

9

SURFACES AND SOLIDS

The architectural design of the buildings shown suggests many of the solids you will encounter in this chapter. The real world is three-dimensional; that is, objects in the real world have length, width, and depth. In this chapter, formal names will be given to objects that you see and informally describe as "boxes" and "tin cans."

Photograph courtesy of Don Manning, Parkland College

CHAPTER OUTLINE

9.1 Prisms and Pyramids

9.2 Cylinders and Cones

9.3 Polyhedrons and Spheres

9.4 A Three-Dimensional Coordinate System

A Look Beyond Historical Sketch of René Descartes

 ## 9.1 PRISMS AND PYRAMIDS

Suppose that two congruent figures lie in parallel planes in such a way that their corresponding sides are parallel. If the corresponding vertices (such as A and A' in figure 9.1a) are joined by segments, then the "solid" that results is a **prism.** The congruent figures that lie in the parallel planes are the **bases** of the prism. Additional terminology is provided with each of the subsequent drawings used for reference. Notice that the parallel planes need not be shown in the drawings of prisms.

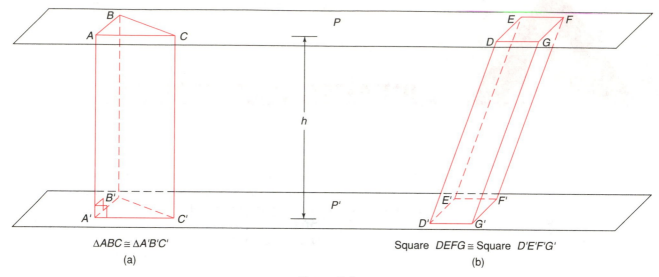

$\triangle ABC \cong \triangle A'B'C'$

(a)

Square $DEFG \cong$ Square $D'E'F'G'$

(b)

Figure 9.1

In figure 9.1a, $\overline{AA'}$, $\overline{BB'}$, and $\overline{CC'}$ are the *lateral edges* of the prism. Because the lateral edges are perpendicular to planes P and P', the **lateral faces** (like quadrilateral $ACC'A'$) are rectangles. The prism in figure 9.1a has vertices A, B, C, A', B', and C' and is known as a **right triangular prism.**

In figure 9.1b, the lateral edges (such as $\overline{DD'}$ and $\overline{GG'}$) are not perpendicular to planes P and P', so we say that each is oblique to these planes. The lateral faces (such as quadrilaterals $DGG'D'$ and $GFF'G'$) are parallelograms. This solid is known as an **oblique square prism.**

Both prisms in figure 9.1 have an **altitude** (the length of a perpendicular segment between the planes containing bases) of length h.

> **DEFINITION:**   The **lateral area L** of a prism is the sum of the areas of all lateral faces.

In the right triangular prism in figure 9.2, a, b, and c are the lengths of sides of the base. These dimensions are used along with the length of the altitude (denoted by h) to calculate the lateral area, which is the sum of the areas of rectangles $ACC'A'$, $ABB'A'$, and $BCC'B'$.

The lateral area L of the right triangular prism is found as follows:

$$L = ah + bh + ch$$
$$= h(a + b + c)$$
$$= h \cdot P$$

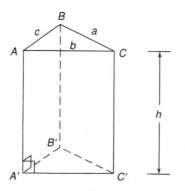

Figure 9.2

where P is the perimeter of a base of the prism.

THEOREM 9.1.1   The lateral area L of a right prism whose altitude has measure h and whose base has perimeter P is given by the formula $L = h \cdot P$.

DEFINITION: For any prism, the **total area T** is the sum of the lateral area and the areas of the bases.

Note: The total area of the prism is also its **surface area**.

Recalling Heron's Formula, we know that the total area of the right triangular prism in figure 9.2 is

$$T = h \cdot P + 2\sqrt{s(s - a)(s - b)(s - c)}$$

in which s is the semiperimeter of the triangular base.

THEOREM 9.1.2   The total area T of any prism whose lateral area has measure L and whose bases have area B is given by the formula $T = L + 2 \cdot B$.

EXAMPLE 1   Find the lateral area L and the surface area T of the right regular hexagonal prism provided in figure 9.3a.

Solution   There are six congruent faces, each rectangular and with dimensions of 4 in. and 10 in. Then

$$L = 6(4 \cdot 10)$$
$$= 240 \text{ in.}^2$$

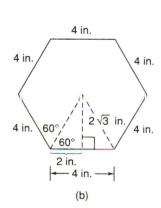

(a)

(b)

Figure 9.3

For the regular hexagonal base (figure 9.3b), the apothem measures $a = 2\sqrt{3}$ in., while the perimeter is $P = 6 \cdot 4 = 24$ in. Then the area B of each base is given by

$$B = \frac{1}{2}a \cdot P$$

$$= \frac{1}{2} \cdot 2\sqrt{3} \cdot 24$$

$$= 24\sqrt{3} \text{ in.}^2$$

Now
$$T = L + 2 \cdot B$$
$$= (240 + 48\sqrt{3}) \text{ in.}^2 \qquad \blacktriangle$$

EXAMPLE 2   The total area of the right square prism in figure 9.4 is 210 cm². Find the length of a side of the square base if the altitude of the prism is 8 cm.

Solution   Let x be the length of the side of the square. Then the area of the base is $B = x^2$ and the area of each of the four lateral faces is $8x$. Therefore

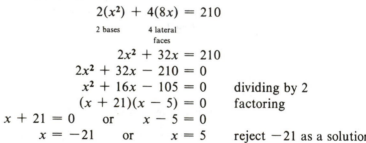

$$\underbrace{2(x^2)}_{\text{2 bases}} + \underbrace{4(8x)}_{\substack{\text{4 lateral}\\\text{faces}}} = 210$$

$$2x^2 + 32x = 210$$
$$2x^2 + 32x - 210 = 0$$
$$x^2 + 16x - 105 = 0 \qquad \text{dividing by 2}$$
$$(x + 21)(x - 5) = 0 \qquad \text{factoring}$$
$$x + 21 = 0 \quad \text{or} \quad x - 5 = 0$$
$$x = -21 \quad \text{or} \quad x = 5 \qquad \text{reject } -21 \text{ as a solution}$$

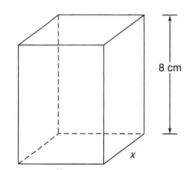

Figure 9.4

Then each side of the square base measures 5 cm. $\blacktriangle$

We may wish to think of the prism as a solid occupying a portion of space, which involves the notion of **volume.** We will not define but merely describe this concept. To begin, we will need a unit for the measuring process. While a meter might be useful for measuring the length of a line segment and a square meter might be used to measure some bounded region of a plane, we need a different type of unit for measuring an enclosed or bounded region of space. That unit is a "cubic unit." The volume occupied by the **cube** (a right square prism in which lateral and base edges are congruent) shown in figure 9.5 is 1 cubic inch or 1 in³.

We assume that any solid in space has a volume that can be expressed as a positive number of cubic units. This is now stated in the form of a postulate.

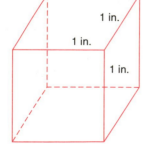

Figure 9.5

POSTULATE 22   **(Volume Postulate)**   Corresponding to every solid is a unique positive number V known as the volume of that solid.

The simplest figure for which we wish to determine volume is the **right rectangular prism.** Such a solid might be described as a box. Since boxes are

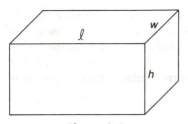

Figure 9.6

used as containers for storage and shipping (like a boxcar), it is important to know their volumes as measures of capacity. A right rectangular prism is shown in figure 9.6; its dimensions are shown as length ℓ, width w, and height (or altitude) h.

The volume of a right rectangular prism of length 4 in., width 3 in., and height 2 in. is easily shown to be 24 in³. The volume is the product of the three dimensions of the given solid. Not only do we see that $4 \cdot 3 \cdot 2 = 24$ but also that in. $\cdot$ in. $\cdot$ in. $=$ in.³.

Figure 9.7a and b emphasizes why the 4 by 3 by 2 box must have the volume 24 cubic units. We see that there are four layers of blocks, each of which is a 2 by 3 configuration of 6 units³. The figure also provides some insight into our next postulate.

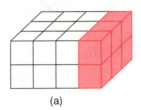
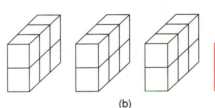

(a)   (b)

Figure 9.7

POSTULATE 23 The volume of a right rectangular prism is given by

$$V = \ell \cdot w \cdot h$$

where ℓ measures the length, w the width, and h the altitude of the prism.

EXAMPLE 3 Find the volume of a box whose dimensions are 1 ft, 8 in., and 10 in.

Solution While it makes no difference which dimension is chosen for ℓ or w or h, it is most important that the units of measure be the same. Thus, 1 ft is replaced by 12 in. in the formula for volume:

$$\begin{aligned} V &= \ell \cdot w \cdot h \\ &= 12 \cdot 8 \cdot 10 \\ &= 960 \text{ in.}^3 \end{aligned}$$

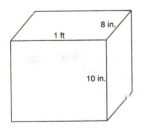

Notice that the formula for the volume of the right rectangular prism, $V = \ell wh$, could be replaced by the formula $V = B \cdot h$, where B is the area of the base of the prism; that is, $B = \ell w$. As stated in our final postulate, this volume relationship is true for right prisms in general.

POSTULATE 24 The volume of a right prism is given by

$$V = B \cdot h$$

in which B is the area of a base and h is the altitude of the prism.

The solids shown in figure 9.8 are **pyramids.** In each pyramid, a point is joined to each vertex of a polygon. The polygon in the plane is the **base** of the pyramid, while the point is the **vertex** of the pyramid. In figure 9.8a, the vertex of the pyramid is point A while the base is square $BCDE;$ this pyramid is a square pyramid. The pyramid in figure 9.8b is a triangular pyramid; its base is $\triangle GHJ$.

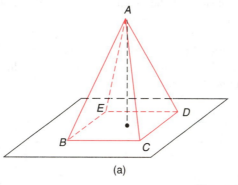

(a) 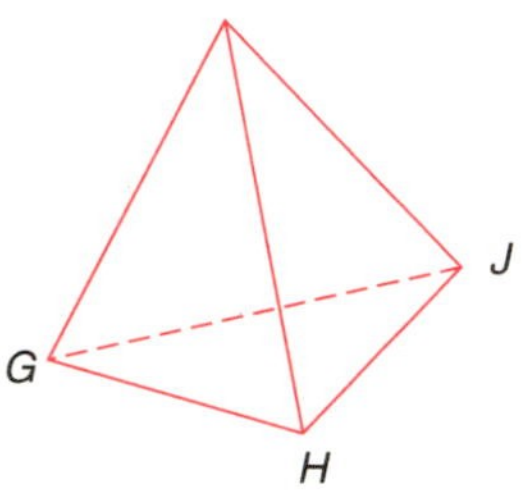   (b)

Figure 9.8

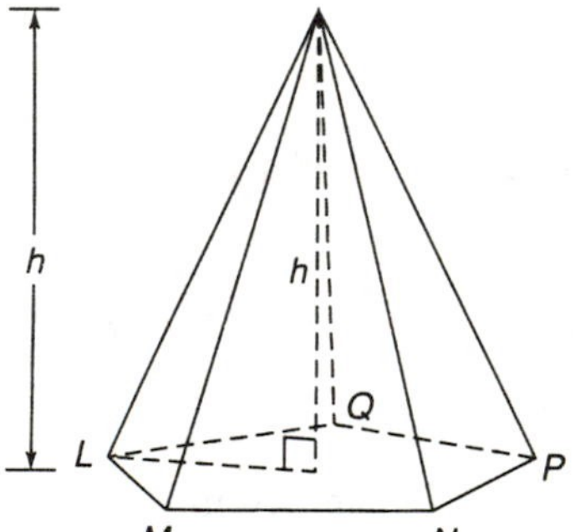

A third pyramid, shown in figure 9.9, is a **pentagonal pyramid.** It has vertex K, pentagon $LMNPQ$ for a base, and **lateral edges** $\overline{KL}$, $\overline{KM}$, $\overline{KN}$, $\overline{KP}$, and $\overline{KQ}$. The sides of the base $\overline{LM}$, $\overline{MN}$, $\overline{NP}$, $\overline{PQ}$, and $\overline{QL}$ are **base edges.** All **lateral faces** of a pyramid are triangles; $\triangle KLM$ is one of the five lateral faces of the pentagonal prism. The **altitude** of the pyramid, of length h, is the segment from the vertex perpendicular to the base.

Figure 9.9

> **DEFINITION:**   A **regular pyramid** is a pyramid whose base is a regular polygon and whose lateral edges are all congruent.

The lateral faces of a regular pyramid are necessarily congruent to each other; in figure 9.9, $\triangle KLM \cong \triangle KMN \cong \triangle KNP \cong \triangle KPQ \cong \triangle KQL$ by SSS. Each lateral face is an isosceles triangle.

> **DEFINITION:**   The **slant height** of a regular pyramid is the altitude of any of the congruent lateral faces of the regular pyramid.
>
> *Note:* Only a regular pyramid has a slant height.

> **THEOREM 9.1.3**   The lateral area L of a regular pyramid with slant height of length ℓ and perimeter P of the base is given by
>
> $$L = \frac{1}{2} \cdot \ell \cdot P$$

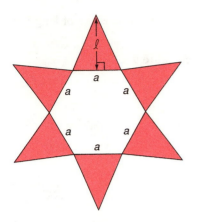

Figure 9.10

While it may not be clear why the formula in Theorem 9.1.3 is correct, we will attempt to justify the result by taking apart one of the regular pyramids and laying it out flat. We will use a regular hexagonal pyramid for this purpose, but the following argument is appropriate if the base has n sides.

When the lateral faces of the regular pyramid are folded down into the plane, as shown in figure 9.10, the shaded lateral area is the sum of areas of the triangular lateral faces. Because the area of each face is $\frac{1}{2} \cdot a \cdot \ell$ (each side of the base of the pyramid has length a, and the slant height is ℓ), the combined areas give the lateral area. Thus

$$L = n \cdot \frac{1}{2} \cdot a \cdot \ell$$

$$= \frac{1}{2} \cdot \ell (n \cdot a)$$

$$= \frac{1}{2} \cdot \ell \cdot P$$

EXAMPLE 4   Find the lateral area of a regular pentagonal pyramid if sides of the base measure 8 cm and the lateral edges measure 10 cm each.

Solution   For the triangular lateral face (figure 9.11b), we find the length of the slant height by applying the Pythagorean Theorem:

$$4^2 + \ell^2 = 10^2$$
$$\ell^2 = 84$$
$$\ell = \sqrt{84} = \sqrt{4 \cdot 21} = \sqrt{4} \cdot \sqrt{21} = 2\sqrt{21}$$

Now $L = \frac{1}{2} \cdot \ell \cdot P$ becomes $L = \frac{1}{2} \cdot 2\sqrt{21} \cdot 40 = 40\sqrt{21}$ cm^2.   ▲

THEOREM 9.1.4   The total area (surface area) T of a pyramid whose lateral area is L and whose base has area B is given by $T = L + B$.

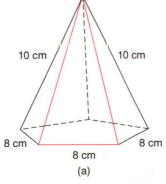

Figure 9.11

EXAMPLE 5

Find the total area of a regular square pyramid having base edges of length 4 ft and lateral edges of length 6 ft.

Solution

First we find the length of the slant height. (See figure 9.12b.)

$$\ell^2 + 2^2 = 6^2$$
$$\ell^2 + 4 = 36$$
$$\ell^2 = 32$$
$$\ell = \sqrt{32} = \sqrt{16 \cdot 2} = \sqrt{16} \cdot \sqrt{2} = 4\sqrt{2}$$

The lateral area is $L = \dfrac{1}{2} \cdot \ell \cdot P$. Therefore

$$L = \frac{1}{2} \cdot 4\sqrt{2}(16) = 32\sqrt{2} \text{ ft}^2$$

Because the area of the square base is 16 ft², the total area is

$$T = 16 + 32\sqrt{2}$$
$$= 16(1 + 2\sqrt{2}) \text{ ft}^2 \qquad \blacktriangle$$

In Example 5, the pyramid was described as a regular square pyramid rather than as a square pyramid. That is, a square pyramid could have a square base and not necessarily have congruent lateral edges, as shown in figure 9.13b.

The final theorem in this section is presented without any attempt to construct the proof. In an advanced course such as calculus, the statement can be proved. The factor "one-third" in the formula for the volume of a pyramid may seem a strange choice, but it provides exact results.

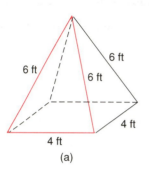

6 ft 6 ft
6 ft
4 ft
4 ft
(a)

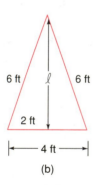

6 ft ℓ 6 ft
2 ft
4 ft
(b)

Figure 9.12

THEOREM 9.1.5   The volume *V* of a pyramid having a base area *B* and an altitude of length *h* is given by

$$V = \frac{1}{3} \cdot B \cdot h$$

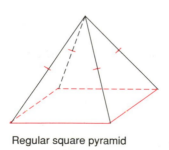

Regular square pyramid
(a)

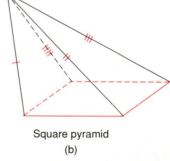

Square pyramid
(b)

Figure 9.13

EXAMPLE 6   Find the volume of the regular square pyramid in Example 5.

Solution   The length of the altitude (of the pyramid) is represented by h, which is determined as follows.

First we see that this altitude meets the diagonals of the square base at their common midpoint (figure 9.14b). Because each diagonal has the length $4\sqrt{2}$ ft, by the 45-45-90 relationship, we have a right triangle whose legs are of length $2\sqrt{2}$ ft and h, while the hypotenuse has length 6 ft (the length of the lateral edge).

Applying the Pythagorean Theorem (figure 9.14c), we have

$$h^2 + (2\sqrt{2})^2 = 6^2$$
$$h^2 + 8 = 36$$
$$h^2 = 28$$
$$h = \sqrt{28} = \sqrt{4 \cdot 7} = \sqrt{4} \cdot \sqrt{7} = 2\sqrt{7}$$

Now we have

$$V = \frac{1}{3} \cdot B \cdot h$$
$$= \frac{1}{3}(16)(2\sqrt{7})$$
$$= \frac{32}{3}\sqrt{7} \text{ ft}^3$$

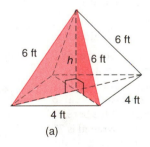
(a)

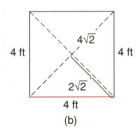
(b)

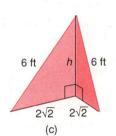
(c)

Figure 9.14

9.1 EXERCISES

1. Consider the solid shown in the accompanying drawing.

  (a) Is it a prism or a pyramid?
  (b) Is it right or oblique?
  (c) What type of base(s) does the solid have?
  (d) Name the type of solid shown.
  (e) What type of figure is each lateral face?

2. Quadrilateral *ABCD* is a square in the solid shown in the accompanying drawing.

  (a) Name the vertex of the pyramid.

  (b) Name the lateral edges.
  (c) Name the lateral faces.
  (d) Name the base.
  (e) Name the base edges.
  (f) Is the solid shown a regular square pyramid?

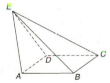

3. A solid is an octagonal prism.

  (a) How many vertices does it have?
  (b) How many lateral edges does it have?
  (c) How many base edges are there in all?

4. A solid is a pentagonal pyramid.

   (a) How many vertices does it have?
   (b) How many lateral edges does it have?
   (c) How many base edges are there in all?

5. Generalize the results found in Exercises 3 and 4 by completing the following table. Assume that the number of sides in each base is *n*.

	Prism	Pyramid
Number of vertices		
Number of lateral edges		
Number of base edges		

6. Find the lateral area *L* of the right prism shown in the accompanying drawing.

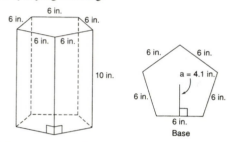

7. Find the total area *T* of the same prism. (*Hint:* The bases are regular pentagons.)

8. Find the volume *V* of the same prism. (*Hint:* The bases are regular pentagons.)

9. Find the lateral area *L* of the right prism shown in the accompanying drawing.

10. Find the total area *T* of the same prism.

11. Find the volume *V* of the same prism.

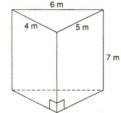

12. Find the lateral area *L* of the regular pyramid shown in the accompanying drawing.

13. Find the total area *T* of the same pyramid.

14. Find the volume *V* of the same pyramid.

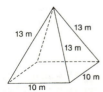

15. Find the lateral area *L* of the regular hexagonal pyramid shown in the accompanying drawing.

16. Find the total area *T* of the same pyramid.

17. Find the volume *V* of the same pyramid.

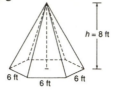

18. A **regular tetrahedron** is a pyramid whose lateral faces and base are congruent equilateral triangles. Given a regular tetrahedron whose edges are 6 cm each, find the length of the

    (a) Slant height ℓ of any face
    (b) Altitude *h* of the pyramid

19. The measures of the sides of the square base of a box are twice the measure of the height of the box. If the volume of the box is 108 in.³, find the dimensions of the box.

20. For a given box, the height measures 4 m. If the length of the rectangular base is 2 m greater than the width of the base and the lateral area *L* is 96 m², find the dimensions of the box.

21. For the box shown in the accompanying drawing, the total area is 94 cm². Determine the value of *x*.

22. For the same box, the volume is 252 in.³. Find the value of *x*.

23. For the regular square pyramid shown in the accompanying drawing, suppose that the altitude has a measure equal to that of the edges of the base. If the volume of the pyramid is 72 in.³, find the total area of the pyramid.

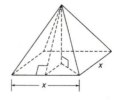

24. For the same regular square pyramid, the slant height of each lateral face has a measure equal to that of the edges of the base. If the lateral area is 200 in.², find the volume of the pyramid.

25. The box shown in the accompanying drawing is to be constructed of materials that cost 1 cent per square inch for the lateral surface and 2 cents per square inch for the bases. What is the total cost of constructing the box?

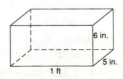

26. A box that is 2 in. by 8 in. by 10 in. contains 16 oz of cereal. What is the volume of a box used for the larger 20-oz size of cereal? Assume that a proportionality exists between weight and volume.

27. The church steeple shown needs to be reshingled. With its measurements as indicated, what is the amount of surface area to be reshingled?

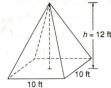

In Exercises 28 and 29, provide a paragraph proof for each claim.

28. The total area T of a right rectangular prism with length ℓ, width w, and height h is given by the formula $T = 2(\ell w + \ell h + wh)$.

29. For the cube whose edges are of length e, the volume is given by $V = e^3$, and the total surface area is given by $T = 6e^2$.

9.2 CYLINDERS AND CONES

Consider the solids in figure 9.15, in which congruent circles lie in parallel planes. If all segments such as $\overline{XX'}$, $\overline{YY'}$, and $\overline{ZZ'}$ are parallel on the left and $\overline{AA'}$, $\overline{BB'}$, and $\overline{CC'}$ are parallel on the right, these solids are known as **circular cylinders.** Because $\overline{XX'}$ is not perpendicular to planes P and P', the solid to the left is an **oblique circular cylinder.** With $\overline{AA'}$ perpendicular to P and P', the solid on the right is a **right circular cylinder.** The distance between planes P and P' is the measure h of the **altitude** of the cylinders. The congruent circles are the **bases** of each cylinder.

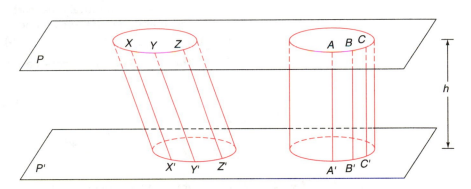

Figure 9.15

As with the right circular cylinder shown in figure 9.16, the parallel planes are generally not pictured. The segment joining the centers of the two circular bases is the **axis** of the cylinder. An easy test to see if the cylinder is a right circular cylinder is that its axis must be perpendicular to the planes of the two circular bases. In that case, the length of the axis equals h, the length of the altitude of the cylinder.

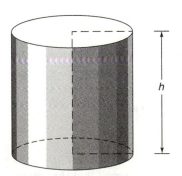

Figure 9.16

> **THEOREM 9.2.1** The lateral area L of a right circular cylinder whose altitude has length h and whose circular bases have radii of length r is given by $L = 2\pi \cdot r \cdot h$.

Think of the right circular cylinder pictured in figure 9.17a as a tin can whose circular bases are the lids of the can and whose lateral surface is the label of the can. If the label were sliced downward by a perpendicular line between the planes, removed, and rolled out flat, it would be rectangular in shape. As shown in figure 9.17b, the dimensions of that rectangle are a length equal to the circumference of the circular base and a width equal to the height of the cylinder. Thus the lateral area is given by $A = b \cdot h$, which becomes $L = (2\pi \cdot r)h$.

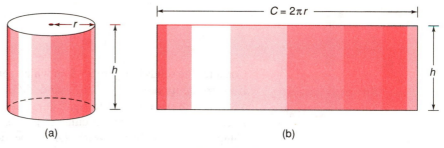

Figure 9.17

THEOREM 9.2.2   The total area (surface area) T of a right circular cylinder whose altitude measures h and whose circular base has a radius of measure r is given by $T = 2\pi r(r + h)$.

Note: This formula may be more easily understood and recalled in the form

$$T = 2\pi r^2 + 2\pi rh$$

EXAMPLE 1

For the right circular cylinder shown in figure 9.18, find the
(a) Exact lateral area L
(b) Exact surface area T

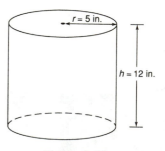

Solution

(a) $L = 2\pi rh$
  $= 2 \cdot \pi \cdot 5 \cdot 12$
  $= 120\pi$ in.^{2}

(b) $T = 2\pi r(r + h)$
  $= 2 \cdot \pi \cdot 5(5 + 12)$
  $= 10\pi(17)$
  $= 170\pi$ in.^{2}

Figure 9.18

▲

In considering the volume of a right circular cylinder, recall that the volume of a prism is given by $V = B \cdot h$, where B is the area of the base. Suppose that

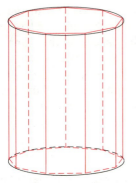

Figure 9.19

the prism is regular and that the number of sides in the base becomes larger and larger. (See figure 9.19.) Thus the base approaches a circle in this limiting process. The area of the polygonal base also approaches the area of the circle, while the volume of the prism approaches that of the right circular cylinder. Our conclusion is stated without proof in the following theorem.

> **THEOREM 9.2.3**   The volume *V* of a right circular cylinder whose height measures *h* and whose radius of the circular base measures *r* is given by $V = \pi r^2 \cdot h$.

EXAMPLE 2   Find the approximate volume of the right circular cylinder shown in figure 9.20, if $d = 4$ cm and $h = 3.5$ cm. Let $\pi \approx \dfrac{22}{7}$.

Solution
$$V = \pi r^2 \cdot h$$
$$= \frac{22}{7} \cdot 2^2(3.5)$$
$$= \frac{22}{\cancel{7}} \cdot \cancel{4}^{\,2} \cdot \frac{7}{\cancel{2}}$$
$$= 44 \text{ cm}^3$$

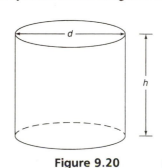

Figure 9.20   ▲

EXAMPLE 3   In the right circular cylinder shown in figure 9.20, the height equals the diameter of the circular base. If the exact volume is 128π in.³, find the lateral area *L* of the cylinder.

Solution
$$h = 2r$$
so
$$V = \pi r^2 \cdot h$$
becomes
$$= \pi r^2(2r)$$
$$= 2\pi r^3$$

Then $2\pi r^3 = 128\pi$; and dividing by 2π, we get
$$r^3 = 64$$
$$r = 4$$
$$h = 8$$
$$L = 2\pi r \cdot h$$
$$= 2 \cdot \pi \cdot 4 \cdot 8$$
$$= 64\pi \text{ in.}^2$$
▲

Consider point *P,* which lies outside the plane containing circle *O.* A surface known as a **cone** results when line segments are drawn from *P* to points on the circle. However, if *P* is joined to points on the circle as well as to points in the interior of the circle, a solid is formed. If $\overline{PO}$ is not perpendicular to the plane of circle *O* in figure 9.21, the cone is an **oblique circular cone.**

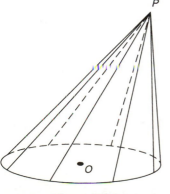

Figure 9.21

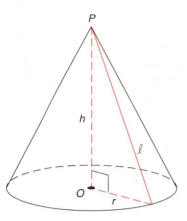

Figure 9.22

In figure 9.22, point P is the **vertex** of the cone while circle O is the **base.** The segment $\overline{PO}$, which joins the vertex to the center of the circular base, is the **axis** of the cone. If the axis is perpendicular to the base or to the plane of the base, the cone is a **right circular cone.** In any cone, the perpendicular segment from the vertex to the plane of the base is the **altitude** of the cone. In a right circular cone, the length h of the altitude equals the length of the axis. For a right circular cone—and only for this type of cone—the segment that joins the vertex to a point on the circle is the **slant height** of the cone; we will denote the length of the slant height by ℓ.

Recall now that the lateral area for the regular pyramid is given by $L = \frac{1}{2} \cdot \ell \cdot P$. For a right circular cone, consider several inscribed pyramids. When the number of sides of the polygonal base grows larger (figure 9.23), the perimeter of that polygon approaches the circumference of the circle as a limit. In addition, the slant height of the congruent triangular faces approaches that of the slant height of the cone. Thus the lateral area of the right circular cone is given by

$$L = \frac{1}{2} \cdot \ell \cdot C$$

in which C is the circumference of the base; the fact that $C = 2\pi r$ leads to the following theorem.

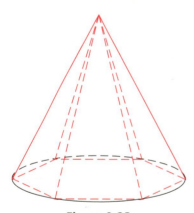

Figure 9.23

> **THEOREM 9.2.4**   For a right circular cone with a slant height of length ℓ and a radius of the base of length r, the lateral area L is given by $L = \pi \cdot r \cdot \ell$.

> **THEOREM 9.2.5**   For a right circular cone with a slant height of length ℓ and a radius of the base of length r, the total area T is given by $T = \pi \cdot r \cdot (r + \ell)$.
>
> *Note:* This formula may be more easily understood and recalled in the form $T = \pi r^2 + \pi r \ell$.

EXAMPLE 4

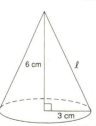

Solution

For the right circular cone for which $r = 3$ cm and $h = 6$ cm, find the
(a) Exact lateral area L
(b) Exact total area T

We need the slant height ℓ for each part, so we apply the Pythagorean Theorem:

$$\ell^2 = r^2 + h^2$$
$$= 3^2 + 6^2$$
$$= 9 + 36 = 45$$
$$\ell = \sqrt{45} = \sqrt{9 \cdot 5} = \sqrt{9} \cdot \sqrt{5} = 3\sqrt{5}$$

(**a**) Now we have

$$L = \pi \cdot r \cdot \ell$$
$$= \pi \cdot 3 \cdot 3\sqrt{5}$$
$$= 9\pi \sqrt{5} \text{ cm}^2$$

(**b**) We also have

$$T = \pi \cdot r \cdot (r + \ell)$$
$$= \pi \cdot 3 \cdot (3 + 3\sqrt{5})$$
$$= (9\pi + 9\pi \sqrt{5}) \text{ cm}^2 \qquad \blacktriangle$$

Recall that the volume of a pyramid is given by the formula $V = \dfrac{1}{3} \cdot B \cdot h$. If we consider a regular pyramid and allow the number of sides to be increased indefinitely, the volume of the pyramid approaches that of a right circular cone. Then the volume of the cone is given by

$$V = \frac{1}{3} \cdot \pi \cdot r^2 \cdot h$$

We state this result as a theorem, without any formal proof.

THEOREM 9.2.6   The volume V of a right circular cone whose height is of length h and whose radius (of the circular base) has length r is given by $V = \dfrac{1}{3} \cdot \pi \cdot r^2 \cdot h$.

EXAMPLE 5   A triangular region having vertices at $(0, 0)$, $(0, 4)$, and $(6, 0)$ is rotated about the x axis of the rectangular coordinate system. Find the volume of the resulting solid.

Solution   The given region is shown in figure 9.24a. When the triangular region is revolved about the x axis, the solid formed is the right circular cone shown in figure 9.24b.

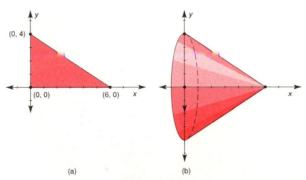

(a)   (b)

Figure 9.24

Because the radius of the circular base measures 4 and the altitude measures 6, we have

$$V = \frac{1}{3} \cdot \pi \cdot r^2 \cdot h$$

$$= \frac{1}{3} \cdot \pi \cdot 4^2 \cdot 6 = 32\pi \text{ units}^3 \qquad \blacktriangle$$

We close by considering two separate issues. The first issue involves the solid that we have described as an oblique circular cylinder. In a later course, this solid will be referred to as an "elliptic cylinder" because any cross section that results from the intersection of the solid and a plane perpendicular to the axis of the cylinder is a figure known as an **ellipse.** (See figure 9.25a and b.)

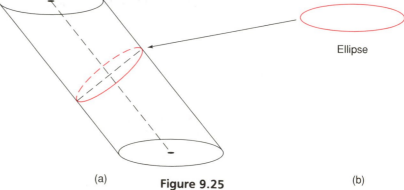

Ellipse

(a)   **Figure 9.25**   (b)

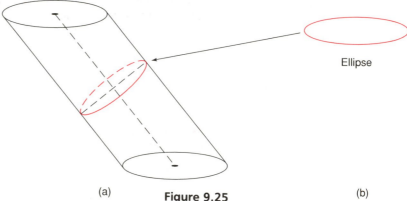

(a)

(b)

Figure 9.26

An elliptic cylinder may have bases that are ellipses. The intersection of the cylinder and a plane perpendicular to the axis of the cylinder will always be an ellipse. (See figure 9.26a and b.)

Our second concern is with volume. For the oblique circular cylinder, the same formula is used to calculate volume as was used for the right circular cylinder. To see why the formula $V = \pi \cdot r^2 \cdot h$ is still appropriate, consider the stacks of pancakes in figure 9.27a and b. The same volume exists regardless of whether the stack is oblique or vertical.

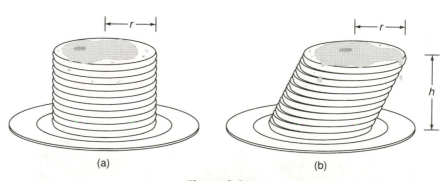

(a)   (b)

Figure 9.27

9.2 EXERCISES

1. For the right circular cylinder shown in the accompanying drawing, suppose that $r = 5$ in. and $h = 6$ in. Find the exact:

   (a) Lateral area
   (b) Total area
   (c) Volume

2. Suppose that $r = 12$ cm and $h = 15$ cm in the same right circular cylinder. Find the exact:

   (a) Lateral area
   (b) Total area
   (c) Volume

3. The tin can in the accompanying drawing has the dimensions shown. Estimate the number of square inches of tin required for its construction. (*Hint:* Include the lid and base in the result. Let $\pi \approx 3.14$.)

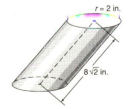

4. What is the volume of the same tin can? If it contains 16 oz of green beans, what is the volume of the can used for 20 oz of green beans? (Assume a proportionality between weight and volume. Let $\pi \approx 3.14$.)

5. If the exact volume of a right circular cylinder is 200π cm³ and its altitude measures 8 cm, what is the measure of the radius of the circular base?

6. Suppose that the volume of an aluminum can is to be 9π in³. Find the dimensions of the can if the diameter of the base is three-fourths the length of the altitude.

7. For an aluminum can, the lateral surface area is 12π in². If the length of the altitude is 1 in. greater than the length of the radius of the circular base, find the dimensions of the cylinder.

8. Find the altitude of a storage tank in the shape of a right circular cylinder that has a circumference measuring 6π m and that has a volume measuring 81π m³.

9. Find the volume of the oblique circular cylinder shown in the accompanying drawing. Notice that the axis meets the plane of the base to form a 45° angle.

10. A cylindrical orange juice container has metal bases of radius 1 in. and a cardboard lateral surface 3 in. high. If the cost of the metal used is 0.5 cents per square inch

and the cost of the cardboard is 0.2 cents per square inch, what is the approximate cost of constructing one container? Let $\pi \approx 3.14$.

11. The oblique circular cone shown in the accompanying drawing has an altitude and a diameter of base that are each of length 6 cm. The line segment joining the vertex to the center of the base is the **axis** of the cone. What is the length of the axis?

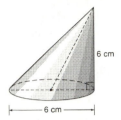

12. For the right circular cone shown in the accompanying drawing, $h = 6$ m and $r = 4$ m. Find the exact:

    (a) Lateral area
    (b) Total area
    (c) Volume

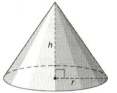

13. Suppose that for the same right circular cone, $h = 7$ in. and $r = 6$ in. Find the approximate:

    (a) Lateral area
    (b) Total area
    (c) Volume

    $\left(\text{Hint: Let } \pi \approx \dfrac{22}{7}. \right)$

14. The accompanying teepee has a circular floor with a radius equal to 6 ft and a height of 15 ft. Using $\pi \approx 3.14$, find the volume of the enclosure.

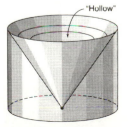

15. If a right circular cone has a circular base with a diameter of length 10 cm and a volume of 100π cm³, find its lateral area.

16. A right circular cone has a slant height of 12 ft and a lateral area of 96π ft². Find its volume.

17. A solid is formed by cutting a conical section away from a right circular cylinder. If the radius measures 6 in. and the altitude measures 8 in., what is the volume of the resulting solid?

18. Find the exact volume of the right circular cone that results when the triangular region with vertices at $(0, 0)$, $(5, 0)$, and $(0, 9)$ is rotated about the

(a) x axis   (b) y axis

19. Find the exact volume of the solid that results when the triangular region with vertices at $(0, 0)$, $(6, 0)$, and $(6, 4)$ is rotated about the

(a) x axis   (b) y axis

20. Find the exact volume of the solid formed when the rectangular region with vertices at $(0, 0)$, $(6, 0)$, $(6, 4)$, and $(0, 4)$ is revolved about the

(a) x axis   (b) y axis

21. Find the exact volume of the solid formed when the region bounded in Quadrant I by the lines $x = 9$ and $y = 5$ is revolved about the

(a) x axis   (b) y axis

22. Find the exact lateral surface of each solid in Exercise 21.

23. Find the volume of the solid formed when the triangular region having vertices at $(2, 0)$, $(4, 0)$, and $(2, 4)$ is rotated about the y axis. (*Hint:* Use the accompanying drawing.)

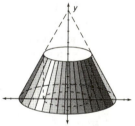

In Exercises 24 to 25, give a paragraph proof for each claim.

24. The total area T of a right circular cylinder whose altitude is of length h and whose circular base has a radius of length r is given by

$$T = 2 \cdot \pi \cdot r \cdot (r + h)$$

25. The volume V of a washer that has an inside radius of length r and an outside radius of length R and an altitude of measure h is given by

$$V = \pi \cdot h \cdot (R + r)(R - r)$$

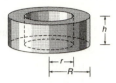

26. For a right circular cone, the slant height has a measure equal to twice that of the radius of the base. If the total area of the cone is 48π in.², what are the dimensions of the cone?

27. For a right circular cone, the ratio of the slant height to the radius is 5:3. If the volume of the cone is 96π in.³, find the lateral area of the cone.

28. If the radius and height of a right circular cylinder are both doubled to form a larger cylinder, what is the ratio of the volume of the larger cylinder to the volume of the smaller? (*Note:* The two cylinders are said to be "similar.")

29. For the two similar cylinders in Exercise 28, what is the ratio of the lateral area of the larger cylinder to that of the smaller cylinder?

30. For a right circular cone, the dimensions are $r = 6$ cm and $h = 8$ cm. If the radius is doubled while the height is made half as large in forming a new cone, will the volumes be equal?

31. A cylindrical storage tank has a depth of 5 ft and a radius measuring 2 ft. If each cubic foot can hold 7.5 gal of gasoline, what is the total storage capacity of the tank (measured in gallons)?

32. If the tank in Exercise 31 needs to be painted and 1 pt of paint covers 50 ft.², how many pints are needed to paint the exterior of the storage tank?

POLYHEDRONS AND SPHERES

When two planes intersect, the angle formed by two half-planes with a common edge (the line of intersection) is a **dihedral angle.** The angle shown in figure 9.28 is such an angle.

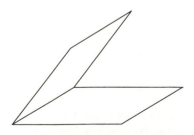

Figure 9.28

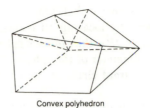

Convex polyhedron

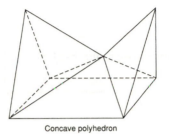

Concave polyhedron

A **polyhedron** is a solid bounded by plane regions. Polygons form the **faces** of the solid, and the segments common to these polygons are the **edges** of the polyhedron. Endpoints of the edges are **vertices** of this type of solid. Because a regular polyhedron is convex, each face determines a half-plane such that all remaining faces lie in that half-plane. The plural of the term *polyhedron* is often given as "polyhedra," although we will use the term "polyhedrons."

> **DEFINITION:** A **regular polyhedron** is a convex polyhedron whose faces are congruent regular polygons arranged in such a way that adjacent faces form congruent dihedral angles.

The five regular polyhedrons are as follows:

1. Regular **tetrahedron,** which has 4 faces
2. Regular **hexahedron** (or **cube**), which has 6 faces
3. Regular **octahedron,** which has 8 faces
4. Regular **dodecahedron,** which has 12 faces
5. Regular **icosahedron,** which has 20 faces

Following are four regular polyhedrons.

Regular Polyhedrons

 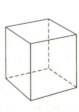 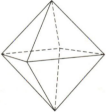 

Tetrahedron   Hexahedron   Octahedron   Dodecahedron

Another type of solid with which you are familiar is the sphere. While the surface of a basketball correctly depicts the sphere, we often use the term **sphere** to mean a solid like a baseball as well.

> In space, the sphere can be described in any of three ways:
>
> 1. A **sphere** is the set of all points at a fixed distance r from a given point O. Point O is known as the **center** of the sphere, even though it is not a part of the spherical surface.
> 2. A **sphere** is the surface determined when a circle is rotated about any of its diameters.
> 3. A **sphere** is the surface that represents the limit of a regular polyhedron whose number of faces is allowed to increase without limit.

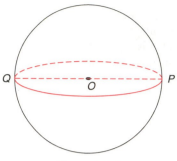

Figure 9.29

Each characterization of a sphere has its advantages, as we shall see.

Figure 9.29 shows a sphere that was produced by locating all points in space that are at distance r from point O. The segment $\overline{OP}$ is the **radius** of sphere O. If points Q, O, and P are collinear, then $\overline{QP}$ is a **diameter** of the sphere. The intersection of a plane and a sphere is known as a **great circle** of the sphere when the plane contains the center of the sphere.

The following theorem is proved in calculus. While the proof is beyond the scope of this textbook, we note that it treats the sphere as a surface of revolution, as described earlier in description 2.

THEOREM 9.3.1   The surface area S of a sphere whose radius has length r is given by $S = 4 \cdot \pi \cdot r^2$.

EXAMPLE 1   Find the surface area of a sphere whose radius has a length of 7 in. Use $\pi \approx \dfrac{22}{7}$.

Solution

$$S = 4 \cdot \pi \cdot r^2$$
$$= 4 \cdot \frac{22}{7} \cdot 7 \cdot 7$$
$$= 616 \text{ in.}^2 \qquad \blacktriangle$$

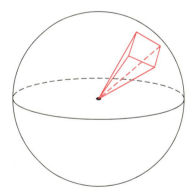

Figure 9.30

The third description of a sphere allows us to reason toward a formula for its volume. Suppose that we consider the sphere to be the limit of a polyhedron whose number of faces is growing without limit. Then the center of the sphere becomes the vertex of a pyramid whose altitude has a length approximately equal to the radius of the sphere it approaches.

We can find the sum of the volumes of all the resulting pyramids and seek the limit of that sum as the volume of the sphere. Figure 9.30 shows one of these pyramids. This sum is given by

$$\frac{1}{3}B_1 h + \frac{1}{3}B_2 h + \frac{1}{3}B_3 h + \cdots + \frac{1}{3}B_n h,$$

in which n is the number of faces and pyramids. This sum equals

$$\frac{1}{3}h(B_1 + B_2 + B_3 + \cdots + B_n)$$

As n increases, $h \to r$ and $(B_1 + B_2 + B_3 + \cdots + B_n) \to S$, which is the surface area of the sphere. Because the surface area of the sphere is given by $S = 4 \cdot \pi \cdot r^2$, the sum of the volumes of all the pyramids approaches the volume of the sphere as a limit. Thus

$$\frac{1}{3}h(B_1 + B_2 + B_3 + \cdots + B_n) \to \frac{1}{3}r(4 \cdot \pi \cdot r^2)$$

which takes us to the following theorem.

> **THEOREM 9.3.2**   The volume V of a sphere whose radius is of length r is given by $V = \dfrac{4}{3} \cdot \pi \cdot r^3$.

EXAMPLE 2   Find the exact volume of a sphere whose length of radius is 1.5 in.

Solution

$$V = \frac{4}{3} \cdot \pi \cdot r^3$$

$$= \frac{4}{3} \cdot \pi \cdot \frac{3}{2} \cdot \frac{3}{2} \cdot \frac{3}{2}$$

$$= \frac{9\pi}{2} \text{ in.}^3 \qquad \blacktriangle$$

EXAMPLE 3   A spherical propane gas storage tank has a volume of $\dfrac{792}{7}$ ft^3. Using $\pi \approx \dfrac{22}{7}$, find the radius of the sphere.

Solution $V = \dfrac{4}{3} \cdot \pi \cdot r^3$, which becomes $\dfrac{792}{7} = \dfrac{4}{3} \cdot \dfrac{22}{7} \cdot r^3$.

Then $\qquad \dfrac{88}{21}r^3 = \dfrac{792}{7} \rightarrow r^3 = 27 \rightarrow r = 3$

The radius of the tank is 3 ft. $\qquad \blacktriangle$

Just as two concentric circles have the same center but different lengths of radii, so also can two spheres be concentric. This fact is the basis for the following example.

EXAMPLE 4   A child's hollow plastic ball has an inside diameter of 10 in. and is approximately $\dfrac{1}{8}$ in. thick. Approximately how many cubic inches of plastic were needed to construct the ball?

Solution   The volume of plastic used is the difference between the inside volume and outside volume. Where R denotes the outside radius and r denotes the inside radius, $R \approx 5.125$ and $r = 5$.

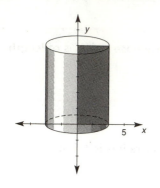

Figure 9.31

$$V = \frac{4}{3} \cdot \pi \cdot R^3 - \frac{4}{3} \cdot \pi \cdot r^3$$

$$\approx \frac{4}{3}(3.14)(5.125)^3 - \frac{4}{3}(3.14)(5)^3$$

$$\approx 563.57 - 523.33 \approx 40.24 \text{ in.}^3$$

The volume of plastic used in construction is approximately 40.24 in.³.   ▲

Let us now recall the notion of a solid of revolution. In figure 9.31, a right circular cylinder is generated when the rectangular region bounded by $(0, 0)$, $(2, 0)$, $(2, 5)$, and $(0, 5)$ is rotated about the y axis.

EXAMPLE 5   Describe the solid that results when the triangular region with vertices at $(0, 0)$, $(4, 0)$, and $(0, 2)$ is rotated about the y axis.

Solution   The resulting solid is a right circular cone whose base has a radius of length 4 and whose altitude has length 2. (See figure 9.32.)   ▲

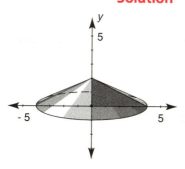

Figure 9.32

It is sometimes easier to define a solid by using the notion of "solid of revolution" than by applying any other method. Suppose that the circle whose equation is

$$(x - 3)^2 + (y - 0)^2 = 1$$

is rotated about the y axis. The solid that results in figure 9.33b has the shape of a doughnut and has the formal name **torus**. Because of the manner in which the solid is generated, the methods of calculus can be used to evaluate the volume or surface area of the torus.

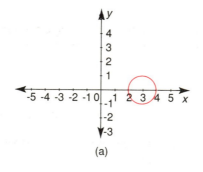

(a)

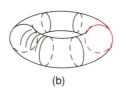

(b)

Figure 9.33

9.3 EXERCISES

1. Which of the two polyhedrons in the accompanying figure is concave? Notice that the interior dihedral angle formed by the planes containing $\triangle EJF$ and $\triangle KJF$ is larger than 180°.

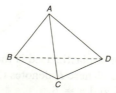

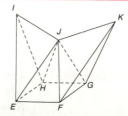

2. For the same figure as in Exercise 1, give the number of faces, vertices, and edges in each polyhedron.

3. Give the number of faces, vertices, and edges in a regular tetrahedron.

4. Give the number of faces, vertices, and edges in a regular hexahedron.

5. In sphere O in the accompanying drawing, the length of radius $\overline{OP}$ is 6 in. Find the length of the chord:

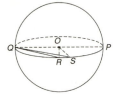

   (a) $\overline{QR}$ if m$\angle QOR = 90°$
   (b) $\overline{QS}$ if m$\angle SOP = 60°$

6. Find the surface area and volume of the sphere in the drawing if $OP = 6$ in. Use $\pi \approx 3.14$.

7. A sphere is inscribed within a right circular cylinder whose altitude and diameter have equal measures.

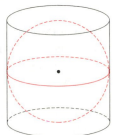

   (a) Find the ratio of the surface area of the cylinder to that of the sphere.
   (b) Find the ratio of the volume of the cylinder to that of the sphere.

8. Given that a right circular cylinder is inscribed within a sphere, what is the least possible volume of the cylinder? (*Hint:* Consider various lengths for radius and altitude.)

9. In calculus, it can be shown that the largest possible volume for the inscribed right circular cylinder in Exercise 8 occurs when its altitude has a length equal to the diameter of the circular base. Find the length of the radius and the altitude of the cylinder of greatest volume if the radius of the sphere is 6 in.

10. Given that a *regular* polyhedron of n faces is inscribed in a sphere of radius 6 in., find the maximum (largest) possible volume for the polyhedron. $\left(\text{Note: Let } \pi \approx \frac{22}{7}. \right)$

11. A right circular cone is inscribed in a sphere, as shown in the accompanying drawing. If the slant height of the cone has a length equal to that of its diameter, find the length of the

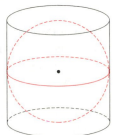

   (a) Radius of the base of the cone
   (b) Altitude of the cone

The radius of the sphere has a length of 6 in.

12. A sphere is inscribed in a cone whose slant height has a length equal to the diameter of its base. What is the length of the radius of the sphere if the slant height and the diameter both measure 12 cm?

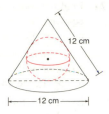

In Exercises 13 to 16, let $\pi \approx \dfrac{22}{7}$.

13. For a sphere whose radius has length 3 m, find the
   (a) Surface area
   (b) Volume

14. For a sphere whose radius has length 7 cm, find the
   (a) Surface area
   (b) Volume

15. A sphere has a volume equal to $\dfrac{99}{7}$ in.^{3}. Determine the length of the radius of the sphere.

16. A sphere has a surface area equal to 154 in.^{2}. Determine the length of the radius of the sphere.

17. The spherical storage tank described in Example 3 had a length of radius of 3 ft. Since it needs to be painted, find its surface area. Also determine the number of pints of rust-proofing paint needed to paint the tank if 1 pt covers approximately 40 ft^2. Use $\pi \approx 3.14$.

18. An observatory has the shape of a right circular cylinder surmounted by a hemisphere, as shown in the accompanying drawing. If the radius of the cylinder is 14 ft and the altitude measures 30 ft, what is the surface area of the observatory? If 1 gal of paint covers 300 ft^2, how many gallons are needed to paint the surface if it needs two coats? (Use $\pi \approx 3.14$.)

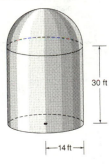

19. A leather soccer ball has an inside diameter of 8.5 in. and a thickness of 0.1 in. Find the volume of leather needed for its construction. (Use $\pi \approx 3.14$.)

20. An ice cream cone is completely filled with ice cream. What is the volume of the ice cream? (Use $\pi \approx 3.14$.)

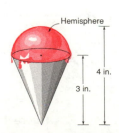

21. Find the volume of the solid formed when a circle whose equation is $x^2 + y^2 = 9$ is rotated about the x axis. Leave an *exact* answer.

22. Find the volume of the solid formed when a circle whose equation is $(x - 2)^2 + y^2 = 4$ is rotated about the x axis. Leave an *exact* answer.

*23. Sketch the torus that results when a circle whose equation is $(x - 0)^2 + (y - 4)^2 = 1$ is rotated about the x axis.

*24. Sketch the solid that results when a circle whose equation is $(x - 0)^2 + (y - 1)^2 = 1$ is rotated about the x axis.

25. Explain how the following formula used in Example 4 was obtained: $V = \dfrac{4}{3} \cdot \pi \cdot R^3 - \dfrac{4}{3} \cdot \pi \cdot r^3$.

26. Derive a formula for the total surface area of the hollow-core sphere shown in the accompanying drawing. (*Note:* Include both interior and exterior surface areas.)

9.4
A THREE-DIMENSIONAL COORDINATE SYSTEM

The French mathematician, René Descartes, is given a great deal of the credit for the development of analytic geometry in the plane. Descartes, who lived from 1596 to 1650, invented the rectangular or Cartesian coordinate system for locating points (x, y) in the plane.

In three-dimensional space, an object can be located by three numbers: latitude, longitude, and altitude. In mathematics, we extend Descartes' two-dimensional system by including a third axis perpendicular to each of the x and y axes. In the resulting **Cartesian space,** we describe a point by the coordinates $x, y,$ and z. Each point in space, such as $(4, 5, 3)$, names the x coordinate, the y coordinate, and the z coordinate, respectively.

While the x axis of the three-dimensional Cartesian system is shown in figure 9.34 to move toward and away from us, the y axis is horizontal. The new axis, the z axis, is vertical. The following chart suggests how to locate points that have the form (x, y, z).

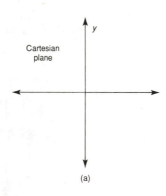

Cartesian plane

(a)

POINT LOCATION			
	x axis	*y* axis	*z* axis
Positive direction	Forward	Right	Upward
Negative direction	Back	Left	Down

In figure 9.35a and b, we locate the point $(4, 5, 3)$ and the point $(2, -3, -1)$. We may better understand the difficulty with this type of point plotting if we recognize that we are representing a three-dimensional system on a two-dimensional surface (like a sheet of paper).

It can be shown (by the Pythagorean Theorem) that the distance between the two points (x_1, y_1, z_1) and (x_2, y_2, z_2) is merely an extension of the Distance Formula in Chapter 8.

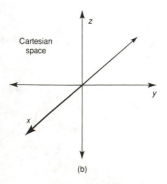

Cartesian space

(b)

Figure 9.34

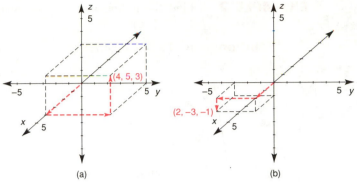

Figure 9.35

THEOREM 9.4.1 The distance between the two points $P_1 (x_1, y_1, z_1)$ and $P_2 (x_2, y_2, z_2)$ is given by

$$d = \sqrt{(x_2 - x_1)^2 + (y_2 - y_1)^2 + (z_2 - z_1)^2}$$

EXAMPLE 1 Find the distance between the points $(5, -7, 2)$ and $(2, 5, 6)$.

Solution Designating points is arbitrary, so we let $P_1 = (5, -7, 2)$ while $P_2 = (2, 5, 6)$. Then Theorem 9.4.1 leads to

$$\begin{aligned} d &= \sqrt{(2 - 5)^2 + (5 - [-7])^2 + (6 - 2)^2} \\ &= \sqrt{(-3)^2 + 12^2 + 4^2} \\ &= \sqrt{9 + 144 + 16} \\ &= \sqrt{169} = 13 \end{aligned}$$

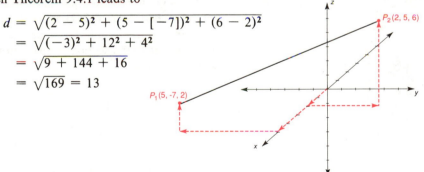

Another formula that can be extended to three dimensions is the Midpoint Formula. The midpoint, in three dimensions, is $M = (x_M, y_M, z_M)$.

THEOREM 9.4.2 The midpoint of the segment joining $P_1 (x_1, y_1, z_1)$ and $P_2 (x_2, y_2, z_2)$ is given by

$$(x_M, y_M, z_M) = \left(\frac{x_1 + x_2}{2}, \frac{y_1 + y_2}{2}, \frac{z_1 + z_2}{2} \right)$$

Equivalently,

$$M = \left(\frac{x_1 + x_2}{2}, \frac{y_1 + y_2}{2}, \frac{z_1 + z_2}{2} \right)$$

EXAMPLE 2   Find the midpoint of the segment that joins the points P_1 (5, −7, 2) and P_2 (2, 5, 6).

Solution   By Theorem 9.4.2,

$$M = \left(\frac{5 + 2}{2}, \frac{-7 + 5}{2}, \frac{2 + 6}{2}\right)$$

$$= \left(\frac{7}{2}, \frac{-2}{2}, \frac{8}{2}\right)$$

$$= (3.5, -1, 4) \qquad \blacktriangle$$

While the plane analytic and solid analytic geometries resemble each other in many ways, they also differ significantly. For instance, the Slope Formula (which involves a quotient) cannot be extended to three variables. Another important difference is that the linear equation in three variables, $Ax + By + Cz = D$, has a plane for its graph, while the graph of $Ax + By = C$ was a line in the xy coordinate system.

EXAMPLE 3   Sketch the graph of the equation $x + 2y + 3z = 12$ in the xyz coordinate system.

Solution   Some points on the graph of the equation $x + 2y + 3z = 12$ are (12, 0, 0), (0, 6, 0), and (0, 0, 4); these points are known, respectively, as the **x intercept,** the **y intercept,** and the **z intercept** of the graph. One more solution, (2, 2, 2), is shown in figure 9.36. This solution is valid because substitution into the given equation leads to $2 + 2(2) + 3(2) = 12$, which is true.

Because the graph is a plane and is infinite, we show only the portion that is in Octant I (the section of space in which all three coordinates of the point are positive). $\qquad \blacktriangle$

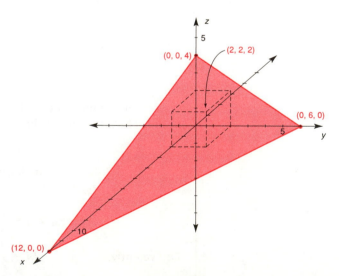

Figure 9.36

The Distance Formula for Cartesian space is used to determine the equation of a sphere. A sphere is the set of all points (x, y, z) that are at a given distance r from a given point (h, k, ℓ), the center of the sphere. When we use the Distance Formula to determine the equation of this sphere, each point (x, y, z) must satisfy the equation

$$\sqrt{(x - h)^2 + (y - k)^2 + (z - \ell)^2} = r$$

Squaring each side of the equation leads to the result

$$(x - h)^2 + (y - k)^2 + (z - \ell)^2 = r^2$$

and to the following theorem:

THEOREM 9.4.3 The sphere that has its center at (h, k, ℓ) and a radius length of r (where $r > 0$) has the equation

$$(x - h)^2 + (y - k)^2 + (z - \ell)^2 = r^2$$

Note: This equation is often simplified to the form

$$x^2 + y^2 + z^2 + Dx + Ey + Fz + G = 0$$

often called the **general form of the equation of a sphere.**

EXAMPLE 4 Find the equation of the sphere that has its center at $(2, -3, 4)$ and has a radius of length $r = 5$.

Solution

$$(x - h)^2 + (y - k)^2 + (z - \ell)^2 = r^2$$
$$(x - 2)^2 + (y - [-3])^2 + (z - 4)^2 = 5^2$$
$$(x - 2)^2 + (y + 3)^2 + (z - 4)^2 = 5^2$$
$$x^2 - 4x + 4 + y^2 + 6y + 9 + z^2 - 8z + 16 = 25$$
$$x^2 + y^2 + z^2 - 4x + 6y - 8z + 4 = 0 \qquad \blacktriangle$$

By the method of completing the squares, we can determine which sphere, if any, is represented by an equation of the form

$$x^2 + y^2 + z^2 + Dx + Ey + Fz + G = 0$$

Recall that the expression $(x^2 + Dx)$ becomes the square of a binomial when $\left(\dfrac{1}{2}D\right)^2$ or $\dfrac{1}{4}D^2$ is added to it.

EXAMPLE 5 Determine which sphere, if any, is represented by the equation

$$x^2 + y^2 + z^2 + 2x - 4y - 2z - 19 = 0$$

Solution We begin by reordering and regrouping terms so that squares can be completed:

$$(x^2 + 2x \qquad) + (y^2 - 4y \qquad) + (z^2 - 2z \qquad) = 19$$

To complete squares, we need

$$\left(\frac{1}{2} \cdot 2\right)^2 \text{ or } 1 \text{ for the } x \text{ term}$$

$$\left[\frac{1}{2}(-4)\right]^2 \text{ or } 4 \text{ for the } y \text{ term}$$

$$\left[\frac{1}{2}(-2)\right]^2 \text{ or } 1 \text{ for the } z \text{ term}$$

> ▲ **WARNING:** *Consider the general form for the equation of the sphere. If the completing of squares leads to an equation in which the value of r² is 0, then the graph of the equation is one point (h, k, ℓ). If r² < 0, then there is no locus (graph).*

Completing squares, we get

$$(x^2 + 2x + 1) + (y^2 - 4y + 4) + (z^2 - 2z + 1) = 19 + 1 + 4 + 1$$
$$(x + 1)^2 + (y - 2)^2 + (z - 1)^2 = 25$$
$$[x - (-1)]^2 + (y - 2)^2 + (z - 1)^2 = 5^2$$

Then the sphere has its center at $(-1, 2, 1)$ and a radius of length $r = 5$. ▲

EXAMPLE 6   For the spheres $x^2 + y^2 + z^2 = 4$ and $x^2 + y^2 + z^2 = 16$, use a ratio to compare
(a) Surface areas (b) volumes

Solution   The first sphere is $(x - 0)^2 + (y - 0)^2 + (z - 0)^2 = 2^2$; it has a radius of length 2. The second sphere is $(x - 0)^2 + (y - 0)^2 + (z - 0)^2 = 4^2$; it has a radius of length 4.

(a) For surface area of a sphere, we have $S = 4\pi r^2$. Denoting the respective surface areas by S_1 and S_2, we have

$$\frac{S_1}{S_2} = \frac{4\pi \cdot 2^2}{4\pi \cdot 4^2} = \frac{16\pi}{64\pi} = \frac{1}{4}$$

Note: In general, the surface areas of two spheres have a ratio equal to the square of the ratio of their corresponding lengths of radii, that is,

$$\frac{S_1}{S_2} = \left(\frac{r_1}{r_2}\right)^2$$

(b) For the volume of a sphere, we have $V = \frac{4}{3}\pi r^3$. Denoting respective volumes by V_1 and V_2, we have

$$\frac{V_1}{V_2} = \frac{\frac{4}{3}\pi \cdot 2^3}{\frac{4}{3}\pi \cdot 4^3} = \frac{\frac{32}{3}\pi}{\frac{256}{3}\pi} = \frac{32}{3} \cdot \frac{3}{256} = \frac{1}{8}$$

Note: In general, the volumes of two spheres have a ratio equal to the cube of the ratio of their corresponding lengths of radii; that is,

$$\frac{V_1}{V_2} = \left(\frac{r_1}{r_2}\right)^3$$

▲

9.4 EXERCISES

1. Identify each point A, B, C, and D shown in the accompanying drawing, by naming an ordered triple (x, y, z).

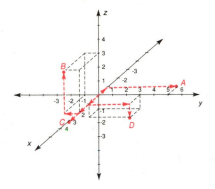

2. Plot each point in a three-dimensional coordinate system: A (2, 0, 0), B (0, 0, 2), C (2, 3, 4), and D (0, −2, 5).

In Exercises 3 to 6, find the distance between points P_1 and P_2.

3. P_1 (0, 0, 0) and P_2 (1, 2, 4)

4. P_1 (−1, 2, 3) and P_2 (2, −2, −9)

5. P_1 (1, −1, 1) and P_2 (2, 1, 3)

6. P_1 (1, 0, 2) and P_2 (3, 4, −1)

For Exercises 7 to 10, find the midpoint M of the line segment that joins the points named.

7. P_1 (0, 0, 0) and P_2 (−6, 4, 14)

8. P_1 (−1, 2, 3) and P_2 (3, 6, 6)

9. P_1 (2, −1, 1) and P_2 (5, 7, −7)

10. P_1 (1, 0, 2) and P_2 (0, 5, 9)

In Exercises 11 to 14, draw the graph of each equation.

11. $x + 2y + z = 6$

12. $x + 2y + 3z = 12$

13. $2x − y + z = 10$

14. $2x − 2y + z = 8$

In Exercises 15 to 18, find the equation of the sphere that has center C and a radius of length r. Leave the answer in the general form

$$x^2 + y^2 + z^2 + Dx + Ey + Fz + G = 0$$

15. C (0, 0, 0); $r = 4$

16. C (−1, 2, 0); $r = 5$

17. C (2, 3, −2); $r = 3$

18. C (4, 2, 3); $r = \sqrt{15}$

19. Which of the following points lie in the plane that has the equation $2x + 3y + 4z = 24$?
$(0, 0, 6)$ $(0, 3, 4)$ $(6, 0, 3)$ $(4, 4, 1)$ $(−2, 6, 3)$

20. Which of the following planes contains the three points $(8, −5, 1)$, $(3, 0, −4)$, and $(−2, 12, −2)$?

$$2x + y − z = 10 \qquad x + 2y + z = −1$$

21. Which of the following points lie on the sphere whose equation is $x^2 + y^2 + z^2 = 9$?
$(0, 0, 3)$ $(1, 2, −2)$ $(1, 3, −2)$ $(0, 0, −3)$

22. If M (2, −3, 4) is the midpoint of $\overline{AB}$, in which A (5, 1, 2) is a known endpoint, find the remaining endpoint B.

23. Without graphing, find the three points at which the plane that has the equation $2x − 3y + 4z = −120$ intersects the axes.

24. Find the points at which the sphere $x^2 + y^2 + z^2 = 16$ intersects the axes.

In Exercises 25 to 32, determine whether the graph of the equation is a plane, a sphere, or a point, or has no locus. If the graph is a sphere, name its center and find the length of its radius. If the graph is a point, name the point (x, y, z).

25. $x + y + z = 4$

26. $x + y + z = −4$

27. $x^2 + y^2 + z^2 = 4$

28. $x^2 + y^2 + z^2 = −4$

29. $x^2 + y^2 + z^2 − 2x − 4y − 8z + 21 = 0$

30. $x^2 + y^2 + z^2 − 2x + 6y + 10z + 36 = 0$

31. $x^2 + y^2 + z^2 − 4x + 6y − 2z − 2 = 0$

32. $2x + 3y + 4z = 12$

*33. In space, the locus of points that are equidistant from two fixed points is a plane. Find the equation of the plane that is equidistant from the points (1, 2, 8) and (−1, −2, 4).

*34. Find the point(s) on the x axis at a distance of 13 units from the point (1, 12, 3).

*35. The part of the plane $x + y + z = 2$ shown in the accompanying drawing forms an equilateral triangle. Find the area of the triangle.

36. Consider the spheres whose equations are $x^2 + y^2 + z^2 = 9$ and $x^2 + y^2 + z^2 = 16$. Find the ratio of the

(a) Surface areas of the spheres
(b) Volumes of the spheres

37. Consider the spheres whose equations are

$$(x - 2)^2 + (y - 3)^2 + (z + 4)^2 = 64 \quad \text{and}$$
$$(x + 1)^2 + (y - 2)^2 + (z + 5)^2 = 16$$

Find the ratio of the

(a) Surface areas of the spheres
(b) Volumes of the spheres

▲ A LOOK BEYOND: HISTORICAL SKETCH OF RENÉ DESCARTES

René Descartes was born in Tours, France, on March 31, 1596, and died in Stockholm, Sweden, on February 11, 1650. He was a contemporary of Galileo, the Italian scientist responsible for many discoveries in the science of dynamics. Descartes was also a friend of the French mathematicians Marin Mersenne (Mersenne Numbers) and Blaise Pascal (Pascal's Triangle).

As a small child, René Descartes was in poor health much of the time. Because he spent so much time reading in bed during his illnesses, he became known as a very intelligent young man. When Descartes was older and stronger, he joined the French army. It was during his time as a soldier that Descartes had three dreams that shaped much of his future. The dreams, dated to November 10, 1619, shaped his philosophy and laid the framework for his discoveries in mathematics.

He resigned his commission with the army in 1621 so that he could devote his life to studies of philosophy, science, and mathematics. In the ensuing years, Descartes came to be highly regarded as a philosopher and mathematician and was invited to the learning centers of France, Holland, and Sweden.

Descartes' chief contribution to mathematics was the development of analytical geometry, using the rectangular (Cartesian) coordinate system as a means of representing points. This convention led, in turn, to the algebraic description (equations) of various geometric figures; subsequently, many conjectured properties concerning those figures could be established through proof. Using the Cartesian system, it was also possible to locate the points of intersection of certain figures such as circles, parabolas, and lines, as well as some combination of these.

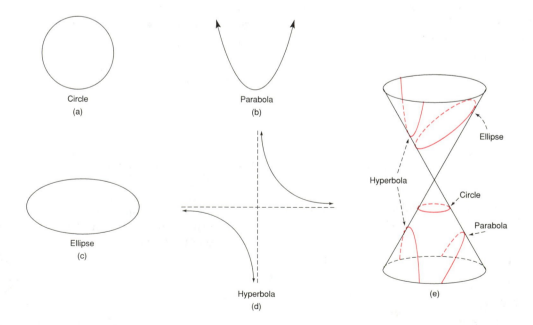

Circle
(a)

Parabola
(b)

Ellipse
(c)

Hyperbola
(d)

Ellipse
Hyperbola
Circle
Parabola
(e)

Figure 9.37

Generally, the phrase "conic sections" refers to four geometric figures: the **circle,** the **parabola,** the **ellipse,** and the **hyperbola.** These figures are shown in figure 9.37 individually and also in relation to the upper and lower nappes of a cone. The conic sections are results found when a plane intersects the nappes of a cone.

Other mathematical works of Descartes were devoted to the study of tangent lines to curves. The notion of a tangent to a curve is illustrated in figure 9.38; this concept is the basis for the branch of mathematics known as **differential calculus.**

Descartes' final contributions to mathematics involved his standardizing the use of many symbols. To mention a few of these, Descartes used (1) a^2 rather than aa and a^3 rather than $aaa;$ (2) ab to indicate multiplication; and (3) $a, b,$ and c as constants and $x, y,$ and z as variables.

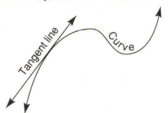

Figure 9.38

▲ Summary

A Look Back at Chapter 9

Our goal in this chapter was to deal with a type of geometry known as solid geometry. We found formulas for the lateral area, the total area (surface area), and the volume of prisms, pyramids, cylinders, cones, and spheres. Some of the formulas used in this chapter were developed using the concept of "limit." The material in Section 9.4 is often described as solid analytic geometry.

A Look Ahead to Chapter 10

In the next chapter, we will again deal with the right triangle. Three trigonometric ratios (sine, cosine, and tangent) will be defined in terms of the sides of a right triangle. An area formula for triangles will be derived using the sine ratio. We will also prove the Law of Sines and the Law of Cosines for acute triangles. Another unit for measuring an angle, called "radian measure," will also be introduced.

Important Terms and Concepts of Chapter 9

9.1 Prisms (Right and Oblique)
    Pyramids and Regular Pyramids

Bases, Lateral Edges, Lateral Faces, Altitude
Vertex and Slant Height of a Pyramid
Lateral Area, Total Area, Volume
9.2 Cylinders (Right and Oblique)
    Cones (Right and Oblique)
    Bases, Altitude
    Vertex and Slant Height of a Cone
    Lateral Area, Total Area, Volume
9.3 Polyhedron (Convex and Concave)
    Faces, Edges, Vertices
    Regular Polyhedron (Tetrahedron, Hexahedron
        Octahedron, Dodecahedron, Icosahedron)
    Sphere (Center, Radius, Diameter, Great Circle)
    Surface Area and Volume
9.4 Three-Dimensional Coordinate System
    Cartesian Space
    Midpoint Formula
    Distance Formula
    Equation of a Sphere
    Equation of a Plane
A Look Beyond Historical Sketch of René Descartes

▲ Review Exercises

1. Each side of the base of a right octagonal prism is 7 in. long. The altitude of the prism is 12 in. Find the lateral area.

2. The base of a right prism is a triangle whose sides measure 7 cm, 8 cm, and 12 cm, respectively. The altitude of the prism is 11 cm. Calculate the lateral area of the right prism.

3. The height of a square box is 2 more than 3 times the length of a side of the base. If the lateral area is 480 in.², find the dimensions of the box and the volume of the box.

4. The base of a right prism is a rectangle whose length is 3 more than its width. If the altitude of the prism is 12 cm and the lateral area is 360 cm², find the total area and the volume.

5. The base of a right prism is a triangle whose sides are 9 in., 15 in., and 12 in. The height of the prism is 10 in. Find the

(a) Lateral area   (b) Total area   (c) Volume

6. The base of a right prism is a regular hexagon whose sides are 8 cm. The altitude of the prism is 13 cm. Find the

(a) Lateral area   (b) Total area   (c) Volume

7. The slant height of a regular square pyramid is 15 in. One side of the base is 18 in. Find the

(a) Lateral area   (b) Total area   (c) Volume

8. The base of a regular pyramid is an equilateral triangle each of whose sides is 12 cm. The altitude of the pyramid is 8 cm. Find the

(a) Lateral area   (b) Total area   (c) Volume

9. The radius of the base of a right circular cylinder is 6 in. The height of the cylinder is 10 in. Find the exact

(a) Lateral area   (b) Total area   (c) Volume

10. (a) For the trough in the shape of the half-cylinder shown in the accompanying drawing, find the volume. (Use $\pi \approx 3.14$ and disregard the thickness.)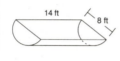
(b) If the trough is to be painted inside and out, find the number of square feet to be painted.

11. The slant height of a right circular cone is 12 cm. The angle formed by the slant height and the altitude is 30°. Find the exact

(a) Lateral area   (b) Total area   (c) Volume

12. The volume of a right circular cone is 96π in.³. If the radius of the base is 6 in., find the length of the slant height.

13. Find the surface area of a sphere if the radius is 7 in. $\left(\text{Use } \pi \approx \dfrac{22}{7}. \right)$

14. Find the volume of a sphere if the diameter is 12 cm. (Use $\pi \approx 3.14$.)

15. The solid shown in the accompanying drawing consists of a hemisphere (half of a sphere), a cylinder, and a cone. Find the exact volume of the solid.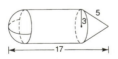

16. If the radius of one sphere is three times as long as the radius of another sphere, how do the surface areas of the spheres compare? How do the volumes compare?

17. Find the volume of the solid that results when the triangular region with vertices at (0, 0), (5, 0), and (0, 7) is rotated about the y axis. $\left(\text{Use } \pi \approx \dfrac{22}{7}. \right)$

18. Find the exact volume of the solid that results when the rectangular region with vertices at (0, 0), (8, 0), (8, 6), and (0, 6) is rotated about the x axis.

19. Find the exact volume of the solid that results when the circle whose equation is $(x - 2)^2 + (y - 0)^2 = 4$ is rotated about the x axis.

20. A plastic pipe is 3 ft long and has an inside radius of 4 in. and an outside radius of 5 in. How many cubic inches of plastic are in the pipe? (Use $\pi \approx 3.14$.)

21. A sphere with a diameter of 14 in. is inscribed in a hexahedron. Find the exact volume of the space inside the hexahedron but outside the sphere.

22. (a) An octahedron has _____ faces that are _____ .
(b) A tetrahedron has _____ faces that are _____ .
(c) A dodecahedron has _____ faces that are _____ .

23. A drug manufacturing company wants to manufacture a capsule that contains a spherical pill inside. The diameter of the pill is 4 mm and the capsule is cylindrical, with hemispheres on either end. The length of the capsule between the two hemispheres is 10 mm. What is the exact volume the capsule will hold, excluding the volume of the pill?

24. Find the volume of cement used in the block shown in the accompanying drawing.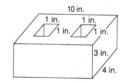

25. Consider the points P_1 (1, 2, −3) and P_2 (−3, 4, 1).
(a) Find the midpoint M of the segment $\overline{P_1P_2}$.
(b) Find the length of the segment $\overline{P_1P_2}$.

26. Sketch the graph of the plane $3x + y + 2z = 12$.

27. Find the equation of the sphere that has center (2, −1, 3) and length of radius $r = 5$. Leave the answer in the form $x^2 + y^2 + z^2 + Dx + Ey + Fz + G = 0$.

28. Determine the graph, if any, for each of these equations:
(a) $x^2 + y^2 + z^2 = 16$
(b) $2x + 2y + 2z = 16$
(c) $x^2 + y^2 + z^2 - 2x - 4y + 2z - 3 = 0$
(d) $(x + 3)^2 + (y - 2)^2 + (z - 5)^2 = 0$
(e) $x^2 + y^2 + z^2 - 6y + 10 = 0$

INTRODUCTION TO TRIGONOMETRY

A surveyor uses a transit to obtain a reasonably accurate angle measurement. The surveyor can then use a trigonometric ratio to find an unknown length. The word "trigonometry" refers to the measurement of triangles. In this chapter, you will discover methods for measuring the angles as well as lengths of sides of right triangles when the measures of other parts of the triangle are known. These techniques can be extended to other types of triangles. Various advanced techniques are described in more intensive courses in trigonometry.

Photographs Courtesy of Don Manning, Parkland College

CHAPTER OUTLINE

10.1 Sine Ratio and Applications

10.2 Cosine Ratio and Applications

10.3 Tangent Ratio and Other Ratios

10.4 More Trigonometric Relationships

A Look Beyond Radian Measure of Angles

 ## SINE RATIO AND APPLICATIONS

In this section we will deal strictly with similar right triangles. In figure 10.1, $\triangle ABC \sim \triangle DEF$ and $\angle C$ and $\angle F$ are right angles. Consider corresponding angles A and D.

    If we compare the length of the side opposite each angle to the length of the hypotenuse of each triangle, we can see that

$$\frac{BC}{AB} = \frac{EF}{DE} \quad \text{or} \quad \frac{3}{5} = \frac{6}{10}$$

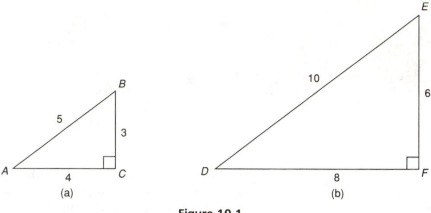

Figure 10.1

In any two similar right triangles, the ratio of this pair of corresponding sides depends on the measures of the corresponding acute angles *A* and *D;* that is, the numerical value of the ratio

$$\frac{\text{Length of side opposite the acute angle}}{\text{Length of hypotenuse}}$$

will become larger for larger measures of ∠*A*. This ratio is unique for each measure of the acute angle, even though the lengths of the two sides of two triangles containing ∠*A* may be different.

In figure 10.2, we name the measures of the angles of the right triangle by *α* (alpha) at vertex *A*, *β* (beta) at vertex *B*, and *γ* (gamma) at vertex *C*. The lengths of the sides opposite vertices *A*, *B*, and *C* are given by *a*, *b*, and *c*, respectively. For convenience, we describe the lengths of the sides of the triangle in the following definition simply as "opposite" and "hypotenuse." The word "opposite" is used to mean the length of the side opposite the angle named; the word "hypotenuse" is used to mean the length of the hypotenuse.

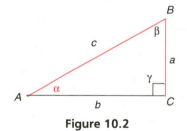

Figure 10.2

DEFINITION: In a right triangle, the **sine ratio** for an acute angle is the ratio $\dfrac{\text{Opposite}}{\text{Hypotenuse}}$.

Note: In right △*ABC*, we have $\sin \alpha = \dfrac{a}{c}$ and $\sin \beta = \dfrac{b}{c}$, in which "sin" is an abbreviated form of the word "sine" (pronounced like the word "sign").

EXAMPLE 1   Find sin *α* and sin *β* for right △*ABC*.

Solution   *a* = 3, *b* = 4, and *c* = 5. Therefore

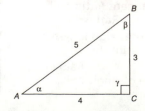

$$\sin \alpha = \frac{a}{c} = \frac{3}{5} \quad \text{and} \quad \sin \beta = \frac{b}{c} = \frac{4}{5}$$ ▲

EXAMPLE 2 Find $\sin \alpha$ and $\sin \beta$ for right $\triangle ABC$.

Solution $a = 5$ and $c = 13$. By the Pythagorean Theorem

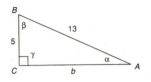

$$c^2 = a^2 + b^2$$
$$13^2 = 5^2 + b^2$$
$$169 = 25 + b^2$$
$$b^2 = 144$$
$$b = 12$$

Therefore $\sin \alpha = \dfrac{a}{c} = \dfrac{5}{13}$ and $\sin \beta = \dfrac{b}{c} = \dfrac{12}{13}$ ▲

Recall the 30-60-90 relationship, in which the side opposite the 30° angle has a length equal to one-half that of the hypotenuse; the remaining leg has a length equal to the product of the length of the shorter leg and $\sqrt{3}$. In this case, we write $\sin 30° = \dfrac{x}{2x} = \dfrac{1}{2}$.

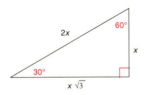

It is also true that $\sin 60° = \dfrac{x\sqrt{3}}{2x} = \dfrac{\sqrt{3}}{2}$. While the value for $\sin 30°$ is exactly 0.5, a calculator would give an approximate value such as 0.8660254 for $\sin 60°$. We will generally round the ratio to four decimal places, so that $\sin 60° = \approx 0.8660$.

EXAMPLE 3 Find exact and approximate values for $\sin 45°$.

Solution Using the accompanying 45-45-90 triangle, we see that $\sin 45° = \dfrac{x}{x\sqrt{2}} = \dfrac{1}{\sqrt{2}}$. Equivalently, $\sin 45° = \dfrac{\sqrt{2}}{2}$. A calculator approximation is given by $\sin 45° \approx 0.7071$. ▲

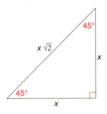

We will now use an earlier result (from Section 5.6) to determine the sine ratios for angles that measure 15° and 75°. Recall that an angle bisector of one angle of a triangle divides the opposite side into two segments that are proportional to the sides forming the bisected angle. Using this fact in a 30-60-90 triangle, we are led to the proportion

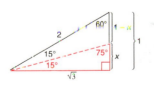

$$\frac{x}{\sqrt{3}} = \frac{1 - x}{2}$$

Clearing this equation of fractions, we have

$$2x = \sqrt{3} - \sqrt{3}x$$
$$2x + \sqrt{3}x = \sqrt{3}$$
$$(2 + \sqrt{3})x = \sqrt{3}$$
$$x = \frac{\sqrt{3}}{2 + \sqrt{3}} \approx 0.46410$$

This result provides us with the length of the side that is opposite the 15° angle of a 15-75-90 triangle. Using the Pythagorean Theorem, we can show that the length of the hypotenuse is equal to approximately 1.79315. Then sin $15° = \dfrac{0.46410}{1.79315} \approx 0.2588$. In the same triangle, we see that sin 75° = $\dfrac{1.73205}{1.79315} \approx 0.9659$.

We now begin to formulate a small table of values of sines. In Table 10.1, the Greek letter θ (theta) designates the angle measure in degrees. The second column has the heading sin θ and provides the ratio for the corresponding angle; this ratio is generally given to four decimal places of accuracy.

Table 10.1   **Sine Ratios**

θ	$sin\ \theta$
15°	0.2588
30°	0.5000
45°	0.7071
60°	0.8660
75°	0.9659

▲ **WARNING:** Notice that $sin\left(\dfrac{1}{2}\theta\right) \neq \dfrac{1}{2} sin\ \theta$ in Table 10.1. For instance, if $\theta = 60°$, $sin\left(\dfrac{1}{2} \cdot 60°\right) \neq \dfrac{1}{2} sin\ 60°$ because sin 30° $\neq \dfrac{1}{2} sin\ 60°$. That is, $0.5000 \neq \dfrac{1}{2} \cdot 0.8660$.

Note: Most values provided in tables (or from a calculator) are approximations. In this chapter, we generally use the equality symbol (=) when reading values from a table (or calculator); but solutions to the problems that follow are generally approximations.

In figure 10.3, let $\angle\theta$ be the acute angle whose measure increases as shown.

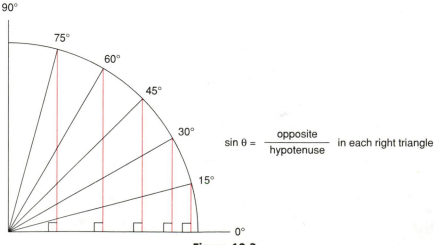

Figure 10.3

Then the side opposite $\angle\theta$ gets larger as θ increases. In fact, as θ approaches 90° ($\theta \to 90°$), the length of the leg opposite $\angle\theta$ approaches the length of the hypotenuse. As θ decreases, $\sin\theta$ also decreases. As θ decreases ($\theta \to 0°$), the length of the side opposite $\angle\theta$ approaches 0. These observations lead to the following definitions.

> **DEFINITION:** $\sin 0° = 0$ and $\sin 90° = 1$.

EXAMPLE 4

Using Table 10.1, find the length of a to the nearest tenth.

Solution

$$\sin 15° = \frac{\text{opposite}}{\text{hypotenuse}}$$

From the table, we have $\sin 15° = 0.2588$.

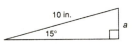

$$\frac{a}{10} = 0.2588$$
$$a = 2.588$$

Therefore $a = 2.6$ in. when rounded to tenths. ▲

    In an application problem, the sine ratio can be used to find the measure of either a side or an angle of a triangle. To find the sine ratio of the angle involved, you may use a table of ratios or a calculator. Table 10.2 provides ratios for many more angle measures than does Table 10.1. Like calculators, tables generally provide only approximations of the sine ratios indicated.

Note: An even more complete table (Table 10.5) is provided at the end of Section 10.3.

EXAMPLE 5

Find $\sin 36°$, using
(a) Table 10.2   (b) A calculator

Solution

(a) Find 36° under the heading θ. Now read the number under the $\sin\theta$ heading:

$$\sin 36° = 0.5878$$

(b) On a calculator that is in degree mode, use the following key sequence:

$$\boxed{3} \to \boxed{6} \to \boxed{\sin} \to \boxed{\textbf{0.5878}}$$

The result is $\sin 36° = 0.5878$, correct to four decimal places. ▲

Note: The boldfaced number in the box represents the final answer.

    The table or a calculator can also be used to find the measure of an angle. This is possible when the sine of the angle is known.

Table 10.2   **Sine Ratios**

θ	$\sin \theta$	θ	$\sin \theta$	θ	$\sin \theta$	θ	$\sin \theta$
0°	0.0000	23°	0.3907	46°	0.7193	69°	0.9336
1°	0.0175	24°	0.4067	47°	0.7314	70°	0.9397
2°	0.0349	25°	0.4226	48°	0.7431	71°	0.9455
3°	0.0523	26°	0.4384	49°	0.7547	72°	0.9511
4°	0.0698	27°	0.4540	50°	0.7660	73°	0.9563
5°	0.0872	28°	0.4695	51°	0.7771	74°	0.9613
6°	0.1045	29°	0.4848	52°	0.7880	75°	0.9659
7°	0.1219	30°	0.5000	53°	0.7986	76°	0.9703
8°	0.1392	31°	0.5150	54°	0.8090	77°	0.9744
9°	0.1564	32°	0.5299	55°	0.8192	78°	0.9781
10°	0.1736	33°	0.5446	56°	0.8290	79°	0.9816
11°	0.1908	34°	0.5592	57°	0.8387	80°	0.9848
12°	0.2079	35°	0.5736	58°	0.8480	81°	0.9877
13°	0.2250	36°	0.5878	59°	0.8572	82°	0.9903
14°	0.2419	37°	0.6018	60°	0.8660	83°	0.9925
15°	0.2588	38°	0.6157	61°	0.8746	84°	0.9945
16°	0.2756	39°	0.6293	62°	0.8829	85°	0.9962
17°	0.2924	40°	0.6428	63°	0.8910	86°	0.9976
18°	0.3090	41°	0.6561	64°	0.8988	87°	0.9986
19°	0.3256	42°	0.6691	65°	0.9063	88°	0.9994
20°	0.3420	43°	0.6820	66°	0.9135	89°	0.9998
21°	0.3584	44°	0.6947	67°	0.9205	90°	1.0000
22°	0.3746	45°	0.7071	68°	0.9272		

EXAMPLE 6   If $\sin \theta = 0.7986$, find θ to the nearest degree, using
(a) Table 10.2   (b) a calculator

Solution   (a) Find 0.7986 under the heading $\sin \theta$. Now look to the left to find the degree measure of the angle in the θ column:

$$\sin \theta = 0.7986 \rightarrow \theta = 53°$$

(b) On some calculators, you can use the following key sequence while in the degree mode:

$$\boxed{.} \rightarrow \boxed{7} \rightarrow \boxed{9} \rightarrow \boxed{8} \rightarrow \boxed{6} \rightarrow \boxed{\text{inv}} \rightarrow \boxed{\text{sin}} \rightarrow \boxed{53}$$

The combination "inv" and "sin" yields the angle, whose sine ratio is known, so $\theta = 53°$.   ▲

In most application problems, a drawing provides a good deal of information and allows some insight into the solution. In the drawings, the phrases "angle of elevation" and "angle of depression" are often used. These angles, which measure from the horizontal, are illustrated in figure 10.4a and b. In figure 10.4a, the angle measured up from the horizontal ray is α, the **angle of elevation.** In figure 10.4b, the angle measured down from the horizontal ray is β, the **angle of depression.**

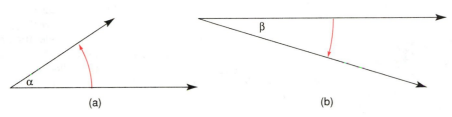

(a) (b)

Figure 10.4

EXAMPLE 7

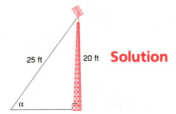

The tower for a television antenna stands 20 ft tall. A guy wire 25 ft long supports the antenna, as shown in the accompanying drawing. Find the measure of the angle of elevation α to the nearest degree.

Solution

$$\sin \alpha = \frac{\text{opposite}}{\text{hypotenuse}} = \frac{20}{25} = \frac{4}{5} = 0.8$$

From Table 10.2 (or from a calculator), we find that the angle whose sine ratio is 0.8 is $\alpha = 53°$. ▲

10.1 EXERCISES

In Exercises 1 to 6, find $\sin \alpha$ and $\sin \beta$ for the triangle shown.

1.
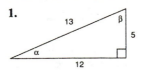

2.
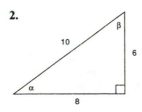

3.
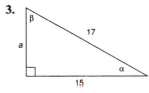

4.
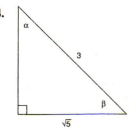

5.

6.
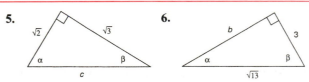

In Exercises 7 to 14, use either Table 10.2 or a calculator to find the sine of the indicated angle to four decimal places.

7. $\sin 90°$ **11.** $\sin 82°$

8. $\sin 0°$ **12.** $\sin 46°$

9. $\sin 17°$ **13.** $\sin 72°$

10. $\sin 23°$ **14.** $\sin 57°$

In Exercises 15 to 20, find the length of the side indicated by the variables. Use either Table 10.2 or a calculator, as needed; and round to the nearest tenth.

15.

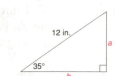

16.

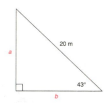

17.

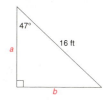

18.

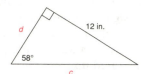

19.

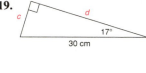

20.

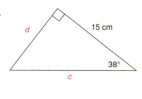

In Exercises 21 to 26, find the measure of the angles named to the nearest degree.

21.

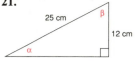

22.

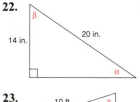

23.

24.

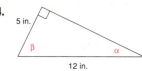

25.

26.

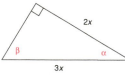

In Exercises 27 to 34, use drawings where provided to solve each problem. Angle measures should be given to the nearest degree; distances should be given to the nearest tenth of a unit.

27. The pitch or slope of a roofline is 5 to 12. Find the measure of angle α.

28. A kite is flying at an angle of elevation of 67° from the ground. If 100 ft of kite string is out, how far is the kite above the ground?

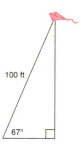

29. Danny sees a balloon that is 100 ft above the ground. If the angle of elevation of the balloon is 75°, how far away is the balloon from Danny?

30. Over a 2000-ft span of highway through a hillside, there is a 100-ft rise in the roadway. What is the measure of the angle formed by the road and the horizontal?

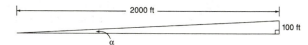

31. From a cliff, a person observes an automobile through an angle of depression of 23°. If the cliff is 50 ft high, how far is the automobile from the person?

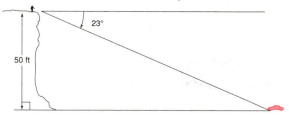

32. A 12-ft rope secures a rowboat to a pier that is 4 ft above the water. What is the angle formed by the rope and the water?

33. A 10-ft ladder is leaning against a vertical wall so that the bottom of the ladder is 4 ft away from the base of the wall. How large is the angle formed by the ladder and the wall?

34. An airplane flying at the rate of 350 ft per sec begins to climb at an angle of 10°. What is the increase in altitude over the next 15 sec?

COSINE RATIO AND APPLICATIONS

Again we deal strictly with similar right triangles, as shown in figure 10.5. While $\overline{BC}$ is the leg opposite angle A, we say that $\overline{AC}$ is the leg adjacent to angle A. In the two triangles, the ratios of the form

$$\frac{\text{Length of the adjacent leg}}{\text{Length of hypotenuse}}$$

are equal; that is,

$$\frac{AC}{AB} = \frac{DF}{DE} \qquad \text{or} \qquad \frac{4}{5} = \frac{8}{10}$$

because corresponding sides of $\sim \Delta$s are proportional.

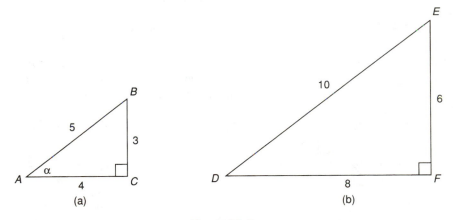

Figure 10.5

As with the sine ratio, this ratio—known as the "cosine" ratio—depends on the measure of the acute angle (A or D). In the following definition, "adjacent" refers to the length of the leg that is adjacent to the angle named.

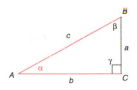

DEFINITION.   In a right triangle, the **cosine ratio** for an acute angle is the ratio $\dfrac{\text{adjacent}}{\text{hypotenuse}}$.

Note: In right $\triangle ABC$, we have $\cos \alpha = \dfrac{b}{c}$ and $\cos \beta = \dfrac{a}{c}$, in which "cos" is an abbreviated form of the word "cosine."

EXAMPLE 1   Find $\cos \alpha$ and $\cos \beta$ for right $\triangle ABC$.

Solution $a = 3$, $b = 4$, and $c = 5$ for the triangle shown in the accompanying drawing.

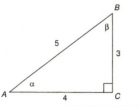

Therefore

$$\cos \alpha = \frac{b}{c} = \frac{4}{5} \quad \text{and} \quad \cos \beta = \frac{a}{c} = \frac{3}{5} \qquad \blacktriangle$$

EXAMPLE 2   Find $\cos \alpha$ and $\cos \beta$ for right $\triangle ABC$.

Solution $a = 5$ and $c = 13$. To find b, we use the Pythagorean Theorem:

$$c^2 = a^2 + b^2$$
$$13^2 = 5^2 + b^2$$
$$169 = 25 + b^2$$
$$b^2 = 144$$
$$b = 12$$

Consequently,

$$\cos \alpha = \frac{b}{c} = \frac{12}{13} \quad \text{and} \quad \cos \beta = \frac{a}{c} = \frac{5}{13} \qquad \blacktriangle$$

Just as the result for the sine ratio of any angle is unique, the result for the cosine ratio of any angle is also unique. Using the most general 30-60-90 and 45-45-90 triangles, we see that

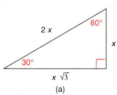

(a)

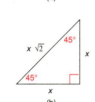

(b)

$$\cos 30° = \frac{x\sqrt{3}}{2x} = \frac{\sqrt{3}}{2}$$

$$\cos 45° = \frac{x}{x\sqrt{2}} = \frac{1}{\sqrt{2}} = \frac{\sqrt{2}}{2}$$

$$\cos 60° = \frac{x}{2x} = \frac{1}{2}$$

Now we use the 15-75-90 triangle shown in figure 10.6 to find $\cos 75°$ and $\cos 15°$. Recall that $\sin 15° = \frac{a}{c}$ and $\sin 15° = 0.2588$. But $\cos 75° = \frac{a}{c}$, so $\cos 75° = 0.2588$. Similarly, because $\sin 75° = \frac{b}{c} = 0.9659$, we have $\cos 15° = \frac{b}{c} = 0.9659$.

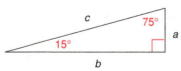

Figure 10.6

In figure 10.7, we see that the cosine ratios become larger as θ decreases and become smaller as θ increases. Consider the definition

$$\cos \theta = \frac{\text{length of adjacent leg}}{\text{length of hypotenuse}}$$

As $\theta \to 0°$, length of adjacent leg $\to$ length of hypotenuse, and we see that $\cos 0° \to 1$. Similarly, $\cos 90° \to 0$ because the adjacent leg grows smaller. Therefore, we have the following definitions.

DEFINITION: cos 0° = 1 and cos 90° = 0.

We now summarize our cosine ratios in Table 10.3.

Table 10.3   **Cosine Ratios**

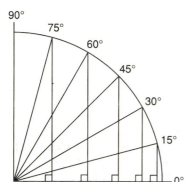

Figure 10.7

θ	$\cos \theta$
0°	1.0000
15°	0.9659
30°	0.8660
45°	0.7071
60°	0.5000
75°	0.2588
90°	0.0000

Table 10.4 (pages 372–373) identifies the ratios for angles from 0° to 90° in increments of 1°.

EXAMPLE 3   Using Table 10.3, find the length of b, correct to the nearest tenth.

Solution $\cos 15° = \dfrac{\text{adjacent}}{\text{hypotenuse}} = \dfrac{b}{10}$ from the triangle. $\cos 15° = 0.9659$ from the table. Then

$$\frac{b}{10} = 0.9659$$
$$b = 9.659$$
Therefore $$b = 9.7 \text{ in.}$$

when rounded to the nearest tenth of an inch.   ▲

   In applications, the cosine ratio can be used to find a missing length or a missing angle measure. While the sine ratio requires that we use "opposite" and "hypotenuse," the cosine ratio requires that we use the "adjacent" and "hypotenuse."

EXAMPLE 4   Which ratio would you use to find:
(**a**) α, if a and c are known?
(**b**) b, if α and c are known?
(**c**) c, if a and α are known?
(**d**) β, if a and c are known?

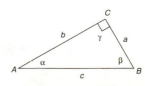

Solution   (a) sine, since $\sin \alpha = \dfrac{a}{c}$

(b) cosine, since $\cos \alpha = \dfrac{b}{c}$

(c) sine, since $\sin \alpha = \dfrac{a}{c}$

(d) cosine, since $\cos \beta = \dfrac{a}{c}$ ▲

   To complete application problems, you generally use either Table 10.4 or a calculator.

EXAMPLE 5 Find $\cos 67°$, using
  (a) Table 10.4   (b) a calculator

Solution (a) Find $67°$ under the heading θ of the table. Then find $\cos 67°$ under the heading $\cos \theta$, which is immediately to the right; $\cos 67° = 0.3907$.
  (b) On a calculator that is in degree mode, use the following key sequence:

$$\boxed{6} \rightarrow \boxed{7} \rightarrow \boxed{\cos} \rightarrow \boxed{\mathbf{0.3907}}$$ ▲

EXAMPLE 6 Find the measure of angle θ if $\cos \theta = 0.5878$.

Solution Using Table 10.4, we look for 0.5878 under the $\cos \theta$ column. Looking to the left, we find 54 under the heading θ. Thus $\theta = 54°$.

Note: If you are using a calculator (in degree mode), follow this key sequence:

$$\boxed{.} \rightarrow \boxed{5} \rightarrow \boxed{8} \rightarrow \boxed{7} \rightarrow \boxed{8} \rightarrow \boxed{\text{inv}} \rightarrow \boxed{\cos} \rightarrow \boxed{\mathbf{54}}$$ ▲

Table 10.4 **Ratios for Sine and Cosine**

θ	$\cos \theta$	$\sin \theta$	θ	$\cos \theta$	$\sin \theta$
0°	1.0000	0.0000	45°	0.7071	0.7071
1°	0.9998	0.0175	46°	0.6947	0.7193
2°	0.9994	0.0349	47°	0.6820	0.7314
3°	0.9986	0.0523	48°	0.6691	0.7431
4°	0.9976	0.0698	49°	0.6561	0.7547
5°	0.9962	0.0872	50°	0.6428	0.7660
6°	0.9945	0.1045	51°	0.6293	0.7771
7°	0.9925	0.1219	52°	0.6157	0.7880
8°	0.9903	0.1392	53°	0.6018	0.7986
9°	0.9877	0.1564	54°	0.5878	0.8090

10°	0.9848	0.1736	55°	0.5736	0.8192
11°	0.9816	0.1908	56°	0.5592	0.8290
12°	0.9781	0.2079	57°	0.5446	0.8387
13°	0.9744	0.2250	58°	0.5299	0.8480
14°	0.9703	0.2419	59°	0.5150	0.8572
15°	0.9659	0.2588	60°	0.5000	0.8660
16°	0.9613	0.2756	61°	0.4848	0.8746
17°	0.9563	0.2924	62°	0.4695	0.8829
18°	0.9511	0.3090	63°	0.4540	0.8910
19°	0.9455	0.3256	64°	0.4384	0.8988
20°	0.9397	0.3420	65°	0.4226	0.9063
21°	0.9336	0.3584	66°	0.4067	0.9135
22°	0.9272	0.3746	67°	0.3907	0.9205
23°	0.9205	0.3907	68°	0.3746	0.9272
24°	0.9135	0.4067	69°	0.3584	0.9336
25°	0.9063	0.4226	70°	0.3420	0.9397
26°	0.8988	0.4384	71°	0.3256	0.9455
27°	0.8910	0.4540	72°	0.3090	0.9511
28°	0.8829	0.4695	73°	0.2924	0.9563
29°	0.8746	0.4848	74°	0.2756	0.9613
30°	0.8660	0.5000	75°	0.2588	0.9659
31°	0.8572	0.5150	76°	0.2419	0.9703
32°	0.8480	0.5299	77°	0.2250	0.9744
33°	0.8387	0.5446	78°	0.2079	0.9781
34°	0.8290	0.5592	79°	0.1908	0.9816
35°	0.8192	0.5736	80°	0.1736	0.9848
36°	0.8090	0.5878	81°	0.1564	0.9877
37°	0.7986	0.6018	82°	0.1392	0.9903
38°	0.7880	0.6157	83°	0.1219	0.9925
39°	0.7771	0.6293	84°	0.1045	0.9945
40°	0.7660	0.6428	85°	0.0872	0.9962
41°	0.7547	0.6561	86°	0.0698	0.9976
42°	0.7431	0.6691	87°	0.0523	0.9986
43°	0.7314	0.6820	88°	0.0349	0.9994
44°	0.7193	0.6947	89°	0.0175	0.9998
			90°	0.0000	1.0000

EXAMPLE 7   For a regular pentagon, the length of the apothem is 12 in. Find the length of the pentagon's radius, to the nearest tenth.

Solution   The central angle of the regular pentagon measures $\frac{360}{5}$ or 72°. An apothem

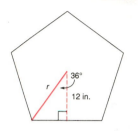

bisects this angle, so the angle formed by the apothem and the radius measures 36°.

In the drawing,

$$\cos 36° = \frac{\text{adjacent}}{\text{hypotenuse}} = \frac{12}{r}$$

From Table 10.4, $\cos 36° = 0.8090$, so $\frac{12}{r} = 0.8090$. Then $0.8090r = 12$, so $r = 14.8$ in. ▲

We will now consider a proof of a statement that is often called an "identity" because it is true for all angles; we will refer to this statement as a theorem. Perhaps you will see that the statement is based entirely on the Pythagorean Theorem.

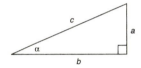

THEOREM 10.2.1   In any right triangle in which α is the measure of an acute angle,

$$\sin^2 \alpha + \cos^2 \alpha = 1$$

Note: $\sin^2 \alpha$ means $(\sin \alpha)^2$ and $\cos^2 \alpha$ means $(\cos \alpha)^2$.

Proof $\sin \alpha = \frac{a}{c}$ and $\cos \alpha = \frac{b}{c}$. Then

$$\sin^2 \alpha + \cos^2 \alpha = \left(\frac{a}{c}\right)^2 + \left(\frac{b}{c}\right)^2$$

$$= \frac{a^2}{c^2} + \frac{b^2}{c^2} = \frac{a^2 + b^2}{c^2}$$

In the right triangle, we know that $a^2 + b^2 = c^2$. Therefore by substitution,

$$\sin^2 \alpha + \cos^2 \alpha = \frac{c^2}{c^2} \qquad \text{and} \qquad \frac{c^2}{c^2} = 1$$

It follows that

$$\sin^2 \alpha + \cos^2 \alpha = 1 \qquad \text{for any angle } \alpha \qquad ▲$$

10.2 EXERCISES

In Exercises 1 to 6, find cos α and cos β.

1.

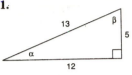

4.

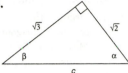

2.

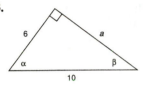

5.

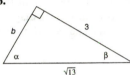

3.

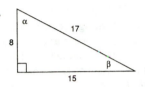

6.

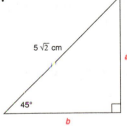

7. In Exercises 1 to 6:

  (a) Why does sin α = cos β?

  (b) Why does cos α = sin β?

8. Using the right triangle from Exercise 1, show that sin² α + cos² α = 1.

In Exercises 9 to 16, use Table 10.4 or a calculator to find the indicated cosine ratio to four decimal places.

9. cos 23° **11.** cos 17° **13.** cos 90° **15.** cos 82°

10. cos 0° **12.** cos 73° **14.** cos 42° **16.** cos 7°

In Exercises 17 to 22, use either the sine ratio or the cosine ratio to find the lengths of the indicated sides of the triangle, correct to the nearest tenth.

17.

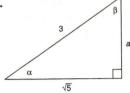

18.

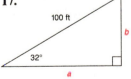

19.

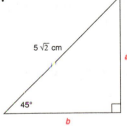

20.

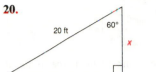

21.

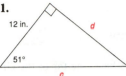

22.

In Exercises 23 to 28, use the sine ratio or cosine ratio as needed to find the measure of each indicated angle to the nearest degree.

23.

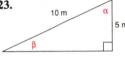

26.

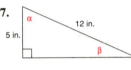

24.

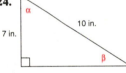

27.

25.

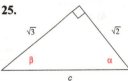

28.

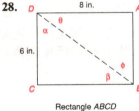

Rectangle ABCD

In Exercises 29 to 37, use the drawings provided to solve each problem. Angle measures should be given to the nearest degree; distances should be given to the nearest tenth of a unit.

29. In building a garage onto his house, Gene wants to use a sloped 12-ft roof to cover an area that is 10 ft wide. Find the measure of angle θ.

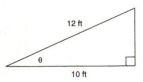

30. Gene redesigned the garage from Exercise 29 so that the 12-ft roof would rise 2 ft as shown. Find the measure of angle θ.

31. When an airplane is descending to land, the angle of depression is 5°. When the plane has a reading of 100 ft on the altimeter, what is its distance x from touchdown?

32. At a point 200 ft from a cliff's base, the top of the cliff is seen through an angle of elevation of 37°. How tall is the cliff?

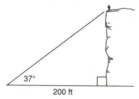

33. Find the length of each apothem in a regular pentagon whose radii measure 10 in. each.

34. Dale looks up to see his friend Lisa waving from her apartment window 30 ft from him. If Dale is standing 10 ft from the building, what is the angle of elevation as Dale looks up at Lisa?

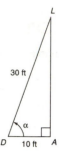

35. Find the length of the radius in a regular decagon for which each apothem has a length of 12.5 cm.

36. In searching for survivors of a boating accident, a helicopter moves horizontally across the ocean at an altitude of 200 ft above the water. If a man clinging to a life raft is seen through an angle of depression of 12°, what is the distance from the helicopter to the man in the water?

*37. What is the size of the angle α formed by a diagonal of a cube and one of its edges?

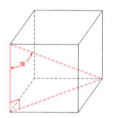

10.3

TANGENT RATIO AND OTHER RATIOS

As in Sections 10.1 and 10.2, we deal here strictly with right triangles. The third trigonometric ratio is the tangent ratio, which we define for an acute angle to be

$$\frac{\text{Length of the leg opposite the acute angle}}{\text{Length of the leg adjacent to the acute angle}}$$

Like the sine ratio, the tangent ratio grows as the measure of the acute angle grows. Unlike the sine and cosine ratios, whose values range from 0 to 1, the value of the tangent ratio is from 0 upward indefinitely; that is, there is no greatest value for the tangent.

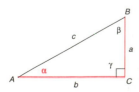

DEFINITION: In a right triangle, the **tangent ratio** for an acute angle is the ratio $\dfrac{\text{opposite}}{\text{adjacent}}$.

Note: In right $\triangle ABC$, we have $\tan \alpha = \dfrac{a}{b}$ and $\tan \beta = \dfrac{b}{a}$, in which "tan" is an abbreviated form of the word "tangent."

EXAMPLE 1 Find the values of tan α and tan β for the triangle shown in the accompanying figure.

Solution

$$\tan \alpha = \frac{a}{b} = \frac{8}{15}$$

$$\tan \beta = \frac{b}{a} = \frac{15}{8} \qquad \blacktriangle$$

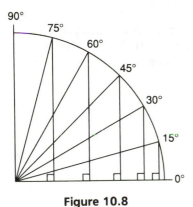

The value of the tangent ratio changes from 0 for a 0° angle to an immeasurably large value as the measure of the acute angle draws near 90°. That the tangent becomes infinitely large follows from the fact that the denominator becomes smaller while the numerator increases in the ratio $\frac{\text{opposite}}{\text{adjacent}}$.

Study figure 10.8 to see why the value of the tangent of an angle grows immeasurably large as the angle approaches 90° in size. We often express this relationship by writing: As $\theta \rightarrow 90°$, tan $\theta \rightarrow \infty$. This means that tan 90° is *undefined*.

Figure 10.8

DEFINITION: tan 0° = 0 and tan 90° is undefined.

Other results for the tangent are found through special triangles. By observing the triangles in figure 10.9, we see that

$$\tan 30° = \frac{\sqrt{3}}{3} \approx 0.5774$$

$$\tan 45° = \frac{1}{1} = 1.0000$$

$$\tan 60° = \sqrt{3} \approx 1.7321$$

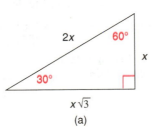
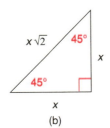

(a)   (b)

Figure 10.9

Table 10.5 at the end of this section includes ratios for the sine, cosine, and tangent of angles in increments of 1°. We will use that table in Example 2.

EXAMPLE 2 A ski lift moves each chair through an angle of 25° as shown on page 378. What vertical change (rise) accompanies a horizontal change (run) of 845 ft?

Solution

$$\tan 25° = \frac{\text{opposite}}{\text{adjacent}} = \frac{a}{845}$$

in the triangle. From the table,

$$\tan 25° = 0.4663$$

$$\frac{a}{845} = 0.4663$$

$$a = 394 \text{ ft} \qquad \blacktriangle$$

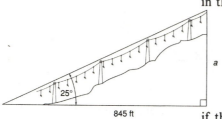

The tangent ratio can also be used to find the measure of an angle if the lengths of the legs of a right triangle are known. This is illustrated in Example 3.

EXAMPLE 3   An airplane is seen flying just over Mission Rock, which is 1 mi away. If Mission Rock is known to be 135 ft high and the airplane is 50 ft above it, then what is the angle of elevation through which the plane is seen?

Solution   From figure 10.10 and the fact that 1 mi = 5280 ft,

θ

Figure 10.10

$$\tan \theta = \frac{\text{opposite}}{\text{adjacent}} = \frac{185}{5280}$$

Then $\tan \theta = 0.0350$, so $\theta = 2°$, to the nearest degree.

Note: This result can be determined by using either a table or a calculator. The typical key sequence in the degree mode is

$$\boxed{0} \rightarrow \boxed{.} \rightarrow \boxed{0} \rightarrow \boxed{3} \rightarrow \boxed{5} \rightarrow \boxed{\text{inv}} \rightarrow \boxed{\text{tan}} \rightarrow \boxed{2} \qquad \blacktriangle$$

For the right triangle in figure 10.11, we now have three ratios that could be used in solving an application problem. These are summarized next.

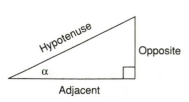

Figure 10.11

$$\sin \alpha = \frac{\text{opposite}}{\text{hypotenuse}}$$

$$\cos \alpha = \frac{\text{adjacent}}{\text{hypotenuse}}$$

$$\tan \alpha = \frac{\text{opposite}}{\text{adjacent}}$$

EXAMPLE 4   Name the ratio that should be used to find:
(a) a, if α and c are known
(b) α, if a and b are known
(c) β, if a and c are known
(d) b, if a and β are known

Solution   (a) sine, since $\sin \alpha = \dfrac{a}{c}$

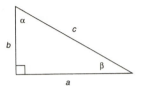

(b) tangent, since $\tan \alpha = \dfrac{a}{b}$

(c) cosine, since $\cos \beta = \dfrac{a}{c}$

(d) tangent, since $\tan \beta = \dfrac{b}{a}$ ▲

EXAMPLE 5   A street and sidewalks separate two apartment buildings by a distance of 40 ft. From a window in her apartment, Vicki can see the top of the other apartment building through an angle of 47°. She can also see the base of the other building through an angle of depression of 33°. Approximately how tall is the other building?

Solution   The height of the building is the sum $x + y$.

$$\tan 47° = \frac{x}{40} \quad \text{and} \quad \tan 33° = \frac{y}{40}$$

$$1.0724 = \frac{x}{40} \quad \text{and} \quad 0.6494 = \frac{y}{40}$$

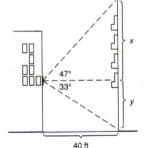

40 ft

Then $x = 43$ and $y = 26$, so the building stands *approximately* 69 ft tall.   ▲

There are a total of six trigonometric ratios. We define the remaining three for completeness; however, we will be able to solve all application problems in this chapter by using only sine, cosine, and tangent ratios. The remaining ratios are the **cotangent** (abbreviated "cot"), **secant** (abbreviated "sec"), and **cosecant** (abbreviated "csc"). These are defined in terms of the right triangle shown in figure 10.12.

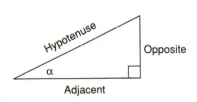

Figure 10.12

$$\cot \alpha = \frac{\text{adjacent}}{\text{opposite}}$$

$$\sec \alpha = \frac{\text{hypotenuse}}{\text{adjacent}}$$

$$\csc \alpha = \frac{\text{hypotenuse}}{\text{opposite}}$$

It is easy to see that cot α is the reciprocal of tan α; sec α is the reciprocal of cos α; and csc α is the reciprocal of sin α. In the following chart, we invert the trigonometric ratio on the left to obtain the reciprocal ratio named on the right.

Trigonometric Ratio	Reciprocal Ratio
sine	cosecant
cosine	secant
tangent	cotangent

Note: Calculators use only the sine, cosine, and tangent ratios. By using the reciprocal key, $\boxed{1/\text{x}}$, you can obtain values for the remaining ratios.

EXAMPLE 6   For the triangle shown, find all six trigonometric ratios for angle θ.

Solution   We need the length of the hypotenuse, which we find by the Pythagorean Theorem.
With c the length of the hypotenuse,

$$c^2 = 5^2 + 6^2$$
$$= 25 + 36$$
$$= 61$$
$$c = \sqrt{61}$$

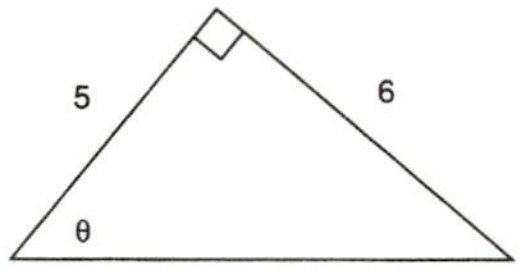

Therefore

$$\sin \theta = \frac{\text{opposite}}{\text{hypotenuse}} = \frac{6}{\sqrt{61}} = \frac{6\sqrt{61}}{61}$$

$$\cos \theta = \frac{\text{adjacent}}{\text{hypotenuse}} = \frac{5}{\sqrt{61}} = \frac{5\sqrt{61}}{61}$$

$$\tan \theta = \frac{\text{opposite}}{\text{adjacent}} = \frac{6}{5}$$

$$\cot \theta = \frac{\text{adjacent}}{\text{opposite}} = \frac{5}{6}$$

$$\sec \theta = \frac{\text{hypotenuse}}{\text{adjacent}} = \frac{\sqrt{61}}{5}$$

$$\csc \theta = \frac{\text{hypotenuse}}{\text{opposite}} = \frac{\sqrt{61}}{6}$$

▲

Note: We are quickly reminded (by the numerical results) which ratios are reciprocals of each other.

EXAMPLE 7   Evaluate the ratio named by using the given ratio:

(**a**) tan θ, if cot $\theta = \dfrac{2}{3}$

(b) $\sin \alpha$, if $\csc \alpha = 1.25$

(c) $\sec \beta$, if $\cos \beta = \dfrac{\sqrt{3}}{2}$

(d) $\csc \gamma$, if $\sin \gamma = 1$

Solution

(a) $\tan \theta = \dfrac{3}{2}$, the reciprocal of $\cot \theta$

(b) $\sin \alpha = \dfrac{4}{5}$, the reciprocal of 1.25 or $\dfrac{5}{4}$

(c) $\sec \beta = \dfrac{2}{\sqrt{3}}$ or $\dfrac{2\sqrt{3}}{3}$, the reciprocal of $\cos \beta$

(d) $\csc \gamma = 1$, the reciprocal of $\sin \gamma$ ▲

EXAMPLE 8

To the nearest degree, how large is θ in the triangle shown in the accompanying drawing?

Solution

Because the lengths "opposite" and "adjacent" are known, we can use the tangent ratio:

$$\tan \theta = \frac{5}{8} = 0.6250$$

From Table 10.5, pages 382–383 (or by a calculator), we find that $\theta = 32°$ ▲

Note: The key sequence for some calculators is

In the application exercises that follow this section, you will have to decide which trigonometric ratio allows you to solve the problem. The Pythagorean Theorem will often be used as well.

EXAMPLE 9

As his fishing vessel moves into the bay, the captain notes that the angle of elevation to the top of the lighthouse is 11°. If the lighthouse is 200 ft tall, how far is the vessel from the lighthouse?

Solution

Again we use the tangent ratio; in figure 10.13,

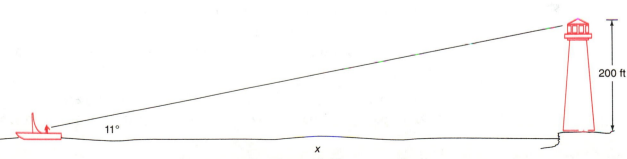

Figure 10.13

$$\tan 11° = \frac{200}{x}$$

$$0.1944 = \frac{200}{x}$$

$$0.1944x = 200$$

$$x = \frac{200}{0.1944} = 1029$$

The vessel is approximately 1029 ft from the lighthouse.   ▲

Table 10.5   **Ratios for Sine, Cosine, and Tangent**

θ	$\tan \theta$	$\cos \theta$	$\sin \theta$	θ	$\tan \theta$	$\cos \theta$	$\sin \theta$
0°	0.0000	1.0000	.0000	45°	1.0000	0.7071	0.7071
1°	0.0175	0.9998	.0175	46°	1.0355	0.6947	0.7193
2°	0.0349	0.9994	.0349	47°	1.0724	0.6820	0.7314
3°	0.0524	0.9986	.0523	48°	1.1106	0.6691	0.7431
4°	0.0699	0.9976	.0698	49°	1.1504	0.6561	0.7547
5°	0.0875	0.9962	.0872	50°	1.1918	0.6428	0.7660
6°	0.1051	0.9945	.1045	51°	1.2349	0.6293	0.7771
7°	0.1228	0.9925	.1219	52°	1.2799	0.6157	0.7880
8°	0.1405	0.9903	.1392	53°	1.3270	0.6018	0.7986
9°	0.1584	0.9877	.1564	54°	1.3764	0.5878	0.8090
10°	0.1763	0.9848	.1736	55°	1.4281	0.5736	0.8192
11°	0.1944	0.9816	.1908	56°	1.4826	0.5592	0.8290
12°	0.2126	0.9781	.2079	57°	1.5399	0.5446	0.8387
13°	0.2309	0.9744	.2250	58°	1.6003	0.5299	0.8480
14°	0.2493	0.9703	.2419	59°	1.6643	0.5150	0.8572
15°	0.2679	0.9659	.2588	60°	1.7321	0.5000	0.8660
16°	0.2867	0.9613	.2756	61°	1.8040	0.4848	0.8746
17°	0.3057	0.9563	.2924	62°	1.8807	0.4695	0.8829
18°	0.3249	0.9511	.3090	63°	1.9626	0.4540	0.8910
19°	0.3443	0.9455	.3256	64°	2.0503	0.4384	0.8988
20°	0.3640	0.9397	.3420	65°	2.1445	0.4226	0.9063
21°	0.3839	0.9336	.3584	66°	2.2460	0.4067	0.9135
22°	0.4040	0.9272	.3746	67°	2.3559	0.3907	0.9205

23°	0.4245	0.9205	.3907	68°	2.4751	0.3746	0.9272
24°	0.4452	0.9135	.4067	69°	2.6051	0.3584	0.9336
25°	0.4663	0.9063	.4226	70°	2.7475	0.3420	0.9397
26°	0.4877	0.8988	.4384	71°	2.9042	0.3256	0.9455
27°	0.5095	0.8910	.4540	72°	3.0777	0.3090	0.9511
28°	0.5317	0.8829	.4695	73°	3.2709	0.2924	0.9563
29°	0.5543	0.8746	.4848	74°	3.4874	0.2756	0.9613
30°	0.5774	0.8660	.5000	75°	3.7321	0.2588	0.9659
31°	0.6009	0.8572	.5150	76°	4.0108	0.2419	0.9703
32°	0.6249	0.8480	.5299	77°	4.3315	0.2250	0.9744
33°	0.6494	0.8387	.5446	78°	4.7046	0.2079	0.9781
34°	0.6745	0.8290	.5592	79°	5.1446	0.1908	0.9816
35°	0.7002	0.8192	.5736	80°	5.6713	0.1736	0.9848
36°	0.7265	0.8090	.5878	81°	6.3138	0.1564	0.9877
37°	0.7536	0.7986	.6018	82°	7.1154	0.1392	0.9903
38°	0.7813	0.7880	.6157	83°	8.1443	0.1219	0.9925
39°	0.8098	0.7771	.6293	84°	9.5144	0.1045	0.9945
40°	0.8391	0.7660	.6428	85°	11.4301	0.0872	0.9962
41°	0.8693	0.7547	.6561	86°	14.3007	0.0698	0.9976
42°	0.9004	0.7431	.6691	87°	19.0811	0.0523	0.9986
43°	0.9325	0.7314	.6820	88°	28.6303	0.0349	0.9994
44°	0.9657	0.7193	.6947	89°	57.2900	0.0175	0.9998
				90°	Undefined	0.0000	1.0000

10.3 EXERCISES

In Exercises 1 to 4, find tan α and tan β for each triangle.

1.
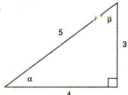

2.
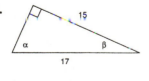

3.
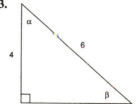

4.
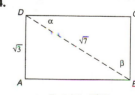

Rectangle *ABCD*

In Exercises 5 to 10, find the value (or expression) for each of the six trigonometric ratios of angle α. Use the Pythagorean Theorem as needed.

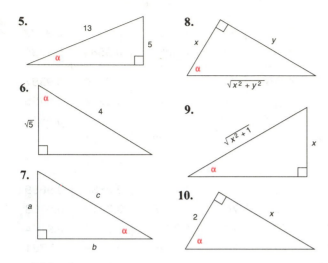

5.

6.

7.

8.

9.

10.

In Exercises 11 to 14, use Table 10.5 or a calculator to find the indicated tangent ratio, to four decimal places.

11. tan 15°   **13.** tan 57°

12. tan 45°   **14.** tan 78°

In Exercises 15 to 20, use the sine, cosine, or tangent ratio to find the length of the indicated sides of the triangle. Give each length correct to tenths.

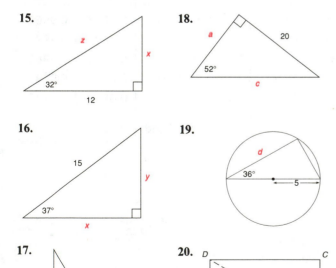

15.

16.

17.

18.

19.

20. Rectangle ABCD

In Exercises 21 to 26, use the sine, cosine, or tangent ratio to find the indicated angles of the triangle to the nearest degree.

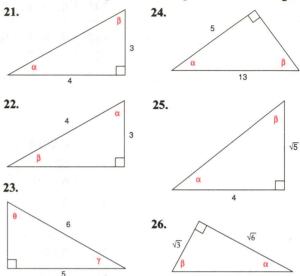

21.

22.

23.

24.

25.

26.

In Exercises 27 to 32, use Table 10.5, a calculator, and reciprocal relationships to find each ratio correct to four decimal places.

27. cot 34°   **30.** cot 67°

28. sec 15°   **31.** sec 42°

29. csc 30°   **32.** csc 72°

In Exercises 33 to 39, use the drawings provided to solve each problem. Angle measures should be given to the nearest degree; distances should be given to the nearest tenth of a unit.

33. When her airplane is descending to land, the pilot notes an angle of depression of 5°. If the altimeter shows an altitude reading of 120 ft, what is the distance x of the plane from touchdown?

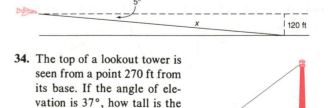

34. The top of a lookout tower is seen from a point 270 ft from its base. If the angle of elevation is 37°, how tall is the tower?

35. Find the length of the apothem to each of the 6-in. sides of a regular pentagon.

*36. What is the angle between the diagonal of a cube and the diagonal of the face of the cube?

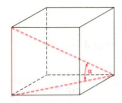

37. Upon approaching a house, Liz hears Lynette shout to her. Liz, who is standing 10 ft from the house, looks up to see Lynette in the third-story window approximately 32 ft away. What is the angle of elevation as Liz looks up at Lynette?

*38. While a helicopter hovers 1000 ft above the water, its pilot spies a man in a lifeboat through an angle of depression of 28°. Along a straight line, a rescue boat can also be seen through an angle of depression of 14°. How far is the rescue boat from the lifeboat?

*39. From atop a 200-ft lookout tower, a fire is spotted due north through an angle of depression of 12°. Firefighters located 1000 ft due east of the tower must work their way through heavy foliage to the fire. By their compasses, through what angle (measured from the north toward the west) must the firefighters travel?

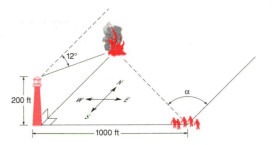

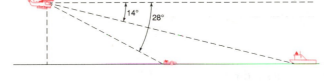

10.4

MORE TRIGONOMETRIC RELATIONSHIPS

Many of the relationships that combine trigonometric ratios are known as *trigonometric identities*. These relationships are true for all angles whenever each ratio involved is defined. For instance, the ratio tan 90° is not defined because it involves division by 0.

The relationships now stated are known as Reciprocal Identities because they involve ratios that are reciprocals of each other. Because

$$\sin \theta = \frac{\text{opposite}}{\text{hypotenuse}} \quad \text{while} \quad \csc \theta = \frac{\text{hypotenuse}}{\text{opposite}}$$

in a right triangle, we say that the sine and cosecant of the same acute angle are reciprocals. For instance, $\sin 30° = \frac{1}{2}$, while $\csc 30° = 2$.

RECIPROCAL IDENTITIES

$$\cot \theta = \frac{1}{\tan \theta} \quad \text{when } \tan \theta \neq 0$$

$$\sec \theta = \frac{1}{\cos \theta} \quad \text{when } \cos \theta \neq 0$$

$$\csc \theta = \frac{1}{\sin \theta} \quad \text{when } \sin \theta \neq 0$$

Some equivalent forms of these identities are found by multiplication. If $\cot \theta = \dfrac{1}{\tan \theta}$, then $\tan \theta \cdot \cot \theta = 1$ when each side of the equation is multiplied by tangent θ. In turn, $\tan \theta = \dfrac{1}{\cot \theta}$ by division.

RECIPROCAL IDENTITIES

$$\tan \theta = \frac{1}{\cot \theta} \quad \text{and} \quad \tan \theta \cdot \cot \theta = 1$$

$$\cos \theta = \frac{1}{\sec \theta} \quad \text{and} \quad \cos \theta \cdot \sec \theta = 1$$

$$\sin \theta = \frac{1}{\csc \theta} \quad \text{and} \quad \sin \theta \cdot \csc \theta = 1$$

EXAMPLE 1   Use the given ratio to find the desired ratio.

(a) Find $\sin \theta$, if $\csc \theta = \dfrac{5}{3}$.

(b) Find $\cos \alpha$, if $\sec \alpha = \sqrt{2}$.

(c) Find $\tan \beta$, if $\cot \beta = \dfrac{a}{b}$.

Solution   (a) $\sin \theta = \dfrac{3}{5}$, the reciprocal of $\csc \theta$.

(b) $\cos \alpha = \dfrac{1}{\sqrt{2}} = \dfrac{\sqrt{2}}{2}$, the reciprocal of $\sec \alpha$.

(c) $\tan \beta = \dfrac{b}{a}$, the reciprocal of $\cot \beta$.   ▲

The quotient relationships that follow are true for any angle θ for which the denominator of the quotient does not equal 0.

QUOTIENT RELATIONSHIPS

$$\tan \theta = \frac{\sin \theta}{\cos \theta} \quad \text{for } \theta \neq 90°$$

$$\cot \theta = \frac{\cos \theta}{\sin \theta} \quad \text{for } \theta \neq 0°$$

We will now prove that $\tan \theta = \dfrac{\sin \theta}{\cos \theta}$. In the right triangle shown,

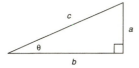

$$\sin \theta = \frac{a}{c} \text{ and } \cos \theta = \frac{b}{c}. \text{ Then}$$

$$\frac{\sin \theta}{\cos \theta} = \frac{\dfrac{a}{c}}{\dfrac{b}{c}} = \frac{a}{c} \div \frac{b}{c} = \frac{a}{c} \cdot \frac{c}{b}$$

$$= \frac{a}{b} = \tan \theta$$

Therefore

$$\tan \theta = \frac{\sin \theta}{\cos \theta}$$ ▲

In Section 10.2, you saw that $\sin^2 \theta + \cos^2 \theta = 1$ is a consequence of the Pythagorean Theorem. Through division, we can add two other Pythagorean relationships to our list.

PYTHAGOREAN RELATIONSHIPS
$\sin^2 \theta + \cos^2 \theta = 1$
$\tan^2 \theta + 1 = \sec^2 \theta$ for $\theta \neq 90°$
$\cot^2 \theta + 1 = \csc^2 \theta$ for $\theta \neq 0°$

We will now prove that $\tan^2 \theta + 1 = \sec^2 \theta$. Using $\sin^2 \theta + \cos^2 \theta = 1$, division by $\cos^2 \theta$ leads to

$$\frac{\sin^2 \theta}{\cos^2 \theta} + \frac{\cos^2 \theta}{\cos^2 \theta} = \frac{1}{\cos^2 \theta}$$

$$\left(\frac{\sin \theta}{\cos \theta} \right)^2 + 1 = \left(\frac{1}{\cos \theta} \right)^2$$

$$(\tan \theta)^2 + 1 = (\sec \theta)^2$$

$$\tan^2 \theta + 1 = \sec^2 \theta$$ ▲

Note: $\tan^2 \theta$ means the square of $\tan \theta$.

EXAMPLE 2   Given that $\tan \theta = \dfrac{3}{4}$ in an acute triangle, find $\sec \theta$.

Solution

$$\tan^2 \theta + 1 = \sec^2 \theta$$

$$\left(\frac{3}{4}\right)^2 + 1 = \sec^2 \theta$$

$$\frac{9}{16} + 1 = \sec^2 \theta$$

$$\frac{25}{16} = \sec^2 \theta$$

$$\sec \theta = \sqrt{\frac{25}{16}} = \frac{\sqrt{25}}{\sqrt{16}} = \frac{5}{4} \qquad \blacktriangle$$

We now turn our attention to some relationships that we will prove for acute triangles. The first of these is an area application.

THEOREM 10.4.1   The area of a triangle equals one-half the product of the lengths of two sides and the sine of the included angle.

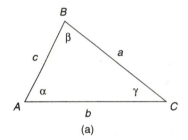

(a)

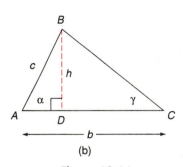

(b)

Figure 10.14

Given: Acute $\triangle ABC$, as shown in figure 10.14a

Prove: $A = \left(\frac{1}{2}\right) \cdot b \cdot c \sin \alpha$

Proof: The area of the triangle is given by $A = \left(\frac{1}{2}\right) \cdot b \cdot h$. With the altitude $\overline{BD}$, as shown in figure 10.14b, we see that $\sin \alpha = \dfrac{h}{c}$ in right $\triangle ABD$.

Then $h = c \cdot \sin \alpha$. Consequently, $A = \left(\frac{1}{2}\right) \cdot b \cdot h$ becomes

$$A = \frac{1}{2} \cdot b \cdot (c \cdot \sin \alpha)$$

so

$$A = \frac{1}{2} \cdot b \cdot c \cdot \sin \alpha \qquad \blacktriangle$$

AREA OF A TRIANGLE

$$A = \left(\frac{1}{2}\right) \cdot b \cdot c \cdot \sin \alpha$$

Equivalently, we could have proved that

$$A = \left(\frac{1}{2}\right) \cdot a \cdot c \cdot \sin \beta$$

$$A = \left(\frac{1}{2}\right) \cdot a \cdot b \cdot \sin \gamma$$

In a more advanced course (trigonometry), this area formula can be proved for obtuse triangles, too.

EXAMPLE 3   Find the area of $\triangle ABC$.

Solution

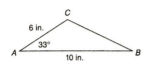

$$A = \left(\frac{1}{2}\right) \cdot b \cdot c \cdot \sin \alpha$$

$$= \left(\frac{1}{2}\right) \cdot 6 \cdot 10 \cdot \sin 33°$$

$$= \left(\frac{1}{2}\right) \cdot 6 \cdot 10 \cdot 0.5446$$

$$\approx 16.3 \text{ in.}^2$$

Note: Use a calculator or Table 10.5 to find sin 33°.   ▲

Because the area of a triangle is unique, we can equate the three area expressions of the previous box as follows:

$$\left(\frac{1}{2}\right) \cdot b \cdot c \cdot \sin \alpha = \left(\frac{1}{2}\right) \cdot a \cdot c \cdot \sin \beta = \left(\frac{1}{2}\right) \cdot a \cdot b \cdot \sin \gamma$$

Dividing each part of this equality by $\left(\frac{1}{2}\right) \cdot a \cdot b \cdot c$, we find

$$\frac{\left(\frac{1}{2}\right) \cdot b \cdot c \cdot \sin \alpha}{\left(\frac{1}{2}\right) \cdot b \cdot c \cdot a} = \frac{\left(\frac{1}{2}\right) \cdot a \cdot c \cdot \sin \beta}{\left(\frac{1}{2}\right) \cdot a \cdot c \cdot b} = \frac{\left(\frac{1}{2}\right) \cdot a \cdot b \cdot \sin \gamma}{\left(\frac{1}{2}\right) \cdot a \cdot b \cdot c}$$

$$\frac{\sin \alpha}{a} = \frac{\sin \beta}{b} = \frac{\sin \gamma}{c}$$

This relationship between the sides of a triangle and the sines of its angles is known as the Law of Sines. The Law of Sines can also be proved true for obtuse triangles.

THEOREM 10.4.2 (Law of Sines)   In any triangle, the three ratios between the sines of the angles and the lengths of the opposite sides are equal. Equivalently,

$$\frac{\sin \alpha}{a} = \frac{\sin \beta}{b} = \frac{\sin \gamma}{c}$$

Generally, only two of the equal ratios described in Theorem 10.4.2 are equated in solving a problem. That is,

$$\frac{\sin \alpha}{a} = \frac{\sin \beta}{b} \qquad \text{or} \qquad \frac{\sin \alpha}{a} = \frac{\sin \gamma}{c} \qquad \text{or} \qquad \frac{\sin \beta}{b} = \frac{\sin \gamma}{c}$$

EXAMPLE 4   Use the Law of Sines to find ST.

Solution   Since we know RT and want ST, we use

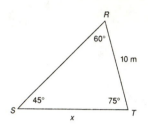

$$\frac{\sin S}{RT} = \frac{\sin R}{ST}$$

$$\frac{\sin 45°}{10} = \frac{\sin 60°}{x}$$

$$\frac{\frac{\sqrt{2}}{2}}{10} = \frac{\frac{\sqrt{3}}{2}}{x}$$

$\left(\text{Recall from Section 10.1 that } \sin 45° = \frac{\sqrt{2}}{2} \text{ and } \sin 60° = \frac{\sqrt{3}}{2}.\right)$ By the Means Extremes Property,

$$\frac{\sqrt{2}}{2} \cdot x = \frac{\sqrt{3}}{2} \cdot 10$$

Multiplying by $\frac{2}{\sqrt{2}}$, we have

$$\frac{2}{\sqrt{2}} \cdot \frac{\sqrt{2}}{2} \cdot x = \frac{2}{\sqrt{2}} \cdot \frac{\sqrt{3}}{2} \cdot 10$$

$$x = \frac{10\sqrt{3}}{\sqrt{2}} = \frac{10\sqrt{3}}{\sqrt{2}} \cdot \frac{\sqrt{2}}{\sqrt{2}} = \frac{10\sqrt{6}}{2} = 5\sqrt{6}$$

Then $ST = 5\sqrt{6}$ m.   ▲

The final relationship we will consider is also proved for an acute triangle. Like the Law of Sines, this relationship (known as the Law of Cosines) can also be used to find unknown measures in a triangle. The Law of Cosines (which can also be established for obtuse triangles in a more advanced course) can be stated in compact form as "The square of one side of a triangle equals the sum of squares of the remaining two sides decreased by twice the product of the other two sides and the cosine of the angle that those sides include."

THEOREM 10.4.3 (Law of Cosines)   In acute $\triangle ABC$,

$$c^2 = a^2 + b^2 - 2 \cdot a \cdot b \cdot \cos \gamma$$
$$b^2 = a^2 + c^2 - 2 \cdot a \cdot c \cdot \cos \beta$$
$$a^2 = b^2 + c^2 - 2 \cdot b \cdot c \cdot \cos \alpha$$

The proof of the first form of the Law of Cosines follows:

Given: Acute $\triangle ABC$

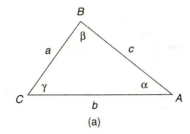

(a)

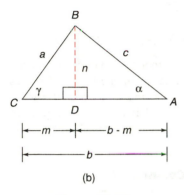

(b)

Figure 10.15

Prove: $c^2 = a^2 + b^2 - 2 \cdot a \cdot b \cdot \cos \gamma$

Proof: In figure 10.15a, draw the altitude $\overline{BD}$ from B to $\overline{AC}$. We designate lengths as shown in figure 10.15b. Now

$$(b - m)^2 + n^2 = c^2 \qquad \text{and} \qquad m^2 + n^2 = a^2$$

by the Pythagorean Theorem.

The second statement is equivalent to $m^2 = a^2 - n^2$. After we expand $(b - m)^2$, the first equation becomes

$$b^2 - 2bm + m^2 + n^2 = c^2$$

Then we replace m^2 by $(a^2 - n^2)$ to obtain

$$b^2 - 2bm + (a^2 - n^2) + n^2 = c^2$$

Simplifying yields

$$c^2 = a^2 + b^2 - 2bm$$

In right $\triangle CDB$,

$$\cos \gamma = \frac{m}{a} \qquad \text{so} \qquad m = a \cdot \cos \gamma$$

Hence we write

$$\begin{aligned} c^2 &= a^2 + b^2 - 2bm \\ &= a^2 + b^2 - 2b(a \cdot \cos \gamma) \\ &= a^2 + b^2 - 2 \cdot a \cdot b \cdot \cos \gamma \end{aligned} \qquad \blacktriangle$$

Note: Similar proofs can be constructed for both remaining forms of the Law of Cosines.

EXAMPLE 5   Find the length of $\overline{AB}$ in the triangle shown in the accompanying figure.

Solution   Referring to the 30° angle as γ, we use the form

$$\begin{aligned} c^2 &= a^2 + b^2 - 2 \cdot a \cdot b \cdot \cos \gamma \\ &= (4\sqrt{3})^2 + 4^2 - 2 \cdot 4\sqrt{3} \cdot 4 \cdot \cos 30° \\ &= 48 + 16 - 2 \cdot 4\sqrt{3} \cdot 4 \cdot \frac{\sqrt{3}}{2} \\ &= 48 + 16 - 48 \\ &= 16 \\ c &= 4 \end{aligned}$$

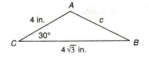

Therefore $AB = 4$ in. $\blacktriangle$

The Law of Cosines can also be used to find an angle of a triangle when its sides are known.

EXAMPLE 6   In acute $\triangle ABC$, find β to the nearest degree.

Solution   The particular form of the Law of Cosines involving β is $b^2 = a^2 + c^2 - 2 \cdot a \cdot c \cdot \cos \beta$. Since $b = 6$ is the length of the side opposite β, we have

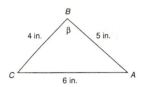

$$6^2 = 4^2 + 5^2 - 2 \cdot 4 \cdot 5 \cdot \cos \beta$$
$$36 = 16 + 25 - 40 \cdot \cos \beta$$
$$36 = 41 - 40 \cdot \cos \beta$$

Therefore $40 \cdot \cos \beta = 5$

$$\cos \beta = \frac{5}{40} = \frac{1}{8} = 0.1250$$

To find β, use Table 10.5 or a calculator. In either case, $\beta = 83°$.   ▲

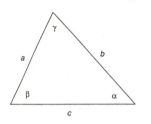

In finding the measure of a side or an angle of an acute triangle, a decision must be made as to which form of the Law of Sines or of the Law of Cosines should be applied. Table 10.6 deals with that question and is based on the acute triangle shown in the accompanying drawing. Note that a, b, and c represent the lengths of the sides, while α, β, and γ represent the measures of the opposite angles, respectively.

Table 10.6   **When to Use the Law of Sines/Law of Cosines**

1. *Three sides are known:* Use the Law of Cosines to find *any* angle.

   Known measures: *a, b,* and *c*
   Desired measure: α

   $\therefore$ *Use:* $a^2 = b^2 + c^2 - 2bc \cdot \cos \alpha$

> ▲ *WARNING: If only the measures of the three angles of the triangle are known, then no side can be determined.*

2. *Two sides and a nonincluded angle are known:* Use the Law of Sines to find the remaining nonincluded angle.

   Known measures: *a, b,* and α
   Desired measure: β

   $\therefore$ *Use:* $\dfrac{\sin \alpha}{a} = \dfrac{\sin \beta}{b}$

3. *Two sides and an included angle are known:* Use the Law of Cosines to find the remaining side.

   Known measures: a, b, and γ
   Desired measure: *c*

   $\therefore$ *Use:* $c^2 = a^2 + b^2 - 2ab \cdot \cos \gamma$

4. *Two angles and a nonincluded side are known:* Use the Law of Sines to find the other nonincluded side.

   Known measures: *a,* α, and β
   Desired measure: *b*

   $\therefore$ *Use:* $\dfrac{\sin \alpha}{a} = \dfrac{\sin \beta}{b}$

EXAMPLE 7   In the design of a child's swing set, the two metal posts that support the top bar each measure 8 ft. At ground level, the posts are to be 6 ft apart. At what angle should the two metal posts be secured?

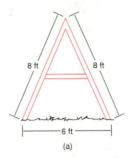
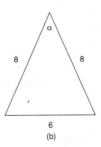

(a)   (b)

Solution   Call the desired angle α. Because the three sides of the triangle are known, we use the Law of Cosines of the form $a^2 = b^2 + c^2 - 2bc \cdot \cos \alpha$.

Because a represents the length of the side opposite the angle α, $a = 6$ while $b = 8$ and $c = 8$. Consequently, we have

$$6^2 = 8^2 + 8^2 - 2 \cdot 8 \cdot 8 \cdot \cos \alpha$$
$$36 = 64 + 64 - 128 \cdot \cos \alpha$$
$$36 = 128 - 128 \cdot \cos \alpha$$
$$-92 = -128 \cdot \cos \alpha$$
$$\cos \alpha = \frac{-92}{-128}$$
$$\cos \alpha = 0.7188$$

Therefore $\alpha = 44°$ ▲

10.4 EXERCISES

In Exercises 1 and 2, use the accompanying figure.

1. Show that $\dfrac{\cos \theta}{\sin \theta} = \cot \theta$.

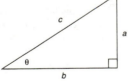

2. Show that $\sec \theta \cdot \sin \theta = \tan \theta$.

In Exercises 3 and 4, use the quotient relationships.

3. If $\sin \theta = 0.6$ and $\cos \theta = 0.8$, find $\tan \theta$ and $\cot \theta$.

4. If $\sin \theta = \dfrac{8}{17}$ and $\cos \theta = \dfrac{15}{17}$, find $\tan \theta$ and $\cot \theta$.

5. If $\sin \theta = \dfrac{3}{5}$, find $\cos \theta$. (*Hint:* Use the fact that $\sin^2 \theta + \cos^2 \theta = 1$.)

6. If $\cos \theta = \dfrac{\sqrt{3}}{2}$, find $\sin \theta$.

7. If $\tan \theta = \dfrac{5}{12}$, find $\sec \theta$.

8. If $\sec \theta = \dfrac{17}{15}$, find $\tan \theta$.

9. Show that $\cot^2 \theta + 1 = \csc^2 \theta$.

10. State whether each equation is true or false:
   (a) $\cos^2 \theta - \sin^2 \theta = 1$
   (b) $\csc^2 \theta - \cot^2 \theta = 1$
   (c) $\tan \theta \cdot \cot \theta = 1$
   (d) $\sin \theta \cdot \cos \theta = 1$

In Exercises 11 to 14, find the area of each triangle shown. Give the answer to the nearest tenth of a square unit.

11.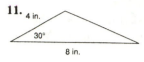

12.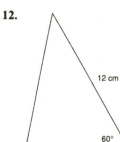

13.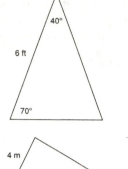

14.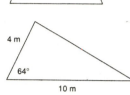

In Exercises 15 and 16, find the area of the given figure. Give the answer to the nearest tenth of a square unit.

15.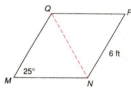

Rhombus *MNPQ*

16.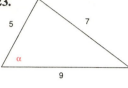

In Exercises 17 to 22, use a form of the Law of Sines to find the measure of the indicated side or angle. Angle measures should be given to the nearest degree; lengths should be given to the nearest tenth of a unit.

17.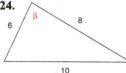

18.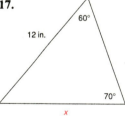

19.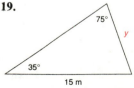

20.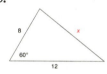

21.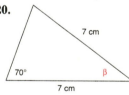

22.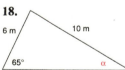

In Exercises 23 to 28, use a form of the Law of Cosines to find the measure of the indicated side or angle. Angle measures should be given to the nearest degree; lengths should be given to the nearest tenth of a unit.

23.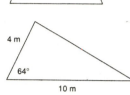

24.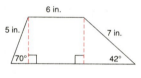

25.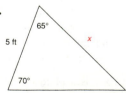

26.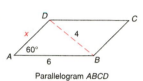

27.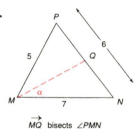

Parallelogram *ABCD*

28.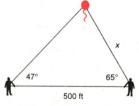

$\overrightarrow{MQ}$ bisects ∠*PMN*

In Exercises 29 to 34, use the Law of Sines or the Law of Cosines to solve each problem. Angle measures should be given to the nearest degree; area and distances should be given to the nearest tenth of a unit.

29. A triangular lot has street dimensions of 150 ft and 180 ft, and an included angle of 80° for these two sides.

(a) Find the length of the remaining side of the lot.
(b) Find the area of the lot, in square feet.

30. Two people observe a balloon. They are 500 ft apart, and their angles of observation are 47° and 65°, respectively. Find the distance *x* from the second observer to the balloon.

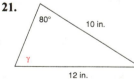

31. A surveillance aircraft at point *C* sights an ammunition warehouse at *A* and enemy headquarters at *B* through the angles indicated. If points *A* and *B* are 10,000 m apart, what is the distance from the aircraft to enemy headquarters?

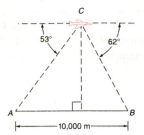

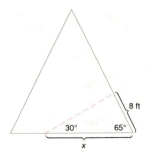

32. Above one room of a house, the rafters meet as shown in the accompanying drawing. What is the measure of the angle α at which they meet?

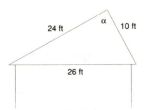

*33. In an A-frame house, a bullet is found embedded at a point 8 ft up the sloped wall. If it was fired at a 30° angle with the horizontal, how far from the base of the wall was the gun fired?

34. Clay pigeons are released at an angle of 30° with the horizontal. A sharpshooter hits one of the clay pigeons when shooting through an angle of elevation of 70°. If the point of release is 120 m from the sharpshooter, how far (*x*) is the sharpshooter from the target when it is hit?

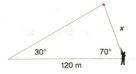

35. Show that the form $c^2 = a^2 + b^2 - 2 \cdot a \cdot b \cdot \cos \gamma$ reduces to the Pythagorean Theorem when $\gamma = 90°$.

▲ A Look Beyond: Radian Measure of Angles

In much of this textbook, we have considered angle measures from 0° to 180°. As you apply geometry, you will find that two things are true:

1. Angle measures do not have to be limited to degree measures from 0° to 180°.

2. The degree is not the only unit used in measuring angles.

We will address the first of these issues in Examples 1, 2, and 3.

EXAMPLE 1

As the time changes from 1 P.M. to 1:45 P.M., through what angle does the minute hand rotate?

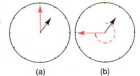

(a)   (b)

Solution

Since the rotation is $\frac{3}{4}$ of a complete circle (360°), the result

is $\left(\frac{3}{4}\right)360°$ or 270°.   ▲

EXAMPLE 2

An airplane pilot is instructed to circle the tower twice during a holding pattern before receiving clearance to land. Through what angle does the airplane move?

Solution

Two circular rotations give 2(360°) or 720°.   ▲

In trigonometry, negative measures for angles are used to distinguish the direction of rotation. A counterclockwise rotation is measured as positive, while a clockwise rotation is measured as negative. The arrowed arcs in figure 10.16 are used to indicate the direction of rotation.

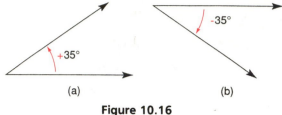

(a)   (b)

Figure 10.16

EXAMPLE 3

To tighten a hex bolt, a mechanic applies rotations of 45° several times. What is the measure of each rotation?

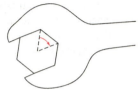

Solution

Tightening occurs if the angle is −45°.

Note: If the angle of rotation is 45° (that is, +45°), the bolt is loosened.   ▲

Our second concern is with an alternative unit of measuring angles—a unit often used in the study of trigonometry and calculus.

> **DEFINITION:**   A **radian** (rad) is the measure of a central angle in a circle that intercepts an arc whose length is equal to the radius of the circle.

In figure 10.17 the length of each radius and the intercepted arc are all equal to r. Thus, the central angle shown measures 1 rad. A complete rotation about the circle corresponds to 360° and to $2\pi r$. Thus, the arc length of 1 radius equals the central angle measure of 1 rad, and the circumference of 2π radii equals the complete rotation of 2π rad.

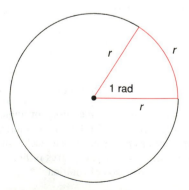

Figure 10.17

This relationship for the complete rotation allows us to equate 360° and 2π radians. As suggested by figure 10.18 there are approximately 6.28 rad (or exactly 2π radians) about the circle. The exact result leads to this important relationship.

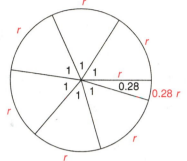

Figure 10.18

$$2\pi \text{ rad} = 360°$$
or
$$360° = 2\pi \text{ rad}$$

Through division by 2, the relationship is often restated as follows:

$$\pi \text{ rad} = 180°$$
or
$$180° = \pi \text{ rad}$$

Because π rad = 180°, we may divide by π to see the following relationship:

$$1 \text{ rad} = \frac{180°}{\pi} \approx 57.3°$$

To compare angle measures, we may also wish to divide each side of the equation 180° = π rad by 180 to get the following relationship.

$$1° = \frac{\pi}{180} \text{ rad}$$

EXAMPLE 4

Using the fact that $1° = \dfrac{\pi}{180}$ rad, find the radian equivalencies for

(a) 30°   (b) 45°   (c) 60°   (d) −90°

Solution

(a) $30° = 30(1°) = 30\left(\dfrac{\pi}{180}\right) \text{rad} = \dfrac{\pi}{6} \text{rad}$

(b) $45° = 45(1°) = 45\left(\dfrac{\pi}{180}\right) \text{rad} = \dfrac{\pi}{4} \text{rad}$

(c) $60° = 60(1°) = 60\left(\dfrac{\pi}{180}\right) \text{rad} = \dfrac{\pi}{3} \text{rad}$

(d) $-90° = -90(1°) = -90\left(\dfrac{\pi}{180}\right) \text{rad} = \dfrac{-\pi}{2} \text{rad}$ ▲

EXAMPLE 5

Using the fact that π rad = 180°, find the degree equivalencies for the following angles measured in radians:

(a) $\dfrac{\pi}{6}$ (b) $\dfrac{2\pi}{5}$ (c) $\dfrac{-3\pi}{4}$ (d) $\dfrac{\pi}{2}$

Solution

(a) $\dfrac{\pi}{6} = \dfrac{180°}{6} = 30°$

(b) $\dfrac{2\pi}{5} = \dfrac{2}{5} \cdot \pi = \dfrac{2}{5} \cdot 180° = 72°$

(c) $\dfrac{-3\pi}{4} = \dfrac{-3}{4} \cdot \pi = \dfrac{-3}{4} \cdot 180° = -135°$

(d) $\dfrac{\pi}{2} = \dfrac{180°}{2} = 90°$ ▲

While we do not pursue this method of measuring angles further in this textbook, you may need to use this method of angle measurement in a more advanced course.

▲ SUMMARY

A LOOK BACK AT CHAPTER 10

One goal of this chapter was to define the sine, cosine, and tangent ratios in terms of the sides of a right triangle. We derived a formula for finding the area of a triangle, given two sides and the included angle. We also proved the Law of Sines and the Law of Cosines for acute triangles. Another unit, called the "radian," was introduced for the purpose of measuring angles.

IMPORTANT TERMS AND CONCEPTS OF CHAPTER 10

10.1 Greek Letters for Angles: $\alpha, \beta, \gamma, \theta$

　　　　Opposite Side
　　　　Sine Ratio
10.2 Adjacent Side
　　　　Cosine Ratio
　　　　Identity
10.3 Tangent Ratio
　　　　Cosecant
　　　　Cotangent
　　　　Secant
10.4 Law of Sines
　　　　Law of Cosines
A Look Beyond Radian Measure of Angles

▲ REVIEW EXERCISES

In Exercises 1 to 4, state the ratio needed, and use it to find the measure of the indicated segment to the nearest tenth of a unit.

1.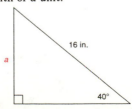

2.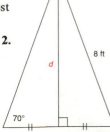

3.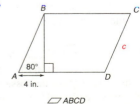

▱ ABCD

4.

Regular pentagon
with radius = 5 ft

In Exercises 5 to 8, state the ratio needed, and use it to find the measure of the angle to the nearest degree. Use Table 10.5 or a calculator.

5.

14 in.   13 in.

α

6.

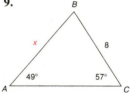

B   10 ft   C

15 ft

θ

A   26 in.   D

Isosceles trapezoid ABCD

7.

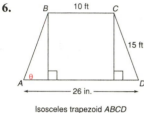

B   C

α 9 cm

12 cm

A   D

Rhombus ABCD

8.

7 in.

24 in.

O

β

Circle O

In Exercises 9 to 12, use the Law of Sines or the Law of Cosines to solve each triangle for the indicated side or angle. Angle measures should be given to the nearest degree; distances should be given to the nearest tenth of a unit.

9.

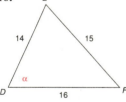

B

x 8

49°   57°

A   C

10.

E

14   15

α

D   16   F

11.

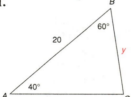

B

60°

20   y

40°

A   C

12.

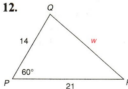

Q

14   w

60°

P   21   R

In Exercises 13 to 17, use the Law of Sines or the Law of Cosines to solve each problem. Angle measures should be given to the nearest degree; distances should be given to the nearest tenth of a unit.

13. A building 50 ft tall is on a hillside. A surveyor at a point on the hill observes that the angle of elevation to the top of the building measures 43° and the angle of depression to the bottom of the building measures 16°. How far is the surveyor from the bottom of the building?

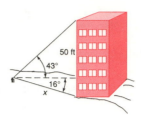

50 ft

43°

16°

x

14. Two sides of a parallelogram are 50 cm and 70 cm long. Find the length of the shorter diagonal if the larger angle is 105°.

15. The sides of a rhombus are 6 in. each and the longer diagonal is 11 in. Find the measure of the acute angle in the rhombus.

16. The area of $\triangle ABC$ is 9.7 in.^{2}. If $a = 6$ in. and $c = 4$ in., find the measure of angle B.

17. Find the area of the rhombus in Exercise 15.

In Exercises 18 to 20, prove each statement without using a table or a calculator. Draw an appropriate right triangle.

18. If $m\angle R = 45°$, then $\tan R = 1$.

19. If $m\angle S = 30°$, then $\sin S = \dfrac{1}{2}$.

20. If $m\angle T = 60°$, then $\sin T = \dfrac{\sqrt{3}}{2}$.

In Exercises 21 to 30, use the drawings where provided to solve each problem. Angle measures should be given to the nearest degree; lengths should be given to the nearest tenth of a unit.

21. In the evening, a tree that stands 12 ft tall casts a long shadow. If the angle of depression from the top of the tree to the top of the shadow is 55°, what is the length of the shadow?

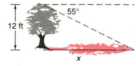

55°

12 ft

x

22. A rocket is shot into the air at an angle of 60°. If it is traveling at 200 ft/s, how high in the air is it after 5 s? (Ignoring gravity, assume that the path of the rocket is in a straight line.)

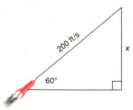

200 ft/s

x

60°

23. A 4-m beam is used to brace a weak wall. If the bottom of the beam is 3 m from the base of the wall, what is the angle of elevation to the top of the wall?

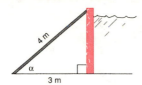

24. A hot air balloon is 300 ft high. The pilot of the balloon sights a stadium 2200 ft away. What is the angle of depression?

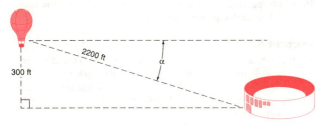

25. The apothem of a regular pentagon is approximately 3.44 cm. What is the approximate length of each side of the pentagon?

26. What is the approximate length of the radius of the pentagon in Exercise 25?

27. The legs of an isosceles triangle are each 40 cm in length. The base is 30 cm in length. Find the measure of a base angle.

28. The diagonals of a rhombus measure 12 in. and 16 in. Find the measure of the obtuse angle in the rhombus.

29. The unit used for measuring the steepness of a hill is the **grade.** A grade of a to b means the hill rises a vertical units for every b horizontal units. If, at some point, the hill is 3 ft above the horizontal and the angle of elevation to that point is 23°, what is the grade of this hill?

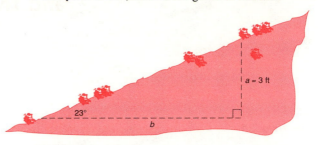

30. An observer in a plane 2500 m high sights two ships below. The angle of depression for one ship is 32°, and the angle of depression for the other ship is 44°. How far apart are the ships?

31. If $\sin \theta = \dfrac{7}{25}$, find $\cos \theta$ and $\sec \theta$.

32. If $\tan \theta = \dfrac{11}{60}$, find $\sec \theta$ and $\cot \theta$.

33. If $\cot \theta = \dfrac{21}{20}$, find $\csc \theta$ and $\sin \theta$.

APPENDIX: Summary of Constructions, Postulates, and Theorems and Corollaries

CONSTRUCTIONS

Section 2.1

1. To construct a segment congruent to a given segment.
2. To construct the midpoint M of a given line segment AB.

Section 2.2

3. To construct an angle congruent to a given angle.
4. To construct the angle bisector of a given angle.

Section 2.3

5. To construct the line perpendicular to a given line at a specified point on the given line.

Section 3.1

6. To construct the line that is perpendicular to a given line from a point not on the given line.

A Look Beyond (Chapter 3)

7. To construct the line parallel to a given line from a point not on that line.

Section 6.4

8. To construct a tangent to a circle at a point on the circle.
9. To construct a tangent to a circle from an external point.

POSTULATES

Section 2.1

1. Through two distinct points, there is exactly one line.
2. (Ruler Postulate) The measure of any line segment is a unique positive number.
3. (Segment-Addition Postulate) If X is a point on $\overline{AB}$ and A-X-B, then $AX + XB = AB$.
4. Through three noncollinear points, there is exactly one plane.
5. If two distinct planes intersect, then their intersection is a line.
6. Given two distinct points in a plane, the line containing these points also lies in the plane.

Section 2.2

7. (Protractor Postulate) The measure of an angle is a unique positive number.
8. (Angle-Addition Postulate) If a point D lies in the interior of angle ABC, then $m\angle ABD + m\angle DBC = m\angle ABC$.

Section 3.1

9. (Parallel Postulate) Through a point not on a line, exactly one line is parallel to the given line.
10. If two parallel lines are cut by a transversal, then the corresponding angles are congruent.

Section 4.1

11. If the three sides of one triangle are congruent to the three sides of another triangle, then the triangles are congruent (SSS).
12. If two sides and the included angle of one triangle are congruent to two sides and the included angle of a second triangle, then the triangles are congruent (SAS).
13. If two angles and the included side of one triangle are congruent to two angles and the included side of a second triangle, then the triangles are congruent (ASA).

Section 5.3

14. If the three angles of one triangle are congruent to the three angles of another triangle, then the triangles are similar (AAA).

Section 6.1

15. (Central Angle Postulate) In a circle, the degree measure of a central angle is equal to the degree measure of its intercepted arc.
16. (Arc-Addition Postulate) If A, B, C lie in that order on a circle, then $m\widehat{AB} + m\widehat{BC} = m\widehat{ABC}$.

A Look Beyond (Chapter 6)

17. The ratio of the circumference of a circle to its length of diameter is a constant.

Section 7.1

18. (Area Postulate) Corresponding to every bounded region is a unique positive number A, known as the area of that region.
19. If two closed plane figures are congruent, then their areas are equal.
20. (Area-Addition Postulate) Let R and S be two regions with a common boundary and for which the intersection is empty (no overlap). Then
    $$A_{R \cup S} = A_R + A_S.$$
21. The area A of a rectangle whose base has length b and whose altitude has length h is given by $A = b \cdot h$.

Section 9.1

22. (Volume Postulate) Corresponding to every solid is a unique positive number V known as the volume of that solid.

23. The volume of a right rectangular prism is given by

$$V = \ell \cdot w \cdot h$$

where ℓ measures the length, w the width, and h the altitude of the prism.

24. The volume of a right prism is given by

$$V = B \cdot h$$

in which B is the area of a base and h is the altitude of the prism.

THEOREMS AND COROLLARIES

2.1.1 The midpoint of a line segment is unique.

2.2.1 There is one and only one angle bisector for a given angle.

2.3.1 There is exactly one line perpendicular to a given line at any point on the line.

2.3.2 The perpendicular bisector of a line segment is unique.

2.4.1 If two lines are perpendicular, then they meet to form right angles.

2.4.2 If two lines meet to form a right angle, the these lines are perpendicular.

2.4.3 If two angles are complementary to the same angle (or to congruent angles), then these angles are congruent.

2.4.4 If two angles are supplementary to the same angle (or to congruent angles), then these angles are congruent.

2.4.5 If two lines intersect, then the vertical angles formed are congruent.

2.4.6 Any two right angles are congruent.

2.4.7 If the exterior sides of two adjacent angles form perpendicular rays, then these angles are complementary.

2.4.8 If the exterior sides of two adjacent angles form a straight line, then these angles are supplementary.

2.4.9 If two segments are congruent, then their midpoints separate these segments into four congruent segments.

2.4.10 If two angles are congruent, then their bisectors separate these angles into four congruent angles.

3.1.1 Exactly one line is perpendicular to a given line from a point not on that line.

3.1.2 If two parallel lines are cut by a transversal, then the alternate interior angles are congruent.

3.1.3 If two parallel lines are cut by a transversal, then the alternate exterior angles are congruent.

3.1.4 If two parallel lines are cut by a transversal, then the interior angles on the same side of the transversal are supplementary.

3.1.5 If two parallel lines are cut by a transversal, then the exterior angles on the same side of the transversal are supplementary.

3.3.1 If two lines are cut by a transversal so that the corresponding angles are congruent, then these lines are parallel.

3.3.2 If two lines are cut by a transversal so that the alternate interior angles are congruent, then these lines are parallel.

3.3.3 If two lines are cut by a transversal so that the alternate exterior angles are congruent, then these lines are parallel.

3.3.4 If two lines are cut by a transversal so that the interior angles on the same side of the transversal are supplementary, then these lines are parallel.

3.3.5 If two lines are cut by a transversal so that the exterior angles on the same side of the transversal are supplementary, then these lines are parallel.

3.3.6 If two lines are each parallel to a third line, then these lines are parallel to each other.

3.3.7 If two lines are each perpendicular to a third line, then these lines are parallel to each other.

3.4.1 In a triangle, the sum of the measures of the interior angles is $180°$.

3.4.2 Each angle of an equiangular triangle measures $60°$.

3.4.3 The acute angles of a right triangle are complementary.

3.4.4 The measure of an exterior angle of a triangle equals the sum of measures of the two nonadjacent interior angles.

3.4.5 If two angles of one triangle are congruent to two angles of another triangle, then the third angles are also congruent.

3.4.6 The sum of the measures of the interior angles of a polygon with n sides is given by $(n - 2) \cdot 180°$. Note that $n > 2$ for any polygon.

3.4.7 The sum of the measures of the exterior angles, one at each vertex, of a polygon is $360°$

4.1.1 If two angles and the nonincluded side of one triangle are congruent to two angles and the nonincluded side of a second triangle, then the triangles are congruent (AAS).

4.2.1 If the hypotenuse and a leg of one right triangle are congruent to the hypotenuse and a leg of a second right triangle, then the triangles are congruent (HL).

4.3.1 Corresponding altitudes of congruent triangles are congruent.

4.3.2   The bisector of the vertex angle of an isosceles triangle separates the triangle into two congruent triangles.

4.3.3   If two sides of a triangle are congruent, then the angles opposite these sides are also congruent.

4.3.4   If two angles of a triangle are congruent, then the sides opposite these angles are also congruent.

4.3.5   An equilateral triangle is also equiangular.

4.3.6   An equiangular triangle is also equilateral.

4.4.1   A diagonal of a parallelogram separates it into two congruent triangles.

4.4.2   Opposite angles of a parallelogram are congruent.

4.4.3   Opposite sides of a parallelogram are congruent.

4.4.4   Diagonals of a parallelogram bisect each other.

4.4.5   Consecutive angles of a parallelogram are supplementary.

4.4.6   If two sides of a quadrilateral are both congruent and parallel, then the quadrilateral is a parallelogram.

4.4.7   If both pairs of opposite sides of a quadrilateral are congruent, then it is a parallelogram.

4.4.8   If the diagonals of a quadrilateral bisect each other, then the quadrilateral is a parallelogram.

4.4.9   A kite has two pairs of congruent adjacent sides.

4.4.10   The segment that joins the midpoints of two sides of a triangle is parallel to the third side and has a length equal to one-half the length of the third side.

4.5.1   All angles of a rectangle are right angles.

4.5.2   The diagonals of a rectangle are congruent.

4.5.3   All sides of a square are congruent.

4.5.4   All sides of a rhombus are congruent.

4.5.5   The diagonals of a rhombus are perpendicular.

4.5.6   The base angles of an isosceles trapezoid are congruent.

4.5.7   The diagonals of an isosceles trapezoid are congruent.

4.5.8   The length of the median of a trapezoid equals one-half the sum of the lengths of the two bases.

4.5.9   The median of a trapezoid is parallel to each base.

4.6.1   (Addition Property of Inequality): If $a > b$ and $c > d$, then $a + c > b + d$.

4.6.2   If one side of a triangle is longer than a second side, then the measure of the angle opposite the first side is greater than the measure of the angle opposite the second side.

4.6.3   If the measure of one angle of a triangle is greater than the measure of a second angle, then the side opposite the first angle is longer than the side opposite the second angle.

4.6.4   The perpendicular segment from a point to a line is the shortest segment that can be drawn from the point to the line.

4.6.5   The perpendicular segment from a point to a plane is the shortest segment that can be drawn from the point to the plane.

4.6.7   (Triangle Inequality) The sum of the lengths of any two sides of a triangle is greater than the length of the third side.

5.3.1   If two angles of one triangle are congruent to two angles of another triangle, then the triangles are similar (AA).

5.4.1   If a line is parallel to one side of a triangle and intersects the other two sides, then it divides these sides proportionally.

5.4.2   When three (or more) parallel lines are cut by a pair of transversals, the transversals are divided proportionally by the parallel lines.

5.4.3   If a ray bisects one angle of a triangle, then it divides the opposite side into segments that are proportional to the two sides which form that angle.

5.5.1   The altitude drawn to the hypotenuse of a right triangle separates the right triangle into two right triangles that are similar to each other and to the original right triangle.

5.5.2   The length of the altitude to the hypotenuse of a right triangle is the geometric mean of the lengths of the segments of the hypotenuse.

5.5.3   The length of each leg of a right triangle is the geometric mean of the length of the hypotenuse and the length of the projection of the leg on the hypotenuse.

5.5.4   (Pythagorean Theorem) The square of the length of the hypotenuse of a right triangle is equal to the sum of the squares of the lengths of the legs.

5.5.5   (Converse of Pythagorean Theorem) If a, b, and c are the lengths of the three sides of a triangle, with c the length of the longest side, and if $c^2 = a^2 + b^2$, then the triangle is a right triangle with the right angle opposite the side of length c.

5.5.6   If the hypotenuse and a leg of a one right triangle are congruent to the hypotenuse and a leg of a second right triangle, then the triangles are congruent (HL).

5.5.7 If *a*, *b*, and *c* are the lengths of the three sides of a triangle and *c* is the length of the longest side, and

1. $$c^2 > a^2 + b^2$$

then the triangle is obtuse and the obtuse angle is opposite the side of length *c*.

or:

2. $$c^2 < a^2 + b^2$$

then the triangle is acute.

5.6.1 (45-45-90 Theorem) In a triangle whose angles measure 45°, 45°, and 90°, the hypotenuse has a length equal to the product of $\sqrt{2}$ and the length of either leg.

5.6.2 (30-60-90 Theorem) In a triangle whose angles measure 30°, 60°, and 90°, the hypotenuse has a length equal to twice the length of the shorter leg, while the length of the longer leg is the product of $\sqrt{3}$ and the length of the shorter leg.

6.1.1 All radii of a circle are congruent.

6.1.2 A radius drawn perpendicular to a chord bisects the chord.

6.1.3 The measure of an inscribed angle of a circle is one-half the measure of its intercepted arc.

6.1.4 In a circle or in congruent circles, congruent minor arcs have congruent central angles.

6.1.5 In a circle or in congruent circles, congruent central angles have congruent arcs.

6.1.6 In a circle or in congruent circles, congruent chords have congruent minor (major) arcs.

6.1.7 In a circle or in congruent circles, congruent arcs have congruent chords.

6.1.8 Chords at the same distance from the center of a circle are congruent.

6.1.9 Congruent chords are at the same distance from the center of a circle.

6.1.10 An angle inscribed in a semicircle is a right angle.

6.1.11 If two inscribed angles intercept the same arc, then these angles are congruent.

6.2.1 If a qudrilateral is inscribed in a circle, then the opposite angles are supplementary.

6.2.2 The measure of an angle formed by two intersecting chords is one-half the sum of measures of the arcs intercepted by the angle and its vertical angle.

6.2.3 The radius (or any other line through the center of a circle) drawn to a tangent at the point of tangency is perpendicular to the tangent at that point.

6.2.4 The measure of an angle formed by a tangent and a chord drawn to the point of tangency is one-half the measure of the intercepted arc.

6.2.5 The measure of an angle formed when two secants intersect at a point outside the circle is one-half the difference of the measures of the two intercepted arcs.

6.2.6 If an angle is formed by a secant and tangent that intersect in the exterior of a circle, then the measure of the angle is one-half the difference of the measures of its intercepted arcs.

6.2.7 If an angle is formed by two intersecting tangents, then the measure of the angle is one-half the difference of the measures of the intercepted arcs.

6.2.8 If two parallel lines intersect a circle, then the intercepted arcs between these lines are congruent.

6.3.1 If a line is drawn through the center of a circle perpendicular to a chord, then it bisects the chord and its arc.

6.3.2 If a line through the center of a circle bisects a chord other than a diameter, then it is perpendicular to the chord.

6.3.3 The perpendicular bisector of a chord contains the center of the circle.

6.3.4 The tangent segments to a circle from an external point are congruent.

6.3.5 If two chords intersect within a circle, then the product of the lengths of the segments (parts) of one chord is equal to the product of the lengths of the segments of the other.

6.3.6 If two secant segments are drawn to a circle from an external point, then the products of the lengths of each secant and its external segment are equal.

6.3.7 If a tangent segment and secant segment are drawn to a circle from an external point, then the square of the length of the tangent equals the product of the lengths of the secant and its external segment.

6.4.1 The line that is perpendicular to the radius of a circle at its endpoint on the circle is a tangent to the circle.

6.4.2 In a circle (or in congruent circles) containing two unequal central angles, the larger angle corresponds to the larger intercepted arc.

6.4.3 In a circle (or in congruent circles) containing two unequal arcs, the larger arc corresponds to a larger central angle.

6.4.4 In a circle (or in congruent circles) containing two unequal chords, the shorter chord is at a greater distance from the center of the circle.

6.4.5 In a circle (or in congruent circles) containing two unequal chords, the chord nearer the center of the circle has a greater length.

6.4.6 In a circle (or in congruent circles) containing two unequal chords, the longer chord corresponds to the greater minor arc.

6.4.7 In a circle (or in congruent circles) containing two unequal minor arcs, the greater minor arc corresponds to the longer of the chords related to these arcs.

6.5.1 The locus of points in a plane and equidistant from the sides of an angle is the angle-bisector.

6.5.2 The three angle bisectors of the angles of a triangle are concurrent.

6.5.3 The three perpendicular bisectors of the sides of a triangle are concurrent.

6.5.4 The three altitudes of a triangle are concurrent.

6.5.5 The three medians of a triangle are concurrent at a point that is two-thirds the distance from any vertex to the midpoint of the opposite side.

7.1.1 The area A of a square whose sides are each of length s is given by $A = s^2$.

7.1.2 The area A of a parallelogram that has a base of length b and whose corresponding altitude has length h is given by

$$A = b \cdot h$$

7.1.3 The area A of a triangle whose base has length b and whose corresponding altitude has length h is given by

$$A = \frac{1}{2} \cdot b \cdot h$$

7.2.1 (Heron's Formula) If the three sides of a triangle have lengths a, b, and c, then the area A of the triangle is given by

$$A = \sqrt{s(s - a)(s - b)(s - c)}$$

where $s = \frac{1}{2}(a + b + c)$

7.2.2 The area A of a trapezoid whose bases have lengths b_1 and b_2 and whose altitude has length h is given by

$$A = \frac{1}{2}h(b_1 + b_2).$$

7.2.3 The area A of a rhombus whose diagonals have lengths d_1 and d_2 is given by

$$A = \frac{1}{2} \cdot d_1 \cdot d_2$$

7.2.4 The area A of a kite whose diagonals have lengths of d_1 and d_2 is given by

$$A = \frac{1}{2} \cdot d_1 \cdot d_2$$

7.3.1 A circle can be circumscribed about (or inscribed in) any regular polygon.

7.3.2 The measure of the central angle of a regular polygon of n sides is given by $c = \dfrac{360}{n}$.

7.3.3 Any radius of a regular polygon bisects the angle at the vertex to which it is drawn.

7.3.4 Any apothem to a side of a regular polygon bisects that side of the regular polygon.

7.3.5 The area A of a regular polygon whose apothem has length a and whose perimeter is P is given by

$$A = \frac{1}{2} \cdot a \cdot P$$

7.4.1 In a circle whose circumference is C, the length ℓ of an arc whose arc measure is m is given by

$$\ell = \frac{m}{360} \cdot C$$

7.4.2 The area A of a circle whose radius is of length r is given by $A = \pi r^2$.

7.4.3 The area A of a sector of a circle whose arc measure is m and whose radius has length r is given by

$$A = \frac{m}{360} \cdot \pi \cdot r^2$$

8.1.1 (Distance Formula) The distance between two points (x_1, y_1) and (x_2, y_2) is given by the formula

$$d = \sqrt{(x_2 - x_1)^2 + (y_2 - y_1)^2}$$

8.1.2 (Midpoint Formula) The midpoint M of the line segment joining (x_1, y_1) and (x_2, y_2) has coordinates x_M and y_M, where

$$(x_M, y_M) = \left(\frac{x_1 + x_2}{2}, \frac{y_1 + y_2}{2} \right)$$

that is, $M = \left(\dfrac{x_1 + x_2}{2}, \dfrac{y_1 + y_2}{2} \right)$

8.2.1 If two nonvertical lines are parallel, then their slopes are equal.

8.2.2 If two lines (neither horizontal nor vertical) are perpendicular, then the product of their slopes is -1.

8.3.1 (Slope-Intercept Form of a Line) The line whose slope is m and whose y intercept is b has the equation $y = mx + b$.

8.3.2 (Point-Slope Form of a Line) The line with slope m and containing the point (x_1, y_1) has the equation

$$y - y_1 = m(x - x_1)$$

8.4.1 The circle whose center is (h, k) and whose radius has length r, where $r > 0$, has the equation

$$(x - h)^2 + (y - k)^2 = r^2$$

8.6.1 The diagonals of a parallelogram bisect each other.

8.6.2 The diagonals of a rhombus are perpendicular.

8.6.3 If the diagonals of a parallelogram are equal in length, then the parallelogram is a rectangle.

8.6.4 The three medians of a triangle are concurrent at a point that is two-thirds the distance from any vertex to the midpoint of the opposite side.

9.1.1 The lateral area L of a right prism whose altitude has measure h and whose base has perimeter P is given by the formula $L = h \cdot P$.

9.1.2 The total area T of any prism whose lateral area has measure L and whose bases have area B is given by the formula $T = L + 2 \cdot B$.

9.1.3 The lateral area L of a regular pyramid with slant height of length ℓ and perimeter P of the base is given by

$$L = \frac{1}{2} \cdot \ell \cdot P$$

9.1.4 The total area (surface area) T of a pyramid whose lateral area is L and whose base has area B is given by $T = L + B$.

9.1.5 The volume V of a pyramid having a base area B and an altitude of length h is given by

$$V = \frac{1}{3} \cdot B \cdot h$$

9.2.1 The lateral area L of a right circular cylinder whose altitude has length h and whose circular bases have radii of length r is given by $L = 2\pi \cdot r \cdot h$.

9.2.2 The total area (surface area) T of a right circular cylinder whose altitude measures h and whose circular base has a radius of measure r is given by $T = 2\pi r(r + h)$ or $T = 2\pi r^2 + 2\pi rh$.

9.2.3 The volume V of a right circular cylinder whose height measures h and whose radius of the circular base measures r is given by $V = \pi r^2 \cdot h$.

9.2.4 For a right circular cone with a slant height of length ℓ and a radius of the base of length r, the lateral area L is given by $L = \pi \cdot r \cdot \ell$.

9.2.5 For a right circular cone with a slant height of length ℓ and a radius of the base of length r, the total area T is given by $T = \pi \cdot r \cdot (r + \ell)$ or $T = \pi r^2 + \pi r\ell$.

9.2.6 The volume V of a right circular cone whose height is of length h and whose radius (of the circular base) has length r is given by

$$V = \frac{1}{3} \cdot \pi \cdot r^2 \cdot h.$$

9.3.1 The surface area S of a sphere whose radius has length r is given by $S = 4 \cdot \pi \cdot r^2$.

9.3.2 The volume V of a sphere whose radius is of length r is given by $V = \frac{4}{3} \cdot \pi \cdot r^3$.

9.4.1 The distance between the two points $P_1(x_1, y_1, z_1)$ and $P_2(x_2, y_2, z_2)$ is given by

$$d = \sqrt{(x_2 - x_1)^2 + (y_2 - y_1)^2 + (z_2 - z_1)^2}$$

9.4.2 The midpoint of the segment joining $P_1(x_1, y_1, z_1)$ and $P_2(x_2, y_2, z_2)$ is given by

$$(x_M, y_M, z_M) = \left(\frac{x_1 + x_2}{2}, \frac{y_1 + y_2}{2}, \frac{z_1 + z_2}{2} \right)$$

Equivalently,
$$M = \left(\frac{x_1 + x_2}{2}, \frac{y_1 + y_2}{2}, \frac{z_1 + z_2}{2} \right)$$

9.4.3 The sphere that has its center at (h, k, ℓ) and a radius length of r (where $r > 0$) has the equation

$$(x - h)^2 + (y - k)^2 + (z - \ell)^2 = r^2$$

10.2.1 In any right triangle in which α is the measure of an acute angle,

$$\sin^2 \alpha + \cos^2 \alpha = 1$$

10.4.1 The area of a triangle equals one-half the product of the lengths of two sides and the sine of the included angle.

10.4.2 (Law of Sines) In any triangle, the three ratios between the sines of the angles and lengths of the opposite sides are equal. Equivalently,

$$\frac{\sin \alpha}{a} = \frac{\sin \beta}{b} = \frac{\sin \gamma}{c}$$

10.4.3 (Law of Cosines) In acute triangle ABC,

$$c^2 = a^2 + b^2 - 2 \cdot a \cdot b \cdot \cos \gamma$$
$$b^2 = a^2 + c^2 - 2 \cdot a \cdot c \cdot \cos \beta$$
$$a^2 = b^2 + c^2 - 2 \cdot b \cdot c \cdot \cos \alpha$$

ANSWERS: Selected Odd-Numbered Exercises and Proofs

CHAPTER 1

1.1 EXERCISES

1. Induction
3. Deduction
5. Intuition
7. None
9. Angle 1 looks equal in measure to angle 2.
11. The three angles in one triangle are equal in measure to the three angles in the other triangle.
13. *A Prisoner of Society* might be nominated for an Academy Award.
15. The instructor is a math teacher.
17. Small children receive a toy from the dentist.
19. Angles 1 and 2 are complementary.
21. Alex has a strange sense of humor.
23. None
25. June Jesse will be in the public eye.
27. Marilyn is a happy person.
29. False; 2 is prime and is not odd.
31. False; the figure is a rhombus.
33. True

1.2 EXERCISES

1. (a) Not a statement
   (b) Statement; true
   (c) Statement; false
   (d) Statement; false
3. (a) Christopher Columbus did not cross the Atlantic Ocean.
   (b) Some jokes are not funny.
5. Conditional
7. Simple
9. Simple
11. H: You go to the game.
    C: You will have a great time.
13. H: The diagonals of a parallelogram are perpendicular.
    C: The parallelogram is a rhombus.
15. H: Two parallel lines are cut by a transversal.
    C: Corresponding angles are congruent.
17. First write the statement in "If, then" form.
    If a figure is a square, then it is a rectangle.
    H: A figure is a square.
    C: It is a rectangle.
19. Conditional; false
    Converse: If your team wins the eight-team tournament, then your team won the first-round game. True.
    Inverse: If your team does not win the first-round game, then your team will not win the eight-team tournament. True.
    Contrapositive: If your team does not win the eight-team tournament, then your team did not win the first-round game. False.
23. True
25. True
27. False

1.3 EXERCISES

1. Undefined terms, definitions, axioms or postulates, and theorems
3. (a) Reflexive
   (b) Transitive
   (c) Substitution
   (d) Symmetric
5. (a) 12   (c) 2
   (b) -2 (d) -12
7. (a) 35   (c) -35
   (b) -35 (d) 35
9. No; Commutative Axiom for Multiplication
11. (a) 9   (c) 8
    (b) -9 (d) -8
13. (a) -4
    (b) -36 (d) $-\frac{1}{4}$
    (c) 18
15. $-\$60$
17. (a) 65   (c) 9
    (b) 16   (d) $8x$
19. (a) 10π (c) $7x^2y$
    (b) $11\sqrt{2}$ (d) $7\sqrt{3}$
21. (a) 14   (c) 14
    (b) 20   (d) 38
23. (a) -1 (c) $\frac{-8}{9}$
    (b) $\frac{1}{9}$ (d) $\frac{-1}{2}$
25. (a) 6
    (b) $12x^2 - 7x - 10$
27. $5x + 2y$
29. $10x + 5y$
31. $10x$

1.4 EXERCISES

1. $5x + 8$

3. $2x - 2$

5. $5x + 8$

7. $x^2 + x + 12$

9. $6x^2 + 11x - 10$

11. $2a^2 + 2b^2$

13. 60

15. 40

17. 148

19. 12π

21. 7

23. -12

25. 12

27. 5

29. 30

31. 8 24

33. 4

35. 32

1.5 EXERCISES

1. The length of $\overline{AB}$ is greater than the length of $\overline{CD}$.

3. The measure of angle ABC is greater than the measure of angle DEF.

5. (a) 4

(b) 10

7. $AB > IJ$

9. (a) False   (c) True

(b) True   (d) False

11. The measure of the second angle must be greater than 148° and less than 180°.

13. (a) $-12 \le 20$ (c) $18 \ge -30$

(b) $-10 \le -2$ (d) $3 \ge -5$

15.

No change	No change
No change	No change
No change	Change
No change	Change

17. $x \le 7$

19. $x < -5$

21. $x < 20$

23. $x \ge -24$

25. $x \le -2$

27. Not true if $c < 0$

29. Not true if $a = -3$ and $b = -2$

31. If $a < b$ and $c < d$, then $a + p_1 = b$ and $c + p_2 = d$, where p_1 and p_2 are both positive. Use the Addition Property to get

$a + p_1 + c + p_2 = b + d$
$a + c + p_1 + p_2 = b + d$

But since p_1 and p_2 are both positive, $p_1 + p_2$ must also be positive. $\therefore a + c < b + d$

REVIEW EXERCISES

1. Undefined terms, defined terms, axioms or postulate, and theorems

3. Intuition

5. Deduction

7. (a) H: The diagonals of a trapezoid are equal in length.

C: The trapezoid is isosceles.

(b) H: The parallelogram is a rectangle.

C: The diagonals of a parallelogram are congruent.

9. (a) Conjunction; false

(b) Simple; true

(c) Disjunction; true

(d) Conditional; false

11. Jody Smithers has a college degree.

13. (a) If we are happy, then the Bears won.

(b) If the Bears do not win, then we will not be happy.

(c) If we are not happy, then the Bears did not win.

15. (a) Associative Axiom for Addition

(b) Multiplication Property of Equality

(c) Multiplication Property of Inequality

(d) Commutative Axiom for Addition

17. (a) -9 (c) -43

(b) 35   (d) -11

19. (a) 5   (c) 3

(b) 8   (d) 30

21. (a) 900   (b) 3

23. The measure of angle 3 is less than 50°.

CHAPTER 2

2.1 EXERCISES

1. (a) A-C-D

(b) $A, B, C,$ or $B, C, D,$ or A, B, D.

3. $\overleftrightarrow{CD}$ means line CD; $\overline{CD}$ means segment CD; CD means the measure or length of $\overline{CD}$; $\overrightarrow{CD}$ means ray CD with endpoint C.

5. (a) m and t

(b) m and $\overleftrightarrow{AD}$, or $\overleftrightarrow{AD}$ and t

7. $x = 3; AM = 7$

9. $x = 7; AB = 38$

11. (a) $\overrightarrow{OA}$ and $\overrightarrow{OD}$

(b) $\overrightarrow{OA}$ and $\overrightarrow{OB}$ (There are other possible answers.)

15. B

17. (a) C   (b) C   (c) H

23. (a) No   (c) No

(b) Yes   (d) Yes

25. Nothing

2.2 EXERCISES

1. (a) Yes   (b) No
3. (a) Obtuse (c) Acute
    (b) Straight (d) Obtuse
5. $\angle FAC$ and $\angle CAD$ are supplementary.

7. 8
9. $\angle CAB \simeq \angle DAB$
11. $x = 128$
    $y = 52$

13. (a) $180 - x$
    (b) $192 - 3x$
    (c) $180 - 2x - 5y$
15. 143

23. It appears that the two sides opposite angles A and B are congruent.

2.3 EXERCISES

3. **1.** $\angle 1 \simeq \angle 2$ and $\angle 2$
        $\simeq \angle 3$
    2. $\angle 1 \simeq \angle 3$
13. No; yes; no; no
15. No; yes; yes; no

17. (a) Perpendicular
    (b) Angles
    (c) Supplementary
    (d) Right
    (e) Measure of angle

(f) Adjacent
(g) Complementary
(h) Ray AB
(i) Is congruent to
(j) Vertical

19. In space, an infinite number of lines perpendicularly bisect a given line segment at its midpoint.

2.3 SELECTED PROOF

11. *Given:* $\overleftrightarrow{AB}$, $\overleftrightarrow{DE}$, and $\overleftrightarrow{CF}$ as shown
        $\angle BCF$ and $\angle CFE$ are supplementary
        $\overrightarrow{CG}$ bisects $\angle BCF$, and $\overrightarrow{FG}$ bisects $\angle CFE$
Prove: $\angle 2$ and $\angle 3$ are complementary

PROOF

Statements	Reasons
1. $\overleftrightarrow{AB}$, $\overleftrightarrow{DE}$, and $\overleftrightarrow{CF}$ as shown; $\angle BCF$ and $\angle CFE$ are supplementary	1. Given
2. $m\angle BCF + m\angle CFE = 180°$	2. The sum of the measures of supplementary $\angle$s $= 180°$
3. $\overrightarrow{CG}$ bisects $\angle BCF$, and $\overrightarrow{FB}$ bisects $\angle CFE$	3. Given
4. $m\angle 1 = m\angle 2$, and $m\angle 3 = m\angle 4$	4. If a ray bisects an angle, two $\angle$s of equal measure are formed
5. $m\angle 1 + m\angle 2 = m\angle BCF$ $m\angle 3 + m\angle 4 = m\angle CFE$	5. Angle-Addition Postulate
6. $m\angle 1 + m\angle 2 + m\angle 3 + m\angle 4 = 180°$	6. Substitution
7. $m\angle 2 + m\angle 2 + m\angle 3 + m\angle 3 = 180°$ or $2 \cdot m\angle 2 + 2 \cdot m\angle 3 = 180°$	7. Substitution
8. $m\angle 2 + m\angle 3 = 90°$	8. Division Property of Equality
9. $\angle 2$ and $\angle 3$ are complementary	9. If the sum of the measures of two angles is $90°$, the angles are complementary

2.4 EXERCISES

1. H: A line segment is bisected.
    C: Each of the equal segments has half the length of the original segment.

3. First write the statement in "If, then" form:
    If a figure is a square, then it is a quadrilateral.
    H: A figure is a square.

C: It is a quadrilateral.
5. H: Each is a right angle.
    C: Two angles are congruent.
7. Statement, Drawing, Given, Prove, and Proof

2.4 SELECTED PROOFS

11. If two lines intersect, the vertical angles formed are congruent.

Given: $\overleftrightarrow{AB}$ and $\overleftrightarrow{CD}$ intersect at E

Prove: $\angle 1 \simeq \angle 2$

PROOF

Statements	Reasons
1. $\overleftrightarrow{AB}$ and $\overleftrightarrow{CD}$ intersect at E	1. Given
2. $\angle 1$ is supplementary to $\angle AED$, and $\angle 2$ is supplementary to $\angle AED$	2. If the exterior sides of two adjacent $\angle$s form a straight line, these $\angle$s are supplementary
3. $\angle 1 \simeq \angle 2$	3. If two $\angle$s are supplementary to the same $\angle$, these $\angle$s are $\simeq$

15. If two angles are congruent, then their bisectors separate these angles into four congruent angles.

Given: $\angle ABC \simeq \angle EFG$

$\overrightarrow{BD}$ bisects $\angle ABC$

$\overrightarrow{FH}$ bisects $\angle EFG$

Prove: $\angle 1 \simeq \angle 2 \simeq \angle 3 \simeq 4$

PROOF

Statements	Reasons
1. $\angle ABC \simeq \angle EFG$	1. Given
2. $m\angle ABC = m\angle EFG$	2. If two $\angle$s are $\simeq$, their measures are $=$
3. $m\angle ABC = m\angle 1 + m\angle 2$ $m\angle EFG = m\angle 3 + m\angle 4$	3. Angle-Addition Postulate
4. $m\angle 1 + m\angle 2 = m\angle 3 + m\angle 4$	4. Substitution
5. $\overrightarrow{BD}$ bisects $\angle ABC$ $\overrightarrow{FH}$ bisects $\angle EFG$	5. Given
6. $m\angle 1 = m\angle 2$ and $m\angle 3 = m\angle 4$	6. If a ray bisects an $\angle$, two $\angle$s of equal measure are formed
7. $m\angle 1 + m\angle 1 = m\angle 3 + m\angle 3$ or $2 \cdot m\angle 1 = 2 \cdot m\angle 3$	7. Substitution
8. $m\angle 1 = m\angle 3$	8. Division Property of Equality
9. $m\angle 1 = m\angle 2 = m\angle 3 = m\angle 4$	9. Substitution
10. $\angle 1 \simeq \angle 2 \simeq \angle 3 \simeq \angle 4$	10. If $\angle$s are $=$ in measure, they are $\simeq$

REVIEW EXERCISES

5. AC; $AC = 3$ **7.** C **9.** 34   **11.** 39   **13.** 28 and 152   **15.** S   **17.** S

19. *Given:* $\overline{AB} \simeq \overline{AE}$

$\overline{BC} \simeq \overline{DE}$

Prove: $\overline{AC} \simeq \overline{AD}$

PROOF

Statements	Reasons
1. $\overline{AB} \simeq \overline{AE}$ and $\overline{BC} \simeq \overline{DE}$	1. Given
2. $AB = AC$ and $BC = DE$	2. If two segments are $\simeq$, their measures are $=$
3. $AB + BC = AE + DE$	3. Addition Property of Equality
4. $AC = AB + BC$ $AD = AE + DE$	4. Segment-Addition Postulate
5. $AC = AD$	5. Substitution
6. $\overline{AC} \simeq \overline{AD}$	6. If two segments are $=$ in measure, they are $\simeq$

23. *Given:* $\angle MOP \simeq \angle MPO$

$\qquad \overrightarrow{OR}$ bisects $\angle MOP$

$\qquad \overrightarrow{PR}$ bisects $\angle MPO$

Prove: $\angle 1 \simeq 2$

PROOF

Statements	Reasons
1. $\angle MOP \simeq \angle MPO$	1. Given
2. $\overrightarrow{OR}$ bisects $\angle MOP$	2. Given
$\quad \overrightarrow{PR}$ bisects $\angle MPO$	
3. $\angle 1 \simeq \angle 2$	3. If two $\angle$s are $\simeq$, their bisectors separate them into four $\simeq \angle$s

CHAPTER 3

3.1 EXERCISES

1. No; yes; yes; no
3. Angle 9 appears to be a right angle.
5. (a) 87°   (c) 93°
   (b) 87°   (d) 87°
7. (a) $\angle 5$ (c) $\angle 8$
   (b) $\angle 5$ (d) $\angle 5$
9. (a) 68°   (c) 112°
   (b) 112°   (d) 34°
11. $x = 10$; m$\angle 4 = 110$
13. $x = 12, y = 4$, m$\angle 7 = 76°$
19. (a) $\angle 4 \simeq \angle 2$, and $\angle 5 \simeq \angle 3$
    (b) 180°
    (c) 180°
23. No

3.1 SELECTED PROOF

15.

PROOF

Statements	Reasons
1. $\overleftrightarrow{CE} \parallel \overrightarrow{DF}$; transversal $\overleftrightarrow{AB}$	1. Given
2. $\angle ACE \simeq \angle ADF$	2. If two $\parallel$ lines are cut by a transversal, the corresponding $\angle$s are $\simeq$
3. $\overrightarrow{CX}$ bisects $\angle ACE$, and $\overrightarrow{DE}$ bisects $\angle CDF$	3. Given
4. $\angle 1 \simeq \angle 3$	4. If two $\angle$s are $\simeq$, their bisectors separate them into four $\simeq \angle$s

3.2 EXERCISES

1. If Juan wins the state lottery, then he will be rich.
Converse: If Juan is rich, then he won the state lottery. (False)
Inverse: If Juan does not win the state lottery, then he will not be rich. (False)
Contrapositive: If Juan is not rich, then he did not win the state lottery. (True)
3. If the sum of the measures of two angles is 90°, then the two angles are complementary.
Converse: If two angles are complementary, then the sum of their measures is 90°. (True)
Inverse: If the sum of the measures of two angles is not 90°, then the two angles are not complementary. (True)
Contrapositive: If two angles are not complementary, then the sum of their measures is not 90°. (True)
5. No conclusion
7. $x = 5$
9. Parts (a), (b), and (e)

3.2 SELECTED PROOFS

11. Assume that $\overleftrightarrow{DC} \parallel \overleftrightarrow{EG}$. If they are $\parallel$, then $\angle AOD \simeq \angle AFE$, because they are corresponding angles. But this contradicts the given information. Therefore, our assumption is false and $\overleftrightarrow{DC} \not\parallel \overleftrightarrow{EG}$.

15. Assume that the angles are vertical angles. If they are vertical angles, then they are congruent. But this contradicts the hypothesis that the two angles are not congruent. Hence, our assumption must be false, and the angles are not vertical angles.

19. Proof: If M is a midpoint of $\overline{AB}$, then $AM = \frac{1}{2} \cdot AB$. Assume that N is also a midpoint of $\overline{AB}$, so $AN = \frac{1}{2} \cdot AB$. By substitution $AM = AN$. By the Segment-Addition Postulate, $AM = AN + NM$. Using substitution again, $AN + NM = AN$. Subtracting gives $NM = 0$. But this contradicts the Ruler Postulate, which states that the measure of a line segment is a positive number. Therefore, our assumption is wrong, and M is the only midpoint for $\overline{AB}$.

3.3 EXERCISES

1. $p \parallel q$ **3.** None   **5.** $\ell \parallel n$ **7.** None   **9.** $\ell \parallel n$ **19.** $x = 9$ **21.** $x = 6$

3.3 SELECTED PROOFS

11.

PROOF

Statements	Reasons
1. $\angle 1$ and $\angle 2$ are complementary $\angle 3$ and $\angle 1$ are complementary	1. Given
2. $\angle 2 \simeq \angle 3$	2. If two $\angle$s are complementary to the same angle, they are $\simeq$
3. $\overline{BC} \parallel \overline{DE}$	3. If two lines are cut by a transversal so that corresponding $\angle$s are $\simeq$, these lines are $\parallel$

15.

PROOF

Statements	Reasons
1. $\overrightarrow{DE}$ bisects $\angle CDA$	1. Given
2. $\angle 2 \simeq \angle 3$	2. If a ray bisects an $\angle$, it forms two $\simeq$ $\angle$s
3. $\angle 3 \simeq \angle 1$	3. Given
4. $\angle 2 \simeq \angle 1$	4. Transitive for $\simeq$
5. $\overline{ED} \parallel \overline{AB}$	5. If two lines are cut by a transversal so that alternate interior $\angle$s are $\simeq$, these lines are $\parallel$

23.

PROOF

Statements	Reasons
1. Lines ℓ and m, and transversal t; $\angle 1 \simeq \angle 2$	1. Given
2. $\angle 1 \simeq \angle 3$	2. If two lines intersect, the vertical $\angle$s formed are $\simeq$
3. $\angle 2 \simeq \angle 3$	3. Transitive for $\simeq$
4. $\ell \parallel m$	4. If two lines are cut by a transversal so that the corresponding angles are $\simeq$, these lines are $\parallel$

27.

PROOF

Statements	Reasons
1. $\ell \perp t$	1. Given
2. $\angle 1$ is a right $\angle$	2. If two lines are $\perp$, they meet to form right $\angle$s
3. $m \perp t$	3. Given
4. $\angle 2$ is a right $\angle$	4. Same as reason 2
5. $\angle 1 \simeq \angle 2$	5. Any two right $\angle$s are $\simeq$
6. $\ell \parallel m$	6. If two lines are cut by a transversal so that the corresponding $\angle$s are $\simeq$, these lines are $\parallel$

31. (a) If the general term of an infinite series does not have a limit of zero, then the infinite series diverges.
(b) Proof of the statement in (a) is true by contraposition. The contrapositive is: If an infinite series does not diverge (that is, if it converges), then the general term of the series has a limit of zero. This is the theorem that was already proved in the calculus class.

3.4 EXERCISES

1. $\angle C \simeq \angle Q$

3. $m\angle 1 = 122°$
 $m\angle 2 = 58°$
 $m\angle 5 = 72°$

5. $m\angle 2 = 56°$
 $m\angle 3 = 82°$
 $m\angle 4 = 42°$

7. $x = 113°$
 $y = 67°$
 $z = 36°$

9. (a) $360°$
 (b) $540°$

11. (a) 7
 (b) 9

(c) $720°$
 (d) $1080°$
 (c) 13
 (d) 15

13. (a) 15
 (b) 20

15. $135°$

17. $x = 45\frac{1}{4}$
 $y = 44\frac{3}{4}$

(c) 8
 (d) 40

19. $y = 20$; $x = 100$;
 $m\angle 5 = 60°$

3.4 SELECTED PROOFS

27.

PROOF

Statements	Reasons
1. Equiangular $\triangle ABC$	1. Given
2. $\angle A \simeq \angle B \simeq \angle C$	2. Equiangular $\triangle$ has three $\simeq \angle$s
3. $m\angle A = m\angle B = m\angle C$	3. If $\angle$s are congruent, their measures are $=$
4. $m\angle A + m\angle B + m\angle C = 180°$	4. The sum of measures of $\angle$s in a $\triangle$ is $180°$
5. $m\angle A + m\angle A + m\angle A = 180°$ or $3 \cdot m\angle A = 180°$	5. Substitution
6. $m\angle A = 60°$	6. Division Property of Equality
7. $m\angle A = m\angle B = m\angle C = 60°$	7. Substitution

31. Assume that a triangle does have more than one right angle. Then the sum of the measures of the interior angles would be greater than 180°. But this contradicts the fact that the sum of the measures of the angles of a triangle is 180°. Therefore our assumption is wrong, and a triangle cannot have more than one right angle.

REVIEW EXERCISES

1. (a) $\overline{BC} \parallel \overline{AD}$
 (b) $\overline{AB} \parallel \overline{CD}$

3. 37

5. $x = 20$; $y = 10$

7. $\overline{AE} \parallel \overline{BF}$

9. $\overline{BE} \parallel \overline{CF}$

11. $\overline{AC} \parallel \overline{DF}$, and
 $\overline{AE} \parallel \overline{BF}$

13. $x = 32°$; $y = 30°$

15. $x = 140°$

17. $m\angle 3 = 69°$
 $m\angle 4 = 67°$
 $m\angle 5 = 44°$

19. S

21. N

23. S

25.

Number of sides	8	12	20	15	10	16	180
Measure of each exterior $\angle$	45°	30°	18°	24°	36°	22.5°	2°
Measure of each interior $\angle$	135°	150°	162°	156°	144°	157.5°	178°

31. *Statement:*   If it is not raining, then I am happy.
 Converse: If I am happy, then it is not raining.
 Inverse: If it is raining, then I am not happy.
 Contrapositive: If I am not happy, then it is raining.

33.

PROOF

Statements	Reasons
1. ∠C ≅ ∠3	1. Given
2. $\overline{CD}$ ∥ $\overline{BE}$	2. If two lines are cut by a transversal so that the corresponding angles are ≅, the lines are ∥
3. ∠CDA ≅ ∠BEA	3. If two ∥ lines are cut by a transversal, the corresponding angles are ≅
4. m∠CDA = m∠BEA	4. If two ∠s are ≅, their measures are =
5. $\overline{BE}$ ⊥ $\overline{DA}$	5. Given
6. ∠BEA is a right ∠	6. If two lines are ⊥, they meet to form right angles
7. m∠BEA = 90°	7. The measure of a right ∠ = 90°
8. m∠CDA = 90°	8. Substitution
9. ∠CDA is a right ∠	9. If the measure of an ∠ is 90°, the ∠ is a right ∠
10. $\overline{CD}$ ⊥ $\overline{DA}$	10. If two lines meet to form a right angle, the lines are ⊥

35.

PROOF

Statements	Reasons
1. ∠A ≅ ∠C	1. Given
2. $\overline{DC}$ ∥ $\overrightarrow{AB}$	2. Given
3. ∠C ≅ ∠1	3. If two ∥ lines are cut by a transversal, the alternate interior ∠s are congruent
4. ∠A ≅ ∠1	4. Transitive Property of Congruence
5. $\overline{DA}$ ∥ $\overline{CB}$	5. If two lines are cut by a transversal so that corresponding ∠s are ≅, these lines are ∥

37. Assume that m ∥ n. Then ∠1 ≅ ∠3 since alternate exterior angles are congruent when parallel lines are cut by a transversal. But this contradicts the given fact that ∠1 ≠ ∠3. Therefore our assumption must be false, and it follows that m ∦ n.

CHAPTER 4

4.1 EXERCISES

1. ∠A; $\overline{AB}$; no; no
3. SAS
5. △AED ≅ △FDE
7. (a) $\overline{AD}$ ≅ $\overline{CE}$
   (b) Not possible

9. (a) ∠N ≅ ∠P
   (b) $\overline{MO}$ ≅ $\overline{MO}$
(c) ∠ABD ≅ ∠CBE
(d) ∠BDA ≅ ∠BEC

(c) Not possible
(d) Not possible
19. Yes; SAS or SSS
21. No

23. (a) △CBE, △ADE, △CDE
   (b) △ADC
   (c) △CBD

4.1 SELECTED PROOFS

11.

PROOF

Statements	Reasons
1. $\overline{AB}$ ≅ $\overline{CD}$; $\overline{AD}$ ≅ $\overline{CB}$	1. Given
2. $\overline{AC}$ ≅ $\overline{AC}$	2. Identity (or Reflexive)
3. △ABC ≅ △CDA	3. SSS

15.

PROOF

Statements	Reasons
1. $\overline{AB} \perp \overline{AC}$ and $\overline{AB} \perp \overline{BD}$	1. Given
2. $\angle ABC$ is a right $\angle$, and $\angle ABD$ is a right $\angle$	2. If two lines are $\perp$, they meet to form a right $\angle$
3. $\angle ABC \simeq \angle ABD$	3. Any two right $\angle$s are $\simeq$
4. $\overline{BC} \simeq \overline{BD}$	4. Given
5. $\overline{AB} \simeq \overline{AB}$	5. Identity
6. $\triangle ABC \simeq \triangle ABD$	6. SAS

27.

PROOF

Statements	Reasons
1. Plane M; C is the midpoint of $\overline{EB}$	1. Given
2. $\overline{EC} \simeq \overline{CB}$	2. The midpoint of a segment divides the segment into two $\simeq$ segments
3. $\overline{AD} \perp \overline{BE}$	3. Given
4. $\angle ACB$ is a right angle, and $\angle DCE$ is a right angle	4. If two lines are $\perp$, they meet to form a right $\angle$
5. $\angle ACB \simeq \angle DCE$	5. Any two right $\angle$s are $\simeq$
6. $\overline{AB} \parallel \overline{ED}$	6. Given
7. $\angle ABC \simeq \angle DEC$	7. If two $\parallel$ lines are cut by a transversal, the alternate interior $\angle$s are $\simeq$
8. $\triangle ABC \simeq \triangle DEC$	8. ASA

4.2 EXERCISES

1. HL
3. SAS
5. AAS
7. SSS

9. $m\angle 2 = 48°$
$m\angle 3 = 48°$
$m\angle 5 = 42°$
$m\angle 6 = 42°$

4.2 SELECTED PROOFS

11.

PROOF

Statements	Reasons
1. $\overline{HJ} \perp \overline{KL}$	1. Given
2. $\angle HJK$ and $\angle HJL$ are right $\angle$s	2. If two lines are $\perp$, they meet to form right $\angle$s
3. $\overline{HK} \simeq \overline{HL}$	3. Given
4. $\overline{HJ} \simeq \overline{HJ}$	4. Reflexive
5. $\triangle HJK \simeq \triangle HJL$	5. HL

15.

PROOF

Statements	Reasons
1. $\angle 1$ and $\angle 2$ are right $\angle$s	1. Given
2. $\angle 1 \simeq \angle 2$	2. Any two right $\angle$s are $\simeq$
3. H is the midpoint of $\overline{FK}$	3. Given
4. $\overline{FH} \simeq \overline{HK}$	4. The midpoint of a segment forms two $\simeq$ segments
5. $\overline{FG} \parallel \overline{HJ}$	5. Given
6. $\angle GFH \simeq \angle JHK$	6. If two $\parallel$ lines are cut by a transversal, the corresponding $\angle$s are $\simeq$
7. $\triangle FHG \simeq \triangle HKJ$	7. ASA

19.

PROOF

Statements	Reasons
1. E is the midpoint of $\overline{FG}$	1. Given
2. $\overline{FE} \simeq \overline{EG}$	2. The midpoint of a segment forms two $\simeq$ segments
3. $\overline{DF} \simeq \overline{DG}$	3. Given
4. $\overline{DE} \simeq \overline{DE}$	4. Reflexive
5. $\triangle FDE \simeq \triangle GDE$	5. SSS
6. $\angle DEF \simeq \angle DEG$	6. CPCTC
7. $\overline{DE} \perp \overline{FG}$	7. If two lines meet to form $\simeq$ adjacent $\angle$s, the lines are $\perp$

23.

PROOF

Statements	Reasons
1. $\overrightarrow{RW}$ bisects $\angle SRU$	1. Given
2. $\angle SRW \simeq \angle URW$	2. If a ray bisects an $\angle$, two $\simeq$ $\angle$s are formed
3. $\overline{RS} \simeq \overline{RU}$	3. Given
4. $\overline{RW} \simeq \overline{RW}$	4. Identity
5. $\triangle RSW \simeq \triangle RUW$	5. SAS
6. $\angle RSW \simeq \angle RUW$	6. CPCTC
7. $\angle TRV \simeq \angle VRS$	7. Identity
8. $\triangle TRU \simeq \triangle VRS$	8. ASA
9. $\angle T \simeq \angle V$	9. CPCTC

27.

PROOF

Statements	Reasons
1. Line m, with point P on m $\overline{PQ} \simeq \overline{PR}$ and $\overline{QS} \simeq \overline{RS}$	1. Given
2. $\overline{SP} \simeq \overline{SP}$	2. Identity
3. $\triangle SQP \simeq \triangle SRP$	3. SSS
4. $\angle SPQ \simeq \angle SPR$	4. CPCTC
5. $\overleftrightarrow{SP} \perp m$	5. If two lines form $\simeq$ adjacent $\angle$s, the lines are $\perp$

4.3 EXERCISES

1. Underdetermined
3. Overdetermined
5. Determined
7. $55°$

11. $m\angle 2 = 68°$; $m\angle 1 = 44°$
13. $124°$
15. $m\angle A = 52°$; $m\angle B = 64°$; $m\angle C = 64°$

35. Point D is on the bisector of angle BCA.
37. $\triangle PMN$, $\triangle PBM$, $\triangle PAN$, $\triangle MQN$

4.3 SELECTED PROOFS

19.

PROOF

Statements	Reasons
1. $\overline{AB} \simeq \overline{AC}$	1. Given
2. $\angle 2 \simeq \angle 1$	2. If two sides of a $\triangle$ are $\simeq$, the $\angle$s opposite these sides are also $\simeq$
3. $\angle 2$ is supplementary to $\angle 6$ $\angle 1$ is supplementary to $\angle 7$	3. If the exterior sides of two adjacent angles form a straight line, these $\angle$s are supplementary
4. $\angle 6 \simeq \angle 7$	4. If two $\angle$s are supplementary to $\simeq$ $\angle$s, these $\angle$s are $\simeq$

23.

PROOF

Statements	Reasons
1. Isosceles $\triangle ABC$ with vertex B	1. Given
2. $\overline{AB} \simeq \overline{BC}$	2. Isosceles $\triangle$ has two $\simeq$ sides
3. $\overline{BD}$ is the altitude to $\overline{AC}$	3. Given
4. $\overline{BD} \perp \overline{AC}$	4. An altitude is a segment drawn from a vertex $\perp$ to opposite side
5. $\angle BDA \simeq \angle BDC$	5. If two lines are $\perp$, they meet to form $\simeq$ adjacent $\angle$s
6. $\angle A \simeq \angle C$	6. Base $\angle$s of an isosceles $\triangle$ are $\simeq$
7. $\triangle ABD \simeq \triangle CBD$	7. AAS
8. $\overline{AD} \simeq \overline{DC}$	8. CPCTC
9. D is the midpoint of $\overline{AC}$	9. If a point of a segment divides the segment into two $\simeq$ segments, the point is the midpoint
10. $\overline{BD}$ is the median to $\overline{AC}$	10. If a segment is drawn from a vertex to the midpoint of the opposite side, the segment is a median

4.4 EXERCISES

1. $AB = DC = 8$; $BC = AD = 9$
3. m$\angle A =$ m$\angle C = 83°$; m$\angle B =$ m$\angle D = 97°$
19. $\triangle DEC$ is a right triangle; $\triangle ADE$ and $\triangle BCE$ are isosceles triangles.
23. $y = 6$; $MN = 9$; $ST = 18$
25. $x = 5$; $RM = 11$; $ST = 22$

4.4 SELECTED PROOFS

7.

PROOF

Statements	Reasons
1. $\angle 1 \simeq \angle 2$ and $\angle 3 \simeq \angle 4$	1. Given
2. $\overline{NQ} \simeq \overline{NQ}$	2. Identity
3. $\triangle NMQ \simeq \triangle NPQ$	3. ASA
4. $\overline{MN} \simeq \overline{NP}$	4. CPCTC
5. $\overline{NO} \simeq \overline{NO}$	5. Identity
6. $\triangle MNO \simeq \triangle PNO$	6. SAS
7. $\angle NOM \simeq \angle NOP$	7. CPCTC
8. $\overline{MP} \perp \overline{NQ}$	8. If two lines meet to form $\simeq$ adjacent $\angle$s, the lines are $\perp$

11.

PROOF

Statements	Reasons
1. Quadrilateral $HJKL$ with M the midpoint of $\overline{LJ}$	1. Given
2. $\overline{LM} \simeq \overline{MJ}$	2. The midpoint of a segment forms two $\simeq$ segments
3. $\overline{LJ} \perp \overline{HK}$	3. Given
4. $\angle HML \simeq \angle HMJ$	4. If two lines are $\perp$, they meet to form $\simeq$ adjacent $\angle$s
5. $\overline{HM} \simeq \overline{HM}$	5. Identity
6. $\triangle LHM \simeq \triangle JHM$	6. SAS
7. $\angle LHM \simeq \angle JHM$	7. CPCTC
8. $\overrightarrow{HM}$ bisects $\angle LHJ$	8. If a ray divides an $\angle$ into two $\simeq$ $\angle$s, the ray bisects the $\angle$
9. $\angle LMK \simeq \angle JMK$	9. Same as reason 4
10. $\overline{MK} \simeq \overline{MK}$	10. Identity
11. $\triangle LMK \simeq \triangle JMK$	11. SAS
12. $\overline{LK} \simeq \overline{JK}$	12. CPCTC

15.

	PROOF	

Statements	Reasons
1. Quadrilateral $ABCD$ with $\overline{AB} \simeq \overline{CD}$ and $\overline{BC} \simeq \overline{AD}$	1. Given
2. Draw in $\overline{AC}$	2. Through two points there is exactly one line
3. $\overline{AC} \simeq \overline{AC}$	3. Identity
4. $\triangle ABC \simeq \triangle CDA$	4. SSS
5. $\angle 1 \simeq \angle 4$ and $\angle 2 \simeq \angle 3$	5. CPCTC
6. $\overline{AB} \parallel \overline{CD}$ and $\overline{BC} \parallel \overline{AD}$	6. If two lines are cut by a transversal so that alternate interior $\angle$s are $\simeq$, the lines are $\parallel$
7. $ABCD$ is a parallelogram	7. If a quadrilateral has both pairs of opposite sides $\parallel$, the quadrilateral is a parallelogram

4.5 EXERCISES

1. $m\angle A = 60°$
$m\angle ABC = 120°$

3. The parallelogram is a rectangle.

5. The trapezoid is an isosceles trapezoid.

7. $\triangle WMA \simeq \triangle ZMD \therefore WA = DZ$ and $\triangle XNB \simeq \triangle YNC \therefore BX = YC$. In rectangle $WZYX$, $WX = MN = ZY$ or $MN = \frac{1}{2}(WX + ZY)$.

But if WA and BX are subtracted from WX and added to ZY, then

$$MN = \frac{1}{2}(AB + DC).$$

9. $7x + 2$

4.5 SELECTED PROOFS

11.

	PROOF	

Statements	Reasons
1. $ABCD$ is an isosceles trapezoid	1. Given
2. $\angle A \simeq \angle B$	2. Lower base angles of an isosceles trapezoid are $\simeq$
3. $\overline{EB} \simeq \overline{EA}$	3. If two $\angle$s of a $\triangle$ are $\simeq$, the sides opposite these $\angle$s are also $\simeq$
4. $\triangle ABE$ is isosceles	4. If a $\triangle$ has two $\simeq$ sides, it is an isosceles $\triangle$

15.

	PROOF	

Statements	Reasons
1. $ABCD$ is a rhombus with $\overline{AB} \simeq \overline{BC}$	1. Given
2. $ABCD$ is a parallelogram	2. A rhombus is a parallelogram with two $\simeq$ adjacent sides
3. $\overline{AB} \simeq \overline{CD}$	3. Opposite sides of a parallelogram are $\simeq$
4. $\overline{BC} \simeq \overline{CD}$	4. Transitive Property for $\simeq$
5. $\overline{BC} \simeq \overline{AD}$	5. Same as reason 3
6. $\overline{CD} \simeq \overline{AD}$	6. Transitive Property for $\simeq$

19.

	PROOF	

Statements	Reasons
1. $ABCD$ is a rhombus	1. Given
2. $\overline{AB} \simeq \overline{BC} \simeq \overline{CD} \simeq \overline{AD}$	2. All sides of a rhombus are $\simeq$
3. $ABCD$ is a parallelogram	3. A rhombus is a parallelogram with two $\simeq$ adjacent sides
4. $\angle B \simeq \angle D$	4. Opposite angles of a parallelogram are $\simeq$
5. $\triangle ABC \simeq \triangle ADC$	5. SAS
6. $\angle BAC \simeq \angle DAC$, and $\angle BCA \simeq \angle DCA$	6. CPCTC
7. $\overrightarrow{AC}$ bisects $\angle BAD$, and $\overrightarrow{CA}$ bisects $\angle BCD$	7. If a ray divides an $\angle$ into two $\simeq$ $\angle$s, the ray bisects the $\angle$

27. $6°$

4.6 EXERCISES

1. False

3. True

5. True

7. False

9. True

17. $BC < EF$

19. $2 < x < 10$

21. $x + 2 < y < 5x + 12$

4.6 SELECTED PROOFS

15.

PROOF

Statements	Reasons
1. Quadrilateral $RSTU$ with diagonal $\overline{US}$; $\angle R$ and $\angle TUS$ are right angles	1. Given
2. $TS > US$	2. Shortest distance from a point to a line is the $\perp$ distance
3. $US > UR$	3. Same as reason 2
4. $TS > UR$	4. Transitive Property of Inequality

23. Assume that $PM = PN$. But that means that $\triangle MPN$ is isosceles. But that contradicts the hypothesis. Thus, our assumption must be wrong, and $PM \neq PN$.

REVIEW EXERCISES

1.

PROOF

Statements	Reasons
1. $\angle AEB \cong \angle DEC$	1. Given
2. $\overline{AE} \cong \overline{ED}$	2. Given
3. $\angle A \cong \angle D$	3. If two sides of a $\triangle$ are $\cong$, the $\angle$s opposite these sides are also $\cong$
4. $\triangle AEB \cong \triangle DEC$	4. ASA

3.

PROOF

Statements	Reasons
1. $\overline{AD}$ bisects $\overline{BC}$	1. Given
2. $\overline{BE} \cong \overline{EC}$	2. If a segment is bisected, two $\cong$ segments are formed
3. $\overline{AB} \perp \overline{BC}$ and $\overline{DC} \perp \overline{BC}$	3. Given
4. $\angle B$ is a right $\angle$, and $\angle C$ is a right $\angle$	4. If two lines are $\perp$, they meet to form a right $\angle$
5. $\angle B \cong \angle C$	5. Any two right $\angle$s are $\cong$
6. $\angle AEB \cong \angle DEC$	6. If two lines intersect, the vertical $\angle$s formed are $\cong$
7. $\triangle ABE \cong \triangle DCE$	7. ASA
8. $\overline{AE} \cong \overline{ED}$	8. CPCTC

5.

PROOF

Statements	Reasons
1. $\overline{BE}$ is the altitude to $\overline{AC}$, and $\overline{AD}$ is the altitude to $\overline{CE}$	1. Given
2. $\overline{BE} \perp \overline{AC}$ and $\overline{AD} \perp \overline{CE}$	2. An altitude is a line segment drawn from a vertex $\perp$ to the opposite side
3. $\angle CBE$ is a right $\angle$, and $\angle CDA$ is a right $\angle$	3. If two lines are $\perp$, they meet to form a right $\angle$
4. $\angle CBE \cong \angle CDA$	4. Any two right $\angle$s are $\cong$
5. $\overline{BC} \cong \overline{CD}$	5. Given
6. $\angle C \cong \angle C$	6. Identity
7. $\triangle CBE \cong \triangle CDA$	7. ASA
8. $\overline{BE} \cong \overline{AD}$	8. CPCTC

7.   **PROOF**

Statements	Reasons
1. $\overline{AB} \simeq \overline{DE}$	1. Given
2. $\overline{AB} \parallel \overline{DE}$	2. Given
3. $\angle A \simeq \angle D$	3. If two $\parallel$ lines are cut by a transversal, the alternate interior $\angle$s are $\simeq$
4. $\overline{AF} \simeq \overline{DC}$ on $\overline{AD}$	4. Given
5. $AF = DC$	5. If two segments are $\simeq$, their lengths are $=$
6. $FC = FC$	6. Reflexive Property of Equality
7. $AC + FC = DC + FC$	7. Addition Property of Equality
8. $AC = AF + FC$ and $FD = DC + FC$	8. Segment-Addition Postulate
9. $AC = FD$	9. Substitution
10. $\overline{AC} \simeq \overline{FD}$	10. If two segments are $=$ in length, they are $\simeq$
11. $\triangle BAC \simeq \triangle EDF$	11. SAS
12. $\angle BCA \simeq \angle EFD$	12. CPCTC
13. $\overline{BC} \parallel \overline{FE}$	13. If two lines are cut by a transversal so that alternate interior $\angle$s are $\simeq$, the lines are $\parallel$

9.   **PROOF**

Statements	Reasons
1. $\overline{JM} \perp \overline{GM}$ and $\overline{GK} \perp \overline{KJ}$	1. Given
2. $\angle M$ is a right $\angle$, and $\angle K$ is a right $\angle$	2. If two lines are $\perp$, they meet to form a right $\angle$
3. $\angle M \simeq \angle K$	3. Any two right $\angle$s are $\simeq$
4. $\angle GHM \simeq \angle JHK$	4. If two lines intersect, the vertical $\angle$s formed are $\simeq$
5. $\angle G \simeq \angle J$	5. If two $\angle$s of one $\triangle$ are $\simeq$ to two $\angle$s of another $\triangle$, then the third $\angle$s are also $\simeq$

11.   **PROOF**

Statements	Reasons
1. $\overline{AC} \simeq \overline{AE}$	1. Given
2. $\angle C \simeq \angle E$	2. If two sides of a $\triangle$ are $\simeq$, the $\angle$s opposite these sides are also $\simeq$
3. $\angle CBD \simeq \angle EFD$	3. Given
4. D is the midpoint of $\overline{CE}$	4. Given
5. $\overline{CD} \simeq \overline{DE}$	5. Midpoint of a segment forms two $\simeq$ segments
6. $\triangle CBD \simeq \triangle EFD$	6. AAS
7. $\overline{BD} \simeq \overline{DF}$	7. CPCTC

13.   **PROOF**

Statements	Reasons
1. $\overline{YZ}$ is the base of an isosceles triangle	1. Given
2. $\angle Y \simeq \angle Z$	2. Base $\angle$s of an isosceles $\triangle$ are $\simeq$
3. $\overrightarrow{XA} \parallel \overline{YZ}$	3. Given
4. $\angle 1 \simeq \angle Y$	4. If two $\parallel$ lines are cut by a transversal, the corresponding $\angle$s are $\simeq$
5. $\angle 2 \simeq \angle Z$	5. If two $\parallel$ lines are cut by a transversal, the alternate interior $\angle$s are $\simeq$
6. $\angle 1 \simeq \angle 2$	6. Transitive Property for $\simeq$

15.   PROOF

Statements	Reasons
1. $\overline{AB} \parallel \overline{DC}$	1. Given
2. $\angle B \simeq \angle DCE$	2. If two $\parallel$ lines are cut by a transversal, the corresponding $\angle$s are $\simeq$
3. $\overline{AB} \simeq \overline{DC}$	3. Given
4. C is the midpoint of $\overline{BE}$	4. Given
5. $\overline{BC} \simeq \overline{CE}$	5. Midpoint of a segment forms two $\simeq$ segments
6. $\triangle ABC \simeq \triangle DCE$	6. SAS
7. $\angle ACB \simeq \angle E$	7. CPCTC
8. $\overline{AC} \parallel \overline{DE}$	8. If two lines are cut by a transversal so that the corresponding $\angle$s are $\simeq$, the lines are $\parallel$

17.   PROOF

Statements	Reasons
1. $ABEF$ is a rectangle	1. Given
2. $ABEF$ is a parallelogram	2. A rectangle is a parallelogram with a right $\angle$
3. $\overline{AF} \simeq \overline{BE}$	3. Opposite sides of a parallelogram are $\simeq$
4. $BCDE$ is a rectangle	4. Given
5. $\angle F$ is a right $\angle$, and $\angle BED$ is a right $\angle$	5. Same as reason 2
6. $\angle F \simeq \angle BED$	6. Any two right $\angle$s are $\simeq$
7. $\overline{FE} \simeq \overline{ED}$	7. Given
8. $\triangle AFE \simeq \triangle BED$	8. SAS
9. $\overline{AE} \simeq \overline{BD}$	9. CPCTC
10. $\angle AEF \simeq \angle BDE$	10. CPCTC
11. $\overline{AE} \parallel \overline{BD}$	11. If lines are cut by a transversal so that the corresponding $\angle$s are $\simeq$, the lines are $\parallel$

19.   PROOF

Statements	Reasons
1. $\triangle FAB \simeq \triangle HCD$	1. Given
2. $\overline{AB} \simeq \overline{DC}$	2. CPCTC
3. $\triangle EAD \simeq \triangle GCB$	3. Given
4. $\overline{AD} \simeq \overline{BC}$	4. CPCTC
5. $ABCD$ is a parallelogram	5. If a quadrilateral has both pairs of opposite sides $\simeq$, the quadrilateral is a parallelogram

21.   PROOF

Statements	Reasons
1. $\overrightarrow{AC}$ bisects $\angle BAD$	1. Given
2. $m\angle 1 = m\angle 2$	2. If a ray bisects an $\angle$, it forms two $\angle$s of $=$ measure
3. $m\angle ACD > m\angle 1$	3. The measure of an exterior $\angle$ of a $\triangle$ is greater than the measure of either of the nonadjacent interior angles
4. $m\angle ACD > m\angle 2$	4. Substitution
5. $AD > CD$	5. If the measure of one angle of a $\triangle$ is greater than the measure of a second angle, the side opposite the first angle is longer than the side opposite the second angle

23. A
25. N
27. S
29. A

31. A
33. $\overline{BC}$, $\overline{AC}$, $\overline{AB}$
35. $\overline{AD}$
37. 5 and 35

39. 115°
41. $m\angle A = 100°$
   $m\angle D = 80°$
43. Isosceles

45. $\overline{AC}$
47. 30
49. $x = 3$; $MN = 15$;
   $JH = 30$

CHAPTER 5

5.1 EXERCISES

1. a, c, d, and f
3. (a) $2\sqrt{2}$ (d) 30
   (b) $3\sqrt{5}$ (e) 3
   (c) 9   (f) $3\sqrt{2}$
5. (a) $\dfrac{3}{4}$ (d) $\dfrac{\sqrt{6}}{3}$
   (b) $\dfrac{5}{7}$ (e) $\dfrac{\sqrt{6}}{3}$
   (c) $\dfrac{\sqrt{7}}{4}$ (f) $\dfrac{\sqrt{10}}{4}$

7. $x = 4$ or $x = 2$
9. $x = 12$ or $x = 5$
11. $x = \dfrac{-2}{3}$ or $x = 4$
13. $x = \dfrac{1}{3}$ or $x = \dfrac{1}{2}$
15. $x = 5$ or $x = 2$
17. $x = \dfrac{7 \pm \sqrt{13}}{2}$

19. $x = 2 \pm 2\sqrt{3}$
21. $x = \dfrac{3 \pm \sqrt{149}}{10}$
23. $x = \pm\sqrt{7}$
25. $x = \pm\dfrac{5}{2}$
27. $x = 0$ or $x = \dfrac{b}{a}$
29. The rectangle is 5 by 8.

5.2 EXERCISES

1. (a) $\dfrac{4}{5}$
   (b) $\dfrac{4}{5}$
   (c) $\dfrac{2}{3}$
   (d) Incommensurable
3. (a) $\dfrac{5}{8}$
   (b) $\dfrac{1}{3}$
   (c) $\dfrac{4}{3}$

   (d) Incommensurable
5. (a) 3   (b) 8
7. (a) 6   (b) 4
9. (a) $\pm 2\sqrt{7}$
   (b) $\pm 3\sqrt{2}$
11. (a) 4
   (b) $\dfrac{-5}{6}$ or 3
13. (a) $\dfrac{3 \pm \sqrt{33}}{4}$
   (b) $\dfrac{7 \pm \sqrt{89}}{4}$

15. $10\dfrac{1}{2}$
17. ≈ 24 outlets
19.
   (a) $4\sqrt{3}$ (b) $4\dfrac{1}{2}$
21. Secretary's salary is
   $18,500; salesperson's
   salary is $27,750; vice
   president's salary is
   $46,250.
23. 40° and 50°
27. (1) If your x value is
   incorrect, then

your y value will
also be incorrect.
   (2) If you use 12.5,
   you are working
   with a decimal
   within a fraction.
29. $a = 12$, $b = 16$

5.3 EXERCISES

1. (a) $\triangle ABC \sim \triangle XTN$ (b) $\triangle ACB \sim \triangle NXT$
3. Yes; yes; spheres have the same shape, and one is
   generally an enlargement of the other.
5. $\triangle RST \sim \triangle UVW$; $\dfrac{WU}{TR} = \dfrac{WV}{TS} = \dfrac{UV}{RS} = \dfrac{3}{2}$

7. (a) 82°   (b) 42°   (c) $10\dfrac{1}{2}$ (d) 8

9. (a) Yes
   (b) Yes
   (c) Yes
   (d) Yes
23. $4\dfrac{1}{2}$

25. 16   **27.** 12
29. $10 + 2\sqrt{5}$ or $10 - 2\sqrt{5}$
31. 3 ft 9 in.
33. 74 ft

5.3 SELECTED PROOFS

11. With $\overline{MN} \perp \overline{NP}$ and $\overline{QR} \perp \overline{RP}$,
   $\angle N$ and $\angle QRP$ are right $\angle$s and
   are therefore congruent. Both
   triangles have $\angle P$ as a common
   angle. By AA, $\triangle MNP \sim \triangle QPR$.
15. By hypothesis, $\overline{AB} \parallel \overline{DF}$. Thus,
   $\angle A \simeq \angle FEG$ since they are

corresponding angles. The
corresponding angles $\angle BCA$ and
$\angle FGE$ are also congruent, using
$\overline{BD} \parallel \overline{FG}$. Therefore $\triangle ABC \sim$
$\triangle EFG$ by AA.
19. By hypothesis, $\overline{RS} \parallel \overline{UV}$. $\angle R \simeq$
$\angle V$ and $\angle S \simeq \angle U$, since they are

alternate interior angles. $\triangle RTS \sim$
$\triangle VTU$ by AA. It follows that
$\dfrac{RT}{VT} = \dfrac{RS}{VU}$ since corresponding
sides of similar $\triangle$s are
proportional.

5.4 EXERCISES

1. 30 oz of ingredient A; 24 oz of ingredient B; 36 oz of ingredient C.

3. (a) Yes   (b) Yes

5. $EF = 4\frac{1}{6}$; $FG = 3\frac{1}{3}$;

  $GH = 2\frac{1}{2}$

7. $x = 5\frac{1}{3}$; $DE = 5\frac{1}{3}$; $EF = 6\frac{2}{3}$

9. $16\frac{4}{5}$

11. $a = \frac{1}{2}$ or $a = 5$ $AD = 4$

13. (a) No   (b) Yes

15. 9

17. $4\sqrt{6}$

19. (a) R

(b) $\overline{PR}$; the shortest distance from a point to a line is the perpendicular distance.

21. (a) $\overline{AD}$ (b) $\overline{FB}$ (c) $\overline{EC}$

23. $\dfrac{1 + \sqrt{73}}{2}$

5.5 EXERCISES

1. $\triangle RST \sim \triangle RVS \sim \triangle SVT$

3. (a) 15   (b) 3

5. (a) 10   (b) $\sqrt{34}$

7. (a) 8   (b) 4

9. (a) Right   (c) Right
  (b) Acute   (d) Obtuse

11. 15 ft

13. $6\sqrt{5}$ m

15. 12 cm

17. The base is 8; the altitude is 6; the diagonal is 10.

19. $6\sqrt{7}$ in.

21. 12 in.

23. 4

25. $9\frac{3}{13}$

27. $5\sqrt{5}$

33. 60°

5.6 EXERCISES

1. $YZ = 8$, $XY = 8\sqrt{2}$

3. $DF = 5\sqrt{3}$, $FE = 10$

5. $HL = 6$, $HK = 12$, $MK = 6$

7. $AC = 6$, $AB = 6\sqrt{2}$

9. $RS = 6$, $RT = 6\sqrt{3}$

11. $5\sqrt{6}$

13. $6\sqrt{3} + 6$
  or $6(\sqrt{3} + 1)$

15. (a) $5 - \sqrt{3}$
  (b) $7 + \sqrt{2}$

(c) $6 - 2\sqrt{5}$
(d) $\sqrt{5} + \sqrt{3}$

17. (a) $-3(1 - \sqrt{3})$
  (b) $\sqrt{5} + 1$

19. $DC = 2\sqrt{3}$, $DB = 4\sqrt{3}$

21. $6\sqrt{3}$

23. $6(2\sqrt{3} - 3)$

25. $20(\sqrt{2} - 1)$

27. $12(2 - \sqrt{2})$

29. $\dfrac{-1 + \sqrt{5}}{2}$

REVIEW EXERCISES

1. False

3. False

5. True

7. True

19. $AB = 6$, $BC = 12$

21. $4\frac{1}{2}$ **23.** $5\frac{3}{5}$ **25.** 6

9. $3.78

11. $79.20

13. 18

15. 150°

29. (a) $8\frac{1}{3}$ (c) $2\sqrt{3}$
  (b) 21   (d) 3

31. (a) 60   (c) 20
  (b) 24   (d) 16

33. $4\sqrt{2}$ in.

35. 25 cm

37. $4\sqrt{3}$ in.

39. (a) $x = 9\sqrt{2}$, $y = 9$
  (b) $x = 4\frac{1}{2}$, $y = 6$
  (c) $x = 12$, $y = 3$
  (d) $x = 2\sqrt{14}$, $y = 13$

41. (a) Acute   (e) No $\triangle$
  (b) No $\triangle$ (f) Acute
  (c) Obtuse   (g) Obtuse
  (d) Right   (h) Obtuse

43. $8(2\sqrt{3} - 3)$

CHAPTER 6

6.1 EXERCISES

1. (a) 90°   (c) 135°
  (b) 270°   (d) 135°

3. (a) 80°   (f) 160°
  (b) 120°   (g) 10°
  (c) 160°   (h) 50°
  (d) 80°   (i) 30°
  (e) 120°

5. (a) 72°   (d) 72°
  (b) 144°   (e) 18°
  (c) 36°

7. (a) 12
  (b) $6\sqrt{2}$

9. $RQ = 3$

11. $\sqrt{7} + 3\sqrt{3}$

13. Square

15. (a) The measure of an arc equals the measure of its intercepted arc. Therefore, congruent arcs would have to have congruent central angles.

  (b) The measure of a central angle equals the measure of its intercepted arc. Therefore, congruent central angles have congruent arcs.

  (c) Draw the radii to the endpoints of the congruent chords. The two triangles formed are congruent by SSS. The central angles

of each triangle are congruent by CPCTC. Therefore, the arcs corresponding to the central angles are also congruent. Hence, congruent chords have congruent arcs.

  (d) Draw the four radii to the endpoints of the congruent arcs. Also draw the chords corresponding to the congruent arcs. The central angles corresponding to the congruent arcs are also congruent. Therefore, the triangles are congruent by SAS. The chords are

congruent by CPTC. Hence congruent arcs have congruent chords.

(e) Congruent central angles have congruent arcs [from part (b)]. Congruent arcs have congruent

chords [from part (d)]. Hence, congruent central angles have congruent chords.

(f) Congruent chords have congruent arcs [from part (c)]. Congruent arcs have congruent

central angles [from part (a)]. Therefore, congruent chords have congruent central angles.

25. $\triangle STV$ is isosceles.
27. (a) 135° (b) 50°

6.1 SELECTED PROOFS

19. *Given:* A circle with inscribed angles A and D intercepting $\overset{\frown}{BC}$

Prove: $\angle A \simeq \angle D$

Proof: Since both inscribed angles, $\angle A$ and $\angle D$, intercept $\overset{\frown}{BC}$,

$$m\angle A = \frac{1}{2}m\overset{\frown}{BC} \quad \text{and} \quad m\angle D = \frac{1}{2}m\overset{\frown}{BC}$$

Therefore $m\angle A = m\angle D$, which means that $\angle A \simeq \angle D$.

23. Using the chords $\overline{AB}$, $\overline{BC}$, $\overline{CD}$, and $\overline{AD}$ in $\odot O$ as sides of inscribed angles, we find $\angle B \simeq \angle D$ and $\angle A \simeq \angle C$ since they are inscribed angles intercepting the same arc. Then $\triangle ABE \sim \triangle CED$ by AA.

6.2 EXERCISES

1. (a) 8°   (d) 54°
    (b) 46°   (e) 126°
    (c) 38°

3. No. If $\overline{RS}$ is a diameter and $\overline{SW}$ is a tangent, then $\overline{RS} \perp \overline{SW}$. $\overline{TS}$ cannot be $\perp$ to $\overline{SW}$ since there is already a segment $\perp$ to $\overline{SW}$ at S.

5. (a) 22°   (c) 15°
    (b) 7°

7. (a) 136°   (c) 68°
    (b) 224°   (d) 44°

9. (a) $101\frac{1}{2}°$

   (b) $78\frac{1}{2}°$

11. (a) 120°   (c) 60°
    (b) 240°

13. 28°
29. $m\overset{\frown}{QP} = 96°$, $m\overset{\frown}{MN} = 28°$
31. 10
33. $m\angle 1 = 36°$, $m\angle 2 = 108°$

6.2 SELECTED PROOFS

19. $m\angle BCD = m\angle A + m\angle B$; but since $m\angle A = m\angle B$, $m\angle BCD = m\angle B + m\angle B$ or $m\angle BCD = 2 \cdot m\angle B$. Also, $m\angle BCD = \frac{1}{2}m\overset{\frown}{BD}$ since it is an inscribed $\angle$.

Therefore, $\frac{1}{2}m\overset{\frown}{BD} = 2 \cdot m\angle B$ or $m\overset{\frown}{BD} = 4 \cdot m\angle B$. But if $\overline{AB}$ is a tangent to $\odot O$ at B, $m\angle B = \frac{1}{2}m\overset{\frown}{BC}$. By substitution, $m\overset{\frown}{BD} = 4\left(\frac{1}{2}m\overset{\frown}{BC}\right)$ or $m\overset{\frown}{BD} = 2 \cdot m\overset{\frown}{BC}$.

23. If $ABCD$ is a trapezoid, then $\overline{BC} \parallel \overline{AD}$. If two parallel lines intersect a circle, then the intercepted arcs between these lines are congruent. Therefore $\overset{\frown}{AB} \simeq \overset{\frown}{CD}$, which means

that $\overline{AB} \simeq \overline{CD}$. If the nonparallel sides of a trapezoid are congruent, then trapezoid $ABCD$ is an isosceles trapezoid.

27. $\overline{AC}$ is a diameter in $\odot O$ and $\overrightarrow{CD}$ is a tangent.

$m\angle BCD = m\angle BCA + m\angle ACD$ Angle-Addition Postulate

$m\angle BCA = \frac{1}{2}m\overset{\frown}{BA}$ inscribed $\angle$

$m\angle ACD = \frac{1}{2}m\overset{\frown}{AC}$ case 1

$m\angle BCD = \frac{1}{2}m\overset{\frown}{BA} + \frac{1}{2}m\overset{\frown}{AC}$

$= \frac{1}{2}(m\overset{\frown}{BA} + m\overset{\frown}{AC})$ substitution and distributive

$m\overset{\frown}{BA} + m\overset{\frown}{AC} = m\overset{\frown}{BAC}$ Arc-Addition Postulate

$m\angle BCD = \frac{1}{2}m\overset{\frown}{BAC}$ substitution

6.3 EXERCISES

1. 30°
3. $6\sqrt{5}$
7. 3
9. $DE = 12$, $EC = 4$

11. 4
13. $9\frac{2}{5}$

15. 9
17. $5\frac{1}{3}$

19. $3 + 3\sqrt{5}$
25. Yes; $\overline{AE} \simeq \overline{CE}$; $\overline{DE} \simeq \overline{EB}$

27. $AM = 5$, $PC = 7$, $BN = 9$
29. 12   31. 45

6.3 SELECTED PROOFS

23. In $\odot Q$, if tangents $\overline{MN}$ and $\overline{MP}$ are $\perp$, then $\angle M$ is a right $\angle$. $\angle N$ and $\angle P$ are right $\angle$s since a radius drawn to the point tangency is $\perp$ to the tangent. Therefore, $\overline{QN} \parallel \overline{PM}$ and $\overline{NM} \parallel \overline{QP}$; hence $QNMP$ is a parallelogram. But since $MNQP$ has a right angle and two adjacent sides congruent $(\overline{QN} \simeq \overline{QP})$, it is also a square.

35. Given: $\overline{TX}$ is a secant segment intersecting the circle at W
    $\overline{TV}$ is a tangent at V
    Prove: $(TV)^2 = TW \cdot TX$

Proof: With secant $\overline{TX}$ and tangent $\overline{TV}$, draw in $\overline{WV}$ and $\overline{VX}$. $m\angle X = \frac{1}{2} \, m\widehat{WV}$ since $\angle X$ is an inscribed angle. $m\angle TVW = \frac{1}{2} \, m\widehat{WV}$ because it is formed by a tangent and a chord. By substitution, $m\angle TVW = m\angle X$ or $\angle TVW \simeq \angle X$. $\angle T \simeq \angle T$, so $\triangle TVW \sim \triangle TXV$ by AA. It follows that $\frac{TV}{TW} = \frac{TX}{TV}$ or $(TV)^2 = TW \cdot TX$.

6.4 EXERCISES

3. No
5. 60°
7. $\overline{AB}$; $\overline{GH}$; for a circle containing unequal chords, the chord nearest the center has the greatest length, and the chord at the greatest distance from the center has the least length.
9. (a) $m\angle AOB > m\angle BOC$

(b) $AB > BC$
11. (a) $m\widehat{AB} > m\widehat{BC}$
(b) $AB > BC$
13. (a) $\angle C$ (b) $\overline{AC}$
15. $4\sqrt{3} - 4\sqrt{2}$

6.4 SELECTED PROOF

19. Given: $\odot O$ with $m\widehat{AB} > m\widehat{CD}$
    Prove: $m\angle AOB > m\angle COD$

Proof: In $\odot O$, $m\angle AOB = m\widehat{AB}$ and $m\angle COD = m\widehat{CD}$. If $m\widehat{AB} > m\widehat{CD}$, then by substitution, $m\angle AOB > m\angle COD$.

6.5 EXERCISES

9. The locus of points at a given distance from a fixed line is two parallel lines on either side of the fixed line at the same distance from that line.
11. The locus of points at a distance of 3 in. from point O is a circle of radius 3 in.
13. The locus of points equidistant from three noncollinear points D, E, and F is the circumcenter of $\triangle DEF$.
15. The locus of the midpoints of the chords in $\odot Q$ parallel to diameter $\overline{PR}$ is the perpendicular bisector of $\overline{PR}$.
17. The locus of points equidistant from two given intersecting lines are two perpendicular lines that bisect the angles formed by the two intersecting lines.
23. The locus of points at a distance of 2 cm from a sphere whose radius is 5 cm is two concentric spheres with the same center. The radius of one sphere is 3 cm and the radius of the other sphere is 7 cm.
25. The locus is another sphere with the same center and a radius of length 2.5 m.
27. The locus of points equidistant from an 8-ft ceiling and the floor is a parallel plane in the middle.
43. Equilateral triangle
45. No
47. $\frac{10\sqrt{3}}{3}$
49. $RQ = 10$, $SQ = \sqrt{89}$

REVIEW EXERCISES

1. 9 mm
3. $\sqrt{41}$ in.
5. 130°
7. 80°
9. $m\widehat{AC} = m\widehat{DC} = 93\frac{1}{3}°$
   $m\widehat{AD} = 173\frac{1}{3}°$

11. $m\angle 2 = 44°$
    $m\angle 3 = 90°$
    $m\angle 4 = 46°$
    $m\angle 5 = 44°$
13. 24   17. N   21. A
15. A   19. A

23. (a) 70°   (f) 260°
    (b) 28°
    (c) 64°
    (d) 21°
    (e) $m\widehat{AB} = 90°$
        $m\widehat{CD} = 40°$

25. 29
27. If $\overline{DC}$ is tangent to circles B and A at points D and C, then $\overline{BD} \perp \overline{DC}$ and $\overline{AC} \perp \overline{DC}$. $\angle D$ and $\angle C$ are congruent

since they are right angles. $\angle DEB \cong \angle CEA$ because of vertical angles. $\triangle BDE \sim \triangle ACE$ by AA. It follows that $\dfrac{AC}{CE} = \dfrac{BD}{ED}$ since corresponding sides are proportional. Hence $AC \cdot ED = CE \cdot BD$.

29. If $\overline{AP}$ and $\overline{BP}$ are tangent to $\odot Q$ at A and

B, then $\overline{AP} \cong \overline{BP}$. $\overline{AC} \cong \overline{BC}$ since C is the midpoint of $\overline{AB}$. It follows that $\overline{AC} \cong \overline{BC}$; and using $\overline{CP} \cong \overline{CP}$, we have $\triangle ACP \cong \triangle BCP$ by SSS. $\angle APC \cong \angle BPC$ by CPCTC, and hence $\overrightarrow{PC}$ bisects $\angle APB$.

31. $24\sqrt{2}$ cm
33. 14 and 15 cm

35. (a) $AB > CD$
    (b) $QP < QR$
    (c) $m\angle A < m\angle C$
39. The locus of the midpoints of the radii of a circle is another circle.
41. The locus of the centers of a penny that rolls around a half-dollar is a circle.

43. The locus of points equidistant from two parallel planes is a parallel plane midway between the two planes.
51. $BF = 6$; $AE = 9$

CHAPTER 7

7.1 EXERCISES

1. No; two triangles with equal areas are not necessarily congruent. Yes; two squares with equal areas must be congruent because the sides will be congruent.
3. 37 units²

5. The altitudes to $\overline{PN}$ and to $\overline{MN}$ are congruent.
7. 54 cm²
9. 9 m²
11. 72 in.²
13. 100 in.²
15. 126 in.²

17. 264 units²
19. 144 units²
21. $\dfrac{27}{4}\sqrt{3}$ units²
23. 192 ft²
25. (a) 60 ft²
    (b) 1 gal
    (c) $15.50

27. $(156 + 24\sqrt{10})$ ft²
35. 8
37. (a) 12 in.
    (b) 84 in.²
39. 56 percent
43. $ac + ad + bc + bd$
45. $4\dfrac{8}{13}$ in.

47. 8
49. (a) 10
    (b) 26
    (c) 18
    (d) No

7.1 SELECTED PROOF

31. Draw $\overline{ST} \perp \overline{MN}$. $\overline{QM} \perp \overline{MN}$ since $MNPQ$ is a rectangle. Because $\overline{QP} \parallel \overline{MN}$, it follows that $QM = ST$ since parallel lines are everywhere equidistant.

$$A_{MNS} = \frac{1}{2}b \cdot h$$

$$= \frac{1}{2}(MN)(ST)$$

$$= \frac{1}{2}(MN)(QM)$$

$$= \frac{1}{2} \cdot A_{MNPQ}$$

7.2 EXERCISES

1. 30 in.
3. $4\sqrt{29}$ m
5. $7(\sqrt{6} + \sqrt{3}) + 41$
7. 38
9. 84 in.²
11. 40 ft²
13. 80

15. $36 + 36\sqrt{3}$
17. 16 in., 32 in., and 28 in.
19. 15 cm
25. $24 + 4\sqrt{21}$
27. 96
29. 6 yd by 8 yd

31. (a) 770 ft
    (b) $454.30
33. 624 ft²
35. Square with sides of length 10 in.

7.2 SELECTED PROOF

23. Using Heron's Formula, we get

$$A = \sqrt{s(s-a)(s-b)(s-c)} \text{ where } s = \frac{1}{2}(a+b+c). \text{ Also,}$$

$$A = \frac{1}{2}bh \quad \rightarrow \quad A = \frac{1}{2} \cdot c(CD)$$

$$\therefore \frac{1}{2}c(CD) = \sqrt{s(s-a)(s-b)(s-c)}$$

$$c(CD) = 2\sqrt{s(s-a)(s-b)(s-c)}$$

$$CD = \frac{2\sqrt{s(s-a)(s-b)(s-c)}}{c}$$

7.3 EXERCISES

3. Draw the diagonals (angle bisectors) $\overline{JL}$ and $\overline{MK}$. These determine center O of the inscribed circle.

Now construct the line segment $\overline{OR} \perp \overline{MJ}$. Use OR as the length of the radius of the inscribed circle.

9. $a = 5$ in., $r = 5\sqrt{2}$ in.
11. $16\sqrt{3}$ ft; 16 ft
13. (a) 120°   (c) 72°
       (b) 90°   (d) 60°
15. $54\sqrt{3}$ cm²

17. $75\sqrt{3}$ in.²
19. $(24 + 12\sqrt{3})$ in.²
21. $\dfrac{2}{1}$
23. 156°

7.4 EXERCISES

1. $C = 16\pi$ cm
    $A = 64\pi$ cm²
3. $C = 66$ in.
    $A = 346\dfrac{1}{2}$ in.²
5. (a) $r = 22$ in.
       $d = 44$ in.
    (b) $r = 30$ ft
       $d = 60$ ft

7. (a) $r = 5$ in.
       $d = 10$ in.
    (b) $r = 1.5$ cm
       $d = 3.0$ cm
9. $\dfrac{8}{3}\pi$ in.
11. 9π cm   **13.** 30.56 in.
15. $P = (12 + 4\pi)$ in.;
    $A = (24\pi - 36\sqrt{3})$ in.²

17. 16 in.²
19. $5 < AN < 13$
21. $\left(25\sqrt{3} - \dfrac{25}{2}\pi\right)$ cm²
23. $(144 - 36\pi)$ m²
25. $(600 - 144\pi)$ ft²
27. 7 cm
29. 8 in.
31. $(12\pi - 9\sqrt{3})$ in.²

33. 36π
37. 3 in. and 4 in.
39. (a) 56.52 ft
       (b) 10
       (c) $95.40
41. 38.2 yd
43. 154 cm²

7.4 SELECTED PROOF

35. $A = A_{\text{lge. circle}} - A_{\text{sm. circle}}$
    $= \pi R^2 - \pi r^2$
    $= \pi(R^2 - r^2)$

But $R^2 - r^2$ is a difference of two squares, so

$A = \pi(R + r)(R - r)$

REVIEW EXERCISES

1. 480 units²
3. 50 units²
5. 336 units²
7. (a) $(24\sqrt{2} + 18)$ units²
    (b) $(24 + 9\sqrt{3})$ units²
    (c) $33\sqrt{3}$ units²
9. (a) 19,000 ft²
    (b) four bags
    (c) $72
11. (a) $\left(\dfrac{289}{4}\sqrt{3} + 8\sqrt{33}\right)$ units²
    (b) $(50 + \sqrt{33})$ units

13. 5 cm by 7 cm
15. 36 units²
17. 20 units²
23. (a) No; $\perp$ bisectors of sides of a parallelogram are not necessarily concurrent.
    (b) Yes; $\perp$ bisectors of sides of a rhombus are concurrent.
    (c) Yes; $\perp$ bisectors of sides of a rectangle are concurrent.
    (d) Yes; $\perp$ bisectors of sides of a square are concurrent.

19. $96\sqrt{3}$ ft²
21. $162\sqrt{3}$ in.²
27. $(64 - 16\pi)$ units²
29. $\left(\dfrac{8}{3}\pi - 4\sqrt{3}\right)$ units²
31. $\left(25\sqrt{3} - \dfrac{25}{3}\pi\right)$ units²
33. (a) 21 ft
    (b) $346\dfrac{1}{2}$ ft²
35. $(9\pi - 18)$ in.²
39. (a) 28 yd²
    (b) 21.2 ft²

CHAPTER 8

8.1 EXERCISES

3. (a) 4   (c) 5
    (b) 8   (d) 9
5. $b = 3.5$ or $b = 10.5$
7. (a) 5   (c) $2\sqrt{5}$
    (b) 10   (d) $\sqrt{a^2 + b^2}$
9. (a) $\left(2, \dfrac{-3}{2}\right)$ (b) $(1, 1)$

(c) $(4, 0)$ **(d)** $\left(\dfrac{a}{2}, \dfrac{b}{2}\right)$
11. (a) $(-3, 4)$
    (b) $(0, -2)$
    (c) $(-a, 0)$
    (d) $(-b, -c)$
13. (a) $(-3, -4)$ (c) $(-a, 0)$
    (b) $(-2, 0)$ (d) $(-b, c)$

15. $(2.5, -13.7)$
17. $(2, 3)$; 16 units²
19. (a) Isosceles
    (b) Equilateral
    (c) Isosceles right triangle
21. $x + y = 6$

23. $(a, a\sqrt{3})$ or $(a, -a\sqrt{3})$
25. $(0, 1 + 3\sqrt{3})$ and $(0, 1 - 3\sqrt{3})$
27. 17 units²

8.2 EXERCISES

1. $(4, 0)$ and $(0, 3)$

3. $(5, 0)$ and
$$\left(0, \frac{-5}{2}\right)$$

5. $(-3, 0)$

7. $(6, 0)$ and $(0, 3)$

9. (a) 4
  (b) Undefined
  (c) -1

(d) 0

(e) $\dfrac{d - b}{c - a}$

(f) $-\dfrac{b}{a}$

11. (a) 10
  (b) 15

13. (a) Collinear
  (b) Noncollinear

15. (a) $\dfrac{3}{4}$

  (b) $-\dfrac{5}{3}$

  (c) -2

  (d) $\dfrac{a - b}{c}$

17. None of these

19. Perpendicular

21. $\dfrac{3}{2}$

23. 23

31. Right triangle

33. $(4, 7)$; $(0, -1)$; $(10, -3)$

8.2 SELECTED PROOFS

35. $m_{\overline{VT}} = \dfrac{e - e}{(c - d) - (a + d)} = \dfrac{0}{c - a - 2d} = 0$

$m_{\overline{RS}} = \dfrac{b - b}{c - a} = \dfrac{0}{c - a} = 0$

$\therefore \overline{VT} \parallel \overline{RS}$

$RV = \sqrt{[(a + d) - a]^2 + (e - b)^2}$
$= \sqrt{d^2 + (e - b)^2}$
$= \sqrt{d^2 + e^2 - 2be + b^2}$

$ST = \sqrt{(c - [c - d])^2 + (b - e)^2}$
$= \sqrt{(d)^2 + (b - e)^2}$
$= \sqrt{d^2 + b^2 - 2be + e^2}$

$\therefore RV = ST$

$RSTV$ is an isosceles trapezoid.

39. If $\ell_1 \parallel \ell_2$, then $\angle A \simeq \angle D$. With $\overline{CB} \perp \overline{AB}$ and $\overline{FE} \perp \overline{DE}$, $\overline{CB} \parallel \overline{FE}$ (both $\perp$ to the x axis). Then $\angle B \simeq \angle E$ (all right $\angle$s are $\simeq$). By AA, $\triangle ABC \sim \triangle DEF$. Then $\dfrac{CB}{FE} = \dfrac{AB}{DE}$ since corresponding sides of $\sim \triangle$s are proportional. By a property of proportions, $\dfrac{CB}{AB} = \dfrac{FE}{DE}$ (means were interchanged). But $m_1 = \dfrac{CB}{AB}$ and $m_2 = \dfrac{FE}{DE}$. Then $m_1 = m_2$, and the slopes are equal.

8.3 EXERCISES

1. $x + 2y = 6$; $y = -\dfrac{1}{2}x + 3$

3. $-x + 3y = -40$; $y = \dfrac{1}{3}x - \dfrac{40}{3}$

11. $2x + 3y = 15$

13. $x + y = 6$

15. $-2x + 3y = -3$

17. $-x + y = -2$

19. $5x + 2y = 5$

21. $4x + 3y = -12$

23. $-x + 3y = 2$

25. $(6, 0)$

27. $(5, -4)$

29. $(6, -2)$

31. $a = 2$; $b = 3$

33. $(3, -2)$; $k = 5$

8.3 SELECTED PROOF

35. For $B \neq 0$, the equation $Ax + By = C$ is equivalent to $By = -Ax + C$. Therefore

$$y = -\dfrac{A}{B}x + \dfrac{C}{B}$$

For $B \neq 0$, the equation $Ax + By = D$ is equivalent to $y = -\dfrac{A}{B}x + \dfrac{D}{B}$.

Since $m_1 = m_2 = -\dfrac{A}{B}$, the graphs (lines) are parallel.

Note: If $B = 0$, both graphs (lines) are vertical and are parallel.

8.4 EXERCISES

1. Circle with center $(0, 0)$ and $r = 5$

3. Circle with center $(2, -3)$ and $r = 3$

5. Circle with center $(1, -2)$ and $r = 2$

7. Circle with center $(3, 2)$ and $r = 4$

9. Parabola that opens up; vertex $(0, 0)$

11. Parabola that opens up; vertex $(2, -9)$

13. Parabola that opens down; vertex $(1, 2)$

15. $(-2, 4)$ and $(1, 1)$

17. $(-4, 3)$ and $(5, 0)$

19. No point of intersection

21. $(-2, 4)$ and $(1, 1)$

23. $(5, 0)$ and $(-4, 3)$

25. No points of intersection

27. $x^2 + y^2 - 4x + 10y + 13 = 0$

29. $x^2 + y^2 + 6x + 2y - 15 = 0$

31. $x^2 + y^2 - 2x + 2y - 23 = 0$

33. $x^2 + y^2 + 10x - 10y + 25 = 0$

35. 2

37. $-2x + y = -1$

39. $3x + 4y = 25$

41. Perpendicular

43. $(-2, 5)$ and $(2, 5)$

8.5 EXERCISES

1. **(a)** $a\sqrt{2}$ if $a > 0$

(b) $\dfrac{d - b}{c - a}$

3. **(a)** $y = -1x + a$

(b) $y = \dfrac{-b}{a}x + b$

5. $A = (0, 0)$
   $B = (a, 0)$
   $C = (a, b)$

7. $M = (0, 0)$
   $N = (r, 0)$
   $P = (r + s, t)$

9. $A = (0, 0)$
   $B = (a, 0)$
   $C = (a - c, d)$

11. **(a)** $A = (0, 0)$
   $B = (a, 0)$
   $C = (a, a)$
   $D = (0, a)$

(b) $E = (0, 0)$
   $F = (2a, 0)$
   $G = (2a, 2a)$
   $H = (0, 2a)$

13. **(a)** $A = (0, 0)$
   $B = (a, 0)$
   $C = (a + b, c)$
   $D = (b, c)$
   Note: D chosen before C

(b) $E = (0, 0)$
   $F = (2a, 0)$
   $G = (2a + 2b, 2c)$

$H = (2b, 2c)$
   Note: H chosen before G

15. **(a)** $R = (0, 0)$
   $S = (2a, 0)$
   $T = (a, b)$

(b) $X = (0, 0)$
   $Y = (4a, 0)$
   $Z = (2a, 2b)$

17. $r^2 = s^2 + t^2$

19. $c^2 = a^2 - b^2$

21. $b^2 = 3a^2$

25. **(a)** Slope
   (b) Distance
   (c) Midpoint
   (d) Slope

31. $\left(c, \dfrac{ac}{b}\right)$

8.6 EXERCISE

23. True

8.6 SELECTED PROOFS

3. The diagonals of a square are perpendicular bisectors of each other.
Proof: Let square $RSTV$ have the vertices shown.

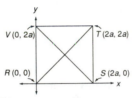

Then the midpoints of the diagonals are

$$M_{\overline{RT}} = \left(\dfrac{0 + 2a}{2}, \dfrac{0 + 2a}{2}\right) = (a, a)$$

$$M_{\overline{VS}} = \left(\dfrac{0 + 2a}{2}, \dfrac{2a + 0}{2}\right) = (a, a)$$

Because the two diagonals share the common midpoint (a, a) they bisect each other.

7. The segments that join the midpoints of the consecutive sides of a quadrilateral form a parallelogram.

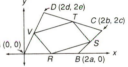

Proof: The midpoints, as shown, of the sides of quadrilateral $ABCD$ are

$$R = \left(\dfrac{0 + 2a}{2}, \dfrac{0 + 0}{2}\right) = (a, 0)$$

$$S = \left(\dfrac{2a + 2b}{2}, \dfrac{0 + 2c}{2}\right) = (a + b, c)$$

$$T = \left(\dfrac{2d + 2b}{2}, \dfrac{2e + 2c}{2}\right) = (d + b, e + c)$$

and

$$V = \left(\dfrac{0 + 2d}{2}, \dfrac{0 + 2e}{2}\right) = (d, e)$$

Now we determine slopes as follows:

$$m_{\overline{RS}} = \dfrac{c - 0}{(a + b) - a} = \dfrac{c}{b}$$

$$m_{\overline{ST}} = \dfrac{(e + c) - c}{(d + b) - (a + b)} = \dfrac{e}{d - a}$$

$$m_{\overline{TV}} = \dfrac{(e + c) - e}{(d + b) - d} = \dfrac{c}{b}$$

and

$$m_{\overline{VR}} = \dfrac{e - 0}{d - a} = \dfrac{e}{d - a}$$

Because $m_{\overline{RS}} = m_{\overline{TV}}$, $\overline{RS} \parallel \overline{TV}$; also, $m_{\overline{ST}} = m_{\overline{VR}}$, so that $\overline{ST} \parallel \overline{VR}$. Then $RSTV$ is a parallelogram.

11. The midpoint of the hypotenuse of a right triangle is equidistant from the three vertices of the triangle.

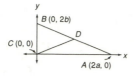

Proof: Let right $\triangle ABC$ have vertices as shown. Then D, the midpoint of the hypotenuse, is given by

$$D = \left(\frac{0 + 2a}{2}, \frac{2b + 0}{2}\right) = (a, b)$$

Now

$$BD = DA = \sqrt{(2a - a)^2 + (0 - b)^2}$$
$$= \sqrt{a^2 + (-b)^2} = \sqrt{a^2 + b^2}$$

Also,

$$CD = \sqrt{(a - 0)^2 + (b - 0)^2}$$
$$= \sqrt{a^2 + b^2}$$

Then D is equidistant from A, B, and C.

---

15. If the diagonals of a parallelogram are perpendicular, then the parallelogram is a rhombus.
Proof: Let parallelogram $ABCD$ have vertices as shown. If $\overline{AC} \perp \overline{DB}$, then $m_{\overline{AC}} \cdot m_{\overline{DB}} = -1$. Because

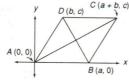

$$m_{\overline{AC}} = \frac{c - 0}{(a + b) - 0}$$
$$= \frac{c}{a + b}$$

and $$m_{\overline{DB}} = \frac{0 - c}{a - b} = \frac{-c}{a - b}$$

it follows that

$$\frac{c}{a + b} \cdot \frac{-c}{a - b} = -1$$

or $$\frac{-c^2}{a^2 - b^2} = -1$$

Then
$$-c^2 = -1(a^2 - b^2)$$
$$-c^2 = -a^2 + b^2$$
$$a^2 = b^2 + c^2 \text{(*)}$$

For $ABCD$ to be a rhombus, we must show that two adjacent sides are congruent.

$$AB = |a - 0| = a$$
$$AD = \sqrt{(b - 0)^2 + (c - 0)^2}$$
$$= \sqrt{b^2 + c^2}$$

Because $a^2 = b^2 + c^2$(*), it follows that

$$a = \sqrt{b^2 + c^2}$$

and $$AB = AD$$

Then parallelogram $ABCD$ is a rhombus.

---

19. The altitudes of a triangle are concurrent.
Proof: For $\triangle ABC$, let $\overline{CH}$, $\overline{AJ}$, and $\overline{BK}$ name the altitudes. Because $\overline{AB}$ is horizontal ($m_{\overline{AB}} = 0$), $\overline{CH}$ is vertical and has the equation $x = b$.

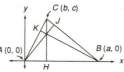

Because $m_{\overline{BC}} = \frac{c - 0}{b - a} = \frac{c}{b - a}$, the slope of altitude $\overline{AJ}$ is $m_{\overline{AJ}} = -\frac{b - a}{c} = \frac{a - b}{c}$. Since $\overline{AJ}$ contains $(0, 0)$, its equation is $y = \frac{a - b}{c}x$.

The intersection of altitudes $\overline{CH}$ ($x = b$) and $\overline{AJ}$ $\left(y = \frac{a - b}{c}x\right)$ is at $x = b$, so that

$$y = \frac{a - b}{c} \cdot b = \frac{b(a - b)}{c} = \frac{ab - b^2}{c}$$

That is, $\overline{CH}$ and $\overline{AJ}$ intersect at $\left(b, \frac{ab - b^2}{c}\right)$. The remaining altitude is $\overline{BK}$. Since $m_{\overline{AC}} = \frac{c - 0}{b - 0} = \frac{c}{b}$, $m_{\overline{BK}} = -\frac{b}{c}$. Because $\overline{BK}$ contains $(a, 0)$, its equation is $y - 0 = -\frac{b}{c}(x - a)$ or $y = \frac{-b}{c}(x - a)$.

For the three altitudes to be concurrent, $\left(b, \frac{ab - b^2}{c}\right)$ must lie on the line $y = \frac{-b}{c}(x - a)$. Substitution leads to

$$\frac{ab - b^2}{c} = \frac{-b}{c}(b - a)$$

$$\frac{ab - c^2}{c} = \frac{-b(b - a)}{c}$$

$$\frac{ab - b^2}{c} = \frac{-b^2 + ab}{c}$$

which is true. Then the three altitudes are concurrent.

REVIEW EXERCISES

1. (a) 7   (c) 13
    (b) 6   (d) 5

3. (a) $\left(6, \dfrac{1}{2}\right)$

    (b) $(-2, 4)$

    (c) $\left(1, \dfrac{-1}{2}\right)$

    (d) $\left(\dfrac{2x - 3}{2}, y\right)$

5. (a) Undefined
    (b) 0

    (c) $\dfrac{-5}{12}$

    (d) $\dfrac{-4}{3}$

7. $(-4, -8)$

9. $\dfrac{4}{3}$

11. (a) Perpendicular
    (b) Parallel
    (c) Neither
    (d) Perpendicular

13. 4

15. $y = \dfrac{4}{3}x - 3$

17. (a) $3x + 5y = 24$
    (b) $3x + y = -7$
    (c) $-2x + y = -8$

(d) $y = 5$
(e) $x^2 + y^2 - 6x +$
    $2y - 6 = 0$
(f) $x^2 + y^2 - 4x +$
    $6y + 9 = 0$

23. $(3, 5)$
25. $(1, 4)$ and $(3, 12)$
27. $(3, 5)$
29. $(1, 4)$ and $(3, 12)$
31. $x^2 + y^2 - 10x -$
    $6y + 9 = 0$
33. $(16, 11)$, $(4, -9)$,
    $(-4, 5)$

35. $C = (-5, 6)$;
    $r = \sqrt{51}$
37. $A = (-a, 0)$
    $B = (0, b)$
    $C = (a, 0)$
39. $R = (0, 0)$
    $U = (0, a)$
    $T = (a, a + b)$
41. (a) $\sqrt{(a + c)^2 + (b + d - 2e)^2}$

    (b) $-\dfrac{a}{b - e}$ or $\dfrac{a}{e - b}$

    (c) $y - 2d = \dfrac{a}{c - b}(x - 2c)$

SELECTED PROOF

19. Let $A = (3, 6)$, $B = (-6, 4)$ and $C = (1, -2)$.

$$AB = \sqrt{(-6 - 3)^2 + (4 - 6)^2}$$
$$= \sqrt{(-9)^2 + (-2)^2} = \sqrt{81 + 4} = \sqrt{85}$$
$$BC = \sqrt{[1 - (-6)]^2 + [(-2) - 4]^2}$$
$$= \sqrt{7^2 + (-6)^2} = \sqrt{49 + 36} = \sqrt{85}$$
$$AC = \sqrt{(1 - 3)^2 + ([-2] - 6)^2}$$
$$= \sqrt{(-2)^2 + (-8)^2} = \sqrt{4 + 64} = \sqrt{68}$$
$$= \sqrt{4 \cdot 17} = 2\sqrt{17}$$

Because $AB = BC$, the triangle is isosceles.

CHAPTER 9

9.1 EXERCISES

1. (a) Prism
    (b) Oblique
    (c) Hexagon
    (d) Oblique hexagonal
      prism
    (e) Parallelogram

3. (a) 16 vertices
    (b) 8 lateral edges
    (c) 16 base edges
7. 423 in.²
9. 105 m²

11. $\dfrac{105\sqrt{7}}{4}$ m³

13. 340 m²
15. $18\sqrt{91}$ ft²
17. $144\sqrt{3}$ ft³

19. 6 in. by 6 in. by 3 in.
21. 3
23. $(36\sqrt{5} + 36)$ in.²
25. $4.44
27. 260 ft²

9.2 EXERCISES

1. (a) 60π in.²
    (b) 110π in.²
    (c) 150π in.³
3. 54.17 in.²
5. 5 cm
7. Radius = 2 in. and
    altitude = 3 in.

9. 32π in.³
11. $3\sqrt{5}$ cm
13. (a) 173.85 in.²
    (b) 286.99 in.²
    (c) 264 in.³
15. 60π cm² (exact)
17. 192π in.³ (exact)

19. (a) 32π units³ (exact)
    (b) 96π units³ (exact)
21. (a) 225π unit³ (exact)
    (b) 405π units³ (exact)

23. $\dfrac{64}{3}\pi$ in.³ (exact)

27. 60π in.² (exact)
29. 4:1
31. ≈ 471 gal

9.3 EXERCISES

1. Polyhedron *EFGHIJK* is concave.
3. A regular tetrahedron has four faces, four vertices, and six edges.
5. (a) $6\sqrt{2}$ in.

(b) $6\sqrt{3}$ in.
7. (a) 3:2   (b) 3:2
9. The radius measures $\sqrt{3}$ in. and the altitude measures $2\sqrt{3}$ in.
11. (a) $3\sqrt{3}$ in.   (b) 9 in.

13. (a) $\approx \dfrac{792}{7}$ m²

(b) $\approx 113\dfrac{1}{7}$ m³
15. $\dfrac{3}{2}$ in.

17. 339.12 ft²; 8.48 pt (9 pt of paint would have to be purchased)
19. ≈ 23.23 in.³
21. 36π units³

9.4 EXERCISES

1. $A = (-1, 5, 0)$ $C = (3, 0, 0)$
   $B = (2, -1, 3)$ $D = (1, 3, -1)$
3. $\sqrt{21}$
5. 3
7. $(-3, 2, 7)$
9. $\left(\dfrac{7}{2}, 3, -3\right)$
15. $x^2 + y^2 + z^2 - 16 = 0$
17. $x^2 + y^2 + z^2 - 4x - 6y + 4z + 8 = 0$
19. $(0, 0, 6)$, $(6, 0, 3)$, and $(4, 4, 1)$

21. $(0, 0, 3)$, $(1, 2, -2)$, and $(0, 0, -3)$
23. $(-60, 0, 0)$, $(0, 40, 0)$, and $(0, 0, -30)$
25. Plane
27. Sphere with center $(0, 0, 0)$ and $r = 2$
29. Point $(1, 2, 4)$
31. Sphere with center $(2, -3, 1)$ and $r = 4$
33. $x + 2y + 2z = 12$
35. $2\sqrt{3}$ units²
37. (a) 4:1   (b) 8:1

REVIEW EXERCISES

1. 672 in.²
3. Length = Width = $\dfrac{15}{2}$ in.
   Height = $\dfrac{49}{2}$ in.
   $V = 1378\dfrac{1}{8}$ in.³

5. (a) 360 in.²
   (b) 468 in.²
   (c) 540 in.³
7. (a) 540 in.²
   (b) 864 in.²
   (c) 1296 in.³

9. (a) 120π in.²
   (b) 192π in.²
   (c) 360π in.³
11. (a) 72π cm²
    (b) 108π cm²
    (c) $72\pi\sqrt{3}$ cm³

13. ≈ 616 in.²
15. 120π units³
17. $183\dfrac{1}{3}$ units³
19. $\dfrac{32\pi}{3}$ units³

21. $\left(2744 - \dfrac{1372\pi}{3}\right)$ in.³
23. 40π mm³
25. (a) $(-1, 3, -1)$
    (b) 6
27. $x^2 + y^2 + z^2 - 4x + 2y - 6z - 11 = 0$

CHAPTER 10

10.1 EXERCISES

1. $\sin \alpha = \dfrac{5}{13}$; $\sin \beta = \dfrac{12}{13}$
3. $\sin \alpha = \dfrac{8}{17}$; $\sin \beta = \dfrac{15}{17}$
5. $\sin \alpha = \dfrac{\sqrt{15}}{5}$; $\sin \beta = \dfrac{\sqrt{10}}{5}$
7. 1
9. 0.2924
11. 0.9903
13. 0.9511
15. $a \approx 6.9$ in.; $b \approx 9.8$ in.
17. $a \approx 10.9$ ft; $b \approx 11.7$ ft
19. $c \approx 8.8$ cm; $d \approx 28.7$ cm
21. $\alpha \approx 29°$; $\beta \approx 61°$
23. $\alpha \approx 17°$; $\beta \approx 73°$
25. $\alpha \approx 19°$; $\beta \approx 71°$
27. $\approx 23°$
29. ≈ 103.5 ft
31. ≈ 128.0 ft
33. $\approx 24°$

10.2 EXERCISES

1. $\cos \alpha = \dfrac{12}{13}$; $\cos \beta = \dfrac{5}{13}$
3. $\cos \alpha = \dfrac{3}{5}$; $\cos \beta = \dfrac{4}{5}$
5. $\cos \alpha = \dfrac{\sqrt{10}}{5}$; $\cos \beta = \dfrac{\sqrt{15}}{5}$

7. (a) $\sin \alpha = \dfrac{\text{leg opposite } \alpha}{\text{hypotenuse}}$

   $= \dfrac{a}{c}$

   $\cos \beta = \dfrac{\text{leg adjacent to } \beta}{\text{hypotenuse}}$

   $= \dfrac{a}{c}$

   $\therefore \sin \alpha = \cos \beta$

   (b) $\cos \alpha = \dfrac{\text{leg adjacent to } \alpha}{\text{hypotenuse}}$

   $= \dfrac{b}{c}$

   $\sin \beta = \dfrac{\text{leg opposite } \beta}{\text{hypotenuse}}$

   $= \dfrac{b}{c}$

   $\therefore \cos \alpha = \sin \beta$

9. 0.9205

11. 0.9563

13. 0

15. 0.1392

17. $a \approx 84.8$ ft; $b \approx 53.0$ ft

19. $a = b = 5$ cm are *exact values*

21. $c \approx 19.1$ in.; $d \approx 14.8$ in.

23. $\alpha = 60°$ (exact); $\beta = 30°$ (exact)

25. $\alpha \approx 51°$; $\beta \approx 39°$

27. $\alpha \approx 65°$; $\beta \approx 25°$

29. $\approx 34°$

31. ≈ 1146.8 ft

33. ≈ 8.1 in.

35. ≈ 13.1 cm

37. $\alpha \approx 55°$

10.3 EXERCISES

1. $\tan \alpha = \dfrac{3}{4}$; $\tan \beta = \dfrac{4}{3}$

3. $\tan \alpha = \dfrac{\sqrt{5}}{2}$; $\tan \beta = \dfrac{2\sqrt{5}}{5}$

5. $\sin \alpha = \dfrac{5}{13}$; $\cos \alpha = \dfrac{12}{13}$;

$\tan \alpha = \dfrac{5}{12}$; $\cot \alpha = \dfrac{12}{5}$;

$\sec \alpha = \dfrac{13}{12}$; $\csc \alpha = \dfrac{13}{5}$

7. $\sin \alpha = \dfrac{a}{c}$; $\cot \alpha = \dfrac{b}{a}$;

$\cos \alpha = \dfrac{b}{c}$; $\sec \alpha = \dfrac{c}{b}$;

$\tan \alpha = \dfrac{a}{b}$; $\csc \alpha = \dfrac{c}{a}$

9. $\sin \alpha = \dfrac{x\sqrt{x^2 + 1}}{x^2 + 1}$;

$\cos \alpha = \dfrac{\sqrt{x^2 + 1}}{x^2 + 1}$; $\tan \alpha = \dfrac{x}{1}$;

$\cot \alpha = \dfrac{1}{x}$; $\sec \alpha = \sqrt{x^2 + 1}$;

$\csc \alpha = \dfrac{\sqrt{x^2 + 1}}{x}$

11. 0.2679

13. 1.5399

15. $x \approx 7.5$; $z \approx 12.8$

17. $y \approx 5.3$; $z \approx 8.5$

19. ≈ 8.1

21. $\alpha \approx 37°$; $\beta \approx 53°$

23. $\theta \approx 56°$; $\gamma \approx 34°$

25. $\alpha \approx 29°$; $\beta \approx 61°$

27. ≈ 1.4826

29. 2.0000 (exact)

31. ≈ 1.3456

33. ≈ 1147.4 ft

35. ≈ 4.1 in.

37. $\approx 72°$

39. N 47° W

10.4 EXERCISES

1. $\dfrac{\cos \theta}{\sin \theta} = \dfrac{\dfrac{b}{c}}{\dfrac{a}{c}} = \dfrac{b}{c} \cdot \dfrac{c}{a} = \dfrac{b}{a} = \cot \theta$

3. $\tan \theta = 0.75$; $\cot \theta \approx 1.3333$

5. $\dfrac{4}{5}$

7. $\dfrac{13}{12}$

11. 8 in.²

13. ≈ 11.6 ft²

15. ≈ 15.2 ft²

17. ≈ 11.3 in.

19. ≈ 8.9 m

21. $\approx 55°$

23. $\approx 51°$

25. ≈ 10.6

27. ≈ 6.9

29. (a) ≈ 213.4 ft

(b) $\approx 13{,}295$ ft²

31. ≈ 8812 m

33. ≈ 15.4 ft

35. Let $\gamma = 90°$. Then

$$c^2 = a^2 + b^2 - 2ab \cos \gamma$$

becomes

$$c^2 = a^2 + b^2 - 2 \cdot a \cdot b \cdot \cos 90°$$

But $\cos 90° = 0$.

$$\therefore c^2 = a^2 + b^2 - 2 \cdot a \cdot b \cdot 0$$
$$= a^2 + b^2$$

REVIEW EXERCISES

1. Sine; ≈ 10.3 in.

3. Cosine; ≈ 23 in.

5. Tangent; $\approx 43°$

7. Sine; $\approx 46°$

9. ≈ 8.9 units

11. ≈ 13.1

13. ≈ 42.7 ft

15. $\approx 47°$

17. ≈ 26.3 in.²

19. If $m\angle S = 30°$, then the sides of $\triangle RQS$ can be represented by $RQ = x$, $RS = 2x$, and $SQ = x \cdot \sqrt{3}$. Then

$$\sin S = \sin 30° = \frac{x}{2x} = \frac{1}{2}$$

21. ≈ 8.4 ft

23. $\approx 41°$

25. ≈ 4.0 cm

27. $\approx 68°$

29. 3:7

31. $\cos \theta = \dfrac{24}{25}$; $\sec \theta = \dfrac{25}{24}$

33. $\csc \theta = \dfrac{29}{20}$; $\sin \theta = \dfrac{20}{29}$

▲ INDEX

AA (AAA), 163
AAS, 103
Absolute value, 282
Acute angle, 45
Acute triangle, 85
Addition, 13
Addition Property of Equality, 22, 301
Addition Property of Inequality, 135
Adjacent angles, 45
Adjacent leg, 369
Alternate exterior angles, 69
Alternate interior angles, 69
Altitude, 21
   of cone, 342
   of cylinder, 21, 339
   of parallelogram, 248
   of prism, 330
   of pyramid, 334
   of trapezoid, 257
   of triangle, 113
Analytic geometry, 281
Analytic proof, 313, 315, 320, 321
Angle, 3, 44, 45, 119
Angle bisector. *See* Bisector
Angle of depression, 367
Angle of elevation, 367
Angle-Addition Postulate, 45
Apothem, 267, 268, 272
Arc, 41, 199, 200, 202, 271
Arc-Addition Postulate, 202
Archimedes, 63, 140, 240
Area
   of circle, 272
   of kite, 258
   of parallelogram, 248
   of rectangle, 247
   of region, 245
   of regular polygon, 268
   of rhombus, 258
   of sector, 273
   of segment, 274
   of square, 248
   of trapezoid, 25, 257

   of triangle, 250, 388
Area-Addition Postulate, 246
Area Postulate, 245
Argument, 4
ASA, 102
Associative Axiom for Addition, 14
Associative Axiom for Multiplication, 14
Assumption, 32
Auxiliary line, 115
Axis:
   of cone, 342
   of cylinder, 339
   of symmetry, 306
Axiom, 12, 36

Banach-Tarski paradox, 325
Base angles:
   of isosceles triangle, 113
   of trapezoid, 129
Base:
   of cone, 342
   of cylinder, 21
   of isosceles triangle, 113
   of parallelogram, 248
   of prism, 329
   of pyramid, 334
   of trapezoid, 129, 257
   of triangle, 250
Bisector:
   of angle, 46, 113
   of arch, 202
   of line segment, 38, 55, 113
Bolyai, Johann, 94
Boundary, 244, 246

Calculus, 336, 350, 359
Cartesian coordinate system, 281, 352
Center:
   of circle, 41, 198
   of regular polygon, 267
   of sphere, 347
Central angle:
   of circle, 199

   of regular polygon, 267
Central Angle Postulate, 200
Centroid of triangle, 236
Chord, 198
Circle, 41, 197–199, 207, 230, 272, 274, 303, 304, 359
Circumcenter of triangle, 235
Circumference, 239, 270, 272
Circumscribe, 262
Circumscribed circle, 207
Circumscribed polygon, 208
Coincide, 98
Coincident lines, 297
Collinear, 37
Commensurable, definition of, 153
Common chord, 206
Common external tangent, 218
Common internal tangent, 218
Common tangent, 218
Commutative Axiom for Addition, 13
Commutative Axiom for Multiplication, 14
Compass, 40
Complement, 46
Complementary angles, 7, 46
Completing the square, 304, 305
Compound statement, 8
Concave polygon, 88
Concave polyhedron, 347
Concentric circles, 199, 217
Concentric spheres, 349
Conclusion, 8, 58
Concurrent lines, 234, 324
Conditional statement, 8, 9, 74
Cone, 341–343, 358
Conic sections, 358
Congruent angles, 45
Congruent arcs, 201
Congruent circles, 198
Congruent line segments, 38
Congruent polygons, 159
Congruent triangles, 98, 99
Conjugate, 188

Conjunction, 10
Constant, 20
Constant of proportionality, 162, 163
Contraposition, 81, 82
Contrapositive, 9, 74
Construction, 40
Converse, 9, 60, 74
Convex polygon, 88
Convex polyhedron, 347
Coordinate, 281, 282, 283
Coplanar points, 40
Corollary, 87
Corresponding angles:
   of congruent triangles, 98, 99
   of parallel lines, 69
   of similar polygons, 160
Corresponding sides:
   of congruent triangles, 98, 99
   of similar triangles, 160
Corresponding vertices:
   of congruent triangles, 99
   of similar triangles, 160
Cosecant ratio, 379
Cosine ratio, 369
Cotangent ratio, 379
Counterexample, 4
CPCTC, 107
Cube, 332
Cylinder, 21, 339–341

Decagon, 88
Deduction, 1, 4
Definition, 32, 36
Degree:
   of equation, 303
   of variable, 303
Degree measure, 3
Descartes, René, 281, 352, 358
Diagonal, 2, 88
Diameter:
   of circle, 198, 239
   of sphere, 348
Dihedral angle, 346
Disjunction, 10
Distance, 283

Distance Formula, 283, 352, 353
Distributive Axiom, 16
Divided proportionally phrase, 168
Division, 14, 15
Division Property of Equality, 22
Dodecahedron, 347

Edge:
    of polyhedron, 347
    of prism, 330
    of pyramid, 334
Ellipse, 344, 359
Endpoint, 3
Equation, 22, 289, 301, 308
    of circle, 303, 304
    of line, 297
    of parabola, 306, 307
    of plane, 354
    of sphere, 355
Equiangular polygon, 88
Equiangular triangle, 85, 117
Equilateral polygon, 88
Equilateral triangle, 85, 117
Equivalence relation, 54, 98, 167
Equivalent equations, 22, 297
Equivalent inequalities, 29
Euclid, 63, 140
Euclidean geometry, 68, 93
Even integer, 7, 33
Extended proportion, 161, 162
Extended ratio, 155
Exterior:
    of angle, 45
    of circle, 198
    of triangle, 87
Exterior angles:
    of polygon, 265
    of triangle, 87
    of two lines, 69
External tangent, 218
Externally tangent circles, 217
Extremes of a proportion, 153

Face:
    of polyhedron, 347
    of prism, 330
    of pyramid, 334
FOIL method, 18, 188
Formula, definition of, 21

Garfield, James A., 276, 277
Geometry, 36

Geometric mean, 155, 177
Graph of equation, 289
Graph of inequality, 30
Great circle, 94, 348
Greater than symbol (>), 26

Height. See Altitude
Heptagon, 88
Heron of Alexandria, 63, 256
Heron's Formula, 256, 257
Hexagon, 88
Hexahedron, 347
HL, 109, 181
Horizontal line, 53, 283
Horizontal plane, 40
Hyperbola, 359
Hyperbolic geometry, 94, 95
Hyperbolic paraboloid, 94, 95
Hypotenuse, 108, 362
Hypothesis, 8, 58

Icosahedron, 347
Identity, 102
Incenter of triangle, 234
Included angle, 101
Included side, 101
Incommensurable, definition of, 153
Incomplete quadratic equation, 147
Indirect proof, 75, 76
Induction, 1, 2
Inequality, 26, 30, 134
Infinity, 272
Inscribing, 262
Inscribed angle, 202
Inscribed circle, 208
Inscribed polygon, 207
Intercepted arc, 199
Intercepts, 289
Interior:
    of angle, 45
    of circle, 198
    of triangle, 245
Interior angles:
    lines, 69
    polygon, 88, 265
    triangle, 85
Internal tangent, 218
Internally tangent circles, 217
Intersecting lines, 39, 301
Intersecting planes, 40
Intersection, 246, 301
Intuition, 1, 2
Invalid argument, 6
Inverse of statement, 9, 74

Inverse operations, 22, 150
Isosceles trapezoid, 130
Isosceles triangle, 32, 85, 113

Kite, 123, 258

Lateral area:
    of cone, 342
    of cylinder, 339
    of prism, 330
    of pyramid, 334
Law of Cosines, 390
Law of Sines, 389
Least common denominator, (LCD), 24
Legs:
    of isosceles triangle, 113
    of right triangle, 108
    of trapezoid, 129
Lemma, 135
Length:
    of arc, 271
    of line segment, 38
Less than symbol (<), 26, 27
Like terms, 16
Limit, 271, 272
Line, 2, 37, 53, 66, 69, 294, 297,
Line of centers, 218
Line segment, 2, 38, 55, 113
Linear equation, 23, 285, 289
Lobachevskian Postulate, 94
Locus (pl. loci), 229
Logic, 4

Major arc, 199
Mathematical system, 12, 36
Mean proportional, 155
Means-Extremes Property, 153
Means of proportion, 153
Measure:
    of angle, 44, 45
    of arch, 200
    of line segment, 2, 38
    of volume, 332
Median:
    of trapezoid, 129
    of triangle, 113, 235, 236, 324
Midpoint:
    of arc, 202
    of line segment, 38
Midpoint Formula, 285, 286, 353
Minor arc, 199
Multiplication, 13

Multiplication Property of Equality, 22

Negation, 8, 9
Nonagon, 88
Noncollinear points, 40
Noncoplanar points, 40
Non-Euclidean geometry, 93

Obtuse angle, 45
Obtuse triangle, 85
Octagon, 88
Octahedron, 347
Odd integer, 7, 33
Opposite angles, 119
Opposite leg, 362
Opposite rays, 39
Opposite sides, 117
Order of operations, 17
Ordered pair, 282
Ordered triple, 352
Origin, 282
Orthocenter of triangle, 235

Parabola, 237, 306, 307, 359
Paradox, 325
Parallel lines, 39, 67, 69, 293
Parallel planes, 40, 68
Parallel Postulate, 68
Parallelogram, 120, 248, 256
Pentagon, 88
Perimeter, 20, 117, 255, 256, 268
Perpendicular lines, 53, 66, 294
Perpendicular bisector of line segment, 55, 113
Pi (π), 239, 270
Plane, 39, 354
Plot, 282
Point, 37
Point of tangency (contact), 207
Point-Slope Form of line, 299, 300
Polygon, 3, 88, 265
Polyhedron, 347
Postulate, 12, 36
Prime number, 3, 4
Prism, 329, 330–332
Product, 13
Product Property of Square Roots, 148
Projection:
    of point on line, 172
    of line segment on line, 172
Proof, 50

Property, 22
Proportion, 153
Proportional segments, 145
Protractor, 3, 32, 44
Protractor Postulate, 45
Pyramid, 334, 335, 336, 348
Pythagoras, 63, 178, 277, 284
Pythagorean relationships, 387
Pythagorean Theorem, 178, 179, 180, 276
Pythagorean Triple, 182

Quadrant, 282
Quadratic equation, 145, 146, 150, 151
Quadratic Formula, 147
Quadrilateral, 88, 120, 256
Quotient Property of Square Roots, 150
Quotient relationships, 386

Radian, 396
Radical, 148, 149, 150, 151
Radicand, 148
Radius (*pl.* radii):
   of circle, 41, 198
   of cone, 342
   of cylinder, 339
   of regular polygon, 267, 268
   of sphere, 348
Ratio, 152, 239
Rationalizing the denominator, 188
Ray, 38
Reciprocal ratios, 380
Reciprocal Identities, 385, 386
Rectangle, 2, 127, 247, 256
Rectangular coordinate system, 281
Reflection, 187, 193
Reflexive Property of Congruence, 99, 102
Reflexive Property of Equality, 13
Reflexive Property of Similarity, 167
Region, 244, 245
Regular polygon, 88, 265, 267, 268
Regular polyhedron, 347
Regular prism, 331
Regular pyramid, 334
Rhombus, 128, 256, 258

Riemannian Postulate, 94
Right angle, 45
Right circular cone, 342
Right circular cylinder, 21
Right triangle, 85, 108
Rotation, 227
Ruler, 244
Ruler Postulate, 38

SAS, 101
Scalene triangle, 85
Secant of circle, 207
Secant ratio, 379
Sector of circle, 273
Segment-Addition Postulate, 38
Segment of circle, 274
Segments divided proportionally, 169, 170, 171, 173
Semicircle, 199
Semiperimeter, 256, 331
Sides:
   of angle, 44
   of polygon, 88
   of triangle, 85
Similar polygons, 145, 159, 160–161
Similar triangles, 7, 145, 160, 163, 361
Sine ratio, 362
Skew quadrilateral, 120
Slant height:
   of regular pyramid, 334
   of right circular cone, 342
Slope, 291
Slope Formula, 290, 291
Slope-Intercept Form of a Line, 297
Solid, 329
Solution of equation, 22
Solution of inequality, 30
Space, 40
Sphere, 93, 230, 347–349, 355
Spherical geometry, 94, 95
Square, 7, 128, 248, 256
Square of number, 148
Square root, 147
Square Roots Property, 150
SSS, 100
Standard form of quadratic equation, 145, 146
Statement, 8

Straight angle, 45
Straightedge, 40
Subscript, 246, 257
Substitution Property of Equality, 13
Subtraction, 14
Subtraction Property of Equality, 22, 301
Sum, 13
Supplement, 46
Supplementary angles, 46
Surface, 329, 331
Surface of revolution, 349, 350
Surface area:
   of cone, 342
   of cylinder, 340
   of prism, 331
   of pyramid, 335
   of sphere, 348
Symmetry, 16, 306
Symmetric Property of Congruence, 99
Symmetric Property of Equality, 13
Symmetric Property of Similarity, 167
System of Equations:
   linear, 300, 301
   nonlinear, 308

Tangent:
   of circle, 207
   to curve, 359
Tangent circles, 217
Tangent ratio, 376
Terms of proportion, 153
Tetrahedron, 338, 347
Theorem, 12, 22, 36, 58
Torus, 350
Total area of solid. *See* Surface area
Transitive Property of Congruence, 99
Transitive Property of Equality, 13
Transitive Property of Inequality, 28
Transitive Property of Similarity, 167
Transversal, 69
Trapezoid, 25, 129, 256, 257
Triangle, 3, 85, 87, 88, 113, 234–236, 250, 288, 324

Triangle Inequality, 138
Trigonometric identity, 372, 385
Trigonometry, 361
Truth table, 10

Undefined term, 12, 36
Underdetermined, 115
Union, 246
Unique, 42, 43

Valid argument, 4, 6
Variable, 16, 20
Vertex:
   of angle, 44
   of cone, 342
   of isosceles triangle, 113
   of parabola, 306
   of polygon, 88
   of polyhedron, 347
   of pyramid, 334, 348
   of triangle, 85
Vertex angle, 113
Vertical line, 53, 283
Vertical plane, 40
Vertical angles, 7, 47
Volume:
   of cone, 343
   of cube, 332
   of cylinder, 341
   of prism, 332
   of pyramid, 336
   of sphere, 349
Volume Postulate, 332

x axis, 282, 352
x intercept, 289, 354
x value (coordinate), 282, 352

y axis, 282, 352
y intercept, 289, 354
y value (coordinate), 282, 352

z axis, 352
z coordinate, 352
z intercept, 354
Zero Product Property, 146

ABBREVIATIONS

AA	angle-angle (proves △s similar)
ASA	angle-side-angle (proves △s congruent)
AAS	angle-angle-side (proves △s congurent)
add.	addition
adj.	adjacent
alt.	altitude, alternate
ax.	axiom
cm	centimeters
cm²	square centimeters
cm³	cubic centimeters
comp.	complementary
corr.	corresponding
cos	cosine
cot	cotangent
CPCTC	Corresponding parts of congruent triangles are congruent.
csc	cosecant
diag.	diagonal
exs.	exercises
ext.	exterior
eq.	equality
ft	foot (or feet)
gal	gallon
HL	hypotenuse-leg (proves △s congruent)
h	hour
in.	inch (or inches)
ineq.	inequality

int.	interior
isos.	isosceles
km	kilometers
m	meters
mi	miles
mm	millimeters
n—gon	polygon of n sides
opp.	opposite
pent.	pentagon
post.	postulate
proj.	projection
prop.	property
pt.	point
quad.	quadrilateral
rect.	rectangle
rt.	right
SAS	side-angle-side (proves △s congruent)
sec, sec.	secant, section
sin	sine
SSS	side-side-side (proves △s congruent)
st.	straight
supp.	supplementary
tan	tangent
trans.	transversal
trap.	trapezoid
vert.	vertical (angles)
yd	yards